# APPLIED STRESS ANALYSIS OF PLASTICS

## A Mechanical Engineering Approach

# APPLIED STRESS ANALYSIS OF PLASTICS

## A Mechanical Engineering Approach

S.I. Krishnamachari

L.J. Broutman & Associates, Ltd.
Chicago, IL

VNR SPRINGER SCIENCE+BUSINESS MEDIA, LLC

*To my mother.*


Originally published by Van Nostrand Reinhold in 1993
Softcover reprint of the hardcover 1st edition 1993
Library of Congress Catalog Card Number 92-27825
ISBN 978-1-4613-6360-6



16 15 14 13 12 11 10 9 8 7 6 5 4 3 2 1

**Library of Congress Cataloging-in-Publication Data**

Krishnamachari, S. I. (Sadagopa I.), 1944-
Applied stress analysis of plastics: a mechanical engineering approach / S. I. Krishnamachari
p. cm.
Includes bibliographical references and index..
ISBN 978-1-4613-6360-6 ISBN 978-1-4615-3110-4 (eBook)
DOI 10.1007/978-1-4615-3110-4
1. Plastics—Testing. 2. Stresses and strains. I. Title.
TA455.P5K7 1992
620.1' 9233' 0287—dc20 92-27825
CIP

# Contents

# Preface

This book is a product of the understanding I developed of stress analysis applied to plastics, while at work at L. J. Broutman and Associates (LJBA) and as a lecturer in the seminars on this topic co-sponsored by LJBA and Society of Plastics Engineers. I believe that by its extent and level of treatment, this book would serve as an easy-to-read desktop reference for professionals, as well as a text book at the junior or senior level in undergraduate programs.

The main theme of this book is what to do with computed stress. To approach the theme effectively, I have taken the "stress category approach" to stress analysis. Such an approach is being successfully used in the nuclear power field. In plastics, this approach helps in the prediction of long term behavior of structures. To maintain interest I have limited derivations and proofs to a minimum, and provided them, if at all, as flow charts. In this way, I believe that one can see better the connection between the variables, assumptions, and mathematics.

I intend this book as a guide for the mechanical engineer working in plastics product design. Therefore, the use, rather than the description of mechanical behavior is treated preferentially. A working knowledge of the strength of materials, differential calculus, and ordinary differential equations should be adequate for most of the text, except that some knowledge of Laplace transforms may be helpful for Chapter 5. A cursory review of this topic is provided in Appendix 3. Wherever appropriate, I have provided interpretive comments, and have explained the use of existing material test practices based on the mechanical behavior of the materials. Examples in this book perform two major tasks. The first one is to illustrate a certain principle, and the second is to provide commentaries about the differences between theory and practice. It is my hope that the student engineer will find these commentaries valuable.

Chapter 1 deals with the overall structure of the theory of elasticity, and in particular, with the fact that stresses caused by loads are basically different from stresses caused by displacements. Understanding this difference is fundamental to understanding plastics. It also introduces the stress category concept. Chapter 2 is an overview of how much can be done in plastics design by using existing techniques in metallics. The techniques used in this chapter range from simple press-fits to matrix structural analysis (using an example in piping flexibility). I should admit that in this chapter, the progression of complexity is rather fast. However, I was able to discuss a wider range of practical problems more compactly. Chapter 3 is basically about incipience of plastic flow. No flow rules are discussed. Chapter 4 returns to the stress category, and its relation to the failure modes known to occur in plastics. The concept of factor of safety is also discussed, as are the facts that it is a number having more details to it than meets the eye, and that rational product quality specifications decide what to look for in an analysis. Chapter 5 is a discussion of the application of viscoelasticity to static problems. I have presented a view that the so-called "pseudoelasticity" is a consequence of the correspondence principle. The examples in Laplace transforms are for illustration of the correspondence principle. If one were to assume that the Poisson's ratio remains constant with time, and express all elastic properties in terms of $E$ and $\nu$ only, the Laplace transform approach will give results identical to those of pseudoelasticity. Thus, the Laplace transform has been used in this book as a pointer to pseudoelasticity technique. A proposal for ranking plastics is also included in this chapter. Chapter 6 concerns linear elastic fracture mechanics and applications. The basic four approaches to crack stability are discussed, but the application to the stress intensity approach is dealt with in detail because it permits computation. Calculation of the fatigue life of plastics is presented as a problem in the progressive increase of $K_I$. The crazing mechanism is explained using the background of stress intensity at crack tip. Chapter 7 discusses the micro- and macro-mechanics of fiber-reinforced plastics. A variety of results have been compiled in a manner that is suitable for computer coding. This chapter, with its exercises, may be suitable for a three-credit course.

Chapters 8 and 9 deal with finite elements. What do finite elements have to do with plastics? Finite elements are replacing classical methods of analysis in the same way that calculators replaced addition and multiplication tables. The design engineer invariably is a user of some finite element software. Covering an extensive topic such as this in one or even two chapters is presumptuous. Only the elementary ideas are discussed and I hope the reader will gain enough confidence to refer to more detailed books on the subject. Chapter 8 discusses the steps involved in the finite element analysis, by taking a linear triangular element as the basic entity. Extensions to higher order elements, or more nodes per element are just indicated. A few exercises stand in for detailed discussion of the topic. Chapter 9 contains cautionary advice to any user of finite elements.

I have used a casual style in algebraic manipulations. At several places, I have used dots such as $(\cdots)$ to mean an exact repetition of the quantity in the previous step. This is not formal, but the reader who wants to get to the bottom line quickly will find this notation agreeable. In Chapter 5, I have used $E'$ and $E''$ with subscripts $\sigma$ and $\epsilon$, to denote the manner in which the test data were obtained. This once again is nonstandard. I have used such shortcuts consciously with the hurried student in mind, the kind of student who might pick a formula from the text without pausing to look for the applicable restrictions. In Chapter 7, for example, the Tsai-Halpin formulas are summarized by the use of an unreadable symbol $(\bullet)$, a symbolism which, I believe, communicates the generality of the formulas.

Some curves in this text are reproductions taken from other texts, trade brochures or company manuals. I gratefully acknowledge the permissions from the organizations to reproduce them.

# Acknowledgments

The origins of this book are in the seminars I worked on with Dr. L. J. Broutman who co-sponsored them with the Society of Plastics Engineers. I gratefully acknowledge the support, resources, and encouragement I received from him, without which this book would not have been possible. Thanks are due to Prof. S. Kalpakjian and Dr. S. Nair, both of Illinois Institute of Technology, Chicago, and to my colleague Dr. Paul K. So, whose counsel and technical discussions were valuable to me all along the writing. I sincerely appreciate the advice on the style and grammar I received from my colleagues Mr. Ken E. Hofer, and Mr. Michael R. Roop. I thank the editors of Van Nostrand Reinhold, a little suggestion from whom sparked off this project. My wife Lalitha and sons Sriram and Parashar gave me loving support, cooperation, and understanding for the last three years of writing.

# 1

# Basic Concepts: Perspectives in Elasticity Theory

## 1.1 INTRODUCTION

The common use of the term *stress analysis* includes any kind of structural analysis. In the field of thermoplastics design, there is a growing awareness of the importance of stress analysis. In recent years, structural plastics have been used for applications in load-bearing structural components in the automotive, aerospace, sporting goods, and construction industries. Hence, design engineers are increasingly concerned about stress-related problems, typically with the strength, stiffness, and life expectancy of their products. About twenty years ago, these problems were primarily associated with the metallic components.

Stress analysis has always been interdisciplinary, because an effective analysis needs to bring together a thorough knowledge of the operating characteristics of the product, material behavior, structural behavior, and solid mechanics. Structural plastics design is a field that is evolving in the same manner as did the aero-space and nuclear power industries. That is, a sequence of product innovations, and better methods of design and analysis continuously reinforce each other and lead to the optimum design of the product. Stress analysis is a vital activity in this process.

The development efforts in mature industries like the aerospace and the nuclear power industries have led to a variety of product standards which are recognized worldwide. In the plastics industry, however, such comprehensive standards have not yet been established.

This book addresses the needs of the practicing engineer, and of the student plastics–designer. Hence a limited amount of prior knowledge of basic strength of materials is assumed. It is assumed that the reader is aware of, and knows to use, handbooks of formulas, material data, and desktop reference books, [1, 2, 3] and hence a few "forward references" have been made in this chapter. Examples in this chapter use well-known formulas. Emphasis is placed more on the use and interpretations of formulas rather than their derivation. In as many examples as possible, interpretive commentaries are provided.

## 1.2 UNIQUENESS OF PLASTICS STRESS ANALYSIS

From the point of view of stress analysis, are thermoplastics very different from metals? Yes and no. Yes, because a few types of behaviors of thermoplastic materials call for advanced techniques of analysis, because such behaviors are encountered only in specialty applications of metals. No, because several calculations and test procedures for characterizing the mechanical properties of thermoplastics are very similar to those of metals; thus such stress analysis is also similar. Material properties of plastics such as elastic modulus, yield point, tensile strength, and fracture toughness are understood, measured, and used in a manner similar to those for metals. Many structural plastics designs may be performed using the familiar strength of materials approach. Likewise detailed stress analyses of plastic components are performed assuming linear elastic behavior. Here, we discuss the aspects of stress analysis that are unique to thermoplastics components.

*Low modulus:* The modulus of any plastic material is about two orders of magnitude lower than that of metals. This leads to the need for a variety of stiffening elements in the mechanical design that result in unconventional shapes. Therefore, the familiar formulas of stress and stiffness are not applicable.

*Nonlinearity:* The stress-strain relationship of all plastics exhibits nonlinearity. However, for many design objectives, a linear approximation is acceptable. However, stresses and strains which are calculated using such procedures have a different significance in plastics. From a materials engineering standpoint, failures are related to the strain in a microscopic scale, be it metal or plastic. For a metallic component failure criteria and mechanisms can be expressed in terms of stress, since it is a linear multiple of strain anyway. For plastics this is not true, because of their nonlinear behavior. In other words, design criteria for plastics have to be expressed in terms of limiting values of stress, as well as strain, which are in general independent of each other. In any case, more detailed analysis is needed for plastics to conform to these limits.

*Viscoelasticity:* The aspect of finite life of the plastic product is highlighted more by viscoelasticity than in the case of metals. As we shall see later, viscoelastic behavior is recognized by its two principal aspects, namely *creep* and *relaxation*. Plastics exhibit this behavior almost at any useful operating temperature. Metals exhibit this behavior only at high temperatures, usually of the order of a half or a third of their melting point. Thus, in metals, creep and relaxation affect only specialty applications. Even so, with metals, the design objective is usually to *avoid* creep by using, for example, thermal shields, or by selecting materials of low creep. In plastics, the objective is to *cope with it*.

*Manufacturing methods:* Molding techniques, leading to integrated product design of plastics, have replaced "machine-and-fabricate" techniques of metallic components. This results in some effect on the stress analysis. In integrated component designs, the mechanism of load transfer from one component to another is no longer simple. New complex shapes arise from molding innovations, for which no pre-engineered stress solutions are available. Such shapes are difficult to model and to stress-analyze even with a powerful computer. Similar difficulties exist in the analysis of metal components of complex shapes. Valve body castings and automotive engine blocks are examples in the context. Such components are designed by rules rather than by analysis.

*Economics:* Stress analysis of structural plastic products needs to consider nonlinearity and viscoelasticity as opposed to the consideration of simple linear elastic behavior in metals. The finite element method, which is very well-known and has been a widely available tool of stress analysis since the 1980s, does not provide cost-effective answers to many stress analysis questions. It is still uneconomical to model the true material behavior of plastics in finite elements, especially when considering the final cost of the product. Currently, only very large industries are able to afford it.

*Material property data:* There is a paucity of material property data for plastics. Long term property data are crucial to many designs, but are not available. When available, they are generally not in a form that is useable directly in a large scale computation. (On the other hand, published data are available for metals and are backed by national standards.)

*Material standards:* Metals are classified according to national standards, and carry with them guaranteed properties, including those related to finite product life. Plastics are engineered materials the final properties of which result from the interaction of many variables. Hence they have not achieved a standardization.

## 1.3 SIMILARITIES BETWEEN PLASTICS AND METALS

Table 1.1 shows the areas of similarity between metals and plastics. *Essentially all the similarities arise from modeling plastics as linear elastic materials*. When a linearized modulus is adopted for plastics, a whole range of stress analysis techniques is also adopted. The only exceptions are as below.

In Table 1.1, we observe that the yield strength of plastics is placed under the "yes/no" category, because plastics do not possess a clear definition as do metals. Measurement of yield strength is made difficult because of its closeness to the ultimate strength and the time-dependent effects inherent in the tests. Also, we shall see that the time-dependent crack propagation at constant applied stress is unique to plastics and does not exist in the metals.

Lastly, the concept of pseudo-elasticity for the analysis of long-term effects is not applicable to metals.

## 1.4 SIGNIFICANCE OF CALCULATED STRESS AND STRAIN

The engineer needs to be aware that stress calculations are subject to numerous assumptions, and the calculated stress is dependent on the way the problem is formulated. Therefore, caution is to be exercised in interpreting the results of the analysis.

Consider, for example, a ribbed plate subjected to uniform pressure load. The classical way to deal with this problem is to suppose that the moment of inertia of the ribs is distributed uniformly over the entire plate. This way, the ribbed plate can be considered equivalent in stiffness to another flat unribbed plate of a greater thickness, or a flat unribbed plate of the same thickness, but having a higher modulus. Using this equivalent thickness or modulus, it is possible to compute the deflections and stresses of the real plate. A review of the theory of this approach, however, would show *that it is good for maximum deflection, but not for stress*. Also, if a detailed map of deflection or stress is required, the approach is not good for either of them. Furthermore, the details of stress concentrations at the

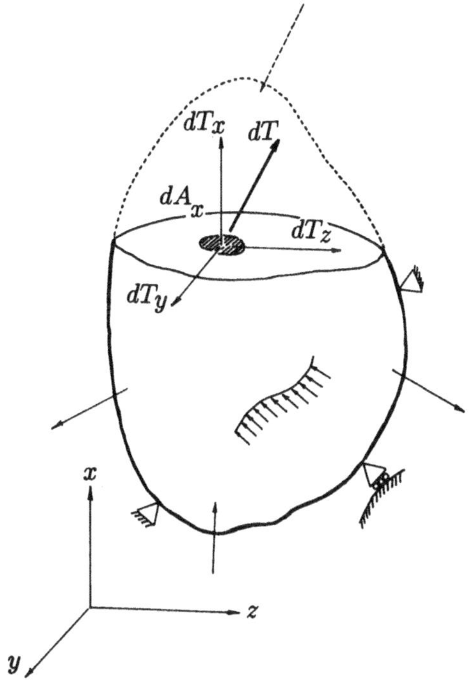


$$\sigma xx = \frac{dTx}{dA_x}$$
$$\tau xy = \frac{dTy}{dA_x}$$
$$\tau xz = \frac{dTz}{dA_x}$$

**FIGURE 1.1.** Definition of stress.

vector $d\vec{T}$ and a different area $dA_y$. Consequently, another independent set of three stress components is obtained. In particular, for a section normal to the $y$ axis, we obtain the definition of $\sigma_{yy}$, $\tau_{yx}$, and $\tau_{yz}$. Similarly by passing a $xy$ plane, we would obtain $\sigma_{zz}$, $\tau_{zx}$, and $\tau_{zy}$.

### 1.6.2 Properties of Stress

It may appear that there can be a total of nine stress components at a point, namely, $\sigma_{xx}$, $\sigma_{yy}$, $\sigma_{zz}$, $\tau_{xy}$, $\tau_{yx}$, $\tau_{xz}$, $\tau_{zx}$, $\tau_{yz}$, and $\tau_{zy}$. However, because of the "symmetry" of shear stresses (to be discussed later in Section 1.9), the number of shear components reduce from six to three. As a result, there are in fact only six independent stress components.

It is also seen that we have used two Greek letters $\sigma_{..}$ and $\tau_{..}$ to represent stresses, and that we have used two subscripts. The components of stress denoted by $\tau_{..}$ are tangential to the plane of the section, as can be seen from Figure 1.1, and are called *shear stresses*. Likewise, the components represented by $\sigma_{..}$ are normal to the plane of the section and are accordingly called *normal stresses*. The normal and shear stresses do not arise solely from the directions of forces and the geometry of planes of the section. All materials, thermoplastics inclusive, exhibit different stiffnesses and strengths under the influence of these two kinds of stresses. It is therefore a fundamental classification of stresses.

Stresses are not vectors, although they have magnitude and direction. They belong to a more general category of physical quantities called *tensors*. A consequence of this is that they do not project like vectors from one direction to another.[2] The rules of transforming stresses from one direction to another are discussed later in Section 1.14.

Stresses are completely identified by two subscripts. The first subscript identifies the plane on which the stress acts. That is, the normal to the plane is directed along the axis denoted by the first subscript. The second subscript denotes the direction of the stress itself.

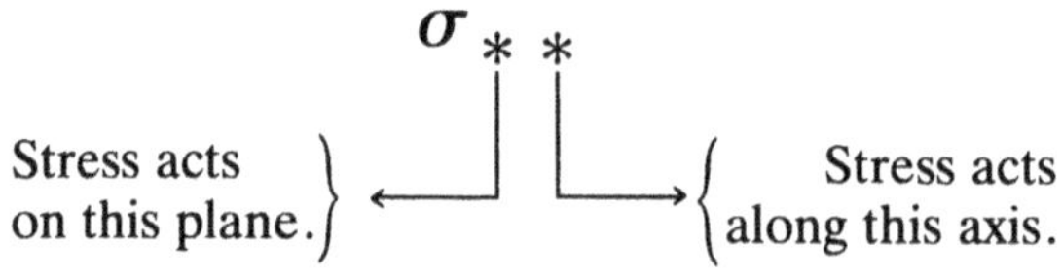


It is easy to see that when the first and second subscripts are the same then the stress is a normal stress. If the subscripts are different, it is a shear stress. Double subscripts are necessary only for the purposes of definition and where there is a genuine need to identify the planes. In normal situations, the context makes clear which component is being discussed. In particular, the use of two different Greek symbols for shear and normal stresses helps in clarifying what stress is being discussed.

> A special property of shear stresses, not mentioned explicitly in text books, is that they cannot be measured by any experimental method. They can only be computed from measurements of normal stresses.[3]

The definition of stress, giving six independent components, is a complete description of the state of stress at a point. That is, the six stress components defined in any one coordinate system is necessary and sufficient to determine the six components of stress along any other coordinate system.

---

[2] In fact, tensors are defined and identified by the rules they obey under transformation from one set of axes to another.

[3] From a more rigorous viewpoint, no stress can be measured. All stress-measuring instruments are designed to measure displacements and display such displacements as equivalent stress.

viz., $\epsilon_{zz}$, $\gamma_{yz}$, $\gamma_{zy}$, $\gamma_{xz}$, and $\gamma_{zx}$. We can complete the definitions of strains by writing the equations simply by cyclic permutation of the subscripts.

$$\epsilon_{xx} = \left( \frac{\partial u}{\partial x} \right) \tag{1.2}$$

$$\epsilon_{yy} = \left( \frac{\partial v}{\partial y} \right) \tag{1.3}$$

$$\epsilon_{zz} = \left( \frac{\partial w}{\partial z} \right) \tag{1.4}$$

$$\gamma_{xy} = \gamma_{yx} = \left( \frac{\partial v}{\partial x} + \frac{\partial u}{\partial y} \right) \tag{1.5}$$

$$\gamma_{yz} = \gamma_{zy} = \left( \frac{\partial v}{\partial z} + \frac{\partial w}{\partial y} \right) \tag{1.6}$$

$$\gamma_{zx} = \gamma_{xz} = \left( \frac{\partial w}{\partial x} + \frac{\partial u}{\partial z} \right) \tag{1.7}$$

The above six equations are commonly known as the *strain-displacement relations* and constitute an essential element of the theory of elasticity.

### 1.7.2 Properties of Strain

Strains, like stresses, are completely identified by two subscripts. The physical meaning of the subscripts for strains is as below.

$$\epsilon_{**}$$

Original length is along this axis. ← (first subscript) | (second subscript) → Displacement is along this axis.

It is easy to see that, as in the case of stresses, when the two subscripts are the same, then it is a normal strain. If they are different, then it is a shear strain. By convention, the Greek letter $\epsilon$ is used to represent normal strains and $\gamma$ is used to represent shear strains.

Like stresses, strains are also tensors.[4] All the strain components at a

---

[4] The engineering shear strain defined here is not a tensor. In Section 1.12.4 we discuss the means for making it a tensor.

point constitute the state of strain at that point. The magnitudes of the individual components depend on the choice of the coordinate system.

It has been mentioned that the symmetry of shear strains is implicit in its definition. However, it is also possible to *prove* it by considering the kinematics of displacements.

Like shear stresses, shear strains also elude measurement by any experimental technique. They can only be computed from observation of normal strains along different directions.

Rigid body displacements and rotations do not cause strains. To produce strains, relative displacement of points within the body is required. To add physical feel for the concept, the reader may substitute the word "particles" for the word "points" everywhere in the definitions.

Intuitively, strain can be visualized as,

- Distribution of displacements
- Intensity of stretch and squeeze
- Ratio of the change in length to original length—$dl/l_0$
- Derivative of displacement, $du/dx$, where $u$ = displacement.

## 1.8 COMMENTARY ON THE DEFINITIONS

Both stress and strain are defined as limiting ratios of infinitesimals. Thus they are quantities associated locally to a point, and hence may be considered as *intensive quantities*. On the other hand, quantities like moments of inertia, bending moments, force resultants, and strain energy are associated with a thickness, a cross section, or a volume. They are obtained by integrating a property over the appropriate length, area, or volume. Hence, they are not local properties. Such quantities may be considered *extensive quantities*.[5]

The definition of stress does not inquire into why and how the load was distributed in a particular way. Likewise the definition of strain does not inquire why and how the displacements were distributed that way.

> It is important to note that there is no a priori notion in these definitions that stress causes or is caused by strain.

Up to this point, stress is understood to be a mechanical quantity and strain to be a kinematic quantity. The definitions of stress and strain given in the foregoing discussions are due to Cauchy and are sometimes referred

---

[5] This classification hardly matters for the analysis of stress. However, for associating a failure mode with stress, this helps.

to as the Cauchy stress and the Cauchy strain. A careful review of the definitions in Sections 1.6.1 and 1.7.1 will show that the Cauchy stress and strains are defined without using the fact that one is associated with the other. However, Cauchy's definition is not the only way they can be defined. The reader may already be aware of the terms "true stress" and "true strain" which recognize the fact that application of stress causes the area to change and therefore, the stress itself changes.

## 1.9 EQUILIBRIUM

The definition of stress as in Equation 1.1 used the fact that the body under consideration is already in overall equilibrium under the action of external forces and constraints. Besides the overall static equibrium, we can also consider the equilibrium of every point in the interior and on the surface of the body. Furthermore, the body may be subjected to body-forces, such as gravity or centrifugal acceleration, which act on the entire volume of the body.

For equilibrium of an interior point, we require

$$\text{Sum of the Forces: } \sum F = 0, \text{ along } x, y, \text{ and } z \text{ axes.}$$

$$\text{Sum of the Moments: } \sum M = 0, \text{ along } x, y, \text{ and } z \text{ axes.}$$

Let us derive the force-equilibrium about the $x$ direction, and the moment-equilibrium in the $z$ direction only. Figure 1.5 shows an elemental volume on which the relevant stresses are shown. The stresses marked on

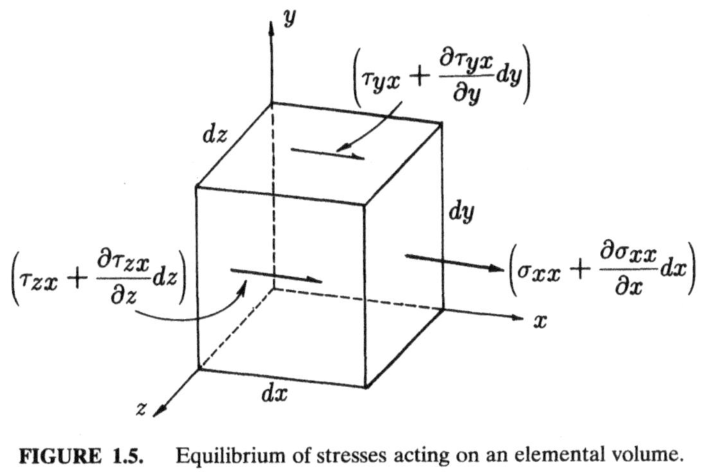


**FIGURE 1.5.** Equilibrium of stresses acting on an elemental volume.

the figure are those acting on the positive planes (those that are seen in the foreground). For the sake of clarity of the figure, those acting on the negative planes (those that are hidden) are not shown. The stresses on the negative planes are a differential magnitude less than those on the positive planes. Gravity force is the only body force considered.

*For force equilibrium, we must convert the stresses to forces by multiplying them by the corresponding areas.* Along the positive $x$ axis, we have

$$\left(\sigma_{xx} + \frac{\partial \sigma_{xx}}{\partial x}\, dx\right) dy\, dz - (\sigma_{xx})\, dx\, dz$$

$$+ \left(\tau_{yx} + \frac{\partial \tau_{yx}}{\partial y}\, dy\right) dx\, dz - (\tau_{yx})\, dy\, dz$$

$$+ \left(\tau_{zx} + \frac{\partial \tau_{zx}}{\partial z}\, dz\right) dy\, dz - (\tau_{zx})\, dy\, dz + \rho g_x\, dx\, dy\, dz = 0$$

Further simplification of the equation gives

$$\frac{\partial \sigma_{xx}}{\partial x} + \frac{\partial \tau_{xy}}{\partial y} + \frac{\partial \tau_{xz}}{\partial z} + \rho g_x = 0$$

We can also consider the moment equilibrium about the $z$ axis, passing through the center of the element. Note that there are no "body moments" present. That is, there are no external moments acting on the volume.

$$\left(\tau_{xy} + \frac{\partial \tau_{xy}}{\partial x}\, dx\right) dy\, dz(dx/2) + \tau_{xy}\, dy\, dz(dx/2)$$

$$- \left(\tau_{yx} + \frac{\partial \tau_{yx}}{\partial y}\, dy\right) dx\, dz(dy/2) - \tau_{yx}\, dx\, dz(dy/2) = 0$$

Upon simplification and cancellation of a common factor $(dx\, dy\, dz)$, we obtain

$$\tau_{xy} = \tau_{yx}.$$

By repeating the exercise for other directions, we can get two more force-equilibrium and two more moment-equilibrium equations. All the resulting equations are listed below.

$$\frac{\partial \sigma_{xx}}{\partial x} + \frac{\partial \tau_{xy}}{\partial y} + \frac{\partial \tau_{xz}}{\partial z} + \rho g_x = 0 \tag{1.8}$$

$$\frac{\partial \tau_{yx}}{\partial x} + \frac{\partial \sigma_{yy}}{\partial y} + \frac{\partial \tau_{yz}}{\partial z} + \rho g_y = 0 \tag{1.9}$$

$$\frac{\partial \tau_{zx}}{\partial x} + \frac{\partial \tau_{zy}}{\partial y} + \frac{\partial \sigma_{zz}}{\partial z} + \rho g_z = 0 \tag{1.10}$$

$$\tau_{xy} = \tau_{yx} \tag{1.11}$$

$$\tau_{yz} = \tau_{zy} \tag{1.12}$$

$$\tau_{zx} = \tau_{xz} \tag{1.13}$$

The above equations are referred to as *equilibrium equations*. The first three of the above equations are the ones useful for the computations in the theory of elasticity. The second set of three equations represent the symmetry of shear stresses, and they reduce the number of stresses from nine to six. Shear symmetry is usually taken for granted in the computations, and effectively, have only three equilibrium equations available at a point.

## 1.10 HOOKE'S LAW

Hooke's law describes linear elastic behavior of a solid. It is a way of formalizing facts of observation and represents one of the many types of material behavior. Linear elastic behavior is the oldest known, most researched, and most used material-behavior type. Hooke's law connects the six strains at a point to the six stresses at that point. *Hence, Hooke's law is local to a point*. In essence, the law states that

- stress and strain are related linearly,
- any one of them can cause or be caused by the other, meaning thereby that application of force is not the only means of producing stresses; application of displacements can also produce stress,
- the stress-strain response is instantaneous,
- normal stresses can cause, or be caused only by normal strains,
- shear stresses can cause, or be caused only by shear strains, and
- there are only *two* independent material constants in linear elasticity.

There are six stress strain equations which together are called Hooke's law. They are as follows:

$$\epsilon_{xx} = \frac{1}{E}\sigma_{xx} - \frac{\nu}{E}\sigma_{yy} - \frac{\nu}{E}\sigma_{zz} \tag{1.14}$$

$$\epsilon_{yy} = -\frac{\nu}{E}\sigma_{xx} + \frac{1}{E}\sigma_{yy} - \frac{\nu}{E}\sigma_{zz} \tag{1.15}$$

$$\epsilon_{zz} = -\frac{\nu}{E}\sigma_{xx} - \frac{\nu}{E}\sigma_{yy} + \frac{1}{E}\sigma_{zz} \tag{1.16}$$

$$\gamma_{xy} = \frac{1}{G}\tau_{xy} \tag{1.17}$$

$$\gamma_{yz} = \frac{1}{G}\tau_{yz} \tag{1.18}$$

$$\gamma_{zx} = \frac{1}{G}\tau_{zx} \tag{1.19}$$

In the above, $E$ is the Young's modulus and $\nu$ is the Poisson's ratio of the material. The shear modulus, denoted by $G$, is not an independent material property, but is connected to the other two by the relation,

$$G = \frac{E}{2(1 + \nu)} \tag{1.20}$$

Sometimes, it is useful to express stresses in terms of strains by just inverting Hooke's law. We obtain the following equations.

$$\lambda = \frac{\nu E}{(1 + \nu)(1 - 2\nu)} \tag{1.21}$$

$$e = \epsilon_{xx} + \epsilon_{yy} + \epsilon_{zz} \tag{1.22}$$

$$\sigma_{xx} = \lambda e + 2G\epsilon_{xx}$$

$$\sigma_{yy} = \lambda e + 2G\epsilon_{yy}$$

$$\sigma_{zz} = \lambda e + 2G\epsilon_{zz} \tag{1.23}$$

$$\tau_{xy} = G\gamma_{xy}$$

$$\tau_{yz} = G\gamma_{yz}$$

$$\tau_{zx} = G\gamma_{zx} \tag{1.24}$$

Note that the quantity $e$ represents the volumetric expansion at a point. The case of one dimensional stress strain relation, say, along $x$ axis, is evident once we note that $\epsilon_{yy} = \epsilon_{zz} = -\nu\epsilon_{xx}$, and substitute in the equation for $\sigma_{xx}$.

Another point to note carefully is the term $(1 - 2\nu)$ in the denominator of $\lambda$. This term sets a theoretical upper limit to the value of Poisson's ratio as 0.5, since at that value of $\nu$, all normal stresses tend to infinity, unless $e$ is held zero. Rubber and other elastomeric materials have a Poisson's ratio quite close to 0.5. Such materials require that $e = 0$. That is, they do not undergo a change in volume. They need a different type of stress-strain relationship.[6] Hooke's laws in the above form do not work for such materials.

## 1.11 PLANE STRESS AND PLANE STRAIN

Many two dimensional problems in elasticity can be seen to belong in one or the other of the two extreme situations, namely *plane stress*, or *plane strain*. In two dimensions it is possible to reduce the number of equations in strain-displacement relations (Equations 1.2 through 1.7), Hooke's law (Equations 1.14 through 1.19), and equilibrium equations (Equations 1.8, 1.9, and 1.10). The modifications needed in the equilibrium equations and the strain-displacement equations are quite straightforward. In this section, we look into only the changes needed in Hooke's law. We can show, that if the stresses lie all in one plane, say, the $x$-$y$ plane, then there is a normal strain along the $z$ direction. Likewise, if the strains all lie in one plane, say the $x$-$y$ plane, then there is a normal stress along the $z$ direction. This can be seen easily from Hooke's law, Equations 1.14 through 1.19. For the case of plane stress, $\sigma_{zz} = \tau_{zx} = \tau_{yz} = 0$. Using these in Hooke's law leads to the following.

$$\epsilon_{xx} = \frac{1}{E}[\sigma_{xx} - \nu\,\sigma_{yy}]$$

$$\epsilon_{yy} = \frac{1}{E}[\sigma_{yy} - \nu\sigma_{xx}]$$

$$\epsilon_{zz} = -\frac{\nu}{E}[\sigma_{xx} + \sigma_{yy}]$$

$$\gamma_{xy} = \frac{1}{G}\tau_{xy}$$

$$\gamma_{yz} = \gamma_{zx} = 0$$

[6] There are several types of stress strain relations possible. The simplest and most commonly used ones are called the Mooney-Rivlin equations.

In the case of plane strain, $\epsilon_{zz} = 0$, and $\tau_{zx} = \tau_{yz} = 0$. Substituting these in Hooke's law gives the following relations for plane strain:

$$\epsilon_{xx} = \frac{(1 - \nu^2)}{E}\left[\sigma_{xx} - \left(\frac{\nu}{1 + \nu}\right)\sigma_{yy}\right]$$

$$\epsilon_{yy} = \frac{(1 - \nu^2)}{E}\left[\sigma_{yy} - \left(\frac{\nu}{1 + \nu}\right)\sigma_{xx}\right]$$

$$\sigma_{zz} = -\nu[\sigma_{xx} + \sigma_{yy}]$$

$$\gamma_{xy} = \frac{1}{G}\tau_{xy}$$

$$\gamma_{yz} = \gamma_{zx} = 0$$

Plane stress situations exist in thin laminae subjected to in-plane loads. Plane strain situations exist in the middle of long prismatic objects subjected to two dimensional loading all along their length. The former case represents the complete absence of $z$-constraint, and the latter represents full $z$-constraint.

## 1.12 ANALYSIS OF STRESS AT A POINT

### 1.12.1 Transformation of Stresses in Two Dimensions

Consider an element of size ($dx \times dy$), which is subjected to a general two dimensional stress state, namely two normal stresses and one shear stress (Figure 1.6(a)). Consider the $x'$ plane, which is inclined to the $x$ plane at an angle $\theta$. The positive normal stress $\sigma_{x'}$ and shear stress $\tau'$ on this plane are shown in Figure 1.6(b). Here, we drop the double subscript notation. Being a two dimensional situation, there is only one shear stress. The normal stresses are denoted by the Greek letter $\sigma$ and just a single subscript. Because every differential element of the body must be in equilibrium, we may consider the equilibrium of the wedge-shaped element $bce$, along and normal to the $x'$ plane. This element is chosen because its sides lie along the axes of both coordinate systems.

The following equations are written for the forces, acting on the faces $bc$, $ce$, and $eb$ for a unit width of the wedge. Equilibrium is considered along the $x'$ axis.

$$\sigma_{x'}(be) = \sigma_x(bc)\cos\theta + \tau(bc)\sin\theta + \sigma_y(ec)\sin\theta + \tau(ec)\cos\theta$$

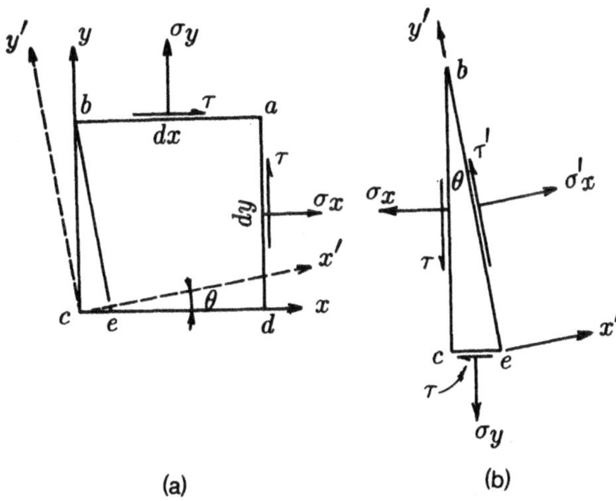


**FIGURE 1.6.** Transformation of stress on to an arbitrary plane.

Dividing both sides by *be*,

$$\sigma_{x'} = \sigma_x \cos^2\theta + \sigma_y \sin^2\theta + \tau \sin 2\theta$$

$$= \left(\frac{\sigma_x + \sigma_y}{2}\right) + \left(\frac{\sigma_x - \sigma_y}{2}\right)\cos 2\theta + \tau \sin 2\theta \tag{1.25}$$

In the same manner, by writing the equilibrium equation along the plane *eb*, we can obtain the following.

$$\tau' = -\left(\frac{\sigma_x - \sigma_y}{2}\right)\sin 2\theta + \tau \cos 2\theta \tag{1.26}$$

No separate equation for $\sigma_{y'}$ is needed, because it is obtained simply by changing $\theta$ to $[\theta + (\pi/2)]$.

### 1.12.2 Mohr's Circle

The transformation rules obtained in Equations 1.25 and 1.26 are easily seen to be quite different from the way vectors would project. This difference is due to the fact that stress is (force)/(area), where both force and area are vectors.

If we consider a $\sigma$-$\tau$ plane in which the normal stress $\sigma$ is laid out along the horizontal axis and $\tau$ is along the vertical axis, then Equations 1.25 and 1.26 suggest that the transformed stresses would be represented

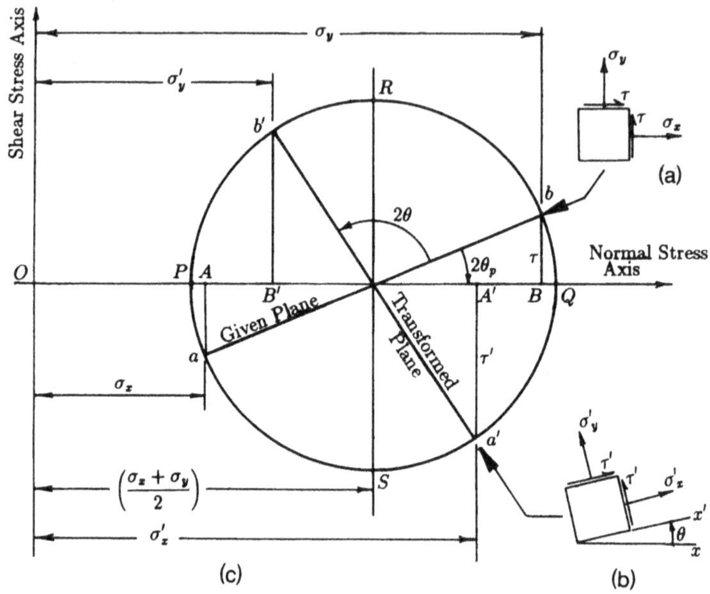


**FIGURE 1.7.** Construction of Mohr's circle.

by points lying on a circle. The center of the circle is at $[(\sigma_x + \sigma_y)/2, 0]$, and its radius is $\sqrt{[(\sigma_x - \sigma_y)/2]^2 + \tau^2}$.

Note that the horizontal and vertical axes of $\sigma$ and $\tau$ have nothing to do with the $x$ and $y$ axes of the problem. Further, an angle of inclination $\theta$ of a plane with respect to the $x$ axis corresponds to an angle of $2\theta$ on the circle. *The stress at a point, and its components on any plane can be located uniquely on the circle*. This approach, called the *Mohr's circle*, elucidates several properties of the stress as a tensor. The procedure helps visualize the problem of projection of stresses as a problem in simple geometry.

The first prerequisite for constructing the Mohr's circle is to have a layout of the stresses and the directions. Since we are considering two dimensions, we have $\sigma_x$, $\sigma_y$, and $\tau$, as shown in Figure 1.7(a). Let us suppose that the components of the stresses are required along the directions $x'$ and $y'$, which are inclined at an angle $\theta$ to the $x$ and $y$ axes respectively, as in Figure 1.7(b). The following is a step by step procedure to draw the Mohr's circle for the problem. Note that the Mohr's circle is a semigraphical procedure, because it is not necessary to draw it exactly to scale.

1. Choose a horizontal axis for normal stresses, positive (or tensile) to the right; choose a vertical axis for shear stresses, direction undefined.

2. Mark $A$ corresponding to $\sigma_x$, and $B$ corresponding to $\sigma_y$.
3. Direction of $\tau$ is decided as follows. On the $x$ plane, the direction of $\tau$ causes a counterclockwise moment about the element centroid in Figure 1.7(a). Draw the ordinate $Aa$ downwards. On the $y$ plane, the direction of $\tau$ causes a clockwise moment and accordingly, the ordinate $Bb$ is drawn upwards corresponding to $\tau$. Thus, the direction of $\tau$ is decided by the direction of moment. (Beer and Johnston [4], suggest a "kitchen rule." *In the kitchen, clock is up and counter is down.*)
4. Join $ab$, which bisects $AB$ at $C$. With $C$ as center, and $Cb$ as radius draw a circle. This is the Mohr's circle for the given state of stress.
5. Corresponding to the counterclockwise rotation of axes by $\theta$, push the point $a$ to $a'$ and $b$ to $b'$, so that the line $a'b'$ makes $2\theta$, counterclockwise, with $ab$.
6. The ordinates $A'a'$ and $B'b'$ represent the $\tau'$ in the new axes. $OA'$ represents $\sigma_{x'}$ and $OB'$ represents $\sigma_{y'}$.
7. The points $P$ and $Q$ represent the maximum normal stresses possible for the given state of stress. Based on the angle $2\theta_p$ in the Figure 1.7(c), we conclude that the stresses attain the extreme values of normal stresses on planes at an angle of $\theta_p$.
8. Points $R$ and $S$ on the Mohr's circle, represent the highest possible value of shear stress, but are also accompanied by normal stresses corresponding to $OC$.

### 1.12.3 Principal Stresses

Many interesting properties associated with the transformation of stresses at a point can be deduced from the Mohr's circle.

- It is obvious from the Mohr's circle, that the ends of the horizontal diameter represent the maximum and minimum values of normal stresses. These extreme values are also known as "principal stresses."
- From the Mohr's circle, it can be seen that the values of principal stresses are given by

$$\sigma_1, \sigma_2 = \text{Abscissa of center} \pm \text{radius}$$

$$\sigma_1, \sigma_2 = \left(\frac{\sigma_x + \sigma_y}{2}\right) \pm \sqrt{\left(\frac{\sigma_x - \sigma_y}{2}\right)^2 + \tau^2} \tag{1.27}$$

- Principal stresses are independent of the coordinate system used to observe or calculate the stress. Thus they are important in coordinate-independent concepts, such as yielding, failure theories, etc.

- The principal stresses lie on planes which are inclined to the $x$ axis at angles $\theta_p$ and $(\pi/2) + \theta_p$, where $\theta_p$ is given by

$$\theta_p = \arctan\left(\frac{2\tau}{\sigma_x - \sigma_y}\right) \tag{1.28}$$

- Maximum shear stress is seen to be simply the value of the radius of the circle, and is given by

$$\tau_{\max} = \sqrt{\left(\frac{\sigma_x - \sigma_y}{2}\right)^2 + \tau^2} \tag{1.29}$$

- On the principal planes, there is only normal stress, and no shear stress.
- However, on the planes of maximum shear stress, there are normal stresses also.
- The sum $(\sigma_x + \sigma_y)$ for any direction is a constant, because it is equal to $2OC$ for all cases. This quantity is called *stress invariant*.
- The maximum shear stress is also a stress invariant, because it corresponds to the radius of the circle.

### 1.12.4 Transformation of Strain in Two Dimensions

As might be expected, the rules of transformation of strain result in equations exactly similar to those of stress. It was mentioned earlier that the engineering shear strain is not an exact tensor. See Footnote 4. In other words, it does not exactly fit into the rules for transformation of tensors. It can be shown that if we multiply the engineering shear strains by a factor of 1/2, then the strain components follow the tensor transformation rules. The need for this reconciliation arises from the fact that a formal mathematical treatment was made long after the engineering definitions were institutionalized. Thus, in Equations 1.25 and 1.26, we need only to replace $\sigma_{\ldots}$ by $\epsilon_{\ldots}$ and $\tau_{\ldots}$ by $\gamma_{\ldots}/2$. The formulas for strain transformation are therefore,

$$\epsilon_{x'} = \left(\frac{\epsilon_x + \epsilon_y}{2}\right) + \left(\frac{\epsilon_x - \epsilon_y}{2}\right)\cos 2\theta + \frac{\gamma}{2}\sin 2\theta$$

$$\gamma' = -\left(\frac{\epsilon_x - \epsilon_y}{2}\right)\sin 2\theta + \frac{\gamma}{2}\cos 2\theta.$$

Principal strains are given by

$$\epsilon_1, \epsilon_2 = \left(\frac{\epsilon_x + \epsilon_y}{2}\right) \pm \sqrt{\left(\frac{\epsilon_x + \epsilon_y}{2}\right)^2 + \left(\frac{\gamma}{2}\right)^2} \tag{1.30}$$

Directions of the principal strains are given by

$$\theta_p = \arctan\left(\frac{\gamma}{\epsilon_x - \epsilon_y}\right) \tag{1.31}$$

It is interesting to note that for isotropic, linear elastic materials, the principal stresses and principal strains have coincident directions. That is, the $\theta_p$ obtained from Equations 1.28 and 1.31 are identical. This is easily seen to be true from the fact that by Hooke's law, we have

$$\left(\frac{\gamma}{\epsilon_x - \epsilon_y}\right) = \left(\frac{2\tau}{\sigma_x - \sigma_y}\right)$$

However, for anisotropic materials this is not true.

Mohr's circle can be used to transform strains exactly in the same manner as the stresses.

## 1.13 REPRESENTATION BY MATRICES

### 1.13.1 Matrix Representation of Stress

It was seen earlier that two subscripts are needed to completely describe a stress. We can use this to represent the stress components as a matrix of size 3 × 3 as shown below.

$$[\boldsymbol{\sigma}] = \begin{bmatrix} \sigma_{xx} & \tau_{xy} & \tau_{xz} \\ \tau_{yx} & \sigma_{yy} & \tau_{yz} \\ \tau_{zx} & \tau_{zy} & \sigma_{zz} \end{bmatrix}$$

Because of the symmetry of shear stresses, the stress matrix $[\boldsymbol{\sigma}]$ is always symmetric.

### 1.13.2 Matrix Representation of Strain

Strain, like stress, is completely identified by two subscripts. Hence, strains can also be represented in the form of a symmetric matrix of size 3 × 3.

Recall the factor of $1/2$ for shear strains in the calculation of principal strains. It is needed to make strain an exact tensor. Accordingly, in the matrix representation of strain, we introduce the factor $1/2$ for shear strains.

$$[\mathbf{e}] = \begin{bmatrix} \epsilon_{xx} & \frac{1}{2}\gamma_{xy} & \frac{1}{2}\gamma_{xz} \\ \frac{1}{2}\gamma_{yx} & \epsilon_{yy} & \frac{1}{2}\gamma_{yz} \\ \frac{1}{2}\gamma_{zx} & \frac{1}{2}\gamma_{zy} & \epsilon_{zz} \end{bmatrix}$$

### 1.13.3 Hooke's Law in Matrix Form

Hooke's law can be expressed in the form of matrices also. However, some modifications are needed in the form of stress and strain matrices. They need to be written as *column vectors* rather than as square matrices. This is required because moduli belong with the fourth order tensors, and as such cannot be represented as a two dimensional array of rows and colums. It can be verified that the following equation is the same as Hooke's law.

$$\begin{Bmatrix} \epsilon_{xx} \\ \epsilon_{yy} \\ \epsilon_{zz} \\ (\gamma_{xy})/2 \\ (\gamma_{yz})/2 \\ (\gamma_{zx})/2 \end{Bmatrix} = \frac{1}{E} \begin{bmatrix} 1 & -\nu & -\nu & 0 & 0 & 0 \\ -\nu & 1 & -\nu & 0 & 0 & 0 \\ -\nu & -\nu & 1 & 0 & 0 & 0 \\ 0 & 0 & 0 & (1+\nu) & 0 & 0 \\ 0 & 0 & 0 & 0 & (1+\nu) & 0 \\ 0 & 0 & 0 & 0 & 0 & (1+\nu) \end{bmatrix} \begin{Bmatrix} \sigma_{xx} \\ \sigma_{yy} \\ \sigma_{zz} \\ \tau_{xy} \\ \tau_{yz} \\ \tau_{zx} \end{Bmatrix}$$

## 1.14 TRANSFORMATION USING MATRICES

We may, at this stage, look at a purely formal procedure for transformation of stress and strain in two dimensions. This would provide the groundwork for proceeding to the rules in three dimensions. Consider the coordinate transformation from $xy$ system to the $x'y'$ system as shown in Figure 1.8.

$$\begin{Bmatrix} x' \\ y' \end{Bmatrix} = \begin{bmatrix} \cos\theta & \sin\theta \\ -\sin\theta & \cos\theta \end{bmatrix} \begin{Bmatrix} x \\ y \end{Bmatrix}$$

Or more briefly,

$$\{\mathbf{x}'\} = [\mathbf{L}]\{\mathbf{x}\},$$

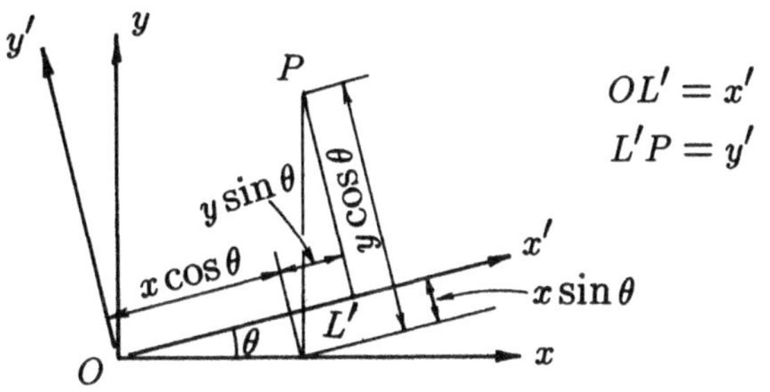


**FIGURE 1.8.** Transformation of coordinates in two dimensions.

where [**L**] is the matrix of "direction cosines." The reader can easily verify that the rules of stress transformation of Equations 1.25 and 1.26 can be summarized in a single matrix equation.

$$\begin{bmatrix} \sigma_{x'} & \tau' \\ \tau' & \sigma_{y'} \end{bmatrix} = \begin{bmatrix} \cos\theta & \sin\theta \\ -\sin\theta & \cos\theta \end{bmatrix} \begin{bmatrix} \sigma_x & \tau \\ \tau & \sigma_y \end{bmatrix} \begin{bmatrix} \cos\theta & -\sin\theta \\ \sin\theta & \cos\theta \end{bmatrix}$$

Or, alternatively $[\boldsymbol{\sigma}'] = [\mathbf{L}][\boldsymbol{\sigma}][\mathbf{L}]^{\mathrm{T}}$, where $[\mathbf{L}]^{\mathrm{T}}$ represents the transpose of the matrix [**L**]. The concept of principal stress can be introduced here with a different perspective. We have earlier seen that the principal stresses are pure normal stresses, unaccompanied by shear stresses. We can utilize this property here. Instead of asking for the maximum normal components of stress, we ask the question "*Is there a set of axes along which the stress has only normal components, and no shear components? If so, what are their magnitudes?*" Stated in the language of the matrix theory, we ask "*For what values of λ is the following equation valid?*"

$$\begin{bmatrix} \sigma_x & \tau \\ \tau & \sigma_y \end{bmatrix} = \begin{bmatrix} \lambda & 0 \\ 0 & \lambda \end{bmatrix}$$

Transposing, we observe that we get a pair of homogeneous equations (that is, equations with only zero terms on the right hand side). They have a solution if, and only if, the determinant of the coefficient matrix is zero. We therefore require,

$$\text{Det}\begin{bmatrix} (\sigma_x - \lambda) & \tau \\ \tau & (\sigma_y - \lambda) \end{bmatrix} = 0$$

Expanding the determinant, we get a quadratic in $\lambda$ which gives the two roots which are identical to the principal stresses in Equation 1.27. Verification of this statement is left to the reader as an exercise.

In the language of matrix algebra, we say that the principal stresses are the *eigenvalues* of the stress matrix $[\boldsymbol{\sigma}]$. Direction cosines of the principal directions are given by the *eigenvectors* of the stress matrix.

### 1.14.1 Transformation of Stress in Three Dimensions

In order to obtain a transformation of stress (or strain) in three dimensions from an *xyz* system to an $x'y'z'$ system, we need in the first place a rule for transformation of coordinates. This is shown in Figure 1.9. We construct the [**L**] matrix in three dimensions. If the direction cosines of the $x'$ axis are $l_1$, $m_1$, and $n_1$ respectively, those of the $y'$ axis are $l_2$, $m_2$, and $n_2$, and finally those of the $z'$ axis are $l_3$, $m_3$, and $n_3$ respectively, then

$$\begin{Bmatrix} x' \\ y' \\ z' \end{Bmatrix} = \begin{bmatrix} l_1 & m_1 & n_1 \\ l_2 & m_2 & n_2 \\ l_3 & m_3 & n_3 \end{bmatrix} \begin{Bmatrix} x \\ y \\ z \end{Bmatrix}$$

Or, as before, the matrix of direction cosines in three dimensions can be denoted by [**L**]. The actual direction cosines in the matrix [**L**] have to satisfy conditions of orthogonality, that is, conditions to reflect the fact that the new set of axes are mutually perpendicular. The conditions are as follows.

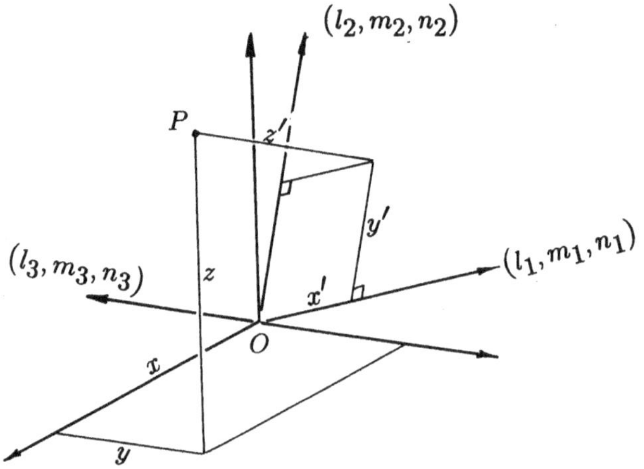


**FIGURE 1.9.** Coordinate transformation in three dimensions.

Since all are unit vectors,

$$l_1^2 + m_1^2 + n_1^2 = 1$$

$$l_2^2 + m_2^2 + n_2^2 = 1$$

$$l_3^2 + m_3^2 + n_3^2 = 1$$

Since any two axes are normal to each other,

$$l_1 l_2 + m_1 m_2 + n_1 n_2 = 0$$

$$l_2 l_3 + m_2 m_3 + n_2 n_3 = 0$$

$$l_3 l_1 + m_3 m_1 + n_3 n_1 = 0$$

Matrices of direction cosines that satisfy all the above conditions are called "orthogonal matrices." Their most important property is that the inverse of the matrix is simply its transpose. The reader can verify this for the two dimensional matrices of the previous section.

The transformation of stresses in three dimensions can be written in terms of the orthogonal [L] matrix, in the same form as for two dimensions. We have,

$$[\boldsymbol{\sigma}'] = [\mathbf{L}][\boldsymbol{\sigma}][\mathbf{L}]^{\mathrm{T}}$$

When written in full, the above equation has the form,

$$\begin{bmatrix} \sigma_{x'x'} & \tau_{x'y'} & \tau_{x'z'} \\ \tau_{y'x'} & \sigma_{y'y'} & \tau_{y'z'} \\ \tau_{z'x'} & \tau_{z'y'} & \sigma_{z'z'} \end{bmatrix}$$

$$= \begin{bmatrix} l_1 & m_1 & n_1 \\ l_2 & m_2 & n_2 \\ l_3 & m_3 & n_3 \end{bmatrix} \begin{bmatrix} \sigma_{xx} & \tau_{xy} & \tau_{xz} \\ \tau_{yx} & \sigma_{yy} & \tau_{yz} \\ \tau_{zx} & \tau_{zy} & \sigma_{zz} \end{bmatrix} \begin{bmatrix} l_1 & l_2 & l_3 \\ m_1 & m_2 & m_3 \\ n_1 & n_2 & n_3 \end{bmatrix} \tag{1.32}$$

### 1.14.2 Principal Stress and Direction in Three Dimensions

Principal stresses in three dimensions are also obtained in exactly the same way as in two dimensions. We require that the determinant below be zero.

That is,

$$\text{Det}\begin{bmatrix} (\sigma_{xx} - \lambda) & \tau_{xy} & \tau_{xz} \\ \tau_{yx} & (\sigma_{yy} - \lambda) & \tau_{yz} \\ \tau_{zx} & \tau_{zy} & (\sigma_{zz} - \lambda) \end{bmatrix} = 0 \tag{1.33}$$

This equation leads to a cubic in $\lambda$, giving three roots $\lambda_1$, $\lambda_2$, and $\lambda_3$, which are the principal stresses. The symmetry of stress matrix $[\boldsymbol{\sigma}]$ guarantees that the roots of the cubic equation will all be real. The eigenvectors associated with each root give the direction cosines for the corresponding principal stresses. Since it is beyond the scope of this book to discuss the computation of the eigenvalues and eigenvectors, it is not explicitly discussed in this section. Examples below illustrate the techniques of calculating the eigenvalues and eigenvectors. The reader must refer to standard text books on matrix algebra [5] for the procedures of extracting eigenvalues.

Associated with each eigenvalue (principal stress), is an eigenvector (direction). Eigenvectors are more complex to calculate than are the eigenvalues. When two or three principal stresses are equal, then the procedure becomes still more complex. The discussions of coincident eigenvalues having different eigenvectors can be found in References [5] and [6], and several other comparable text books.

### 1.14.3 Mohr's Circle in Three Dimensions

There is, of course, a procedure for constructing Mohr's circle in three dimensions to determine the normal and shear stresses on a plane. Many text books on advanced strength of materials, such as, for example, Faupel and Fisher [2], and Seeley and Smith [7] describe the procedure. We will not discuss the procedure here. The procedure is *not* completely sufficient, especially when compared to what can be done on programmable calculators, and personal computers which are currently available. The main disadvantages of the Mohr's circle in three dimensions are:

- It is no longer a semigraphical procedure as in two dimensions; the diagram must be drawn to scale.
- It is not good for projecting from any one set of axes to any other; it is good for projecting from the *principal planes* to any other plane. This leaves upfront the task of determining the principal stresses and directions. Hence, there is practically no saving in labor.

- In terms of intuitive understanding of the transformation, the matrix method appears better than the three-dimensional Mohr's circle.

### 1.14.4 Properties of Transformation

There are several properties associated with the process of transformation of stress and strain, in two as well as in three dimensions. Knowing those properties may not be indispensable for regular calculations, but it certainly improves the understanding of the nature of stress and strain as physical quantities.

- If stress and strain are both transformed from the *xyz* system to the $x'y'z'$ system, they will satisfy Hooke's law in the $x'y'z'$ coordinates automatically. For isotropic elasticity, the two constants $E$ and $\nu$ transform into themselves. In anisotropic elasticity, however, the transformation reveals the "fourth-order-tensor" character of the moduli.[7] In brief, the transformed moduli will fit into Hooke's law together with the transformed stresses and transformed strains.
- The quantities $(\sigma_x + \sigma_y + \sigma_z)$ and $(\epsilon_x + \epsilon_y + \epsilon_z)$ remain constant under transformation. The sum of the strains represents the volumetric expansion at a given point.
- If the principal stresses are such that $\sigma_1 = \sigma_2 = \sigma_3 = \sigma_o$, then the components along any set of orthogonal axes are equal to $\sigma_o$, a constant. Such a state is called the "hydrostatic" stress state.
- If the $\sigma_x + \sigma_y + \sigma_z = 0$, then there exists a coordinate system along which there are only shear stresses, and no normal stresses at all.
- If stresses are projected along the principal planes, shear stresses are guaranteed to turn out to be zero. Text books in matrix algebra refer to this process (of projecting on to the principal axes) as "diagonalization" of a stress/strain matrix.

**Example 1.1**

At a point in a component, there is a two dimensional stress state shown schematically in Figure 1.10.

Find (a) the components of stresses on planes inclined at $\theta = 20^o$ to *xy* axes and (b) the principal stresses and principal directions.

*Solution* We can do this problem by all the three approaches, namely (i) direct application of transformation formulas, Equation 1.27, (ii) using Mohr's circle and (iii) lastly, by using the matrices.

---

[7] The transformation rules for moduli will be discussed in Chapter 7.

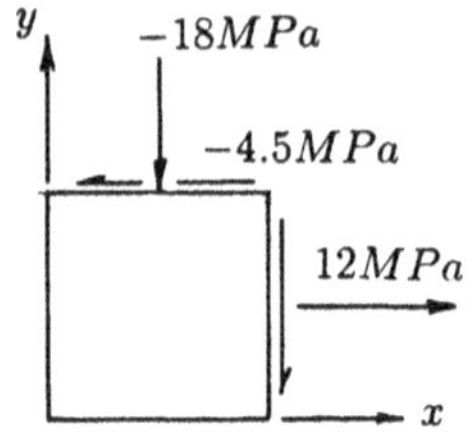


**FIGURE 1.10.** Example 1.1.

***By Direct Application of Formulas***

$$\sigma_x = 12 \text{ MPa}; \ \sigma_y = -18 \text{ MPa and } \tau = -4.5 \text{ MPa}; \ \theta = 20^o$$

$$\sigma_{x'} = \left(\frac{\sigma_x + \sigma_y}{2}\right) + \left(\frac{\sigma_x - \sigma_y}{2}\right)\cos 2\theta + \tau \sin 2\theta$$

$$\sigma_{y'} = \left(\frac{\sigma_x + \sigma_y}{2}\right) - \left(\frac{\sigma_x - \sigma_y}{2}\right)\cos 2\theta + \tau \sin 2\theta$$

$$\tau' = \left(\frac{\sigma_x - \sigma_y}{2}\right)\sin 2\theta + \tau \cos 2\theta$$

Substituting the values of $\sigma_x$, $\sigma_y$, and $\tau$ in the above equations, we obtain,

$$\sigma_{x'} = \left(\frac{12 - 18}{2}\right) + \left(\frac{12 + 18}{2}\right)\cos 40 - 4.5 \sin 40$$

$$= -3.0 + 15.0 \cos 40 - 4.5 \sin 40$$

$$= 5.598 \text{ MPa}$$

$$\sigma_{y'} = \left(\frac{12 - 18}{2}\right) - \left(\frac{12 + 18}{2}\right)\cos 40 - 4.5 \sin 40$$

$$= -3.0 - 15.0 \cos 40 - 4.5 \sin 40$$

$$= -11.60 \text{ MPa}$$

$$\tau = \left(\frac{12 + 18}{2}\right)\sin 40 - 4.5 \cos 40$$

$$= -13.08 \text{ MPa}$$

The principal stresses can be found from Equations 1.27.

$$\sigma_1 = \left(\frac{\sigma_x + \sigma_y}{2}\right) + \sqrt{\left(\frac{\sigma_x - \sigma_y}{2}\right)^2 + \tau^2}$$

$$= \left(\frac{12 - 18}{2}\right) + \sqrt{\left(\frac{12 + 18}{2}\right)^2 + (-4.5)^2}$$

$$= 12.66 \text{ MPa}$$

$$\sigma_2 = \left(\frac{\sigma_x + \sigma_y}{2}\right) - \sqrt{\left(\frac{\sigma_x - \sigma_y}{2}\right)^2 + \tau^2}$$

$$= \left(\frac{12 - 18}{2}\right) - \sqrt{\left(\frac{12 + 18}{2}\right)^2 + (-4.5)^2}$$

$$= -18.66 \text{ MPa}$$

***By Mohr's Circle***

The procedure for drawing the Mohr's circle (Figure 1.11) is as follows.

1. Mark a horizontal axis for normal stresses $\sigma_*$, positive to the right and

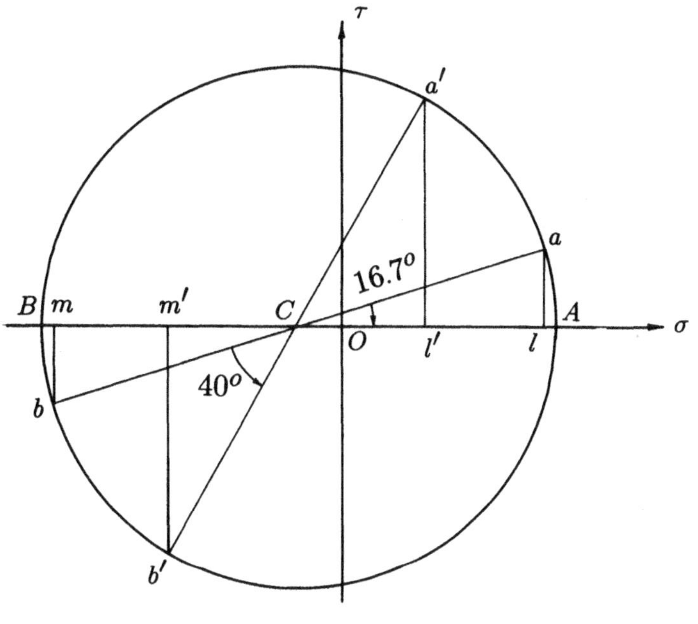


**FIGURE 1.11.** Use of Mohr's circle for stress transformation.

negative to the left of the origin $O$. Draw a vertical axis for shear stress $\tau$, the direction to be decided as below.
2. Mark $Ol = 12$ MPa, for $\sigma_x$; Mark $Om = -18$ MPa, for $\sigma_y$.
3. To determine the direction of the ordinate at $l$, we note that the direction of the shear is such as to produce a clockwise moment about the center of the element being considered. By the "kitchen rule," the ordinate is to be "up." Automatically, the ordinate at $m$, corresponding to $\sigma_y$ must be drawn downwards.
4. Draw the ordinates, $la$ and $mb$; join $ab$, which cuts the $x$-axis at $C$, such that $OC = (\sigma_x + \sigma_y)/2 = -3.0$ MPa.
5. With $C$ as center, and $Ca$ as radius, draw the Mohr's circle.
6. To find the components on the plane $20^o$ from the $xy$ planes, simply turn the diameter $ab$ by $2 \times 20^o$ counterclockwise, to the new position $a'b'$, its projections on the horizontal being $l'$ and $m'$ respectively.
7. We have

$$Ca = \text{radius of the circle},$$

$$R = \sqrt{(Cl^2 + al^2)} = \sqrt{(15.0^2 + 4.5^2)} = 15.66 \text{ MPa}.$$

Further, the angle $a\hat{C}A$ is easily seen to be equal to $16.7^o$.
8. Components on the $x'y'$ planes are simply given by

$$Ol' = \sigma_{x'} = R\cos(40 + 16.7) - OC = 5.598 \text{ MPa}$$

$$Om' = \sigma_{y'} = -R\cos(40 + 16.7) - OC = -11.6 \text{ MPa}$$

$$a'l' = \tau = -R\sin(40 + 16.7) = -13.08 \text{ MPa}$$

In the above, the sign of $\tau$ was assigned by using the "kitchen rule" in reverse.
9. Principal stresses are given by $OA$ and $OB$, which are seen to be

$$\sigma_1 = OA = R - OC = 15.66 - 3.0 = 12.66 \text{ MPa}$$

$$\sigma_2 = OB = R + OC = -(15.66 + 3.0) = -18.66 \text{ MPa}$$

The principal direction is simply given by one half of the angle $a\hat{C}A$. Therefore, the principal directions are at $8.35^o$, clockwise to the $xy$ planes.

***By Matrix Transformation***

The [L] matrix of coordinate transformation is given by

$$[\mathbf{L}] = \begin{bmatrix} \cos 20 & \sin 20 \\ -\sin 20 & \cos 20 \end{bmatrix}$$

The stress transformation equation is $[\boldsymbol{\sigma}'] = [\mathbf{L}][\boldsymbol{\sigma}][\mathbf{L}]^T$. Written out in full, the transformation equation becomes,

$$\begin{bmatrix} \sigma_{x'} & \tau' \\ \tau' & \sigma_{y'} \end{bmatrix}$$

$$= \begin{bmatrix} \cos 20 & \sin 20 \\ -\sin 20 & \cos 20 \end{bmatrix} \begin{bmatrix} 12 & -4.5 \\ -4.5 & -18 \end{bmatrix} \begin{bmatrix} \cos 20 & -\sin 20 \\ \sin 20 & \cos 20 \end{bmatrix}$$

$$= \begin{bmatrix} 0.9397 & 0.3420 \\ -0.3420 & 0.9397 \end{bmatrix} \begin{bmatrix} 12 & -4.5 \\ -4.5 & -18 \end{bmatrix} \begin{bmatrix} 0.9397 & -0.3420 \\ 0.3420 & 0.9397 \end{bmatrix}$$

$$= \begin{bmatrix} 5.598 & -13.08 \\ -13.08 & -11.60 \end{bmatrix} \text{MPa}$$

The principal stresses are found by requiring that the determinant of the eigen-matrix be zero. That is,

$$\text{Det}\begin{bmatrix} (\sigma_x - \lambda) & \tau \\ \tau & (\sigma_y - \lambda) \end{bmatrix} = 0$$

$$\text{Det}\begin{bmatrix} (12.00 - \lambda) & -4.5 \\ -4.5 & -(18.00 + \lambda) \end{bmatrix} = 0$$

Expanding the determinant, we get

$$-(12 - \lambda)(18 + \lambda) - 4.5^2 = 0$$

$$\lambda^2 + 6\lambda - 236.25 = 0$$

Solving,

$$\lambda_1 = \frac{-6 + \sqrt{6^2 + 4(236.25)}}{2} = 12.66 \text{ MPa}$$

$$\lambda_2 = \frac{-6 - \sqrt{6^2 + 4(236.25)}}{2} = -18.66 \text{ MPa}$$

**Example 1.2**

At a point in a body that is triaxially stressed, the stresses are given by

$$\sigma_{xx} = 75 \text{ MPa}; \quad \sigma_{yy} = 60 \text{ MPa}; \quad \sigma_{zz} = 50 \text{ MPa};$$

$$\tau_{xy} = 25 \text{ MPa}; \quad \tau_{yz} = -25 \text{ MPa}; \quad \tau_{zx} = 0 \text{ MPa};$$

Figure 1.12 shows this stress distribution. Find the components of this stress state on a $x'y'z'$ coordinate axes system defined by the followng direction cosines.

| | $x$ | $y$ | $z$ |
|---|---|---|---|
| $x'$ | 2/3 | 2/3 | −1/3 |
| $y'$ | −2/3 | 1/3 | −2/3 |
| $z'$ | −1/3 | 2/3 | 2/3 |

Find also the magnitudes of the principal stresses, and the directions of the principal axes.

*Solution* Being a three dimensional stress state, we resort to the matrix method. We have the following matrices for [**L**] and [$\boldsymbol{\sigma}$].

$$[\mathbf{L}] = \begin{bmatrix} 2/3 & 2/3 & -1/3 \\ -2/3 & 1/3 & -2/3 \\ -1/3 & 2/3 & 2/3 \end{bmatrix}; \quad [\boldsymbol{\sigma}] = \begin{bmatrix} 75 & 25 & 0 \\ 25 & 60 & -25 \\ 0 & -25 & 50 \end{bmatrix}$$

We use the transformation rule $[\boldsymbol{\sigma}'] = [\mathbf{L}][\boldsymbol{\sigma}][\mathbf{L}]^{\mathrm{T}}$. Taking a common

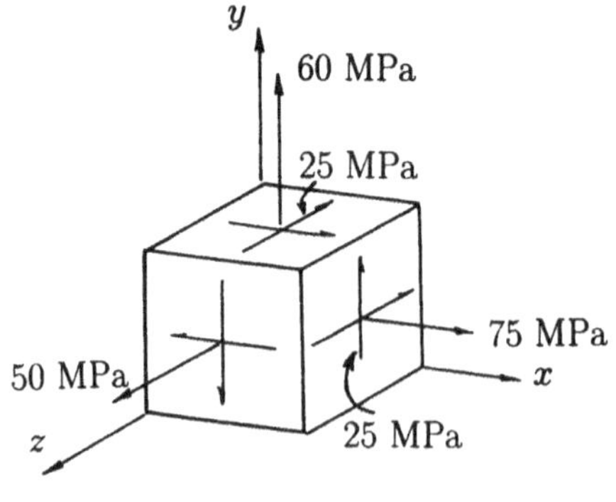


**FIGURE 1.12.** Example 1.2.

factor 1/3 out of the two [L] matrices, we get

$$[\boldsymbol{\sigma}'] = \frac{1}{9}\begin{bmatrix} 2 & 2 & -1 \\ -2 & 1 & -2 \\ -1 & 2 & 2 \end{bmatrix}\begin{bmatrix} 75 & 25 & 0 \\ 25 & 60 & -25 \\ 0 & -25 & 50 \end{bmatrix}\begin{bmatrix} 2 & -2 & -1 \\ 2 & 1 & 2 \\ -1 & -2 & 2 \end{bmatrix}$$

$$= \frac{1}{9}\begin{bmatrix} 2 & 2 & -1 \\ -2 & 1 & -2 \\ -1 & 2 & 2 \end{bmatrix}\begin{bmatrix} 200 & -125 & -25 \\ 195 & 60 & 45 \\ -100 & -125 & 50 \end{bmatrix}$$

$$= \frac{1}{9}\begin{bmatrix} 890 & -5 & -10 \\ -5 & 560 & -5 \\ -10 & -5 & 215 \end{bmatrix} \text{MPa}$$

$$= \begin{bmatrix} 98.89 & -0.56 & -1.11 \\ -0.56 & 62.22 & -0.56 \\ -1.11 & -0.56 & 23.89 \end{bmatrix} \text{MPa}$$

The near-zero values of shear stresses suggest that the principal directions are close to the $x'y'z'$ axes, and that the principal stresses are also nearly equal to the diagonal terms in the $[\boldsymbol{\sigma}']$ matrix above.

***Principal Stresses***

Principal stresses are determined as the eigenvalues of the stress matrix. In other words, the principal stresses are the same as the roots of the following "characteristic equation."

$$\text{Det}\begin{bmatrix} (75-\lambda) & 25 & 0 \\ 25 & (60-\lambda) & -25 \\ 0 & -25 & (50-\lambda) \end{bmatrix} = 0$$

Expanding the determinant, we get,

$$\lambda^3 - 185\lambda^2 + 10000\lambda - 146875 = 0$$

We use the clue that the principal stresses are close to the components on $x'y'z'$ planes. Thus, we choose the first approximation of the roots between 23 and 24, between 62 and 63, and between 98 and 99 respectively. By an iterative process of refinements the roots are found to be

$$\sigma_1 = \lambda_1 = 98.914 \text{ MPa}$$

$$\sigma_2 = \lambda_2 = 62.222 \text{ MPa}$$

$$\sigma_3 = \lambda_3 = 23.864 \text{ MPa}$$

***Directions of Principal Planes***

The method of determining the principal directions (or the eigenvectors) is a little more complex than is that for the principal stresses. In the following, we proceed to determine the eigenvectors, using only first principles.

In order to determine the eigenvectors, we substitute for $\lambda_1$, in the characteristic matrix, and ask for the nonzero solutions $l_1$, $m_1$, and $n_1$ for the following equation.

$$\begin{bmatrix} (\sigma_x - \lambda) & \tau_{xy} & \tau_{xz} \\ \tau_{yx} & (\sigma_y - \lambda) & \tau_{yz} \\ \tau_{zx} & \tau_{zy} & (\sigma_z - \lambda) \end{bmatrix} \begin{Bmatrix} l \\ m \\ n \end{Bmatrix} = \begin{Bmatrix} 0 \\ 0 \\ 0 \end{Bmatrix}$$

***Case 1 ($\lambda_1$ = 98.914 MPa)***

Substituting for $\lambda_1$, we get the following equation.

$$\begin{bmatrix} -23.914 & 25 & 0 \\ 25 & -38.914 & -25 \\ 0 & -25 & -48.914 \end{bmatrix} \begin{Bmatrix} l_1 \\ m_1 \\ n_1 \end{Bmatrix} = \begin{Bmatrix} 0 \\ 0 \\ 0 \end{Bmatrix}$$

This must have a nonzero solution because the determinant is zero. (That is the condition used for determining the eigenvalues!) Also, there must be one constant in the solution, to be selected arbitrarily. We will use the condition that the direction vector has unit magnitude. From the first equation, we get, $l_1 = c_1$; and $m_1 = -0.95656c_1$. From the third equation, we get, $n_1 = -0.5111m_1 = -0.4889c_1$. Combining both, we see that

$$\frac{l_1}{1} = \frac{m_1}{0.95656} = \frac{n_1}{-0.4889} = \frac{\sqrt{l_1^2 + m_1^2 + n_1^2}}{\sqrt{1^2 + 0.95656^2 + 0.4889^2}} = \frac{1}{1.46769}$$

From this we obtain,

$$\begin{Bmatrix} l_1 \\ m_1 \\ n_1 \end{Bmatrix} = \begin{Bmatrix} 0.6813 \\ 0.6517 \\ -0.3331 \end{Bmatrix}$$

***Case 2 ($\lambda_2$ = 62.222 MPa)***

In this case the characteristic equation becomes

$$\begin{bmatrix} 12.778 & 25 & 0 \\ 25 & -2.222 & -25 \\ 0 & -25 & -12.222 \end{bmatrix} \begin{Bmatrix} l_2 \\ m_2 \\ n_2 \end{Bmatrix} = \begin{Bmatrix} 0 \\ 0 \\ 0 \end{Bmatrix}$$

As before, from the first equation, we get

$$l_2 = c_2; \quad \text{and} \quad m_2 = -0.51112c_2.$$

The third equation gives

$$n_2 = -2.0455m_2 = 1.0455c_2.$$

Combining the two relations, we see that

$$\frac{l_2}{1} = \frac{m_2}{-0.51112} = \frac{n_2}{1.0455} = \frac{\sqrt{l_2^2 + m_2^2 + n_2^2}}{\sqrt{1^2 + 0.51112^2 + 1.0455^2}} = \frac{1}{1.5344}$$

From this, we see that

$$\begin{Bmatrix} l_2 \\ m_2 \\ n_2 \end{Bmatrix} = \begin{Bmatrix} 0.6517 \\ -0.3331 \\ 0.6814 \end{Bmatrix}$$

***Case 3 ($\lambda_3$ = 23.864 MPa)***
For the last case, the substitution of $\lambda_3$ gives

$$\begin{bmatrix} 51.136 & 25 & 0 \\ 25 & 36.136 & -25 \\ 0 & -25 & 26.136 \end{bmatrix} \begin{Bmatrix} l_3 \\ m_3 \\ n_3 \end{Bmatrix} = \begin{Bmatrix} 0 \\ 0 \\ 0 \end{Bmatrix}$$

The first equation gives

$$l_3 = c_3; \quad \text{and} \quad m_3 = -2.04544l_3 = -2.04544c_3$$

The third equation gives

$$n_3 = 0.95656m_3 = -1.9566c_3$$

Combining the three equations,

$$\frac{l_3}{1} = \frac{m_3}{-2.04544} = \frac{n_3}{-1.9566} = \frac{\sqrt{l_3^2 + m_3^2 + n_3^2}}{\sqrt{1^2 + 2.04544^2 + 1.9566^2}} = \frac{1}{3.002}$$

From this we obtain

$$\begin{Bmatrix} l_3 \\ m_3 \\ n_3 \end{Bmatrix} = \begin{Bmatrix} 0.3331 \\ -0.6814 \\ -0.6518 \end{Bmatrix}$$

Notice that the first vector corresponding to $\lambda_1$ is close to the $x'$ axis, the second vector corresponding to $\lambda_2$ is close to the negative $y'$ axis, and the third vector corresponding to $\lambda_3$ is close to the negative $z'$ axis. The negative $y'$ and $z'$ axes can be fixed easily by just choosing the arbitrary constants $c_2$ and $c_3$ to be $-1$.

**Example 1.3**

The three-dimensional state of stress at a point in a structure is found to be

$$[\sigma] = \begin{bmatrix} 0 & 1 & 1 \\ 1 & 0 & 1 \\ 1 & 1 & 0 \end{bmatrix} \text{MPa}$$

Find the principal stresses and principal directions.

*Solution* To find the eigenvalues, we require, as before, that the following determinant be zero.

$$\text{Det}\begin{bmatrix} -\lambda & 1 & 1 \\ 1 & -\lambda & 1 \\ 1 & 1 & -\lambda \end{bmatrix} = 0$$

Expanding the determinant, we get

$$\lambda^3 - 3\lambda - 2 = 0$$

The roots of this cubic are the principal stresses and are easily seen to be

$$\lambda_1 = 2 \text{ MPa}; \quad \lambda_2 = -1 \text{ MPa}; \quad \lambda_3 = -1 \text{ MPa}$$

***Principal Direction of $\lambda_1$***

This is a case of coincident eigenvalues. Starting with the first value, we proceed to find its direction, which is the corresponding eigenvector.

$$\begin{bmatrix} -2 & 1 & 1 \\ 1 & -2 & 1 \\ 1 & 1 & -2 \end{bmatrix} \begin{Bmatrix} l_1 \\ m_1 \\ n_1 \end{Bmatrix} = \begin{Bmatrix} 0 \\ 0 \\ 0 \end{Bmatrix}$$

Add (First row) + 2 × (Second row), and we get $m_1 = n_1$. Add 2 × (Second row) + (Third row), and we get $l_1 = m_1$. Hence,

$$l_1 = m_1 = n_1 = \frac{\sqrt{l_1^2 + m_1^2 + n_1^2}}{\sqrt{1 + 1 + 1}} = \frac{1}{\sqrt{3}}$$

***Principal Direction of $\lambda_2$ and $\lambda_3$***

As before, we substitute the value of $\lambda$ in the matrix to obtain

$$\begin{bmatrix} 1 & 1 & 1 \\ 1 & 1 & 1 \\ 1 & 1 & 1 \end{bmatrix} \begin{Bmatrix} l_2 \\ m_2 \\ n_2 \end{Bmatrix} = \begin{Bmatrix} 0 \\ 0 \\ 0 \end{Bmatrix}$$

This is just one equation repeated three times. Therefore, the solution has two arbitrary constants:

$$l_2 = c_1$$

$$m_2 = c_2$$

$$n_2 = -(c_1 + c_2)$$

We can arbitrarily choose $l_2 = m_2 = 1/\sqrt{6}$ and $n_2 = -2/\sqrt{6}$, which makes the magnitude of the direction vector unity.

The third direction can be found by the condition that it has to be normal to the other two. From vector calculus we know that it is simply the cross-product of the other two direction vectors. In this case it turns out to be:

$$l_3 = -1/\sqrt{2}; \quad m_3 = 1/\sqrt{2}; \quad n_3 = 0;$$

***Commentary***

*Note that the first two eigenvectors are perpendicular to each other,* ***although we did not impose it****. Physically, it corresponds to the condition that principal stresses are mutually perpendicular. Analytically, it is a property of eigenvectors themselves.*

*The significance of equal principal stresses (or, coincident eigenvalues) is that in the plane of $\sigma_2$-$\sigma_3$, all directions are just about the same. It is easy to verify this, by constructing the two dimensional Mohr's circle in that plane. Due to two normal stresses being equal, and shear stress being zero, the Mohr's circle degenerates to a single point. Hence, all directions have the same magnitude of stress. See Section 1.14. This can also be verified by considering the transformation Equations 1.25 and 1.26.*

## 1.15 COMPATIBILITY

An important premise of Equations 1.1 through 1.7 is that while sustaining external loads, the body remains intact and in one piece. Every set of neighboring points before deformation is found to be in a one-to-one mappable neighborhood after deformation. This physical fact is expressed in terms of what are called the "compatibility" equations. Technically, compatibility is a result of the fact that six strains were defined (Equations 1.2 through 1.7) in terms of only three displacements. Hence, to be unique, the strains must satisfy additional equations among themselves. These additional equations are called the "compatibility" conditions. See Figure 1.13 for the compatibility equations and their derivation.

By virtue of the strain-displacement relationships, compatibility equations can be expressed in various other forms, (i) in terms of displacements, (ii) in terms of strains, and (iii) in terms of stresses, by using Hooke's law.

## 1.16 FRAMEWORK OF LINEAR ELASTICITY THEORY

A discussion of the structure of the theory of elasticity is presented in this section. We already have discussed all its ingredients, namely, the equilibrium equations, the strain-displacement equations, and Hooke's law. It is only necessary to appropriately combine them. Figure 1.14 illustrates the relationship among them. The kinematic variables, namely, the displacement and strain-displacement equations are on the left side. The static variables, namely the force and stress are shown on the right side. The extraneous requirements on strain and stress, namely compatibility and equilibrium are also shown. Lastly, the connection between the static and kinematic quanitities is established by Hooke's law. A few observations can be made from Figure 1.14.

- The origin of stress can be applied load, or applied displacement.
- In the place of Hooke's law, any other stress-strain relation can be inserted to represent another specific material behavior.
- Any procedure to calculate the stress must satisfy all the requirements, namely, equilibrium, strain-displacement equations, and Hooke's law.

### 1.16.1 Load-Controlled Problems

Problems in which the stress is caused by externally applied loads are called *load-controlled problems*. The path of computation of stresses is shown in Figure 1.15. The suggestion is that while going from stress to

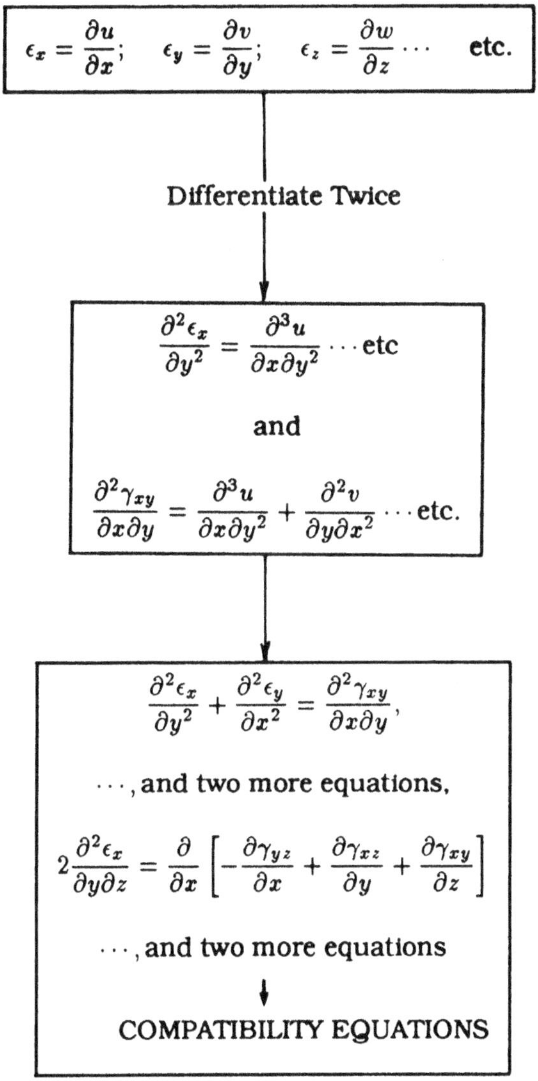


**FIGURE 1.13.** Derivation of compatibility equations.

strains, a factor of $1/E$ is picked up and while coming back a factor of $E$ is picked up. The net effect in *almost all* cases is that *the material property terms are cancelled out*. However, as mentioned before, in certain cases, Poisson's ratio may remain, but as a weak factor.[8] That is, the presence of

[8] This is the case when there are body forces of certain types. For details, see Reference 8 or 9.

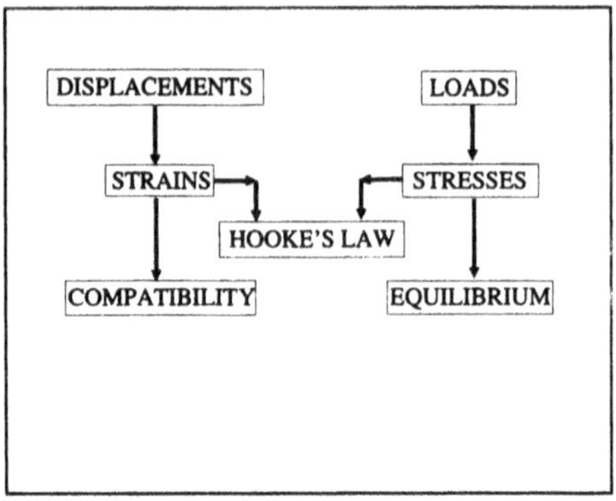


**FIGURE 1.14.** Framework of the elasticity theory.

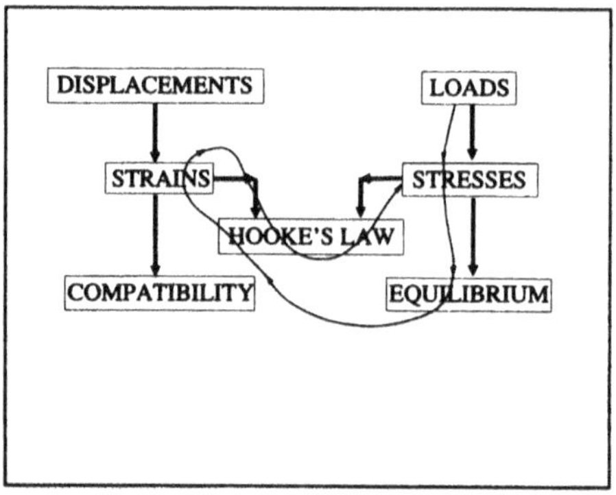


**FIGURE 1.15.** Load controlled problems.

Poisson's ratio accounts for a variability of only about 2 to 3% in the stresses.

We therefore make the important conclusion, that

> the stress solution for load controlled problems is substantially independent of material properties.

**Example 1.4**

Consider the formulas for hoop and longitudinal stresses in a pressurized cylinder given below (Figure 1.16). The pressure is the external load, and the formulas are completely independent of the material.

**Example 1.5**

Consider the more involved formulas for stresses in the vicinity of a

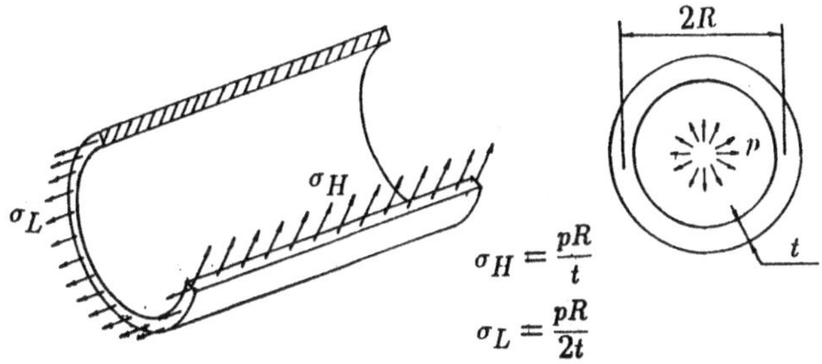


**FIGURE 1.16.** The stresses in a pressurized cylinder are independent of the modulus of the material.

circular hole in a large flat plate shown (Figure 1.17). The plate is subjected to a tensile stress of $\sigma_o$ far away from the hole.

Note that the formulas depend on the location $(r, \theta)$ and on applied stress $\sigma_o$, but not on the material at all!

**Example 1.6**

As a next step consider the radial and tangential stresses developed in a flat circular disk that rotates at uniform angular speed $\omega$ rad/sec (Figure 1.18).

The above formulas depend on Poisson's ratio, but not on Young's modulus. Most materials have Poisson's ratio varying from 0.25 to 0.35; we see that the difference due to Poisson's ratio alone is small, as shown.

| | $\nu = 0.25$ | $\nu = 3.5$ | % Difference |
|---|---|---|---|
| Max $\sigma_r$ | $0.40625C_1\,\rho\omega^2$ | $0.4188C_1\,\rho\omega^2$ | 3% |
| Max $\sigma_\theta$ | $0.4297C_2\,\rho\omega^2$ | $0.4390C_2\,\rho\omega^2$ | 2.2% |
| (based on $a/b = 0.5$) | | | |

## 1.16.2 Displacement-Controlled Problems

Situations in which stresses are caused by the need to accommodate strains, are tentatively called *displacement-controlled problems*. In such problems, the stresses are directly proportional to the modulus of elasticity and hence are highly sensitive to the material. In these problems, the rough path of computation is as shown in Figure 1.19.

Obviously, since the path starts with given displacements, the factor $E$ is carried in the onward direction for calculating stresses. Thermal and residual stresses are the most common examples of this kind. From the point of view of the theory of elasticity, free thermal expansion does not produce any stresses. Constraining thermal expansion does. *In all cases*

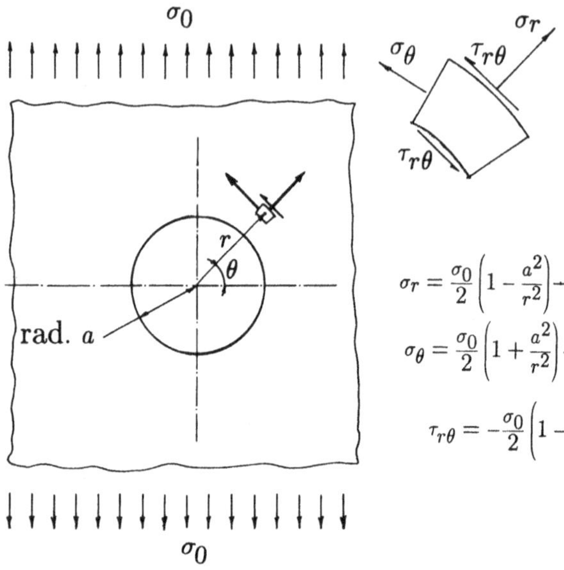


$$\sigma_r = \frac{\sigma_0}{2}\left(1 - \frac{a^2}{r^2}\right) - \frac{\sigma_0}{2}\left(1 + \frac{3a^4}{r^4} - \frac{4a^2}{r^2}\right)\cos 2\theta$$

$$\sigma_\theta = \frac{\sigma_0}{2}\left(1 + \frac{a^2}{r^2}\right) + \frac{\sigma_0}{2}\left(1 + \frac{3a^4}{r^4}\right)\cos 2\theta$$

$$\tau_{r\theta} = -\frac{\sigma_0}{2}\left(1 - \frac{3a^4}{r^4} + \frac{2a^2}{r^2}\right)\sin 2\theta$$

**FIGURE 1.17.** A more complex example of the stress field around a hole also leads to the same observation as in the previous example.

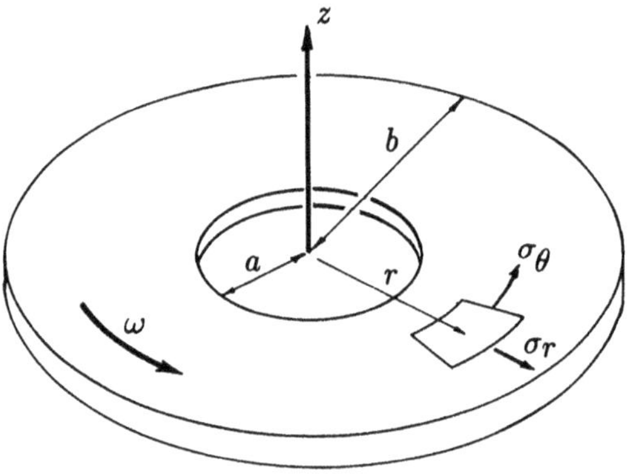


$$\sigma_r = \left(\frac{3+\nu}{8}\right)\rho\omega^2\left(b^2 + a^2 - \frac{a^2b^2}{r^2} - r^2\right)$$

$$\sigma_\theta = \left(\frac{3+\nu}{8}\right)\rho\omega^2\left(b^2 + a^2 + \frac{a^2b^2}{r^2} - \frac{1+3\nu}{3+\nu}r^2\right)$$

**FIGURE 1.18.** Rotating cylinders involve body forces, and hence are marginally dependent on material property, $\nu$.

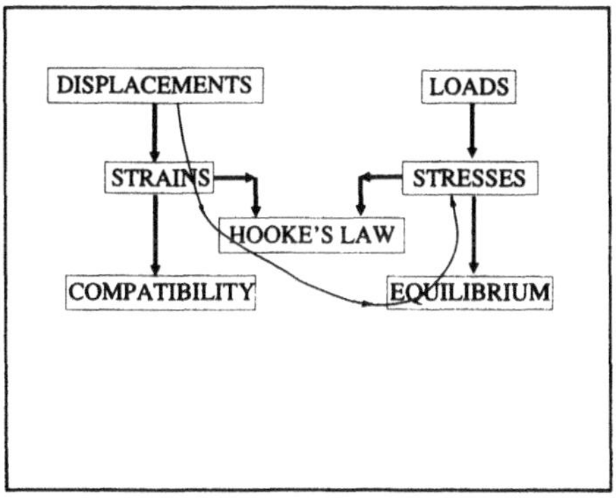


**FIGURE 1.19.** Path of computation in a displacement-controlled problem.

*thermal stresses are proportional to the amount of thermal expansion which has been arrested.*

The important characteristic of the stress distribution in such problems is that across any cross section of the body the stress adds up to zero force, *because there are no externally applied forces.*

We observe that,

> in a displacement-controlled problem, the stress is directly proportional to the modulus.

### Example 1.7

For a straight bar that is fully constrained at its ends, a uniform temperature rise of $\Delta T$ causes an axial stress given by

$$\text{Thermal stress} = \sigma_T = E\alpha\,\Delta T$$

where $\alpha$ = linear thermal expansion coefficient of the material.

### Example 1.8

Consider the press fit of two pipes (Figure 1.20), which have a radial interference of $\delta$. Let the inner and outer radii of the inner pipe be $a$ and $b$, and those of the outer pipe be $b$ and $c$. For simplicity, let both the pipes be of the same material, for which the modulus is $E$.

The pressure developed at the interface of the two pipes is $P$ given by the following.

$$P = E\frac{\delta}{b}\,\frac{(c^2 - b^2)(b^2 - a^2)}{2b^2(c^2 - a^2)}$$

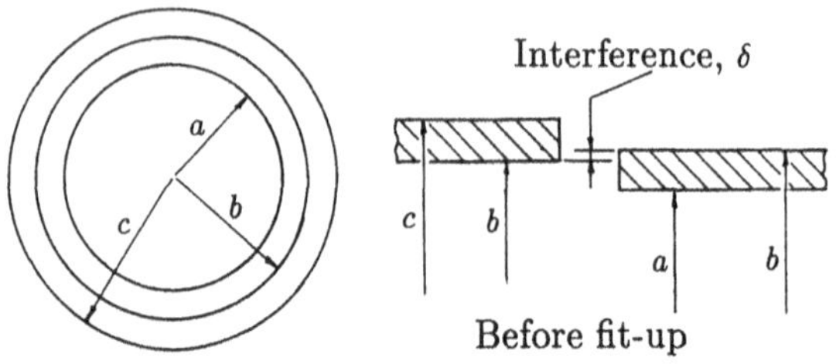


**FIGURE 1.20.** Press-fitted joints are common examples of displacement-controlled problems. The stresses depend on the material property.

Let $\sigma_o$ and $\sigma_i$ be the maximum hoop stresses in the outer and inner cylinders. They are given by,

$$\sigma_o = \frac{(c^2 + b^2)}{(c^2 - b^2)} P$$

$$\sigma_i = -\frac{b^2}{a^2} \frac{(b^2 + a^2)}{(b^2 - a^2)} P$$

Notice that the stresses are proportional to $E$, since $P$ carries it.

### 1.16.3 Principle of Similitude in Elasticity

Since experimental stress analysis on down-scaled (or up-scaled) prototypes is becoming common in the plastics industry, it is useful to know how to relate the stresses observed from down-scaled experimental models to the actual stresses in the component caused by service loads.

We can derive a very useful principle called the *principle of similitude*, as a direct consequence of the definitions of stress and strain. It states that if we consider two geometrically similar bodies made of the same material, which are loaded at corresponding points by forces that are proportional to the square of the linear dimensions, then the stress and strain at corresponding points are equal. Also, if corresponding points are given displacements which are proportional to the linear dimensions, then, again, the stress and strain are equal at corresponding points. This gives a quick rule to scale up or down forces and displacements based on an existing known design.

This principle becomes obvious if we consider the fact that the stress is the limiting ratio of load to area. Areas in the two bodies vary as the square of the linear dimensions. So do the applied forces. Hence the stress at corresponding points must match. A similar argument applies for the case of applied displacements.

A more general form of the principle of similitude is central to the *photoelastic technique* of experimental stress analysis. See also References [10] and [11] for detailed discussions of the technique and the use of similitude. This technique depends on the fact that stresses in scale models are scalable too, even if the model and the prototype are made of different materials. The requirement is that the loads and displacements applied to the model must be in a predetermined relationship to those on the prototype. Let us explain the principle as applied to stress analysis in some detail.

Let

$P$ = Concentrated load, (Force)
$Q$ = Load distributed along a line, (Force/Length)
$q$ = Load distributed over an area, (Force/Area)
$f_b$ = Body forces distributed volumetrically, (Force/Volume)
$U$ = Applied displacements, (Length)
$L$ = A characteristic dimension, (Length)
$\mathcal{S}$ = Scale factor for model dimensions,
$\sigma$ = Stress at a point, (Force/Area)
$E$ = Modulus of elasticity, (Force/Area)
$\nu$ = Poisson's ratio.

By a powerful theorem in similitude, called *Buckingham's $\pi$-theorem*, it is possible to show that the stress state can be expressed in terms of *dimensionless* groups shown below.

$$\frac{\sigma}{E} = f\left(\mathcal{S}, \nu, \frac{P}{EL^2}, \frac{Q}{EL}, \frac{q}{E}, \frac{f_b L}{E}, \frac{U}{L}\right)$$

The functional form of $f(\cdots)$ may be unknown, but is the same between the model and the prototype. The implication, therefore, is that if the dimensionless ratios inside $f(\cdots)$ are maintained the same between model and prototype, then the resulting ratios $\sigma/E$ will be the same between the model and the prototype. Hence, we see that if concentrated loads are in the ratio of $(\mathcal{S})^2$, if lineally distributed loads are in the ratio of $\mathcal{S}$, if pressure loads are the equal, etc., then

$$\frac{\sigma_p}{E_p} = \frac{\sigma_m}{E_m}.$$

There is, of course, a difficulty in maintaining the same $\nu$ between

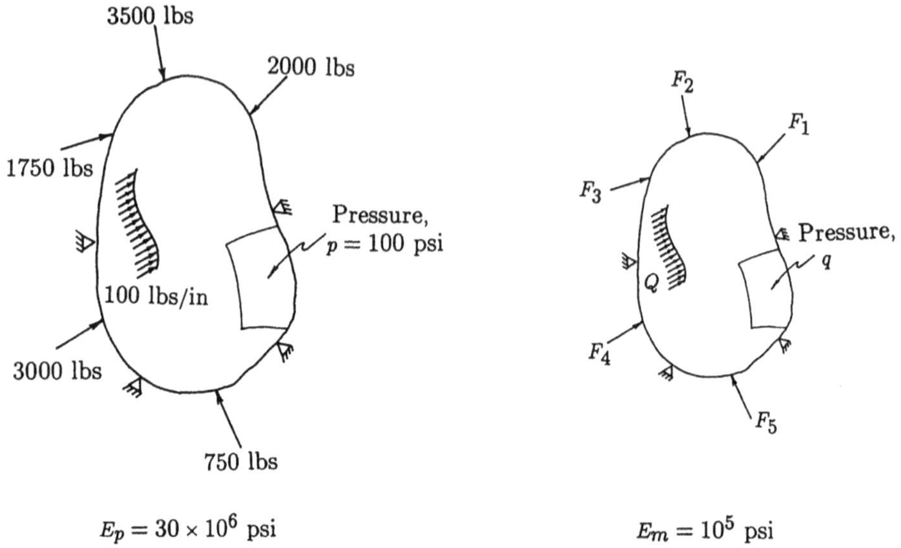


**FIGURE 1.21.** Use of similitude principle for model testing.

model and prototype, because the engineer has no control over this. We had earlier mentioned that $\nu$-dependence arises only under certain types of body forces. However, the practical range of difference in $\nu$ from one material to another causes very little error in the value of stresses. Hence, the error due to different $\nu$ is negligible.

**Example 1.9**

A body, made of steel is acted upon by concentrated forces, lineally distributed forces, and pressure forces, as shown in Figure 1.21. Build a model, one-tenth of the original size, of a material whose modulus is $10^5$ psi. Find out the magnitudes of the concentrated and distributed forces to be applied on the model. If the stress at a point $X'$ in the model is measured to be 35 psi, determine the stress at the corresponding point $X$ in the prototype.

*Solution* The dimensionless numbers corresponding to concentrated and distributed forces are to be kept the same between the model and the prototype. Hence,

$$\left(\frac{F_1}{EL^2}\right)_m = \left(\frac{F_1}{EL^2}\right)_p; \quad \left(\frac{Q}{EL}\right)_m = \left(\frac{Q}{EL}\right)_p; \quad \left(\frac{q}{E}\right)_m = \left(\frac{q}{E}\right)_p$$

Therefore,

$$F_1 = 2000\left(\frac{E_m}{E_p}\right)\left(\frac{L_m}{L_p}\right)^2 = 2000\left(\frac{10^5}{30 \times 10^6}\right)\left(\frac{1}{10}\right)^2 = 0.0667 \text{ lbs.}$$

Similarly,

$$F_2 = \frac{3500}{3 \times 10^4} = 0.117 \text{ lbs.}$$

$$F_3 = \frac{1750}{3 \times 10^4} = 0.0585 \text{ lbs.}$$

$$F_4 = \frac{3000}{3 \times 10^4} = 0.10 \text{ lbs.}$$

$$F_5 = \frac{750}{3 \times 10^4} = 0.025 \text{ lbs.}$$

$$Q = 100\left(\frac{E_m L_m}{E_p L_p}\right) = 0.0333 \text{ lbs./in}$$

$$q = 100 \text{ psi}$$

The stress in the prototype can be obtained by computing the ratio $(\sigma/E)$. Therefore,

$$\sigma_p = \sigma_p \frac{E_p}{E_m}$$

$$= 35 \text{ psi } \frac{30 \times 10^6}{10^5}$$

$$= 10{,}500 \text{ psi}$$

### 1.16.4 Balance Sheet of Elasticity Theory

Referring back to the structure of the theory of elasticity, we find that the key elements of elasticity are: (i) equilibrium, (ii) strain-displacement equations, and (iii) Hooke's law. Together with the boundary conditions, they are sufficient to solve for all the field quantities as functions of coordinates. This fact is expressed in Tables 1.2.

Note that these equations are solvable only in the purely mathematical sense. A real life problem is solved in one of the following ways. The first

**Table 1.2. Equations and Unknowns in Elasticity.**

| Unknowns | | Available Equations | |
|---|---|---|---|
| **General Three Dimensions** | | | |
| Displacements$(u, v, w)$ | 3 | Equilibrium$(x, y, z)$ | 3 |
| Stress $(\sigma_{xx}, \sigma_{yy}, \sigma_{zz}, \tau_{xy}, \tau_{yz}, \tau_{zx})$ | 6 | Stress Strain Equations (Hooke's Law) | 6 |
| Strain $(\epsilon_{xx}, \epsilon_{yy}, \epsilon_{zz}, \gamma_{xy}, \gamma_{yz}, \gamma_{zx})$ | 6 | Strain-Displacement Equations (Equations 1.2 thru 1.7) | 6 |
| TOTAL | 15 | TOTAL | 15 |
| **Plane Stress** $\sigma_{zz} = \tau_{xz} = \tau_{yz} = 0; \epsilon_{zz} = -\nu(\epsilon_{xx} + \epsilon_{yy})$ | | | |
| Displacements$(u, v)$ | 2 | Equilibrium$(x, y)$ | 2 |
| Stress$(\sigma_{xx}, \sigma_{yy}, \tau_{xy})$ | 3 | Stress-Strain Equations | 3 |
| Strain$(\epsilon_{xx}, \epsilon_{yy}, \gamma_{xy})$ | 3 | Strain-Displacement Equations | 3 |
| TOTAL | 8 | TOTAL | 8 |
| **Plane Strain** $\epsilon_{zz} = \gamma_{xz} = \gamma_{yz} = 0; \sigma_{zz} = -\nu(\sigma_{xx} + \sigma_{yy})$ | | | |
| Displacements$(u, v)$ | 2 | Equilibrium$(x, y)$ | 2 |
| Stress$(\sigma_{xx}, \sigma_{yy}, \tau_{xy})$ | 3 | Stress-Strain Equations | 3 |
| Strain$(\epsilon_{xx}, \epsilon_{yy}, \gamma_{xy})$ | 3 | Strain-Displacement Equations | 3 |
| TOTAL | 8 | TOTAL | 8 |

In plane strain conditions, $\sigma_{zz}$ may alternatively be shown as an additional unknown, and the relation $\sigma_{zz} = -\nu(\sigma_{xx} + \sigma_{yy})$ would be added as the extra available equation.

approach is to relegate the theory of elasticity to a lower level theory such as a beam theory, or a plate theory, or a shell theory, and to formulate the problem within one such theory. Quite a few, but not all, problems are solvable this way. A second approach uses a collection of semi-analytical seminumerical methods, called *variational methods*. With these methods, a starting solution is assumed in terms of a polynomial whose coefficients are obtained by minimizing the error. A third approach is to attack the problem entirely numerically, using a finite difference or a finite element approach. In the mathematical treatment of the finite elements, some equivalence of the second and third approaches is established. In general, the compatibility conditions are satisfied right at the stage of assuming a solution, which is usually in the form of assumed displacements. This way, the further algorithms need to address only equilibrium, boundary conditions, and Hooke's law.

**References**


1. Roark, R. J. and Young, W. C. *Formulas for Stress and Strain*, 5th Ed., 1975, McGraw Hill, New York, NY.
2. Faupel, J. H. and Fisher, F. E. *Engineering Design—A Synthesis of Stress Analysis and Materials Engineering*, 2nd Ed. 1981, Wiley Interscience, New York, NY.
3. Peterson, R. E. *Stress Concentration Design Factors*, 1953, John Wiley and Sons, Inc., New York, NY.
4. Beer, F. P. and Johnston, E. R. Jr., *Mechanics of Materials*, 1981, McGraw Hill Book Company, New York, NY.
5. Bronson, R. *Matrix Methods, An Introduction*, 1970, Academic Press, New York, NY.
6. Hildebrand, F. B. *Methods of Applied Mathematics*, 2nd Ed. 1965, Prentice-Hall, Inc., Englewood Cliffs, NJ.
7. Seeley, F. B. and Smith, J. O. *Advanced Mechanics of Materials*, 1967, 2nd Ed. John Wiley and Sons, Inc., New York, NY.
8. Timoshenko, S. P. and Goodier, J. N. *Theory of Elasticity*, 3rd Ed. 1970, McGraw Hill Book Company, New York, NY.
9. Chou, P. C. and Pagano, N. J. *Elasticity—Tensor, Dyadic and Engineering Approaches*, 1967, D. Van Nostrand Company, Inc., Princeton, NJ.
10. Dally, J. W. and Riley, W. F. *Experimental Stress Analysis*, 2nd Ed. 1978, McGraw Hill Book Company, New York, NY.
11. Durelli, A. J., Phillips, E. A., and Tsao, C. H. *Introduction to the Theoretical and Experimental Analysis of Stress and Strain*. 1958, McGraw Hill Book Company, New York, NY.


**Exercises**

Some of the mathematical techniques necessary for the solution of some of the following exercises are not discussed in the text. References cited at the end of this chapter rigorously discuss a variety of applications, and are highly recommended.

**1.1.** For the figures shown below, write the traction vector $\vec{T}$, at the point $P$, with respect to the axis of coordinates.

**1.2.** In the bending of a beam the bending stress is given by

$$\sigma_x = Axy.$$

Find the constant $A$ using equilibrium of moments.

**1.3.** At a point in the plane cross section of a rectangular beam, the components

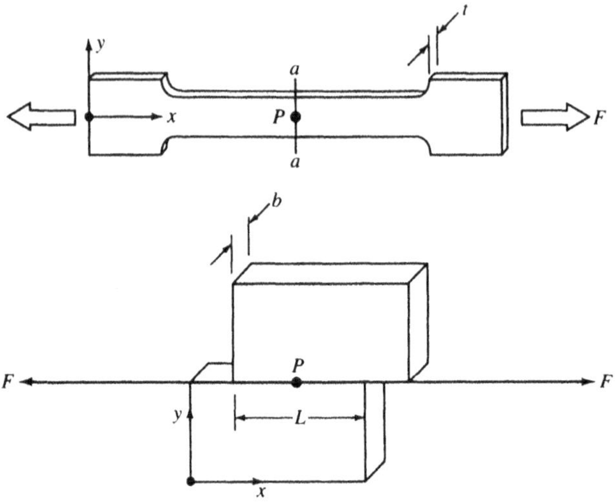


**EXERCISE 1.1.**

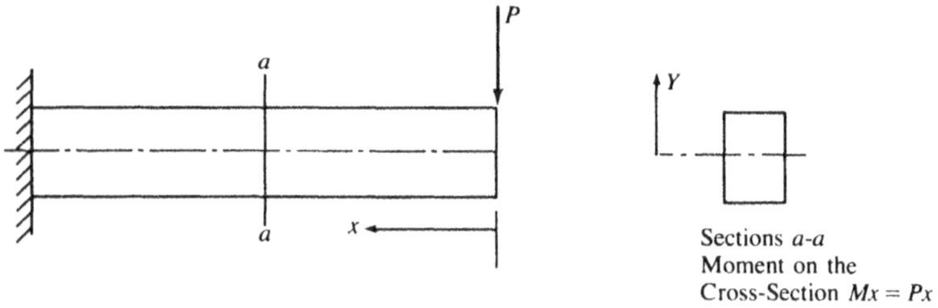


**EXERCISE 1.2.**

of the traction vector $d\vec{T}$ are expressible as

$$d\vec{T} = \left[ Axy\vec{i} + B\left(y - \frac{d}{2}\right)^2 \vec{j} \right] d\vec{A}.$$

The shear force on the section is $P$, and the moment is $M$, as shown. Determine the constants $A$ and $B$, using equilibrium equations.

**1.4.** For the axial load $F$ acting on the specimen shown in Exercise 3 above, find the traction vectors $T_n$ and $T_t$ along an inclined set of axes.

**1.5.** For the bending moment shown, write the traction vector $T_n$ acting on a small area $dA$ on the point $a$, $b$ and $c$.

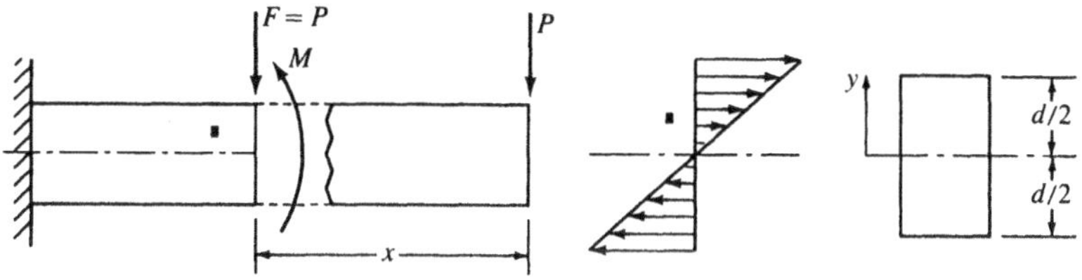


EXERCISE 1.3.

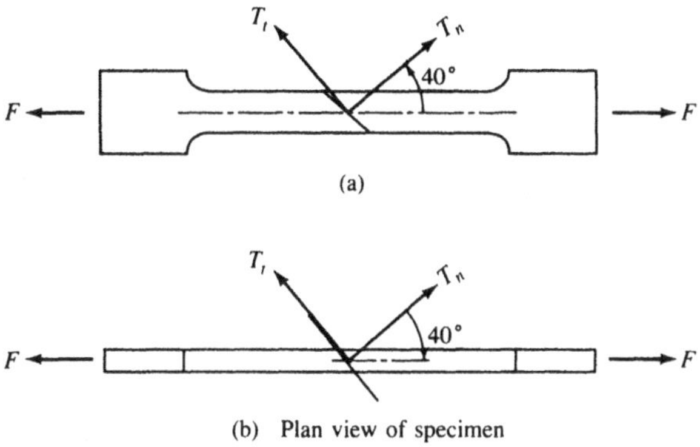


EXERCISE 1.4.

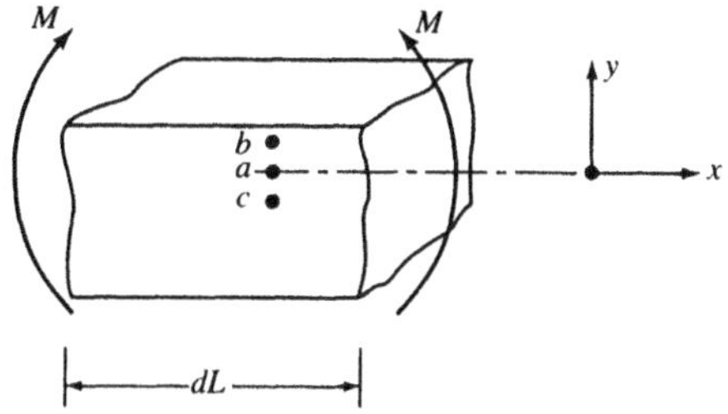


EXERCISE 1.5.

**1.6.** The $x$-displacement of a two dimensional problem is given by

$$u = \frac{q}{2EI}\left[\left(l^2x - \frac{1}{3}x^3\right)y + x\left(\frac{2}{3}y^3 - \frac{2}{5}c^2y\right) + \nu x\left(\frac{1}{3}y^3 - c^2y + \frac{2}{3}c^3\right)\right]$$

Find $v$.

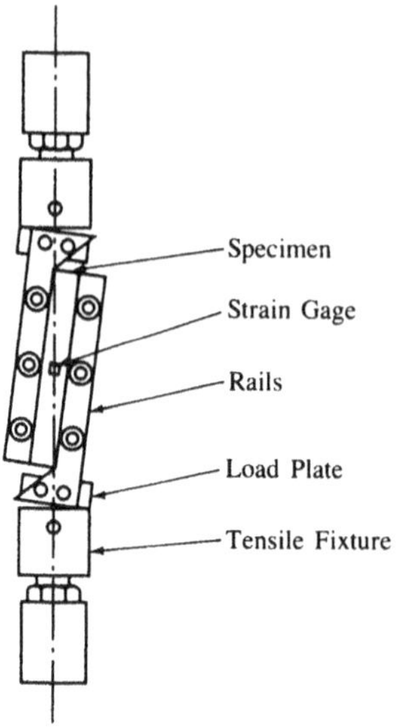


**EXERCISE 1.7.**

**1.7.** By using equilibrium equations, show that a rail shear test (described in ASTM D-4255) can only be in a state of bending plus shear stress.

**1.8.** By applying equations of moment equilibrium to a small element of size $dx \times dy \times dz$ show that for any shear stress, there must be a complementary shear stress acting on a perpendicular plane.

**1.9.** The bending stress in a beam of rectangular cross section is known to vary linearly through the depth, and hence may be expressed as

$$\sigma = Ay; \qquad \left(-\frac{d}{2} \leq y \leq \frac{d}{2}\right).$$

Show that the elements of the beam cannot be in complete equilibrium under this stress state, unless acted upon by a shear stress with a distribution given by

$$\tau = B\left(y - \left|\frac{d}{2}\right|\right)^2.$$

**1.10.** A thin circular disk is in a state of plane stress, and is acted upon by a uniform constant pressure $p$ all around its perimenter. Find $\sigma_{xx}$, $\sigma_{yy}$, and $\epsilon_{zz}$ at any point in the disk.

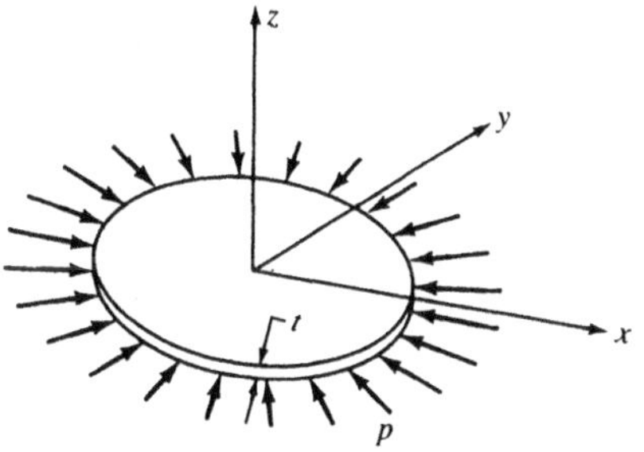


**EXERCISE 1.10.**

**1.11.** A linear elastic cylindrical plug is inserted into a cavity in a body which may be supposed to be infinitely rigid. An axial stress $\sigma_{zz}$ is applied to the plug. Find $\sigma_{xx}$, $\sigma_{yy}$, and $\epsilon_{zz}$.

**1.12.** Prove that any linear elastic solid subjected to a constant normal pressure $p$ all over its bounding surface will have a stress distribution given by

$$\text{All normal stresses} = p$$

$$\text{All shear stresses} = 0$$

**1.13.** Prove that for the plane stress state $(\sigma_x \sigma_y - \tau^2)$ is a constant with respect to coordinate transformations.

**1.14.** Prove that for isotropic, linear elastic materials the principal stresses lie along the same directions as do the principal strains.

**1.15.** For a cylindrical pipe under internal pressure, show that the hoop and axial stresses are the principal stresses.

**1.16.** Express the state of stress shown in the figure in a matrix form. Also find the strains (1) by writing the long hand form of Hooke's law, and (2) by writing it

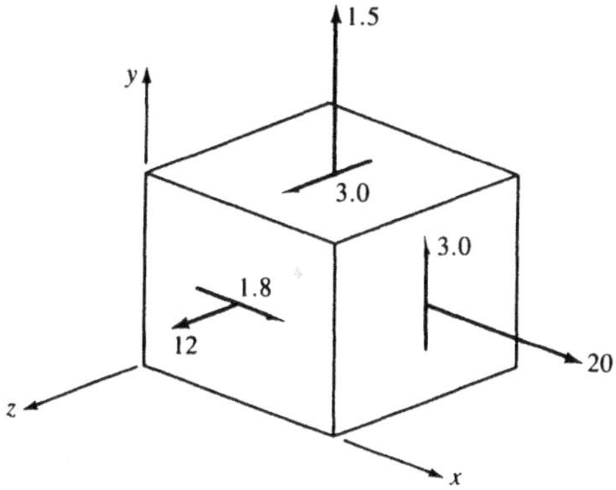


**EXERCISE 1.16.**

in the matrix form. Assume $E = 2200\ \text{kg/mm}^2$, and $\nu = 0.3$. (3) What are the components of the strain tensor?

**1.17.** Prove that the form of the matrix equation in Section 1.13.3 and the elastic constants in Hooke's law (for isotropic materials) remain unchanged under coordinate transformation.

**1.18.** Show that the quantities $(\sigma_{xx} + \sigma_{yy} + \sigma_{zz})$ and $(\epsilon_{xx} + \epsilon_{yy} + \epsilon_{zz})$ are constants at a point with respect to coordinate transformation.

**1.19.** Find the principal stresses and directions for the stress state given in the previous problem.

**1.20.** For the figures shown, write the boundary conditions.

**1.21.** Prove that Equations 1.25 and 1.26 together are identical to the Mohr's circle procedure.

**1.22.** Verify that Equations 1.25 and 1.26 are also given by the expression $[L][\sigma][L]^T$ in two dimensions.

**1.23.** Get a feel for the word "isotropy" in the following way. Assume a 2-dimensional stress state as shown in the figure.

- Find strains along $x - y$.
- Transform stresses to $x' - y'$ axes using Equations 1.25 and 1.26.

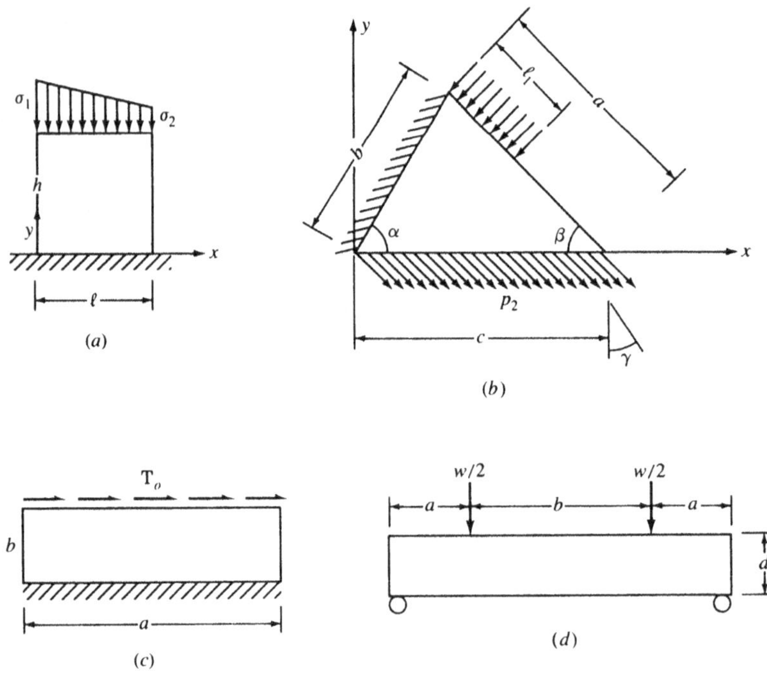


Treat this problem as a planestren elasticity problem, and also as an Euler beam problem.

**EXERCISE 1.20.**

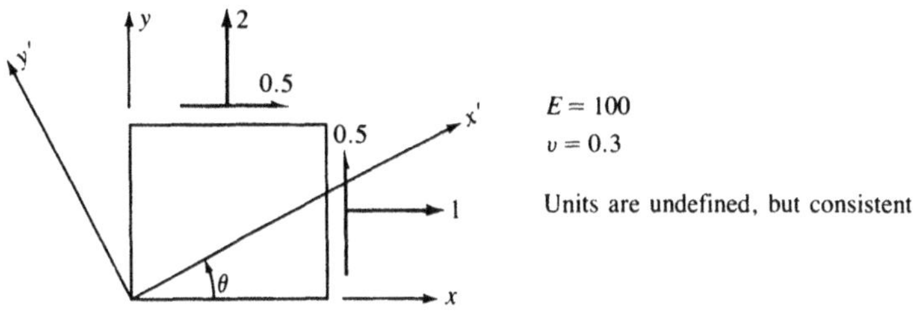


**EXERCISE 1.23.**

- Transform also strains to $x' - y'$ axes.
- Suppose that moduli $E'$ and $\nu'$ are different in the $x' - y'$ axes. Write the stress strain relations in this coordinate system, and verify that $E'$ and $\nu'$ are the same as in the $x - y$ system.

**1.24.** A plastic furniture article is prototyped in wood to check styling. Would you strain gage the wooden model to estimate the service stresses?

**1.25.** A simply supported wooden beam has a cross section 25 mm wide and 100 mm deep, and supports a concentrated load of 70 kgs at the center of the span of 2 m. Beam theory predicts zero bending stress at the support, and 35 kg shear force. Nonetheless, the wood shows markings at the support which give evidence of *normal* stresses. What calculation approach would lead to the prediction and estimation of these stresses.

**1.26.** **Term Paper:** Consider this problem which is very common in metallics. A pin contacts the eye in a lifting hook. Since contact is along a line, the area of load application is zero, leading to infinite normal stress for any load. Explain why infinite stress is a fallacy. Outline a calculation procedure that would lead to an estimation of the actual stresses at the contact.

**1.27.** Why is the stress concentration factor a function of only geometry for linear elastic materials, and not for nonlinear elastic materials? Would the SCF be a function of only geometry for orthotropic materials such as layered composites?

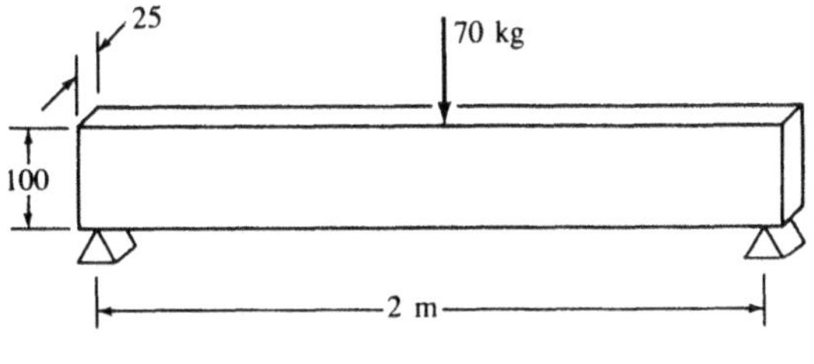


**EXERCISE 1.25.**

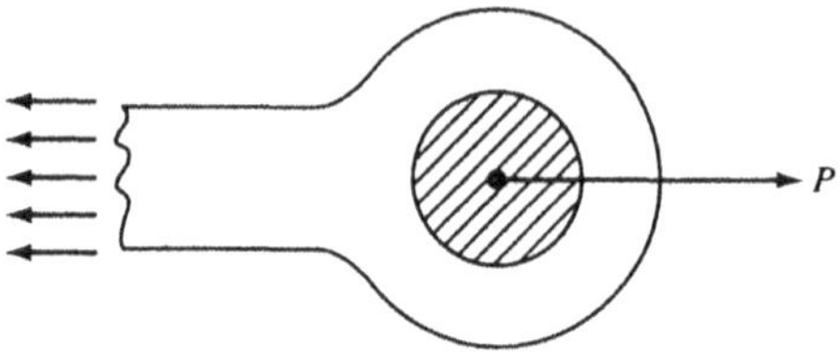


**EXERCISE 1.26.**

**1.28.** The example cited in Section 1.4 states that treating a ribbed plate as a plain, unribbed plate of different thickness (or different modulus) is an artifice that is good for deflection calculation but not for stress. Substantiate this.

**1.29.** Plastics have a modulus that is sensitive to strain rates. Discuss the finesse involved in analyzing, by finite elements, the problem of a plastic component being shatter-tested by dropping a weight on the component. In what aspects would a metallic component analysis would be different?

# 2

# Applications of Linear Elastic Behavior

## 2.1 INTRODUCTION

The theory of linear elasticity is a complete mathematical description of the behavior of solids under the action of mechanical and thermal loads. However, because of the large number of equations to be solved, it is not useful for simple problems. On the other hand, simple theories of beam bending and twisted bars have a proven record of applications. The conventional beam bending theory for example, can be applied to many practical problems. Although historically the beam theory precedes the elasticity theory, we can visualize it as a logical subset of the latter. Table 2.1 presents the theory of beams, the theory of torsion, and the theory of plates and shells as subsets of elasticity theory. The assumptions underlying the formulation of these theories are also briefly mentioned in the table. In this chapter, we discuss the applications of these theories to certain practical situations. In doing so, the topics have been arranged as one and two dimensional problems. Sections 2.2 through 2.9 deal with the one dimensional problems, and the last section deals with two dimensional problems. No effort has been made in this chapter to restate the well known beam theory or plate theory. Instead, this chapter is devoted to examples of applications that are germane to the plastics industry. In-depth development of the theory of beams with examples may be found in References 1, 2, 3, and 4. For plate theory, References 3, 4, 5, 6, and 7 are a few of the well-known texts. Reference 8 is the "cookbook" of stress analysis, and is a large collection of pre-engineered solutions for beams, plates, and shells of various shapes.

The examples in this chapter are supplementary to the material in the text, and provide an elaboration of the theory discussed. The motivation is

**Table 2.1. Scope of Beam and Plate Theories.**

| THEORY | LIMITATIONS | REMARKS |
|---|---|---|
| Beam Theory<br>Thin Plate Theory<br>Thin Shell Theory | Small deflections | 1. There exists a neutral surface on which all strains are zero.<br>2. Plane, normal sections remain so after bending under the action of a bending moment.<br>3. Bending is the predominant mode in which the applied load is supported. Bending deflections are very large in comparison with shear deflections. Hence, thin plate/shell theories calculate shear forces but not shear deflections.<br>4. Stresses and strains in the direction of thickness are negligible.<br>5. In-plane stresses (due to bending) are neglected. |
| Torsion | Small rotations | 6. Out-of-plane displacements of cross sections (called warping) are constant for all cross sections. |
| Thick Plates | Small deflections | 7. Deflection due to shear is accounted for. (This amounts to a modification of assumption 2 and 3 above.)<br>8. Stress variation from inside (bottom) surface to outside (top) surface is nonlinear. |

to show the breadth of applications of elasticity, rather than to give a formal lesson.

## 2.2 BENDING OF BEAMS

The theory of beams, attributed to Euler, is the most elementary specialization of the theory of elasticity, and is valid for long and straight bars or beams of uniform cross section, subjected to transverse loading. The basic assumptions of this theory are as follows.

1. There is a neutral plane at which the bending strains are zero.
2. Plane cross sections normal to the neutral plane remain so after bending under the action of a bending moment.

Euler's bending theory has been dealt with in several excellent text books. Refer to, for example, Beer and Johnston [1]. We provide a brief outline of the theory in Figure 2.1.

From Figure 2.1 we see that the theory provides the equations for obtaining the deflections ($\delta$), slopes ($\theta$), shear forces ($V$), bending moments ($M$), shear stresses ($\tau$), and bending stresses ($\sigma$) at any point along

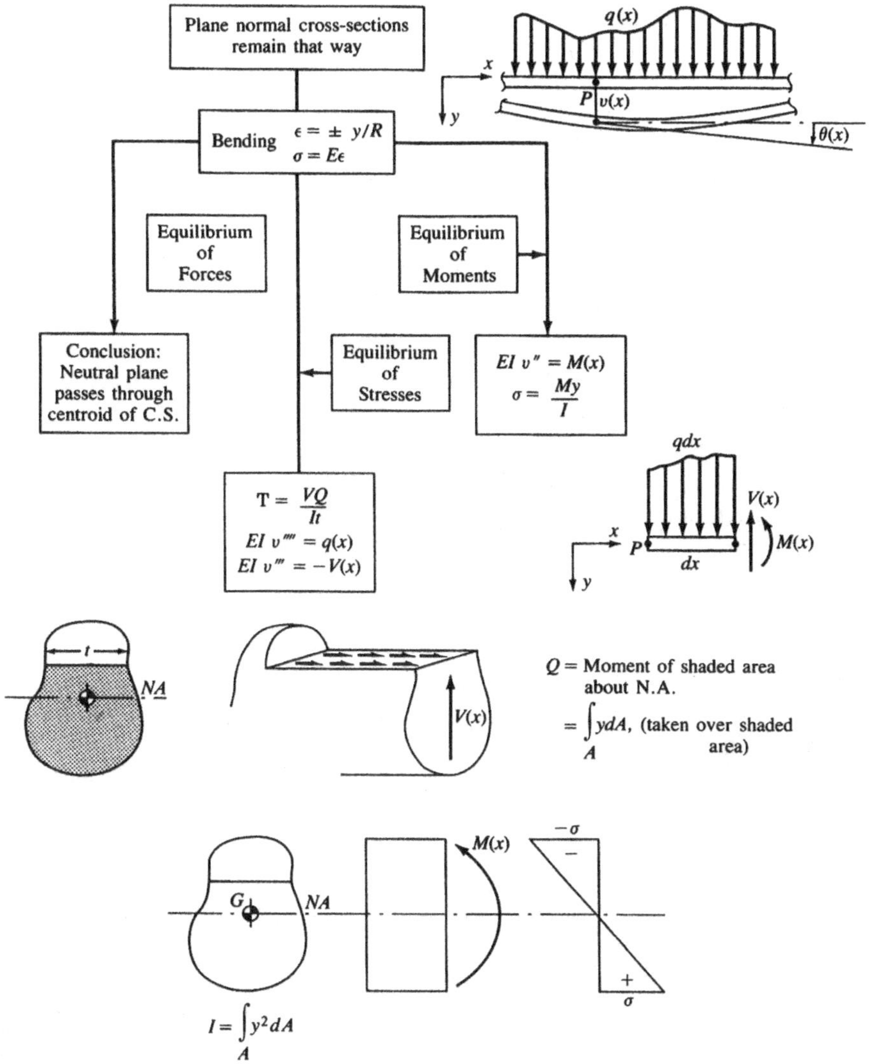


**FIGURE 2.1.** Derivation of the equations of beam theory.

a beam, subjected to general loading. The typical solution process is to apply equilibrium conditions to a segment of a beam—called the *free body* —and obtain the moments $M(x)$ and shear forces $V(x)$ at any cross section. Thereafter, the equations in Figure 2.1 are utilized to formally solve for the other variables, considering the boundary conditions of the problem.

However, $M(x)$ and $V(x)$ are obtainable as above only in a few cases, such cases called the "statically determinate" beams. These are beams that have only two support reactions. It may be two forces, or a force and a moment. In such beams, the external support reactions of the beam can be obtained by using equilibrium equations only. The support reactions are used with the free body to determine $M(x)$ and $V(x)$. This way, the solution of a statically determinate beam is straightforward.

On the other hand, the statically indeterminate beams have more unknown reactions to solve than the number of equilibrium equations. A beam clamped at both ends is an example of such a case. It has two support reactions at the two ends and two support moments. In these cases, the calculation of $M(x)$ and $V(x)$ needs additional equations. The additional equations are usually the continuity of deflections and slopes at as many points as needed. In other words, "compatibility" conditions are utilized to write the required number of equations. Other techniques used for solving indeterminate beam problems, such as the superposition principle, the three-moment theorem, etc., are simply variants of the "compatibility" conditions.

The beam theory has a very successful record with metal structures. The results of this theory are central to advanced analyses such as matrix structural analysis and piping flexibility analysis.

A list of well-known formulas for beam deflections and slopes derived from the beam theory are provided in Appendix B.

**Example 2.1**

A snap fit assembly shown below, is used for permanent closure of a housing in a business machine. The part is made of unfilled PolyCarbonate with a modulus of $E = 245\ \text{kg/mm}^2$, and a flexural strength of 9.5 $\text{kg/mm}^2$. Being a one time closure operation the stresses developed in the snap during assembly can have a factor of safety on the flexural strength as low as 2.0. For the dimensions of the assembly shown, calculate the force needed to push them into assembly, and the maximum stress developed in the part. Do the problem with and without friction. Assume friction coefficient $\mu = 0.3$.

*Solution* This problem can be solved by modeling the snap fit as a cantilever beam. Component #2 is completely rigid and hence the entire deflection is to be taken up by component #1. The force $P$ applied horizontally causes a reactive force system $F$ and $\mu F$ on the inclined surface of the snap as shown.

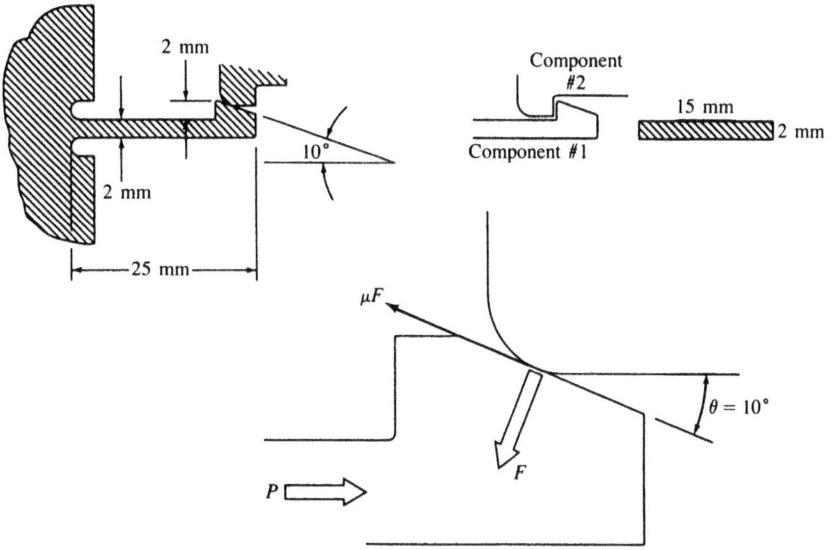


**FIGURE 2.2.** Example 2.1.

For force equilibrium,

$$P = F \sin \theta + \mu F \cos \theta \tag{2.1}$$

While the snap is being pushed left to right, the net vertical downward component $V$, due to $F$ and $\mu F$ causes it to deflect downwards. Net vertical force is given by

$$V = F \cos \theta - \mu F \sin \theta \tag{2.2}$$

Treating the snap as a cantilever,

$$\begin{aligned}\delta &= \frac{VL^3}{3EI} \\ &= \frac{FL^3}{3EI}(\cos \theta - \mu \sin \theta) \\ &= \frac{PL^3}{3EI}\frac{(\cos \theta - \mu \sin \theta)}{(\sin \theta + \mu \cos \theta)}\end{aligned}$$

Hence,
$$P = \frac{3EI\delta}{L^3} \frac{(\sin\theta + \mu\cos\theta)}{(\cos\theta - \mu\sin\theta)} \tag{2.3}$$

$$E = 245\ \text{kg/mm}^2$$

$$I = \frac{15 \times 2^3}{12} = 10\ \text{mm}^3$$

$$\delta = 2\ \text{mm}$$

$$\theta = 10^\circ$$

$$\mu = 0.3$$

$$L = 25\ \text{mm}$$

Substituting, $P = 0.473$ kgs, with friction. In the absence of friction, $\mu = 0$, and hence

$$P = \frac{3EI\delta}{L^3} \tan\theta = 0.166\ \text{kgs}$$

*Stresses*

$$\sigma = \frac{6VL}{bd^2}$$

$$= \frac{6PL}{bd^2} \frac{(\cos\theta - \mu\sin\theta)}{(\sin\theta + \mu\cos\theta)}$$

$$= \frac{6 \times 3EI\delta}{L^3} \frac{L}{bd^2}$$

$$= \frac{3}{2} \frac{Ed\delta}{L^2}$$

$$= 2.35\ \text{kg/mm}^2, \text{ with friction}$$

It can be easily verified that the stress is the same value without friction also, because it depends only on the deflection, which is a result of dimensional differences between parts #1 and #2.

Actual factor of safety is $(9.5\ \text{kg/mm}^2/2.35\ \text{kg/mm}^2) = 4.04 > 2.0$ desired.

**Example 2.2**

Two designs of a shirt hanger are shown below, one made of solid metal wire, and the other having a hollow tubular cross section and made of thermoset polystyrene. Assume that the horizontal portion of the hanger is uniformly loaded with a total of 3 kg weight. Assume the following material properties:

| | | |
|---|---|---|
| Tensile modulus of polystyrene, | $E_p = 200$ | $\text{kg/mm}^2$ |
| Tensile modulus of steel, | $E_s = 21100$ | $\text{kg/mm}^2$ |
| Tensile strength of polystyrene, | $F_p = 2.4$ | $\text{kg/mm}^2$ |
| Yield strength of steel, | $F_{y,s} = 25.5$ | $\text{kg/mm}^2$ |
| Specific gravity of steel, | $SP_s = 7.85$ | |
| Specific gravity of polystyrene, | $SP_p = 1.05$ | |

Dimensions are as shown in the figure. Obtain the stiffness and safety factor available in the thermoplastic hanger, and calculate the weight ratio of the steel to plastic designs, if the steel hanger design is to have at least the same stiffness and safety margin as does the plastic.

*Solution* We may assume that the horizontal portion of the hanger behaves like a simply supported beam. This assumption is justified because of the absence of any special arrangement to clamp it fixed. For such a

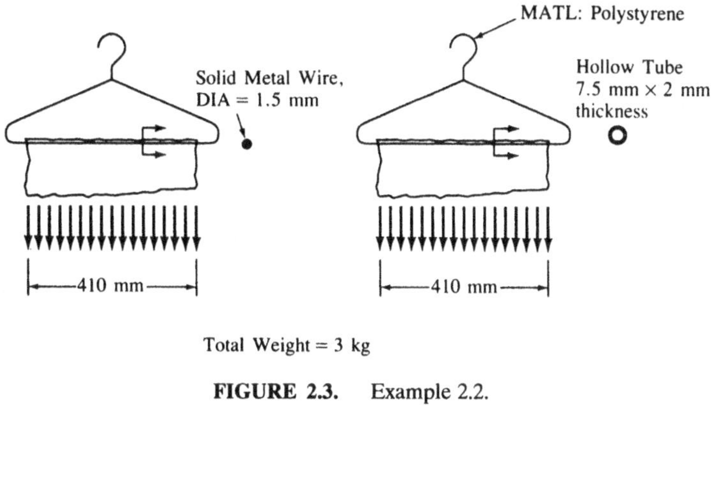


**FIGURE 2.3.** Example 2.2.

beam the maximum deflection is at the center, and is given by

$$\delta_{\max} = \frac{5}{384}\frac{WL^3}{EI}$$

Furthermore, the maximum bending moment of $(WL/8)$ also occurs at the center and consequently the maximum stress occurs at the center too. Maximum stress is given by

$$\sigma_{\max} = \frac{M_{\max}}{z}$$

$$= \frac{M_{\max}}{(2I/d_0)}$$

***Part 1, Polystyrene***

$$I_p = \frac{\pi}{64}(d_o^4 - d_i^4)$$

$$= \frac{\pi}{64}(7.5^4 - 3.5^4)$$

$$= 147.9 \text{ mm}^4$$

$$E_p I_p = 29580 \text{ kg mm}^2$$

$$\text{Stiffness } \frac{W}{\delta} = \frac{384}{5}\frac{E_p I_p}{L^3} = 58.03 \text{ kg/mm}$$

$$\sigma_{\max} = \frac{3 \text{ kg} \times 410 \text{ mm}}{8\left(\dfrac{2 \times 147.9 \text{ mm}^4}{3.75 \text{ mm}}\right)}$$

$$= 1.95 \text{ kg/mm}^2$$

$$\text{Factor of safety} = \frac{2.4 \text{ kg/mm}^2}{1.95 \text{ kg/mm}^2}$$

$$= 1.23$$

***Part 2, Equivalent Steel Design***

Let us determine the diameter of the wire needed for (i) same stiffness and (ii) same safety margin, and use the higher of the two results for calculating the weight ratio.

For equal stiffness we require that $E_s I_s = E_p I_p$. Hence,

$$\frac{\pi}{64} d^4 = \frac{E_p I_p}{E_s} = \frac{29580 \text{ kg mm}^2}{21100 \text{ kg mm}^2}$$

$$= 1.41 \text{ mm}^4$$

Hence,

$$d = 2.31 \text{ mm}.$$

For an equal margin of safety,

$$\sigma_{\max} = \frac{F_{y,s}}{1.23} = \frac{25.5 \text{ kg/mm}^2}{1.23} = 20.73 \text{ kg/mm}^2$$

Or,

$$20.73 \text{ kg/mm}^2 = \frac{3 \text{ kg} \times 410 \text{ mm}}{8z}$$

$$z = \frac{2I_s}{d} = \frac{\pi}{32} d^3 = \frac{3 \text{ kg} \times 410 \text{ mm}}{8 \times 20.73 \text{ kg/mm}^2}$$

$$= 7.42 \text{ mm}^3$$

$$d = 4.23 \text{ mm}.$$

$$\text{Weight ratio} = \frac{SP_s A_s}{SP_p A_p}$$

$$= \frac{7.85 \times \frac{\pi}{4}(4.23 \text{ mm})^2}{1.05 \times \frac{\pi}{4}(7.5^2 - 3.5^2) \text{ mm}^2}$$

$$= 3.01$$

***Stiffness Criterion Development for Automobile Bumpers***

Starting from early 1980s, metal automobile bumpers have been progressively replaced by metal-backed thermoplastic and later by all-thermoplastic bumpers [9]. The merits of thermoplastic bumpers are light weight, streamlined contours for minimizing drag, aesthetics, and structural performance. The development of stiffness criteria for the bumpers is relatively simple, compared to the development of an actual design. An actual design is highly interdisciplinary and complex.

The basic requirement of the Society of Automotive Engineers (SAE) is that bumpers must be able to absorb the kinetic energy of impact that is equivalent to a 1500 lb. mass (680 kg) moving at 5 miles per hour (8 kmph). This corresponds to 1253 ft-lbs. of energy (172 kg m). A further requirement is that the deflection of the bumper be limited such that adjacent components such as head lights, radiator elements, and hood latches must remain functional after impact. A third requirement is that the bumper itself must be reuseable after impact.

In the event of an impact, both force and displacements are "dynamic" quantities. Therefore, the momentary values of the force and the displacements are both higher than the corresponding static quantities. In other words, the apparent stiffness of the bumper in dynamic conditions is less than under the static conditions. Let us assume that the dynamic force developed be $F$ kg, and the dynamic deflection of the bumper be $\delta$ mm. We impose the condition

$$F\delta = 1.72 \times 10^5 \text{ kg mm} \tag{2.4}$$

For the purposes of this discussion, the dynamic deflection of the bumper may be arbitrarily assumed to be limited to about 100 mm, based on the arrangement of the various "adjacent components" in an automobile. The term "functionality of the bumper" may simply be interpreted to mean that the bumper should not undergo large plastic deformations; it is required to remain substantially elastic and hence to regain its original shape.[1]

The bumper is taken to be a simply supported beam with only two supports. For such beams the relationship between static and dynamic deflections is simply a function of the velocity of impact [4], and is given by

$$\delta = \delta_s \left[ 1 + \sqrt{1 + \frac{v^2/g}{\delta_s}} \right]. \tag{2.5}$$

Equations 2.4 and 2.5 may be superimposed on the same plot, as in Figure 2.4. The way to read this figure is as follows. The starting point of the

[1] Obviously, other interpretations are possible. Current philosophy requires that the passenger safety at higher collision velocities be maintained, which entails very advanced analyses.

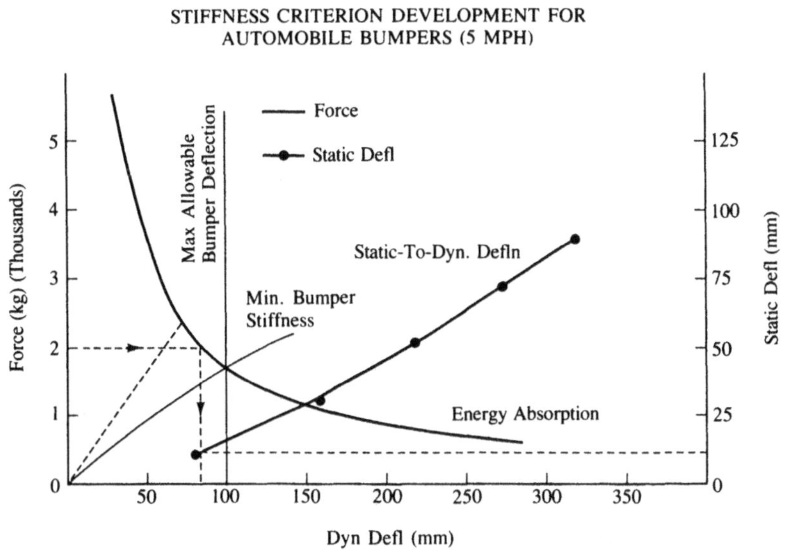


**FIGURE 2.4.** Energy absorption curve for impact loads on bumper.

design is to compute a stiffness required for the bumper. This stiffness corresponds to the static conditions. Hence one must use the force $F$ in conjunction with the corresponding static deflection $\delta_s$. For any assumed allowable $\delta$, we can find the force $F$ from the force curve, and the static deflection $\delta_s$ from the static deflection curve. The stiffness of the bumper needed is given by $F/\delta_s$. Knowing the Young's modulus of the material, one can design a cross section of the bumper to achieve a moment of inertia. Given the constraints on the bumper stiffness, it can be seen that

the dynamic load-deflection curve of the bumper must intersect the $F$-$\delta$ curve in the shaded area.

In practice, a hollow box section of wall thickness varying from 4.5 mm to 6.5 mm is used. Box sections have high torsional stiffness compared to open sections. This is needed because in practice, collisions are unsymmetric and cause torsional moments. For high modulus, dimensional stability, and high impact strength, polycarbonate/polybutylene terephthalate blend is used for bumpers.

The actual static load-deflection curve may be slightly nonlinear due to the stress-strain characteristics. The actual energy absorption and the reuseability of the bumper needs to be assessed on the basis of a full-fledged dynamic impact analysis, which involves among other things, strain-rate dependent stress-strain behavior of the material. On the other hand, the support system of the bumpers has to be designed to minimize the reactions transmitted to the chassis. Considerable improvements can be made to balance the designs of the support system and the bumper by adding damping in the bumper support system.

### 2.2.1 Limitations of Euler's Beam Theory

In its fundamental form, Euler's theory is applicable only to "statically determinate" beams. Other physical principles are to be invoked to solve statically indeterminate forms.

The Euler beam theory is based on the equation for equilibrium of moments about the $z$ direction. Equilibrium is not considered in any other direction. Strictly, equilibrium of transverse forces needs to be considered. This process gives rise to shear stresses which in turn contribute to the deflections. However shear stresses are considered in the beam theory, but not shear deflections. For sufficiently long beams, this procedure is accurate and predicts the stresses and deflections well. For very short beams, the bending stress is small and the shear stress becomes considerable. In this context, we may define a long beam as follows.

> A beam is said to be a *long beam* if the length of the beam exceeds *10* times the depth.

Another limitation of the Euler beam theory is that it is valid strictly only for beams of constant cross section. However, this limitation does not seriously affect the application of Euler's theory to beams of gradually varying cross section. It is only necessary to treat the moment of inertia $I$ as a function of $x$. Text books in strength of materials give a number of interesting examples in optimization of beam sections using this approach.

### Example 2.3

For a uniformly loaded simply supported beam of rectangular cross section the formula for maximum deflection at the center (Figure 2.5), including

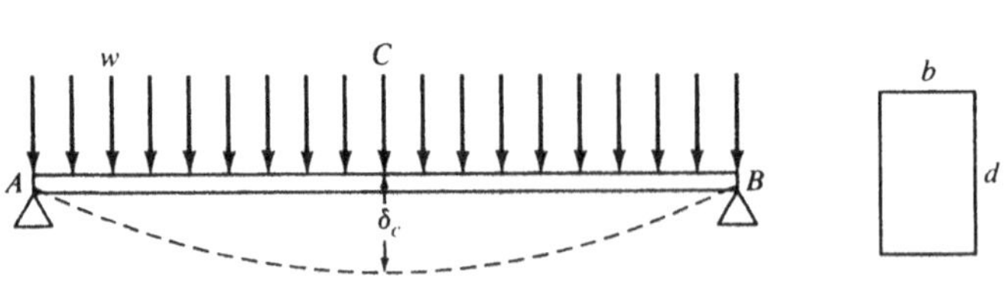


**FIGURE 2.5.** Example 2.3.

the shear effects can be shown to be

$$\delta_c = \frac{5wL^4}{384EI}\left[1 + \frac{48\alpha_s EI}{5\,GAL^2}\right]$$

where $w$ = load per unit length, $L$ = length of the beam, $E$ = Young's modulus, $G$ = shear modulus of the beam, $A$ = area of cross section, and $I$ = second moment of area about the neutral axis. $\alpha_s$ is a factor called the shear coefficient[2] and is equal to $3/2$ for rectangular cross-section. (a) Find the length $L$ of the beam for which the contribution of shear deflection would be less than 5% of the bending deflection. (b) For $L = 10 \times$ (depth of the beam), what would be the contribution of the shear deflection as compared to bending?

*Solution* Comparing with the formula for deflection due to bending only, we see that the second term in the parenthesis denotes the contribution of the shear stress.

(a) We have the following.

$$G = \frac{E}{2(1+\nu)}$$

$$A = bd$$

$$I = bd^3/12$$

$$L = kd, \text{ say, } k \text{ to be determined}$$

To find $k$, we impose the condition that

$$\frac{48\alpha_s EI}{5\,GAL^2} < 0.05$$

$$\frac{48\left(\frac{3}{2}\right)(E)\left(\frac{bd^3}{12}\right)}{5\frac{E}{2(1+\upsilon)}(bd)(k^2d^2)} < 0.05$$

[2] Shear coefficients for other shapes can be obtained systematically. Reference 2 outlines this procedure and also lists the coefficients for a few other cross sectional shapes. Whatever its value, the criterion $L > 10$ (depth) for a long beam is valid for all cross sections.

Simplifying,

$$2.4 < k^2(0.05)$$

Equivalently,
$$k > \sqrt{\frac{2.4}{0.05}} = \sqrt{48} \approx 7.0$$

Length of Beam > 7 × Depth

(b) When $L = 10d$,

$$\frac{48\alpha_s EI}{5\,GAL^2} = \frac{48\frac{3}{2}E\left(\frac{bd^3}{12}\right)}{5\frac{E}{2(1+\upsilon)}bd(10d)^2}$$

$$= 0.024(1+\upsilon)$$

$$= 0.024(1+0.3), \text{ assumed } \upsilon = 0.3$$

$$= 0.0312$$

$$= 3.12\%$$

### 2.2.2 Beams Made of Nonlinear Elastic Materials

Most plastic materials display a nonlinear elastic stress strain relationship over the range of their useful load taking capability. This characteristic renders the measurement of modulus difficult. Fortunately, plastics have low stiffness; besides, it is known that bending loads produce far greater deflections than do membrane loads, thus facilitating easier measurements. Taking advantage of the two facts, ASTM D-790 [10] outlines a procedure for measuring the modulus which utilizes loads in the bending mode. In the process, the standard introduces a term called "flexural modulus" of plastics. The rationale of the test procedure and the significance of "flexural modulus" are discussed in the following section. As a preliminary to that discussion, we need to consider the bending of a beam made of nonlinear elastic material.

The equation governing the deflection for nonlinear elastic beams of uniform cross section can be derived using a procedure analogous to the procedure used for linear materials, Figure 2.6. The derivation is left to the reader as an exercise. In the following example, we solve the governing equation for a beam of rectangular cross section which is the standard

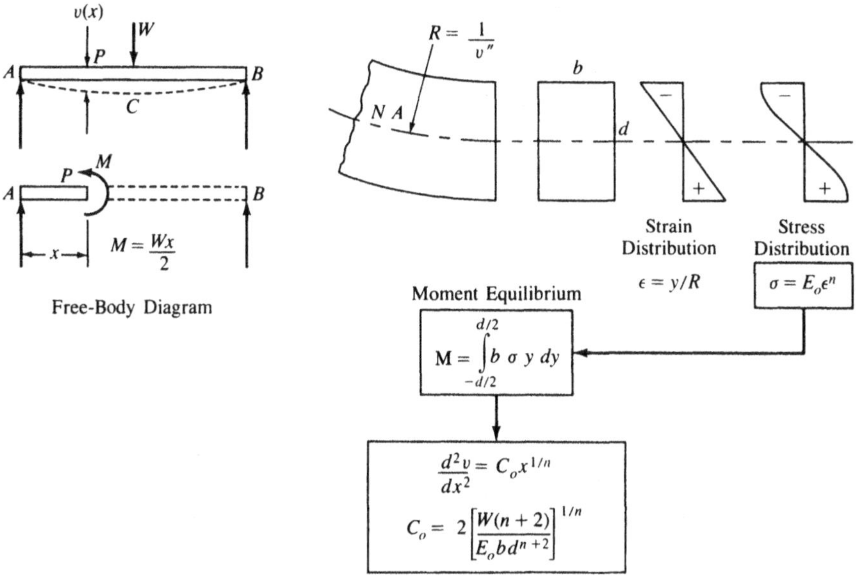


**FIGURE 2.6.** Derivation of governing equation for beams made of nonlinear elastic materials.

shape required for materials testing for flexural modulus as per the ASTM D-790 standard.

### Example 2.4

For the standard 3-point bend test specimen of ASTM D-790, of width $b$, thickness $d$, and length $L$ as shown in Figure 2.7, derive the load-displacement relation. Material stress-strain behavior can be described as $\sigma = E_o \epsilon^n$, where $(0 < n < 1)$.

*Solution* Consider the half $AC$ of the beam, from the left end. In this half, the bending moment $M(x)$ at a point $P$, distance $x$ from the left support is given by $Wx/2$. We have, from Figure 2.6,

$$\frac{d^2 v}{dx^2} = A_1 x^{1/n}$$

where,
$$A_1 = \left[ \frac{2^{(n+1)}(n+2)}{E_o b d^{(n+2)}} \left( \frac{W}{2} \right) \right]^{1/n}$$

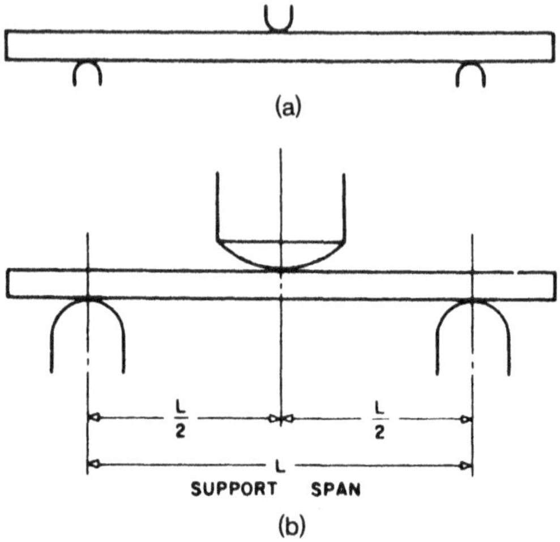


**FIGURE 2.7.** Typical test specimens for 3-point bend tests have a rectangular cross section. Specimens are up to 6.3 mm deep. Restrictions apply to the nose radii of supports and loads.

Solution of this equation for $v$ poses no difficulty. We obtain $v$ by integrating the above equation twice and imposing the conditions that $v' = 0$ at $x = L/2$, and $v = 0$ at $x = 0$.

$$v = 2\left[\frac{W(n+2)}{E_o bd^{(n+2)}}\right]^{1/n}\left(\frac{n}{n+1}\right)x\left[\left(\frac{n}{2n+1}\right)x^{(n+1/n)} - \left(\frac{L}{2}\right)^{(n/n+1)}\right]$$

$$v_{\max} = 2W^{1/n}\left[\frac{(n+2)}{E_o bd^{n+2}}\right]^{1/n}\left(\frac{n}{2n+1}\right)\left(\frac{L}{2}\right)^{(2n+1/n)} \tag{2.6}$$

Note that if we set $n = 1$ in the above, we are led to the conventional beam theory of beams. Following Reference 11 we may write the expressions for $v_{\max}$ in terms of nondimensional groups.

$$\left(\frac{6v_{\max}d}{L^2}\right) = \left(\frac{3n}{2n+1}\right)\left(\frac{n+2}{24}\right)^{1/n}\left(\frac{WLd}{E_o I}\right)^{1/n} \tag{2.7}$$

Now we may substitute the numbers.

***Commentary***

*In general, the partial derivative* $\partial W / \partial v_{\max}$ *gives a measure of the tangent modulus of the material. This idea is basic to the ASTM D-790 standard procedure. Therefore, it is necessary to provide a recording device in the test setup to obtain a graph of load versus deflection. However, if we try to calculate the derivative for this exercise, it tends to infinity, and so does the flexural modulus! This is to be expected, because the stress-strain relation written as a power law leads to the same conclusion anyway. The infinite slope at origin is a consequence of the power law assumed, with* $0 < n < 1$. *In practice, the material exhibits a finite value for the modulus. Furthermore, one will find that the modulus is given by*

$$E_{\text{flex}} = \frac{L^3}{48I} f(n) \left( \frac{L}{d} \right)^{1-n} \left( \frac{L}{v} \right)^{1-n}$$

*The presence of the term* $(L/d)^{1-n}$ *suggests the influence of the depth of the beam on the value of* $E_{\text{flex}}$.

### 2.2.3 A Note on ASTM Standard D-790

ASTM Standard D-790 on "Standard Test Methods for Flexural Properties of Unreinforced and Reinforced Plastics and Electrical Insulating Materials" covers the test methods for the determination of flexural properties of unreinforced plastics and reinforced plastics. The test specimens are simple flat strips conforming to certain dimensional proportions listed in this standard.

This standard states procedures to determine (i) flexural modulus and (ii) flexural strength, if the specimen fails during the test. If the specimen does not fail, the flexural strength cannot be measured.

Salient features of the procedures are as follows.

1. Either three-point bending or four-point bending test methods can be used. For either test method, two different strain rates can be applied. The lower one is 0.01 mm/mm/min, and is advised for materials which fail at a relatively low strain. The higher one is 0.1 mm/mm/min, and is appropriate for materials which undergo large amounts of displacements before failure.

2. Regular beam bending formulas are used to calculate a variety of quantities such as stress, strain, modulus of section, moment of inertia, etc.
3. The standard recognizes and admits a procedure of the so-called "toe correction" of the load deflection curve plotted automatically by the testing machine. In principle, this correction eliminates the spurious inflexion of the curve at startup which is an "artifact caused by takeup of slack and alignment or seating of the specimen."
4. Most important of all is that this standard recognizes that flexural properties are useful in relative comparison of one material with another, or the consistent quality of a material between lots. The following statements in the standard are to be carefully understood.

> Flexural properties determined by (three point bending) are especially useful for quality control and specification purposes....
> Flexural properties may vary with specimen depth, temperature atmospheric conditions, and the difference in rate of straining specified.

The flexural modulus is thus not an intrinsic material property because it is sensitive to the size of the specimen. On the other hand, Young's modulus determined from regular tension tests does not exhibit such dependence. Hence for stress analysis purposes, the Young's modulus must be used and not the flexural modulus.

Why then is the ASTM D-790 method used at all? The D-790 method is an economical alternative to the tension tests. Only simple machinery is needed. Tests can be performed at relatively small loads. For a typical industrial setup adopting a standard size of bend test specimen, the influence of size dependence is eliminated. Typically all the materials in such a setup are compared on a common basis.

## 2.3 THE UNIT LOAD METHOD (ULM)

We return to linear elasticity applications. We introduce a simple procedure called the *unit load method* (ULM), which extends the power of the beam bending theory to more complex shapes. The introduction of ULM in this context is somewhat arbitrary. It is presumed that the reader has enough knowledge of solving simple beam problems, and would therefore be able to appreciate the power of this method. In fact, it is just one of the family of so called "energy methods." The reader is referred to Reference 2 for a complete treatment of the derivation of the ULM. The ULM is a product of a more general principle of mechanics, called the "principle of virtual work."

The unit load method (ULM) consists of the following steps.

1. Apply a fictitious load (or moment) of unit magnitude in addition to the other real loads on the structure. This load must be applied at the location where we wish to calculate the displacement (or rotation). Also it must be applied in the same direction in which displacement (or rotation) is to be calculated.
2. Let $M_L$ be the moment due to the real loads, and $M_U$ be the moment due to the fictitious unit load. Then the required displacement (or rotation) is given by

$$\delta(\text{or } \theta) = \int \frac{M_L M_U}{EI}\, dl \tag{2.8}$$

where $dl$ is a small length of the beam along its axis.

Obviously, this procedure can be used for hand calculation only if the number of points of interest is small, usually only one or two points.

**Example 2.5**

A PVC pipe of size 84 mm o.d. × 4.0 mm thickness is used as a hot burnt gas exhaust in a domestic gas furnace. The pipe is connected to the outlet of the exhaust fan by pipe clamps. For access during repair and replacement of furnace components, the pipe may have to be moved laterally by about 75 mm. For easiness of operation, no more than 5 kg should be required to effect the lateral movement. The normal headroom available in a utility room of a house is about 2.4 m. The fan outlet is at a height of 450 mm above the floor. Determine whether it is possible to accomplish a vertical layout of the pipe consistent with the criteria of lateral flexibility.

*Solution* The actual pipe layout shown in Figure 2.8a can be modeled as a combination of beams shown in Figure 2.8b. The horizontal force $F_H$ is applied at the point $C$. The displacement at $C$ is $\delta_H$. We want to find a relationship between $F_H$ and $\delta_H$. To apply the Unit Load Method to this

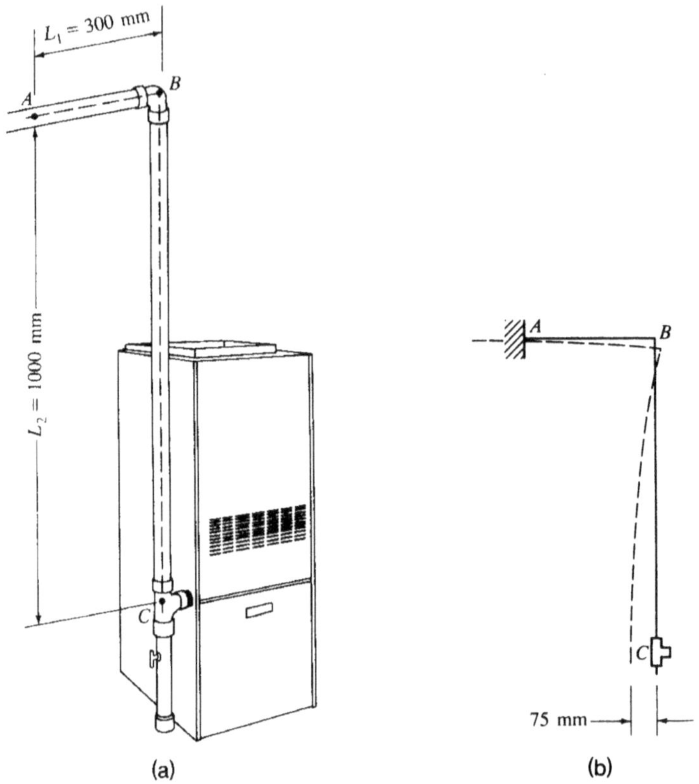


**FIGURE 2.8.** Example 2.5.

problem, we apply a unit horizontal force $U$ at $C$. We can calculate the moments $M_L$ and $M_U$ as follows.

$$M_L = F_H y \text{ (for } P \text{ on } BC)$$

$$= F_H h \text{ (for } P \text{ on } AB)$$

$$M_U = Uy \text{ (for } P \text{ on } BC)$$

$$= Uh \text{ (for } P \text{ on } AB)$$

$$\sigma_U = 1.5 \text{ kg/mm}^2$$

$$E = 210 \text{ kg/mm}^2$$

$$I = \frac{\pi}{64}(84^4 - 76^4) = 0.806 \times 10^6 \text{ mm}^4$$

$$z = I/(42 \text{ mm}) = 1.919 \times 10^4 \text{ mm}^3$$

Using ULM, the displacement at $C$ is obtained as

$$\delta_H = \int_{ABC} \frac{M_U M_L}{EI}\, dl$$

$$= \int_{AB} \frac{F_H U}{EI} h^2\, dl + \int_{BC} \frac{F_H U}{EI} y^2\, dy$$

$$= \frac{F_H U h^2 L_1}{EI} + \frac{F_H U h^3}{3EI}$$

$$= \frac{F_H h^2}{EI}\left(\frac{h}{3} + L_1\right) \quad (\text{since } U = 1)$$

Substituting the values,

$$\delta_H = 75 \text{ mm} \quad F_H = 5 \text{ kg} \quad L_1 = 500 \text{ mm}$$

$$\delta_H = \frac{F_H}{3EI}(h^3 + 3h^2 L_1)$$

$$= \frac{(5 \text{ kg})(h^3 + 1500h^2)}{3(210 \text{ kg/mm}^2)(0.806 \times 10^6 \text{ mm}^4)}$$

To avoid the large sized numbers, we may express $h$ in meters. Accordingly,

$$\delta_H = \frac{F_H}{3EI}(h^3 + 3h^2 L_1)$$

$$0.075 \text{ m} = \frac{(5 \text{ kg})(h^3 + 1.5h^2)}{3(210 \times 10^6 \text{ kg/m}^2)(0.806 \times 10^{-6} \text{ m}^4)}$$

$$0.075 = \frac{5(h^3 + 1.5h^2)}{3(220)(0.806)}$$

Simplifying,

$$h^3 + 1.5h^2 - 5.32 = 0$$

Solving, we obtain

$$h = 1.3625 \text{ m}$$

It is possible to meet the flexibility criterion within the given headroom. In fact, we may set $L_1 = 0$, and re-do the problem to obtain

$$h = 1.746 \text{ m}$$

With the headroom available, this length may be accommodated in a typical utility room, and hence may be considered an acceptable solution.

***Stresses in pipe***
Stress in pipe is calculated from the bending moments. (a) When $L_1 = 500$ mm:

$$\sigma = \frac{F_H h}{z} = \frac{(5 \text{ kg})(1362.5 \text{ mm})}{1.919 \times 10^4 \text{ mm}^4}$$

$$= 0.4547 \text{ kg/mm}^2$$

(b) When $L_1 = 0$

$$\sigma = \frac{F_h h}{z} = \frac{(5 \text{ kg})(1746 \text{ mm})}{1.919 \times 10^4 \text{ mm}^4}$$

$$= 0.355 \text{ kg/mm}^2$$

Comparing with the ultimate strength of 1.5 $\text{kg/mm}^2$ of the pipe material (PVC), we see that the pipe will carry a factor of safety that is higher than the desired value of 2.0.

ULM applies to a more general variety of beam problems. Consider the following example of ovaling of a loaded pail, which appears to be a three dimensional problem. However, it is possible to model it as a combination of one dimensional sub-problems.

**Example 2.6**

A pail of paint weighs 13.5 kg. It is fitted with a handle in the shape of a half circle of radius $R = 225$ mm. The handle is made from a steel wire of diameter 5 mm. The pail is made of PVC, modulus $E_o = 210 \text{ kg/mm}^2$. Calculate the reduction of the diameter of the pail, when the lid is off and the full load of the pail is lifted by the top most point of the handle, as shown in Figure 2.9a. (a) Neglect the stiffness of the pail. (b) Repeat the

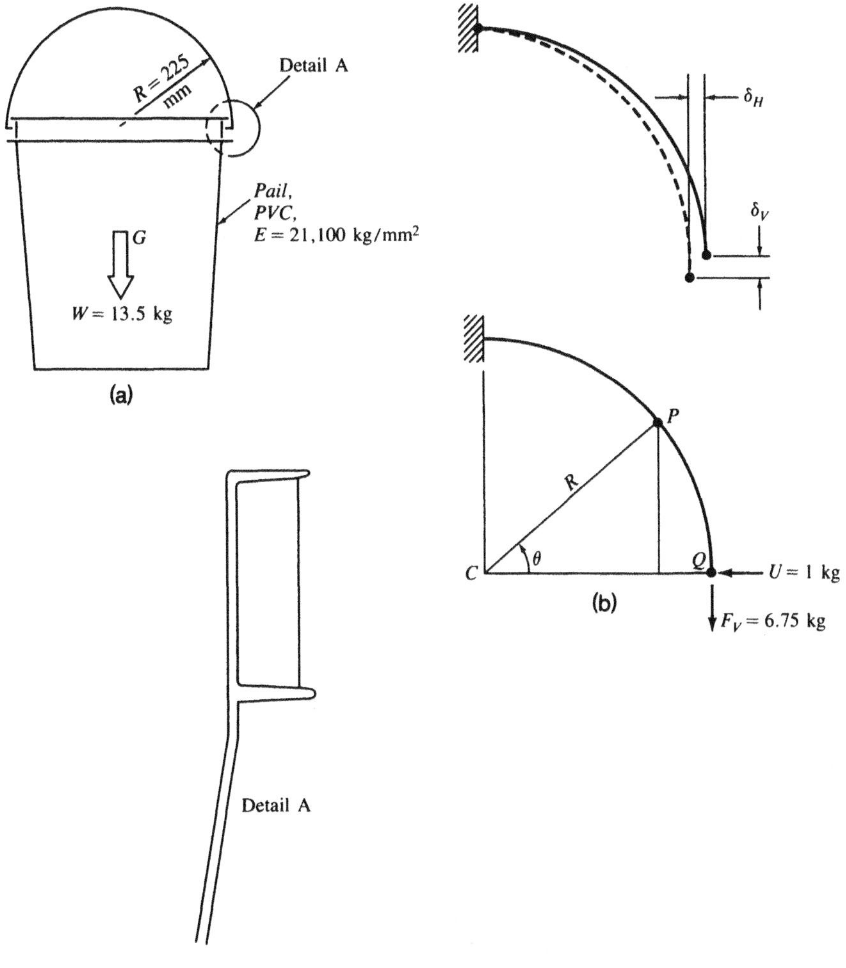


**FIGURE 2.9.** Example 2.6.

problem by assuming that the pail offers a resistance. For this purpose, consider that the ring at the rim of the pail between the two circumferential rings shown in the figure is effective for the stiffness calculation.

*Solution* Part (a): Neglect the stiffness of the pail.

The free body diagram of the pail handle is shown in Figure 2.9b. The Unit Load Method consists of applying a fictitious load $U = 1$ kg to the pail-handle. This load must be applied at $Q$, horizontally acting towards the left. This is so because the displacement is to be calculated at $Q$, horizontal, and towards the left.

$$\text{B.M at } P \text{ due to } F_V = M_L = F_V R(1 - \cos\theta)$$

$$\text{B.M at } P \text{ due to } U = M_U = UR\sin\theta$$

$$= R\sin\theta$$

$$dl = Rd\,\theta$$

$$\delta_H = \int \frac{M_U M_L}{EI}\,dl = \frac{F_V R^3}{EI}\int_0^{\pi/2}(1 - \cos\theta)\sin\theta\,d\theta = \frac{F_V R^3}{2EI}$$

$$F_V = 6.75\text{ Kg}$$

$$R = 225\text{ Kg}$$

$$I = \frac{\pi}{64}d^4 = \frac{\pi}{64}(5\text{ mm})^4 = 30.68\text{ mm}^4$$

$$\delta_H = \frac{(6.75\text{ Kg})(225\text{ mm})^3}{2(21{,}100\text{ Kg/mm}^2)(30.68\text{ mm}^4)}$$

$$= 59.4\text{ mm}$$

Considering the other half of the handle, ovaling = 118.8 mm.
Part (b): Consider the resistance of the pail to ovaling.

Since the pail has its own stiffness, it offers a reaction $F_H$ to the pail handle, at $Q$, acting towards the right. Since its behavior is linear elastic, we can suppose that the reaction $F_H$ is proportional to the ovaling itself, or the reduction in the radius of the pail. Therefore, we can write

$$F_H = k\delta_H$$

where $k$ is the stiffness due to the effective width of the ring at the rim of the pail, for the force $F_H$. The stiffness can be obtained from the load displacement relation for a circular ring. From Reference 8, the following equation is obtained for thin circular rings subjected to two radial forces.

$$\delta_H = \frac{F_H R_o^3}{E_o I_o}\left(\frac{\pi}{4} - \frac{2}{\pi}\right) = \frac{F_H}{k}$$

where $k$ is the stiffness of the ring, and the subscript $_o$ denotes the properties of the ring. From Figure 2.9 we see that

$$R_o = 425 \text{ mm}$$

$$E_o = 210 \text{ kg/mm}^2$$

$$I_o = \frac{60 \text{ mm} \times (2.0 \text{ mm})^3}{12} = 40 \text{ mm}^4$$

$$k = \frac{E_o I_o}{R_o^3\left(\frac{\pi}{4} - \frac{2}{\pi}\right)}$$

$$= \frac{(210 \text{ kg/mm}^2)(40 \text{ mm}^4)}{(212.5 \text{ mm}^3)\left(\frac{\pi}{4} - \frac{2}{\pi}\right)}$$

$$= 5.884 \times 10^{-3} \text{ kg/mm}$$

We may now apply the unit load $U$, in the same manner as before. The moments are calculated as:

$$M_L = F_V R(1 - \cos\theta) - F_H R \sin\theta$$

$$M_U = UR\sin\theta$$

$$\delta_H = \frac{1}{EI}\int_0^{\pi/2}[F_V R(1 - \cos\theta) - F_H R\sin\theta]UR\sin\theta\, R\,d\theta$$

$$= \frac{F_V R^3}{2EI} - \frac{\pi F_H R^3}{4EI}$$

$$= \frac{F_V R^3}{2EI} - \frac{\pi(k\delta_H)R^3}{4EI}$$

Rearranging,

$$\delta_H = \frac{F_V R^3}{2EI\left(1 + \dfrac{\pi}{4}\dfrac{kR^3}{EI}\right)}$$

$$= \frac{F_V R^3}{2EI + \dfrac{\pi}{2}kR^3}$$

$$= \frac{(6.75\text{ kg})(225\text{ mm})^3}{2(21{,}100\text{ kg/mm}^2)(30.68\text{ mm}^4) + \dfrac{\pi}{2}(5.884 \times 10^{-3}\text{ kg/mm})(225\text{ mm})^3}$$

$$= 54.92\text{ mm}$$

Considering both halves of the problem, ovaling = 110 mm.

Note that one can go on to include the effect of the two circumferential ribs near the pail-handle in the calculation of $I_o$, and obtain a higher stiffness for the pail. A further level of refinement of the calculation is to consider the stiffness of the entire pail including its bottom, which would call for numerical techniques.

## 2.4 APPLICATION TO PIPING FLEXIBILITY ANALYSIS

Piping flexibility analysis is a calculation technique applied to large industrial piping, and its objective is to determine the stresses in the pipes which arise from the arresting of the thermal expansion of the pipe. Though simple in its basic form, the piping flexibility analysis is considerably more complex in practice because of several features of piping layout, such as supports, restraints, valves, pipe fittings, pipe branches, and also because of the networking of pipes. In addition to thermal expansion, dead loads, self-weight, internal pressure, start-up shut-down cycles, thermal gradients through the wall thickness, etc., must be considered. All these factors make the total problem an area of specialty in itself. Plastic pipings used for underground applications are further subjected to soil friction which tends to prevent expansions throughout the length of the pipeline. Such pipes may experience considerably higher thermal expansion stress, and may derive no advantage against thermal expansion stress from the shape of their layout. The reader is referred to recent texts such as References 12, 13, and 14. For a simple way to understand the design of underground reinforced or unreinforced plastic pipes, the AWWA Standard C950-88 [15] is an excellent starting point.

### 2.4.1 Use of Plastics For Plumbing and Other Pipelines

Plastics pipes of small sizes, typically less than 1 inch nominal size, made of chlorinated PVC are finding increasing application in domestic plumbing, and are replacing the traditional copper piping which had been used in all newly constructed homes until the mid 1970s. The reason for this is the economics of plastics plumbing material and installation costs, as well as the engineering merits of the material.

Table 2.2 lists the values of modulus $E$, thermal expansion coefficient $\alpha$, and tensile strength $\sigma_y$ for several materials, metals and plastics, and the factor of safety, $n$. The factor $n$ is the ratio of $\sigma_y$ to the product $E\alpha\,\Delta T$ and is based on the fact that the stress created in the piping due to thermal expansion is proportional to this product. In fact, for a straight pipe which is completely constrained at its ends, the thermal expansion stress is equal to $E\alpha\,\Delta T$. For other zigzag piping layouts, the stress in the pipe can be considered to be (a factor $K$, less than 1) $\times\, E\alpha\,\Delta T$. The factor $K$ is dependent upon the layout of the pipe. Usually, if a pipe of length $L$ is connected between two points which are at a distance $r$, ($r < L$), then we may state that the factor $K$ is smaller for larger values for the ratio $r/L$.

Figure 2.10 shows a bar chart of the $n$, for those materials. On the whole, it is seen that plastics possess a larger value of $n$, while having a lower $E\alpha$. This means that plastics used in piping will have lower anchor reactions, and the pipes themselves will carry a larger factor of safety as compared to metals.

Another advantage of plastics as piping materials is that due to their viscoelastic behavior, the stress must relax over time. Thus the initial stress is also the highest over time. However, the strength of plastics also decreases with time, unlike the situation in metals. Thus, we have the problem of ensuring that the strength *at any given time* is adequate against the stresses (combined effects of thermal expansion, operating pressure etc.) at that of time. The $n$ values of Figure 2.10 apply for the initial or short term conditions. It has been demonstrated however, that even after allowing for the decline in strength of the plastics with time, CPVC plastics

**Table 2.2. The Ratio of Yield Strength to Thermal Stress per Unit Temperature Rise. Plastics Show a High Ratio Indicating Suitability for Plumbing Applications.**

| | E (GPa) | $\alpha$ $10^{-6}/°C$ | $\sigma_y$ (MPa) | E$\alpha$ (MPa) | $\eta = \frac{\sigma_y}{E\alpha}$ |
|---|---|---|---|---|---|
| *Aluminum (Pure)* | 70 | 23 | 20 | 1.610 | 12.4 |
| 2014-T6 | 73 | 23 | 3410 | 1.680 | 244 |
| 6061-T6 | 70 | 23 | 270 | 1.610 | 169 |
| 7075-T6 | 72 | 23 | 480 | 1.660 | 289 |
| *Steel* | | | | | |
| Structural | 210 | 12 | 250 | 2.52 | 99 |
| 304-SS | 200 | 17 | 280 | 3.40 | 82 |
| HSLA | 200 | 14 | 350 | 2.80 | 125 |
| *Copper* | | | | | |
| Hard-Drawn | 120 | 17 | 330 | 2.04 | 162 |
| Soft (Annealed) | 120 | 17 | 55 | 2.04 | 27 |
| Beryllium | 120 | 17 | 760 | 2.04 | 372 |
| *Plastics Piping* | | | | | |
| Polybutylene | 0.241 | 128 | 29 | 0.030 | 942 |
| Acetal | 2.830 | 81 | 69 | 0.229 | 301 |
| CPE | 1.030 | 97 | 41 | 0.100 | 413 |
| Flourocarbons | 0.480 | 83 | 21 | 0.040 | 518 |
| Polyamide | 2.830 | 81 | 83 | 0.230 | 361 |
| PMA Grade 8 | 3.100 | 54 | 72 | 0.167 | 432 |
| PMA Grade 6 | 3.100 | 72 | 69 | 0.223 | 309 |
| PMA Grade 5 | 3.100 | 72 | 65 | 0.223 | 294 |
| PE (IpI) | 0.172 | 180 | 13 | 0.031 | 420 |
| Type II | 0.552 | 153 | 16 | 0.084 | 189 |
| Type III | 0.861 | 126 | 22 | 0.108 | 203 |
| High Molecular Wt. Tp (IV) | 0.896 | 108 | 22 | 0.097 | 227 |
| Rigid PVC | 3.070 | 59 | 51 | 0.177 | 288 |
| Polypropylene | 0.930 | 79 | 24 | 0.073 | 327 |

pipes can sustain continued application of temperatures and pressures for plumbing purposes, over an estimated life of 40 years.

Examples 2.5 and 2.6 above belong to the category of "load-controlled problems." The application of the unit load method was straightforward in these cases, but its application becomes less obvious in the case of "displacement-controlled problems." Piping flexibility analysis problems belong in the latter category. The ULM is designed to give the displacements for a given set of loads. On the contrary, the displacement-controlled problems seek to find the loads (and subsequently the stresses) for

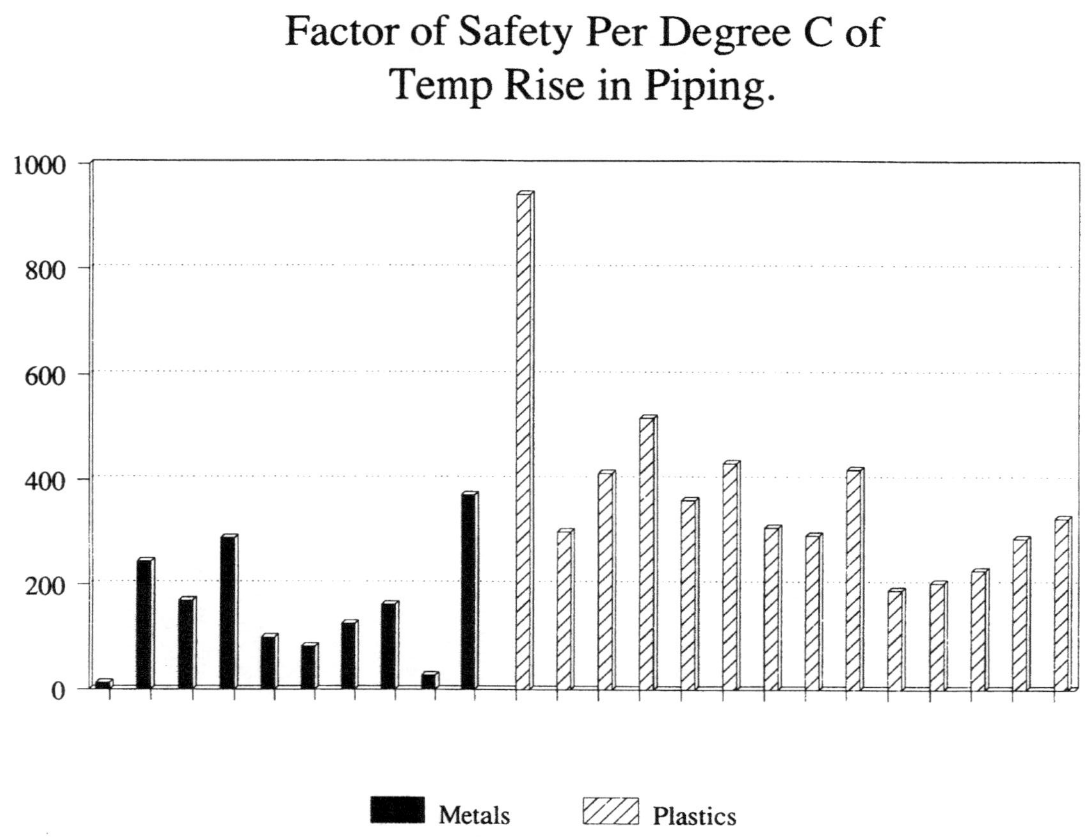


**FIGURE 2.10.** Plastics, when used for piping material, have a much greater factor of safety on their strength than metals do. See also Table 2.2.

a given set of displacements. The objective of the problem being reversed, the application of ULM appears less obvious.

### 2.4.2 A Practical Example of Piping Flexibility Analysis

In this rather long example, we wish to drive home a few points. Firstly, the practical problems dealt with in industry are based on simple theory, but their complexity arises because of the combination of several loads and load-cases which must be analyzed. Secondly, piping flexibility analysis of a relatively simple shape such as is used in the following example is very long and suggests the need for a generalized computer program. The third and the most important point is that

> the stresses in a displacement controlled problem cannot be reduced by increasing the cross sectional area or moment of inertia of the section, etc.; rather, a decrease in stress may be achieved by reducing material!

**Example 2.7**

Reconsider the earlier example of the L-shaped exhaust pipe of a domestic furnace. If the pipe is expected to see a maximum service temperature of $\Delta T$, and if the two end points $A$ and $C$ are fixed rigidly, discuss the resulting stresses in the pipe arising from the prevention of thermal expansion.

*Solution* The standard procedure in all the displacement controlled problems is to calculate the natural displacements that would have left the structure stress-free, apply the negative of these displacements on the structure to simulate the complete restraints.

In this example, if the end $C$ were free, it would move relative to $A$ by an amount

$$\vec{\mathbf{u}} = \alpha \, \Delta T \left( L_1 \vec{i} - L_2 \vec{j} \right)$$

The effect of applying displacements is calculated in the reverse way. That is, by determining the displacements produced by known forces.

***Calculate $\delta_H$ caused by $F_H$, $F_V$ and M***

$$M_L = -F_H L_2 + F_V x + M \text{ (on } AB\text{) and}$$
$$= -F_H y + M \text{ (on } BC\text{)}.$$
$$M_U = -1.L_2 \text{ (on } AB\text{) and}$$
$$= -1.y \text{ (on } BC\text{)}.$$

$$\delta_H = \frac{1}{EI}\int_{AB}(-F_H L_2 + F_V x + M)(-L_2)dx$$
$$+\frac{1}{EI}\int_{BC}(-F_H y + M)(-y)dy \tag{2.9}$$

$$EI\delta_H = F_H L_2^2\left(L_1 + \frac{L_2}{3}\right) - F_V\left(\frac{L_1^2 L_2}{2}\right) - ML_2\left(L_1 + \frac{L_2}{2}\right) \tag{2.10}$$

***Calculate $\delta_V$ caused by $F_H$, $F_V$ and M***

$$M_L = -F_H L_2 + F_V x + M \text{ on } AB \text{ and}$$
$$= -F_H y + M \text{ on } BC$$
$$M_U = 1.x \text{ on } AB$$
$$= 0 \text{ on } BC$$

Hence,
$$EI\delta_V = \int_{AB}(-F_H L_2 + F_V x + M).x\,dx$$

$$EI\delta_V = -F_H L_2\frac{L_1^2}{2} + F_V\frac{L_1^3}{3} + M\frac{L_1^2}{2} \tag{2.11}$$

***Calculate $\theta$ caused by $F_H$, $F_V$ and M***

We may write the equation for $EI\theta$ directly as

$$EI\theta = \int_{AB}(-F_H L_2 + F_V x + M).1.dx + \int_{BC}(-F_H y + M).1.dy$$

$$= -F_H L_2\left(L_1 + \frac{L^2}{2}\right) + F_V\frac{L_1^2}{2} + M(L_1 + L_2) \tag{2.12}$$

It is convenient to arrange Equations 2.10, 2.11, and 2.12 in the form of a matrix equation as

$$\{\mathbf{u}\} = [\mathbf{S}]\{\mathbf{F}\}, \tag{2.13}$$

where, $\{\mathbf{u}\} = \begin{Bmatrix} \delta_H \\ \delta_V \\ \theta \end{Bmatrix}$ $\{\mathbf{F}\} = \begin{Bmatrix} F_H \\ F_V \\ M \end{Bmatrix}$ and,

$$[\mathbf{S}] = \frac{1}{EI}\begin{bmatrix} L_2^2\left(L_1 + \dfrac{L^2}{3}\right) & -\left(\dfrac{L_1^2 L_2}{2}\right) & -L_2\left(L_1 + \dfrac{L_2}{2}\right) \\ -\left(\dfrac{L_1^2 L_2}{2}\right) & \dfrac{L_1^3}{3} & \dfrac{L_1^2}{2} \\ -L_2\left(L_1 + \dfrac{L_2}{2}\right) & \dfrac{L_1^2}{2} & (L_1 + L_2) \end{bmatrix} \tag{2.14}$$

$\{\mathbf{u}\}$ is the free displacement vector, $[\mathbf{S}]$ is the flexibility matrix (and hence the name flexibility analysis), and $\{\mathbf{F}\}$ is the force vector. As mentioned earlier, the complete restraint to thermal expansion is obtained by prescribing

$$\{\mathbf{u}\} = -\alpha\,\Delta T\begin{Bmatrix} L_1 \\ L_2 \\ 0 \end{Bmatrix}$$

We are now ready to substitute the actual numbers. Here, we prefer to do so for lengths only. Note that $L_1 = 0.3$ m and $L_1 = 1.0$ m respectively, Equation 2.14 gives

$$-EI\alpha\,\Delta T\begin{Bmatrix} 0.3 \\ 1.0 \\ 0 \end{Bmatrix} = \begin{bmatrix} 0.633 & -0.045 & -0.345 \\ -0.045 & 0.009 & 0.045 \\ -0.345 & 0.045 & 1.300 \end{bmatrix}$$

Inverting this equation, we can obtain the forces needed to achieve complete restraint.

$$\begin{Bmatrix} F_H \\ F_V \\ M \end{Bmatrix} = -EI\alpha\,\Delta T\begin{bmatrix} 2.532 & 11.247 & 0.283 \\ 11.247 & 184.325 & -3.396 \\ 0.283 & -3.396 & 0.962 \end{bmatrix}\begin{Bmatrix} 0.3 \\ 1.0 \\ 0.0 \end{Bmatrix}$$

$$= EI\alpha\,\Delta T\begin{Bmatrix} 12.01 \\ 187.70 \\ -3.31 \end{Bmatrix}$$

***Calculate Stress***

Stress at any point is calculated by finding $M_L$ at that point using the formulas stated above. For example, the stress at $A$ may be calculated as

$$\begin{aligned} M_A &= -F_H L_2 + F_V L_1 + M \\ &= -EI\alpha\,\Delta T[-(12.01)(1.0) + (187.7)(0.3) - 3.31] \text{ Kg m} \\ &= -EI\alpha\,\Delta T(41.0) \text{ Kg m} \end{aligned}$$

$$\sigma_A = \frac{M_A d}{2I} = E\alpha\,\Delta T \frac{d}{2}(41.0) \text{ Pa} \qquad (2.15)$$

From the formulas for the force system, as well as for stress, we find that increasing the size of the pipe does not improve the situation. The force system is proportional to the moment of inertia, and the stress is proportional to the diameter.

Only the bending moments were considered in the whole exercise. This component represents substantially all the stresses in the pipe. However, there are other effects too. Complexity of the calculations increases if we further consider effects of normal stresses, shear stresses, etc. In real life, one finds that there can be several different pipe sizes in a network and thus there is a need for a general computer program.

## 2.5 PROBLEMS IN POLAR COORDINATES

Polar coordinates are appropriate for problems involving circular shapes. Such problems occur in press fits, rotating shafts, and disks. Intuitively, the stresses, strains, and displacements of these problems are best described in polar coordinates.

Problems in two dimensional elasticity involving circular or cylindrical shapes are best solved using equations of elasticity theory written in polar coordinates. Equations of equilibrium, strain-displacement, and stress-strain equations in polar coordinate system are given below. The derivation of the following equations can be accomplished either directly [16], or by converting their Cartesian form to polar form by using coordinate transformation rules. The quantities referred to in the equations are defined in Figure 2.11.

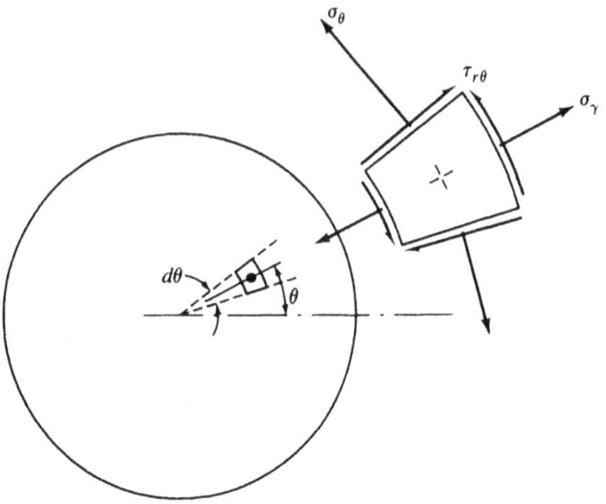


**FIGURE 2.11.** Conventions of measuring displacements, strains, and stresses in polar coordinates. Stresses are positive as marked.

## *Equilibrium Equations in Polar Coordinates*

$$\frac{\partial \sigma_r}{\partial r} + \frac{1}{r}\frac{\partial \tau_{r\theta}}{\partial \theta} + \frac{\sigma_r - \sigma_\theta}{r} + F_r = 0 \tag{2.16}$$

$$\frac{1}{r}\frac{\partial \sigma_\theta}{\partial \theta} + \frac{\partial \tau_{r\theta}}{\partial r} + \frac{2\tau_{r\theta}}{r} + F_\theta = 0 \tag{2.17}$$

## *Strain-Displacement Equations in Polar Coordinates*

$$\epsilon_r = \frac{\partial u_r}{\partial r} \tag{2.18}$$

$$\epsilon_\theta = \frac{u_r}{r} + \frac{1}{r}\frac{\partial u_\theta}{\partial \theta} \tag{2.19}$$

$$\gamma_{r\theta} = \frac{1}{r}\frac{\partial u_r}{\partial \theta} + \frac{\partial u_\theta}{\partial r} - \frac{u_\theta}{r} \tag{2.20}$$

***Stress-Strain Equations—Plane Stress***

$$\epsilon_r = \frac{1}{E}(\sigma_r - \nu\sigma_\theta) \tag{2.21}$$

$$\epsilon_\theta = \frac{1}{E}(\sigma_\theta - \nu\sigma_r) \tag{2.22}$$

$$\gamma_{r\theta} = \frac{1}{G}\tau_{r\theta} \tag{2.23}$$

***Stress-Strain Equations—Plane Strain***

$$\epsilon_r = \frac{1-\nu^2}{E}\left[\sigma_r - \frac{\nu}{1-\nu}\sigma_\theta\right] \tag{2.24}$$

$$\epsilon_\theta = \frac{1-\nu^2}{E}\left[\sigma_\theta - \frac{\nu}{1-\nu}\sigma_r\right] \tag{2.25}$$

$$\gamma_{r\theta} = \frac{1}{G}\tau_{r\theta} \tag{2.26}$$

### 2.5.1 Equations for One-Dimensional Problems

The above equations are two dimensional, since they involve $r$ and $\theta$. Our interest is in the one dimensional and radially symmetric problems, which involve $r$ only, and not $\theta$. Such problems cannot have a shear stress $\tau_{r\theta}$, since its presence violates the radial symmetry of one dimensional problems. Furthermore, the $\theta$-derivative of any quantity must be zero.

$$\tau_{r\theta} = 0$$

$$F_\theta = 0$$

$$u_\theta = 0$$

$$\gamma_{r\theta} = 0$$

$$\frac{\partial(\cdots)}{\partial\theta} = 0$$

Under these restrictions, Equations 2.16 to 2.26 reduce to the following.

***Equilibrium Equations in One Dimension***

$$\frac{\partial \sigma_r}{\partial r} + \frac{\sigma_r - \sigma_\theta}{r} + F_r = 0 \tag{2.27}$$

***Strain-Displacement Equations in One Dimension***

$$\epsilon_r = \frac{\partial u_r}{\partial r} \tag{2.28}$$

$$\epsilon_\theta = \frac{u_r}{r} \tag{2.29}$$

***Stress-Strain Equations in One Dimension***

$$\epsilon_r = \frac{1}{E^*}(\sigma_r - \nu^* \sigma\theta) \tag{2.30}$$

$$\epsilon_\theta = \frac{1}{E^*}(\sigma_\theta - \nu^* \sigma_r) \tag{2.31}$$

where,

$$E^* = \begin{cases} E & \text{for plane stress conditions,} \\ \dfrac{E}{1-\nu^2} & \text{for plane strain conditions.} \end{cases} \tag{2.32}$$

$$\nu^* = \begin{cases} \nu & \text{for plane stress conditions,} \\ \dfrac{\nu}{1-\nu} & \text{for plane strain conditions.} \end{cases} \tag{2.33}$$

## 2.6 THICK PRESSURIZED PIPE

Thin pressurized pipes have hoop stress that is constant over the thickness. Thick pipes do not. Hoop stresses in thick pipes vary with $r$. Furthermore, radial stresses are seldom considered in the design of thin shells. The hoop and radial stresses in a thick pipe caused by internal or external pressure can be obtained by solving the one dimensional equations given above,

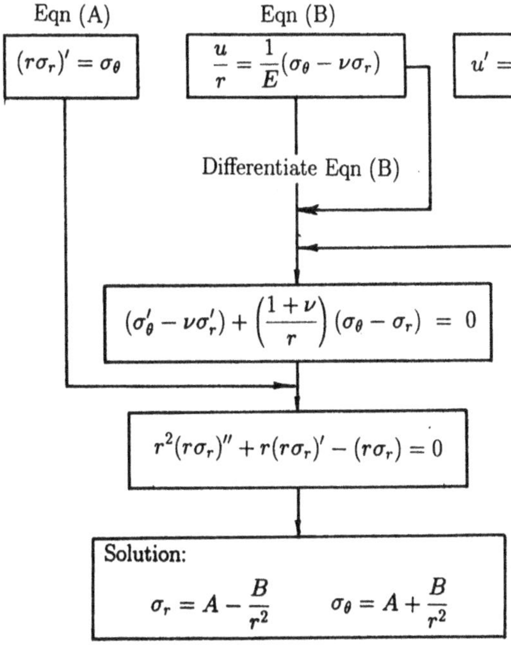


**FIGURE 2.12.** Derivation of elasticity equations in polar coordinates.

according to the procedures indicated in Figure 2.12. The end result of this exercise is the solution for $\sigma_r$ and $\sigma_\theta$ given in these equations.

$$\sigma_r = A - \frac{B}{r^2}$$

$$\sigma_\theta = A + \frac{B}{r^2}$$

The constants $A$ and $B$ are determined from boundary conditions. In particular, for a pipe subjected to both internal and external pressures, we may calculate the hoop stress, the radial stress, and the radial displacements in a serial fashion. The procedures of calculating them are quite straightforward and can be found in References 16, 3, and 17. Only the results are given here for use in later sections.

Considering the general case of an external pressure $p_o$ and an internal pressure $p_i$ acting together, we can summarize the results for all the relevant quantities.

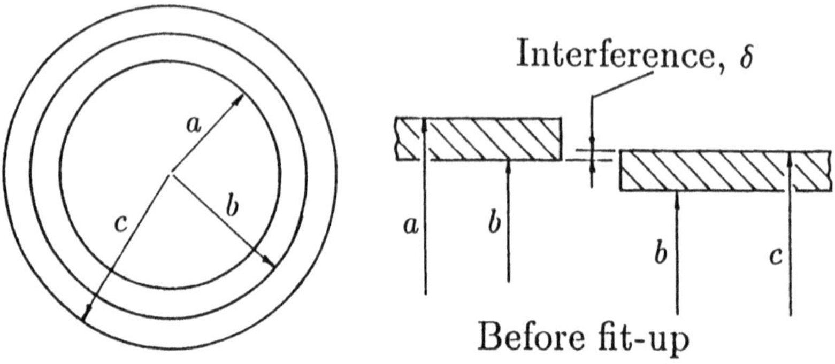


**FIGURE 2.13.** Terminology for press-fit formulas.

***Formulas for Stresses and Displacements***

The symbols and notations used below are with reference to Figure 2.13.

| | |
|---|---|
| Internal radius | $= a$ |
| Any general radius | $= R$ |
| External radius | $= b$ |
| Internal pressure | $= p_i$ |
| External pressure | $= p_o$ |
| Radial displacement at | |
| Inner surface | $= u_i$ |
| Any general radius | $= u_R$ |
| Outer surface | $= u_o$ |

Let the radial stress be $\sigma_r$, hoop stress be $\sigma_h$ and $\beta = (b/a)$.

$$\sigma_{h,R} = \left(\frac{1}{\beta^2 - 1}\right)\left[p_i\left(1 + \frac{b^2}{R^2}\right) + p_o\beta^2\left(1 + \frac{a^2}{R^2}\right)\right] \tag{2.34}$$

$$\sigma_{h,a} = \frac{p_i(\beta^2 + 1) - 2p_o\beta^2}{\beta^2 - 1} \tag{2.35}$$

$$\sigma_{h,b} = \frac{2p_i - p_o(\beta^2 + 1)}{\beta^2 - 1} \tag{2.36}$$

$$\sigma_{r,R} = \left(\frac{1}{\beta^2 - 1}\right)\left[p_i\left(1 - \frac{b^2}{R^2}\right) - p_o\beta^2\left(1 - \frac{a^2}{R^2}\right)\right] \tag{2.37}$$

$$\sigma_{r,a} = -p_i \tag{2.38}$$

$$\sigma_{r,b} = -p_o \tag{2.39}$$

$$u_R = \frac{R}{E(\beta^2 - 1)}\left[(1 - \nu)(p_i - p_o\beta^2) + (1 + \nu)(p_i - p_o)\frac{b^2}{R^2}\right] \tag{2.40}$$

$$u_i = \frac{a}{E(\beta^2 - 1)}\{p_i[(1 - \nu) + (1 + \nu)\beta^2] - 2p_o\beta^2\} \tag{2.41}$$

$$u_o = \frac{b}{E(\beta^2 - 1)}\{2p_i - p_o[(1 + \nu) + (1 - \nu)\beta^2]\} \tag{2.42}$$

The hoop and radial stress distributions in the wall of a thick cylinder are shown in Figure 2.14.

### 2.6.1 Press Fits

An important application of the thick cylinder formulas is in the press fit of wheels and hubs on to shafts. Press fit is an assembly technique for cylindrical components without the use of fasteners or adhesives. An inner tube or shaft is sized slightly larger than the inner diameter of the outer tube. Under these conditions, the outer tube has to be dilated to accommodate the inner tube or shaft. After dilation and fit-up, there is an interfacial pressure set up between the two components. Referring to Figure 2.15 a procedure can be set up for calculating the interfacial pressure $P_o$.

The pressure $P_o$ is required meet the following two conditions.

1. It acts as an internal pressure to the outer pipe, and as external pressure to the internal pipe (or shaft).
2. The radial displacements produced in the outer and inner pipes must together add up to the initial dimensional interference, $\delta$.

The first condition ensures that after fit-up, the pressure $P_o$ will represent a force system in self-equilibrium. Hence the summation of the

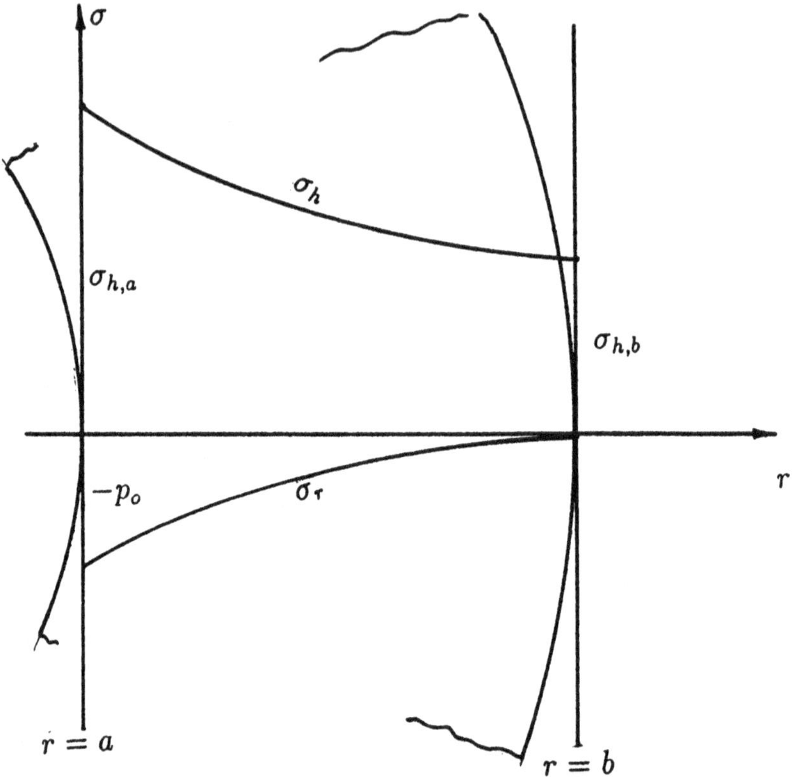


**FIGURE 2.14.** Hoop and radial stress distributions in a thick cylinder under internal pressure.

forces exerted by one component on the other is zero. This reflects the fact that the shaft-wheel assembly has no external forces acting on it. The second criterion ensures that the forces are such that it is just possible to slide the outer pipe over the inner pipe. These two conditions are both necessary and sufficient.

Let

$$P_o = \text{the interfacial pressure,}$$

$$u_s = \text{radially inward displacement of the shaft due to } P_o,$$

$$u_w = \text{radially outward displacement of the wheel due to } P_o,$$

and $$\delta = \text{the interference.}$$

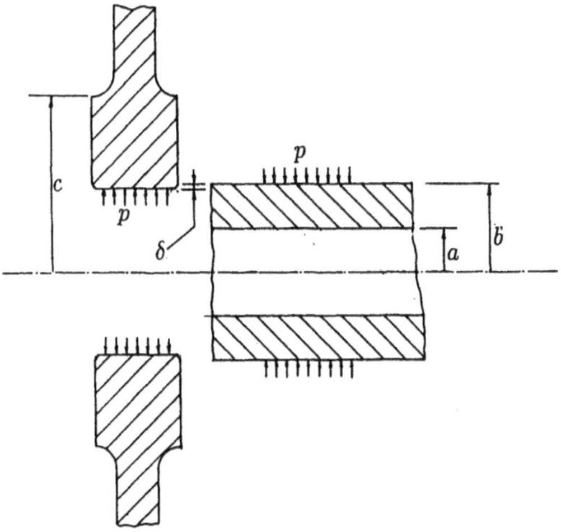


**FIGURE 2.15.** Free body diagrams of two press-fitted components.

Then, we have

$$|u_s| + |u_w| = \delta \tag{2.43}$$

$$\beta_s = \frac{b}{a}$$

$$\beta_w = \frac{c}{b}$$

$$|u_s| = \frac{bP_o}{E(\beta_s^2 - 1)}\left[(1 + \nu) + (1 - \nu)\beta_s^2\right] \tag{2.44}$$

$$|u_w| = \frac{bP_o}{E(\beta_w^2 - 1)}\left[(1 + \nu) + (1 - \nu)\beta_w^2\right] \tag{2.45}$$

The interfacial pressure $P_0$ and the resulting stress distribution are obtained using Equations 2.43, 2.44, and 2.45. Stresses caused by $P_o$ may be obtained from equations 2.34 and 2.37.

$$P_o = E\frac{\delta}{b}\frac{(c^2 - b^2)(b^2 - a^2)}{2b^2(c^2 - a^2)} \tag{2.46}$$

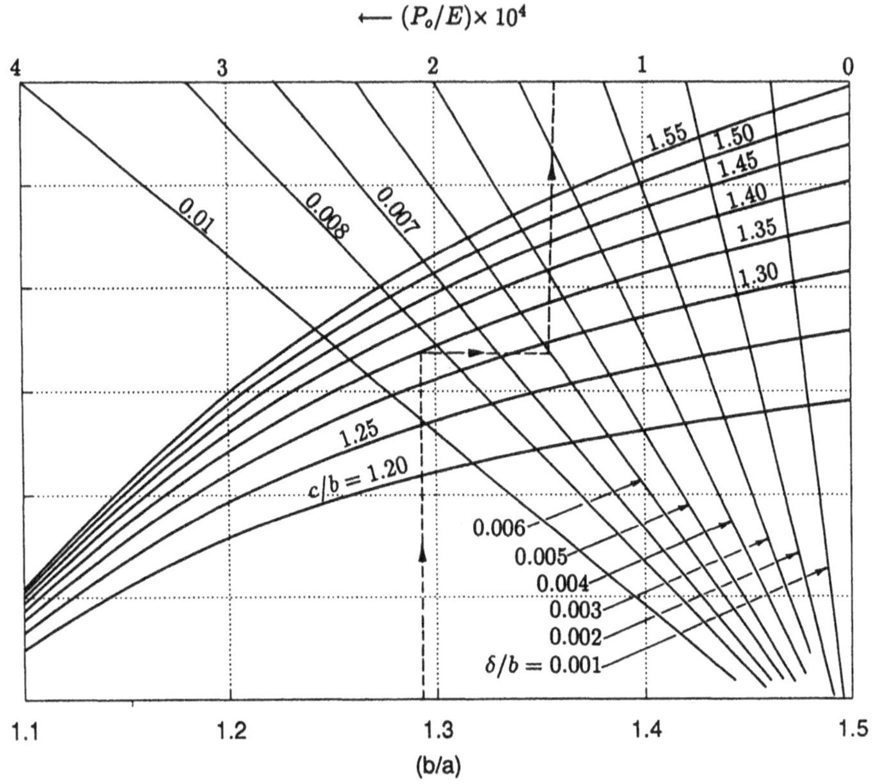


**FIGURE 2.16.** Nomogram for determining the interfacial pressure in a press-fit.

The case of different materials in the assembly poses no problem. We only need to substitute the respective material properties in the Equations 2.40, 2.41, and 2.42.

Equation 2.46 is nondimensionalized and is presented in the form of a nomogram in Figure 2.16. It may be noted that the nomogram uses the ratio $(\delta/b)$ as a parameter. This ratio is commonly known as "interference per inch of radius."

## Example 2.8

Calculate the torque-taking capability of the wheel-shaft assembly shown in Figure 2.17. The shaft is steel, and the wheel is polypropylene. Other dimensions are shown in the table. Coefficient of friction $\mu = 0.2$.

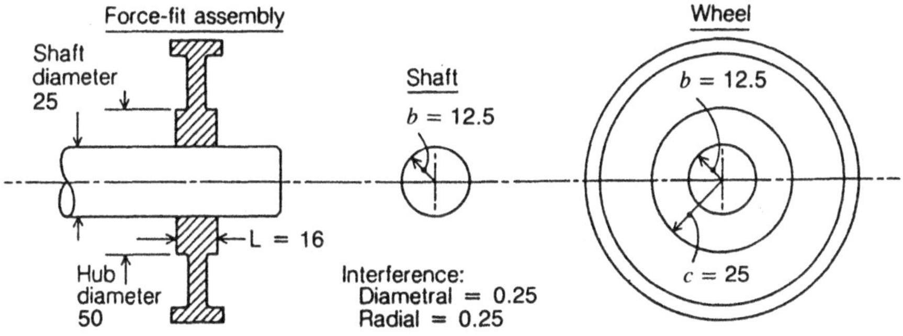


**FIGURE 2.17.** A wheel press-fitted on to a shaft.

Data:

| | Shaft | Wheel |
|---|---|---|
| Material | Steel | Polypropylene |
| Tensile Modulus (Kg/mm$^2$) (Short term) | 21100 | 70 |
| Poisson's ratio | 0.30 | 0.45 |
| Inner Radius (mm) | 0 | 12.5 |
| Outer Radius (mm) | 12.5 | 25.0 |

Radial interference $\delta = 0.25$ mm

Take the axial length of contact, $L = 16$ mm.

*Solution* Since two different materials are used, Equation 2.44 cannot be used as it is. Let us use the subscript $s$ for shaft and $w$ for wheel.

Since $\beta_s \to \infty$, $|u_s|$ is to be calculated as a limit.

$$|u_s| = -\lim_{\beta_s \to \infty} \frac{bP_o\left[(1 + \nu_s) + (1 - \nu_s)\beta_s^2\right]}{E_s(\beta_s^2 - 1)}$$

$$= -\frac{bP_o(1 - \nu_s)}{E_s}$$

$$|u_s| = \frac{12.5\ \text{mm}(P_o\ \text{kg/mm}^2)(1 - 0.3)}{21100\ \text{kg/mm}^2} = 4.150 \times 10^{-4}\ P_o\ \text{mm}$$

$$|u_w| = \frac{a_w}{E_w(\beta_w^2 - 1)} P_o\left[(1 - \nu_w) + (1 + \nu_w)\beta_w^2\right]$$

$$= (12.5\ \text{mm})\frac{P_o[(1 - 0.45) + (1 + 0.45)2^2]}{70\ \text{kg/mm}^2(2^2 - 1)} = 0.378\ P_o\ \text{mm}$$

Since $|u_s| + |u_w| = 0.25$ mm, we get

$$0.3784 \times P_o = 0.25 \text{ mm}$$

Hence,

$$P_o = 0.661 \text{ kg/mm}^2$$

At this point, it is a good idea to check the hoop stress in the wheel caused by this pressure.

$$\sigma_{h,w} = P_o\left[\frac{\beta_w^2 + 1}{\beta_w^2 - 1}\right]$$

$$= 0.661\left[\frac{2^2 + 1}{2^2 - 1}\right]$$

$$= 1.10 \text{ kg/mm}^2$$

Polypropylene has a yield strength of about 2.45 kg/mm$^2$, and hence the hoop stress $\sigma_{h,w}$ has a factor of safety of about 2.23. This is not quite an adequate factor safety.

***Torque Capability***

$$\text{Torque} = \underbrace{(2\pi bL\mu P_o)}_{\text{Friction Force}} \quad \underbrace{b}_{\text{Torque Arm}}$$

$$T = [2\pi \times 12.5 \text{ mm} \times 16 \text{ mm} \times 0.20 \times 0.661 \text{ kg/mm}^2]12.5 \text{ mm}$$

$$= 2.08 \text{ kg m}$$

***Commentary***

- *The value of the torque capability is very low in comparison with what may be needed for a household application—food processor, hand tools, etc.*
- *The effect of the centrifugal force caused by rotation is to separate the shaft and wheel. Consequently, the interface pressure $P_o$ reduces with higher $\omega$. Hence the torque capability calculated above is good only for static conditions (for very low angular velocities for which the centrifugal forces are negligible). We discuss this in a later section.*
- *Interface pressure $P_o$ caused by press fit is a secondary effect. That is, it is proportional to the interference, and hence directly dependent on E. The pressure decays in the same way as the modulus decreases with time. Hence the result in this example is valid only in the short term, because the short term modulus is used in the calculation.*

Three cylindrical parts press-fitted to form a pressure-bearing assembly occurs commonly in practice. The example below is one of a tee fitting and a pipe being assembled together by a metal pipe clamp, which is crimped on the pipe. The nature of this assembly is equivalent to press-fitting, although there was no interference at the beginning. If all the component parts remain within their elastic range, it is unnecessary to consider how the final state of stress was reached. Based on this, it may be seen that the crimping operation is simply an interference fit. Another example for this case is the rubber hose that connects the engine block to the radiator in an automobile.

## Example 2.9

A plastic pipe is pulled over a fitting and they are both held tightly together by a pipe clamp (see Figure 2.18). For the dimensional and material data given, find the stress in the pipe and the fitting caused by the clamping operation.

| | Fitting | Pipe | Clamp |
|---|---|---|---|
| Inner dia (mm) | 50.0 | 55.0 | 58.0 |
| Outer dia (mm) | 55.0 | 58.0 | 60.0 |
| Modulus (kg/mm$^2$) | 422 | 211 | 7035 |
| Poisson's ratio | 0.35 | 0.35 | 0.28 |

$\delta_1$ = Interference between fitting and pipe = 0.05 mm
$\delta_2$ = Interference between pipe and clamp = 0.13 mm

*Solution* The procedure is very similar to the case of two cylinders. However, since we have two interface pressures $p_1$ and $p_2$, we need two conditions to solve the problem. Using the notation in the Figure 2.18b we have the following:

$$\beta_1 = \frac{b}{a}; \quad \beta_2 = \frac{c}{b}; \quad \beta_3 = \frac{d}{c}$$

By noting that both the outer and inner surfaces of the pipe experience displacements because of the pressures $p_1$ and $p_2$, we can obtain the

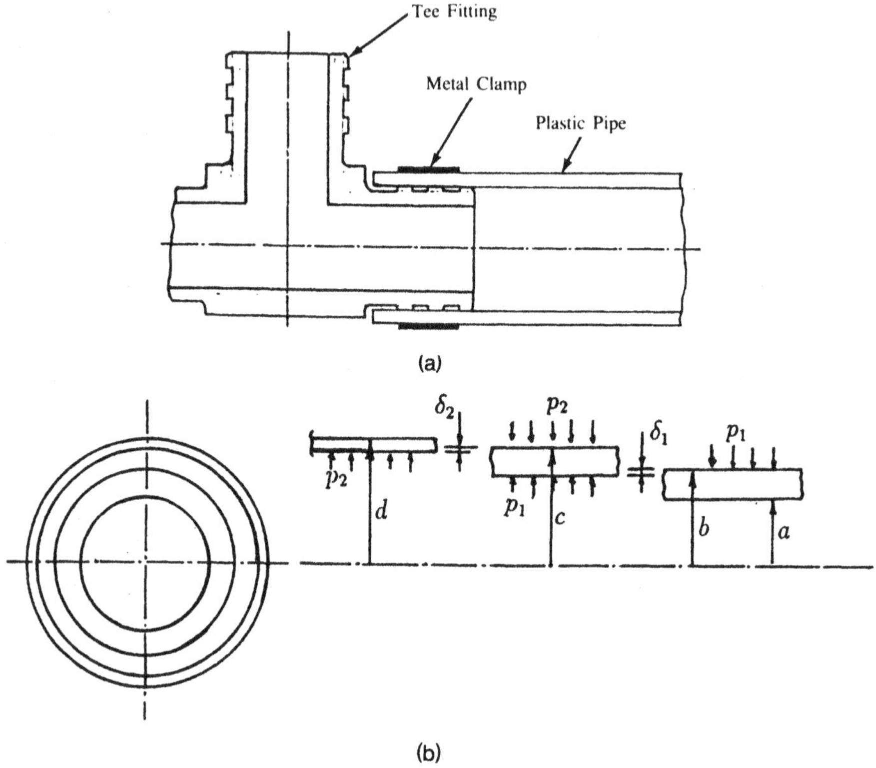


**FIGURE 2.18.** Press-fit of a tee and a pipe with a pipe clamp.

following set of equations. (The details of working are left to the reader as an exercise.)

$$S_{11} = \frac{b}{E_1(\beta_1^2 - 1)}\{(1 + \nu_1) + (1 - \nu_1)\beta_1^2\}$$

$$+ \frac{b}{E_2(\beta_2^2 - 1)}\{(1 - \nu_2) + (1 + \nu_2)\beta_2^2\} \tag{2.47}$$

$$S_{12} = -\frac{2b\beta_2^2}{E_2(\beta_2^2 - 1)} \tag{2.48}$$

$$S_{21} = -\frac{2c}{E_2(\beta_2^2 - 1)} \tag{2.49}$$

$$S_{22} = \frac{c}{E_2(\beta_2^2 - 1)}\{(1 + \nu_2) + (1 - \nu_2)\beta_2^2\}$$
$$+ \frac{2c\beta_3^2}{E_3(\beta_3^2 - 1)} \tag{2.50}$$

$$p_1 S_{11} + p_2 S_{12} = \delta_1 \tag{2.51}$$

$$p_1 S_{21} + p_2 S_{22} = \delta_2 \tag{2.52}$$

$$\Delta = S_{11}S_{22} - S_{12}S_{21} \tag{2.53}$$

$$p_1 = \frac{1}{\Delta}[\delta_1 S_{22} - \delta_2 S_{12}] \tag{2.54}$$

$$p_2 = \frac{1}{\Delta}[\delta_2 S_{11} - \delta_1 S_{21}] \tag{2.55}$$

From the tables, we have $a = 25$ mm, $b = 27.5$ mm, $c = 29$ mm, and $d = 30$ mm. Substituting, we get

$$S_{11} = 3.165$$
$$S_{12} = -2.587$$
$$S_{21} = -2.453$$
$$S_{22} = 2.668$$
$$3.165p_1 - 2.587p_2 = 0.05$$
$$-2.453p_1 + 2.668p_2 = 0.13$$

1. Pressure between the fitting and the pipe, $p_1 = 0.224 \text{ kg/mm}^2$.
2. Pressure between the pipe and the clamp, $p_2 = 0.254 \text{ kg/mm}^2$.

***Stresses in the Components of the Assembly (kg / mm²)***

| Type | Fitting | Pipe | Clamp |
|---|---|---|---|
| Hoop, Inside | −2.58 | −.832 | 7.51 |
| Hoop, Outside | −2.35 | −.802 | 7.25 |

The stresses in the pipe and the fitting are reasonable for many structural plastics. In particular, the properties used for the fitting are for an acetal copolymer, with an ultimate strength of over 9.0 $kg/mm^2$ in tension. The stresses in the fitting pose no problem. Nor are there any problems with the stresses in the pipe, which permit the use of PVC or polybutylene materials with an adequate factor of safety. The metal of the crimp ring will present no difficulty either. A commercial quality aluminum is a good candidate.

**Example 2.10**

Assemblies similar to the ones in the previous example are used in household plumbing. Assuming a water head of about 20 meters (about 0.02 $kg/mm^2$ of water pressure), a pipe clamp width of 20 mm, and a friction factor of 0.2 find whether the assembly will have sufficient pull-out strength against pressure. Assume that the width of the aluminum crimp ring is 10 mm.

*Solution* The pull-out strength is the resistance to the longitudinal pressure force. A pipe has longitudinal stress in it because of the locked up fluid (water in this example) which is under pressure. In the event of pullout of the pipe, it will happen with the pipe clamp riding on the pipe. Hence the pressure that is responsible for holding the pipe against pullout is $p_1$. Let the fluid pressure be denoted by $p_o$. We need to check whether the following inequality is satisfied.

$$2\pi L R_i p_1 \mu > \pi R_i^2 (\text{water pressure})$$

That is,

$$2\pi \times 10\text{ mm} \times 0.2(27.5\text{ mm}) \times (0.224\text{ kg/mm}^2)$$

$$> \pi (27.5\text{ mm})^2 \times (0.02\text{ kg/mm}^2)$$

$$77.32\text{ kg} > 47.52\text{ kg}$$

***Commentary***
*This is barely good enough for the short term. However, in practice the axial force may be close to zero. Because end pressure loads are often taken up by*

*some piping restraints, anchors, clamps, etc. which are provided in adequate numbers in a household plumbing layout.*

*In an actual design process, it is necessary to be aware of the factors that tend to lower the stresses, but it is conservative practice not to take these factors into account in the component design.*

## 2.7 ROTATING CYLINDERS

Rotating cylinders and disks (Figure 2.19) involve centrifugal body forces $F_r$, directed away from the axis of rotation. Stress analysis of such parts is easily handled by stating the problem using polar coordinates. Such problems arise in the mechanical design of hand tools, small fans, small capacity pumps, etc.

### 2.7.1 Formulas for Stresses and Displacements

For a disk of mass density $\rho$, internal radius $a$, outer radius $b$ rotating in its own plane at an angular speed $\omega$ radians/sec, the following are the stress distributions, and radial displacements. As in the case of thick pipes, the solution procedure is presented as a flow chart (Figure 2.20) for the reader to try out as an exercise. Refer to Den Hartog, [3], for a formal derivation of the following equations. Note that the Poisson's ratio $\nu$ gets involved in the solutions, although the problem is a load controlled

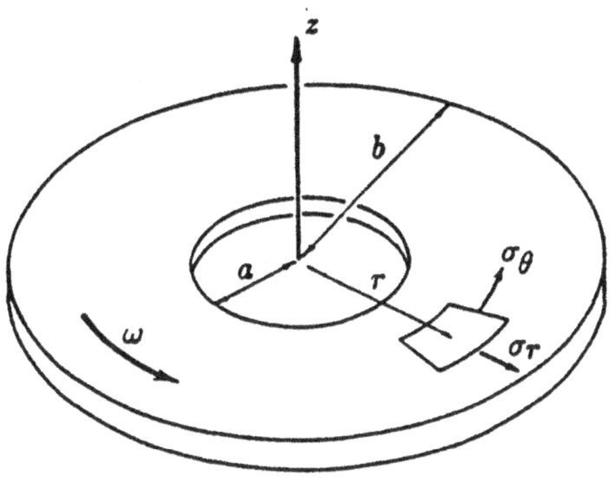


**FIGURE 2.19.** Rotating disk. Stresses in rotating cylinders are governed by the same equations as are pipes. The difference between the two is that rotating cylinders are subject to centrifugal forces instead of static pressure.

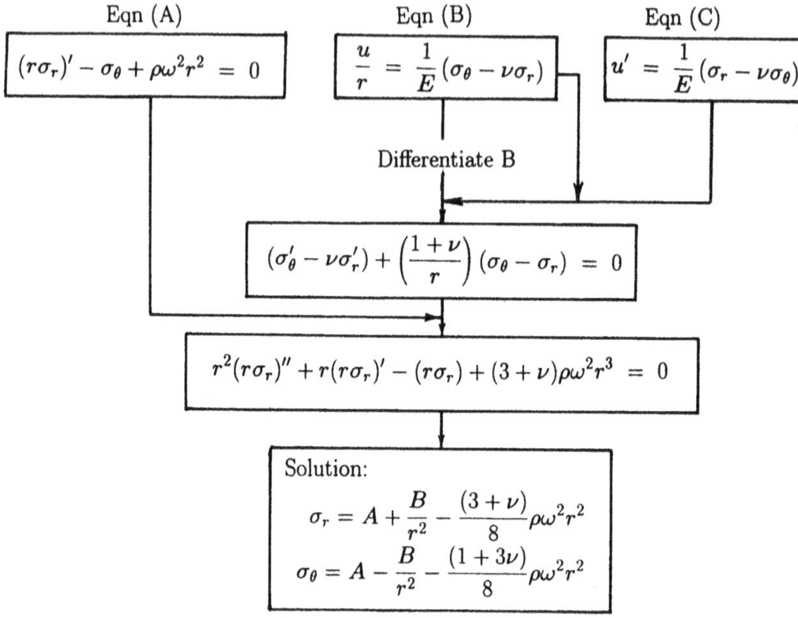


**FIGURE 2.20.** Flow chart for the derivation of stress and displacement in a disk rotating at an angular speed of $\omega$.

problem. This occurs because the rotation causes a body force to be distributed over the entire body.

$$\sigma_r = \left(\frac{3+\nu}{8}\right)\rho\omega^2\left(b^2 + a^2 - \frac{a^2b^2}{r^2} - r^2\right) \tag{2.56}$$

$$\sigma_\theta = \left(\frac{3+\nu}{8}\right)\rho\omega^2\left(b^2 + a^2 + \frac{a^2b^2}{r^2} - \frac{1+3\nu}{3+\nu}r^2\right) \tag{2.57}$$

$$u_r = \rho\omega^2\frac{r}{E}\frac{(3+\nu)(1-\nu)}{8}\left(b^2 + a^2 + \frac{1+\nu}{1-\nu}\frac{b^2a^2}{r^2} + \frac{1+\nu}{3+\nu}r^2\right) \tag{2.58}$$

**Example 2.11**

In the wheel-shaft example of the previous section, let us suppose that a torque of only 0.255 kg m is actually being transmitted. Reduced torque is used to provide adequate factor of safety on the torque capability and in consideration of its reduction over long term. Find the limiting speed of rotation at the end of a period at which only 40% of the modulus is

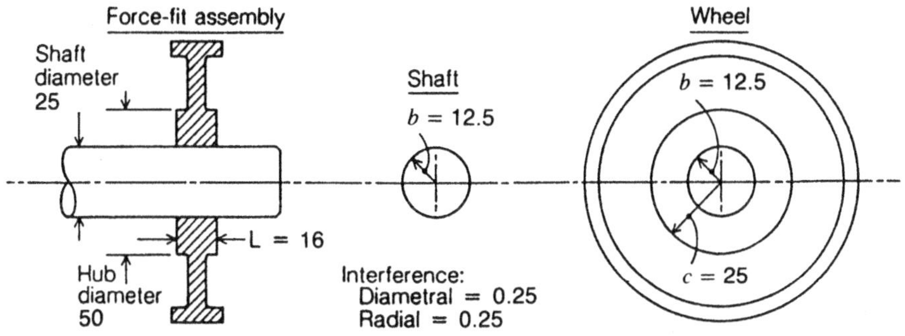


**FIGURE 2.21.** A wheel press-fitted on to a shaft. Centrifugal forces caused by rotation tend to reduce the grip. (Figure repeated.)

retained by the material. Assume that the density of polypropylene is $\rho = 1.2 \times 10^{-6}$ kg/mm$^3$. Consider a FOS of 2 on the long term torque capability.

*Solution* The steel shaft does not lose modulus. Hence $|u_s| = 4.15 \times 10^{-4}P^*$ mm. The polypropylene wheel retains only 40% of its modulus, that is, 28 kg/mm$^2$. Therefore,

$$|u_w| = 0.945P^* \text{ mm}$$

Furthermore, the effect of rotation is to reduce the interference. This happens because the rotation tends to separate the wheel and the shaft by an amount $U_\omega$. Since both the wheel and hub tend to move outward because of rotation by different amounts, the net separation between them $U_\omega$ is the difference,

$$U_\omega = (u_r \text{ of the wheel})_\omega - (u_r \text{ of the shaft})_\omega$$

However, the displacement of the shaft is very small and usually can be neglected. Hence, $U_\omega$ is equal to the displacement of the disk at the inner radius. (Note carefully that notations pertain to the disk.)

$$U_\omega = \rho\omega^2 \frac{a}{E} \frac{(3+\nu)(1-\nu)}{8}\left(b^2 + a^2 + \frac{1+\nu}{1-\nu}b^2 - \frac{1+\nu}{3+\nu}a^2\right)$$

$$= 1.2 \times 10^{-6}\omega^2 \frac{a}{E} \frac{(3+\nu)(1-\nu)}{8}$$

$$\times \left(25^2 + 12.5^2 + \frac{1+\nu}{1-\nu}25^2 - \frac{1+\nu}{3+\nu}12.5^2\right)$$

$$= 2.4 \times 10^{-5}\omega^2 \text{ mm}$$

$P^*$ can be obtained from the required long term torque. We see that for a FOS = 2, we need a torque capability of .255 × 2 = .511 kg mm.

$$\text{Torque} = .511 \text{ kg mm} = (2\pi a L \mu P^*)a$$

from which $$P^* = 0.162 \text{ kg/mm}^2.$$

As a next step, we wish to modify the interference fit equation to account for $\omega$. The equation becomes

$$|u_s| + |u_w| = \delta - U_\omega \tag{2.59}$$

This gives,

$$0.945 \times 0.162 \text{ kg/mm}^2 = 0.25 - 2.4 \times 10^{-5}\omega^2$$

$$\omega^2 = 4018$$

$$\omega = 63.39 \text{ rad/sec or, rpm} = 605$$

This is equivalent to a power transmission of only 160 watts. Considering this, the design feature discussed may be suitable only for low power gadgets, such as a hand-held fan at a dressing table.

### 2.7.2 Tightness Enhancements of Interference Fits

The torque capability and pullout strength of press fits represent their strengths in two common modes. The preceding examples represent fairly practical values of strengths and physical dimensions, and we see that the resulting torque capability of interference joints is not high enough for some common utility items or for long term purposes. Hence it is desirable in a number of cases to enhance the strengths of the press fits by auxiliary means. Ribs, adhesive joints, etc. may be considered for this purpose. Here, we discuss the stress analysis aspects of these alternatives.

The mechanics of enhancement of joint strength can be explained with the help of Figure 2.22. The ribs reduce the area of contact, while maintaining the total force required to achieve the interference, resulting in a higher intensity of contact pressure at the interfaces. A second way the ribs contribute to the enhancement is by the displacement near the edges of ribs. The outer pipe forms a kink which provides a "pinch" effect,

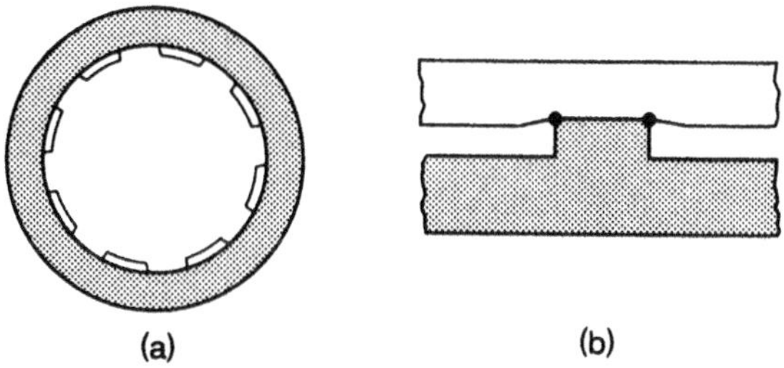


**FIGURE 2.22.** Use of ribs in press fit areas, and their effects.

resulting in a positive transmission of forces. Essentially, friction transmission is replaced by positive contact forces at the pinch areas.

There are, however, disadvantages with this option from the point of view of long term behavior. First of all, we must note that the interface pressure set up by the press-fit is a *secondary* stress, and that required for torque transmission is a *primary* stress. Furthermore, the pinch effect causes a *peak* stress. Over the lifetime of the product, the secondary component will drop down to a low value, because of relaxation. (Theoretically, the terminal value is zero!) Countermeasures such as the use of high initial tightness have a low initial factor of safety against short term yielding. Secondly, the site of the pinch effect is a potential source of craze, creep, and crack initiation. These types of failure must be checked for in the analysis of long term behavior. Because the geometry near the ribs is not amenable to hand calculation, a detailed finite element analysis may be needed to evaluate the design.

Another potential method of enhancement is the adhesively bonded joint. An adhesive joint, properly made, has a high shear strength which is, in all cases, better than what can be achieved by a friction joint. Hence the choice appears to be obvious. However, consider the case of torque transmission by an adhesive joint at a speed of 3000 rpm. The tendency of the shaft and wheel to separate from each other is resisted by the adhesive layer in the tensile (or peel) mode. (See Figure 2.23(b).) Given the fact that adhesive joints have low tensile strengths, the peel stresses are a source of potential failure of the joint.

Further, viewing the adhesive joint in the plane of rotation, we see that the wheel hub tends to shrink in size because of Poisson's effect. Such displacements are low for the shaft but are considerable for the wheel. The relative displacement is entirely taken up by the adhesive layer, the thickness of which is of the same order of magnitude. Thus, the relative displacement may amount to well over 100% strain in the adhesive material, and can cause severe shear stress concentrations. The reader is

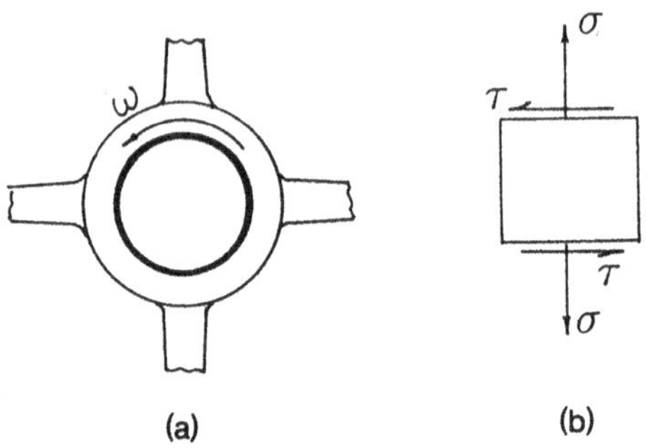


**FIGURE 2.23.** Peel stresses are created in adhesive joints for rotating components.

referred to Reference 18 to appreciate the high magnitudes of shear stress concentration factors (5 to 10 are not uncommon) caused near the tips and at the spew areas of adhesive joints. Except, perhaps, in disposable applications, the combination of dissimilar materials, high angular velocity, and adhesive jointing needs to be evaluated very carefully.

## 2.8 AXISYMMETRIC SHELL PROBLEMS

Axisymmetric shells are those that have an axis of radial symmetry, and whose midsurfaces are generated by revolving a curve about the axis of symmetry. They are an important subset of shell geometries because such shapes are easily produced. A pipe, for instance is an axisymmetric shell. Cylinders, spheres, cones, toroids, and ellipsoids are other common shell shapes used in pipings and vessels. If such shells are loaded in a radially symmetric (axisymmetric) manner, then the deflections, strains, and stresses are also axisymmetric and hence one dimensional. That is, all such forces, stresses, strains, etc., can be expressed in terms of a single variable, usually the axial distance measured along the surface. In the case of cylinders, they are expressed as a function of axial distance $x$; in spheres, they are expressed as a function of the meridional angle $\phi$, and so on. Figure 2.24 illustrates this point.

Usually, shells are subjected to internal pressure loads. There are two modes in which shells resist loads—the membrane and the bending modes. The membrane mode is similar to an inflated balloon withstanding pressure by in-plane tensile stress. The bending mode, on the other hand, is the resistance to applied load in bending, or in other words, by a stress

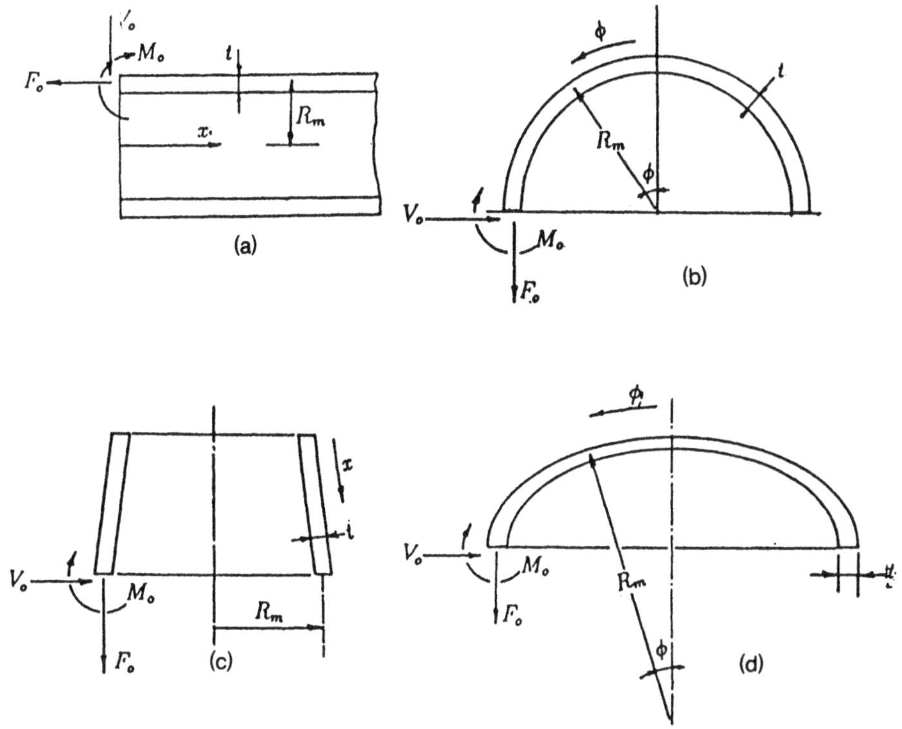


**FIGURE 2.24.** The meridional coordinate for various shell shapes. Meridional coordinates can vary with the shapes. A coordinate may be distance $x$ in cylinders or cones, or an angle $\phi$ in an ellipsoidal or spherical heads. The edge forces $F_o$, $V_o$, and edge moment $M_o$ are the only types that can produce an axisymmetric distribution of stresses.

distribution that varies linearly through the thickness. In practice, a shell is almost always in a combination mode, that is, membrane-plus-bending.

### 2.8.1 Membrane Stresses in Shells of Revolution

Membrane stresses in a shell are obtained by supposing that the whole pressure load is sustained in the membrane mode. Such an approach ignores the capability of shells to develop moments. Shells of revolution have two principal radii of curvature at any point, and the membrane stresses in shells caused by internal pressures are directly related to these radii. The hoop and longitudinal membrane stresses are given for a few common shells of revolution in Table 2.3. The derivation of the expres-

**Table 2.3. Membrane Stresses in Axisymmetric Shells under Pressure.**

| Figure | Membrane Stress |
|---|---|
| 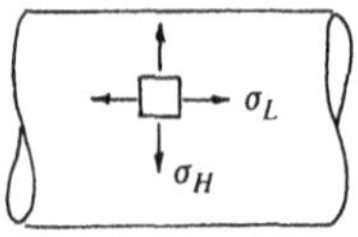  Cylinder | $\sigma_H = \frac{pR}{t}$ <br> $\sigma_L = \frac{pR}{2t}$ |
| 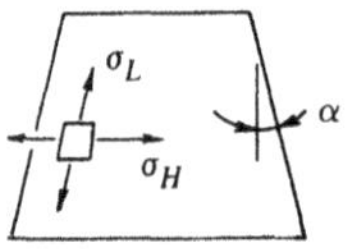  Cone | $\sigma_H = \frac{pR}{t\cos\alpha}$ <br> $\sigma_L = \frac{pR}{2t\cos\alpha}$ <br> $\alpha$ : Semi Cone angle |
| 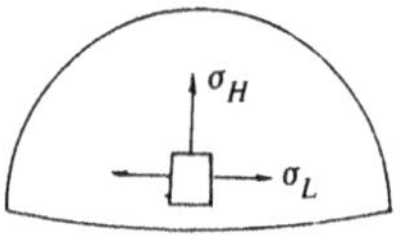  Sphere | $\sigma + H = \sigma_L = \frac{pR}{2t}$ |
| 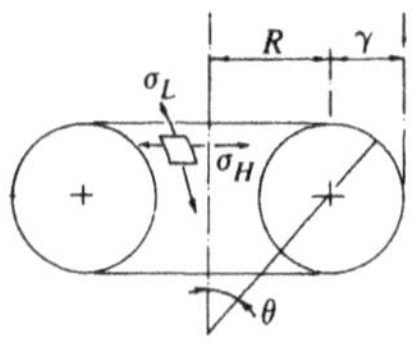  Torus | $\sigma_H = \frac{pr}{2t}\left[\frac{2R + r\sin\theta}{R + r\sin\theta}\right]$ <br> $\sigma_L = \frac{pr}{2t}$ <br> Note the definitions of $r$ and $R$ |
| 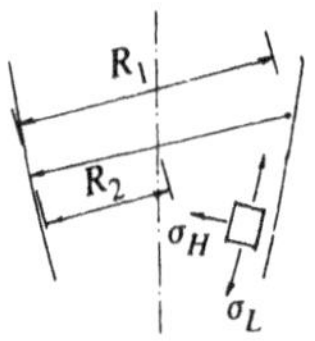  General Shell | $\sigma_H = \frac{pR_2}{2t}\left(2 - \frac{R_2}{R_1}\right)$ <br> $\sigma_L = \frac{pR_2}{2t}$ |

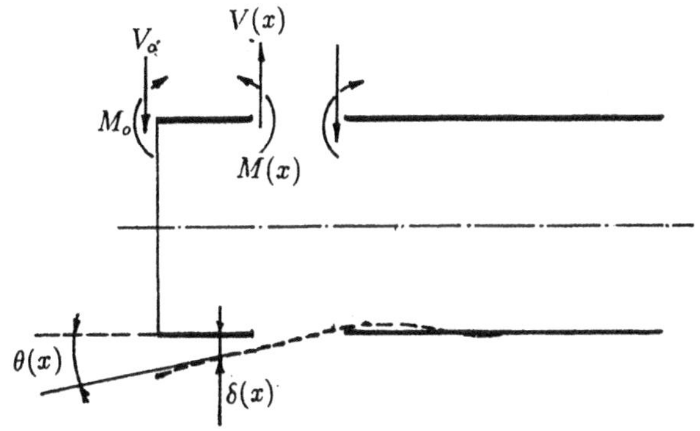


**FIGURE 2.25.** Ring loads on a semi-infinite cylinder.

sions in this table can be found in standard text books in elasticity, or pressure vessel [8, 19].

### 2.8.2 Ring Forces and Moments on Cylindrical Shells

With axisymmetry, there can be no $\theta$-dependence of any quantity. Thus the shell geometry, the loads, the displacements, rotations, strains, stresses, internal shear forces, and moments[3] must all be axisymmetric. The only axisymmetric forces are $F_o$, $V_o$, and $M_o$ shown in Figure 2.25. Any other force or moment would violate radial symmetry. Note that these forces and moments are distributed uniformly around the circumference and that they are expressed as N/mm and Nm/m (or N mm/mm), respectively.

The effect of $F_o$ on the cylindrical shell is trivially simple to calculate. It is simply equal to an axial stress $\sigma_a$ given by

$$\sigma_a = \frac{2\pi R_m F_o}{2\pi R_m t} = \frac{F_o}{t} \tag{2.60}$$

The internal shear force and moment, namely $V_o$ and $M_o$, induce the

[3] The bending moment referred to here produces a stress that varies linearly throught the *thickness* of the cylinder. In piping flexibility analysis we refer to moments that produce a stress that varies linearly across the *entire cross section* of the pipe. The reader should be able distinguish the two based on the context. In shell theory we always refer to the former kind.

following four internal forces and moments, all of them functions of axial distance, and distributed "per unit length." See Figure 2.23.

- Internal shear force $V(x)$, force/unit length,
- Internal moment $M(x)$, moment/unit length,
- Rotation of the tangent to the surface $\theta(x)$,
- Radial displacement $\delta(x)$ .

We list the formulas for the responses of a semi-infinite cylinder.

$$k = \frac{Et}{R_m^2} \tag{2.61}$$

$$\beta = \frac{[3(1-\nu^2)]^{0.25}}{\sqrt{R_m t}} mm^{-1} \tag{2.62}$$

$$A(\beta x) = e^{-\beta x}(\cos \beta x + \sin \beta x) \tag{2.63}$$

$$B(\beta x) = e^{-\beta x} \sin \beta x \tag{2.64}$$

$$C(\beta x) = e^{-\beta x}(\cos \beta x - \sin \beta x) \tag{2.65}$$

$$D(\beta x) = e^{-\beta x} \cos \beta x \tag{2.66}$$

$$V(x) = -P_o C(\beta x) - 2M_o \beta B(\beta x) \text{ N/mm} \tag{2.67}$$

$$M(x) = \frac{P_o}{\beta} B(\beta x) + M_o A(\beta x) \text{ N mm/mm} \tag{2.68}$$

$$\theta(x) = -\frac{2P_o \beta^2}{k} A(\beta x) + \frac{4M_o \beta^3}{k} D(\beta x) \text{ radians} \tag{2.69}$$

$$\delta(x) = \frac{2P_o \beta}{k} D(\beta x) - \frac{2M_o \beta^2}{k} C(\beta x) \text{ mm} \tag{2.70}$$

It can be seen that all of the functions $A(\beta x)$, $B(\beta x)$, $C(\beta x)$, and $D(\beta x)$ see(Figure 2.26)decay with distance, because of the factor $e^{-\beta x}$. So do the internal shear forces and moments, $V(x)$, $M(x)$, as well as the displacements and rotations, $\delta(x)$ and $\theta$. For this reason the quantity is called the *attenuation parameter*.

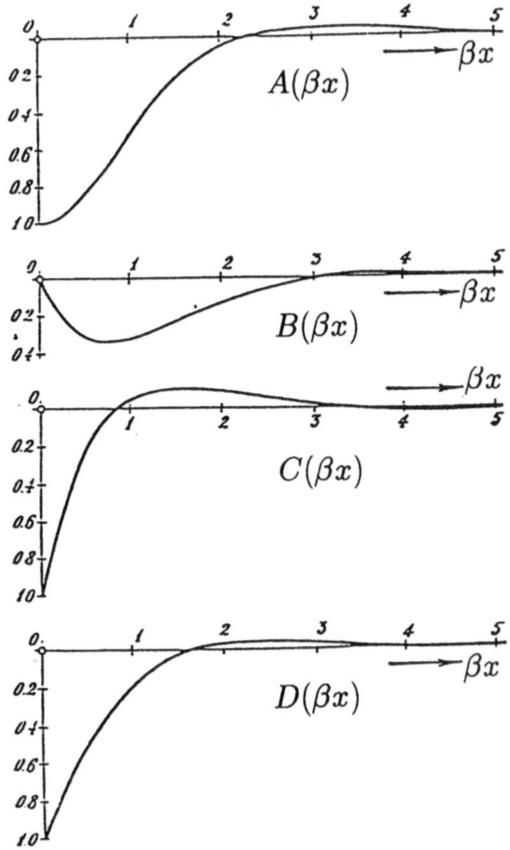


FIGURE 2.26. The functions, $A$, $B$, $C$ and $D$.

At large enough distances, all the internal forces and moments fade out exponentially. This fact is very important in the design of the minimum distance of separation needed between openings, local loads, ribs, and other structural features on pressure vessels.

For an infinite cylinder with a loading as shown in Figure 2.27, the following formulas apply.

$$V(x) = \frac{P_o}{2} D(\beta x) \tag{2.71}$$

$$M(x) = P_o \frac{C(\beta x)}{4\beta} \tag{2.72}$$

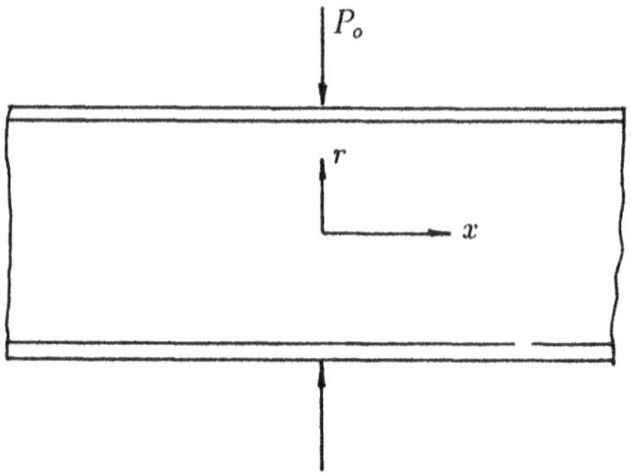


**FIGURE 2.27.** Ring load on an infinite cylinder.

$$\theta(x) = -P_o \frac{\beta^2 B(\beta x)}{k} \tag{2.73}$$

$$\delta(x) = P_o \frac{\beta A(\beta x)}{2k} \tag{2.74}$$

***Effective Length of Cylindrical Shell***

Table 2.4 shows that over a length $x$ of the cylinder for which $\beta x = 2\pi$, all effects fade out to 0.19% of their full magnitude. Hence for practical purposes, this length is equivalent to an infinite length. In other words, two applied loads separated by a distance greater than $2\pi/\beta$ can be treated as totally independent. That is, the effect of one load is not experienced at the location of the other, and hence analysis can be performed for each load independently. This is valuable information for the analysis of shells, especially when modeling individual effects for analysis. Since $\nu$ varies over only a limited range of values, we may develop a rule of a thumb for the effective length of cylindrical shells, by taking $\nu = 0.35$.

$$L_{\text{eff}} = \frac{2\pi}{\beta} = \frac{2\pi\sqrt{R_m t}}{3[1-\nu^2]^{0.25}} = 4.933\sqrt{R_m t} \tag{2.75}$$

**Table 2.4. Exponential Decay of Internal Forces and Moments in a Cylinder.**

| $\beta x$ | 0 | 1 | $\pi/2$ | 2 | 3 | $\pi$ | $2\pi$ |
|---|---|---|---|---|---|---|---|
| Factor of Reduction, $e^{-\beta x}$ | 1 | .368 | .208 | .135 | .05 | .043 | .0019 |

Note that in the above formula, a factor of reduction of 0.19% was treated as negligible. In many practical cases, error tolerance can be increased, say to 5%, in which case

$$L_{\text{eff}} = 2.355\sqrt{R_m t}$$

The ASME/ANSI B & PV Code, [20], adopts a value of $x = 3/\beta$ to be as good as an infinite length.

***Commentary***

*In finite element analyses, the portions of the cylinder beyond its effective length are of no consequence to the analysis. If one wants to study the stresses in a local feature of a cylinder, it would be only necessary to model the effective length of the cylinder. Moreover, the boundary conditions applied at a distance of* $L_{\text{eff}}$*, or greater, will have negligible effect on the local feature being studied. Thus, any convenient boundary condition can be applied at the point at a distance* $L_{\text{eff}}$ *without any effect on the quality of results at the point of interest.*

***An Example of Rib Effectiveness on Pipes***

Large diameter pipes are used underground for several municipal and utility applications. These pipes, which were once made of heavy wall concrete, are now made of flexible pipes, sometimes of corrugated metal, and more often of fiber reinforced plastic. The flexible pipes are really thin walled, the $r/t$ ratio exceeding 40 or even 50. Stiffening ribs are provided at distances of 1.5 to 2 diameters, more for rigidity during handling.

Let us investigate a hypothetical example of a large pipe for water transportation. Consider a pipe 2000 mm diameter, 25 mm thickness and made of composite material.[4]

| | |
|---|---|
| Hoop modulus of pipe material: | 850 kg/mm$^2$ |
| Axial modulus of pipe material: | 425 kg/mm$^2$ |
| Poisson's ratio (major): | 0.28 |

[4] Composite materials have different moduli along two perpendicular directions. That is why two moduli are specified for the pipe along and perpendicular to the axis of the pipe. This is discussed in Chapter 7.

In many practical pipe design environments, it is customary to speak of rib spacing as a multiple of the pipe diameter. This is a sound practice for relatively large $(t/R)$ ratios. However, in the case of thin wall, large diameter pipes as in this example, a more detailed consideration is required.

Assume that the ribs were spaced at 1.5 diameters apart. In this case, it means that the two ribs are 3000 mm apart. We see that $\beta = 4.382 \times 10^{-3}$ mm$^{-1}$, and $L_{\text{eff}} = 720$ mm.

Purely from the point of view of their spacing, the ribs can be shown to be inadequate in design, regardless of the additional moment of inertia they provide. Consider, for instance, a load acting anywhere in the middle one-half of the pipe section between the ribs. Any location in this band is farther away from the ribs than $3/\beta$. Technically, the rib is located at an infinite distance, and hence would be unable to provide any additional stiffness at the location of the load. This observation can be proved formally. However, the finite element method was used in this case.

This hypothetical example is illustrated in Figure 2.28. (See color insert.) The figure shows the results of a finite element analysis of a pipe section analyzed using axisymmetric elements, and two extreme ways of modeling ribs. The pressure load was applied at the middle one-half. In one case the rib stiffness was taken as infinite. This was reflected as a fixed end boundary condition. In the other case, the rib was considered to be absent. No boundary condition was applied at the ends. The deflection distribution throughout the middle one-half is seen to be unaffected by the presence or absence of the ribs.

Stiffness being the relevant parameter for buckling analysis, we may also extend the notion to buckling problems. That is, buckling characteristics under the influence of loads acting at a distance farther than $2\pi/\beta$ from ribs are not affected by the ribs at all. As a result, for pipes with rib spacing as described above, the building strength under local ring loads is as though the pipe is not ribbed at all.

**Example 2.12**

In Example 2.9, press-fit of three components, we used a tee fitting, which is not strictly an axisymmetric shape. Examine whether it is a valid procedure to use Equations 2.34 through 2.42 for this assembly.

*Solution* Recall that the formulas in Equations 2.34 through 2.42 are one-dimensional. In other words, they are valid for infinitely long cylinders. When these formulas are used for components of finite length, and with the $L_{\text{eff}}$ one overlapping on that of the other, the results are not valid in

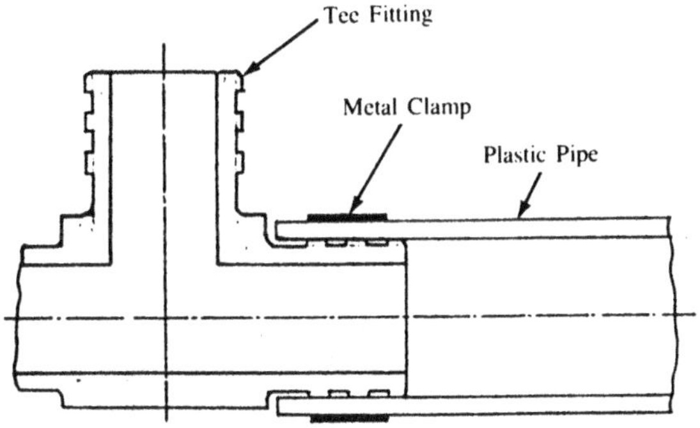


**FIGURE 2.29.** The press-fit of a tee and a pipe with a clamp.

the neighborhood of the overlap. Essentially, the stress distribution is two dimensional, which the one dimensional approach is unable to describe. However, if the geometrical features that depart from axisymmetry are sufficiently far away, then the one-dimensional approach is still a valid procedure. We want to find out if that is true in this case.

We need to find the free length available between the pipe clamp and the branch pipe of the tee.

*Pipe*

$$R_m = 28.25 \text{ mm}$$

$$t = 1.5 \text{ mm}$$

$$\beta = \frac{[3(1 - \nu^2)]^{1/4}}{\sqrt{R_m t}}$$

$$= \frac{[3(1 - 0.35^2)]^{1/4}}{\sqrt{28.25 \times 1.5}}$$

$$= 0.195 \text{ mm}^{-1}$$

The criterion $\beta x = 3$, gives $x = 15.3$ mm. Obviously, the pipe has this length on one side of the assembly, and hence can be considered as a long pipe.

*The Tee-Fitting*

$$R_m = 26.25 \text{ mm}$$

$$t = 2.5 \text{ mm}$$

$$\beta = \frac{[3(1 - 0.35^2)]^{1/4}}{\sqrt{26.25 \times 2.5}}$$

$$= 0.157 \text{ mm}^{-1}$$

For a limit of $\beta x = 3$, we obtain $x = 19.1$ mm.

For this tee fitting, as it is for most other ASTM F845 fittings (see ASTM standard [21]) a length of $3/\beta$ is not available from the branch pipe. We are comparing the length $3/\beta$ with dimension "$L$" in Figure 2.27. The geometry departs from being axisymmetric, within the effective length. This means that the effect of clamping is experienced by the body of the fitting. Thus, the formulas for cylinders are not applicable for the tee-pipe clamped joint. Depending on how critical the application is, a detailed finite element analysis may be desirable.

### 2.8.3 Spherical Shells under Symmetric Edge Loads

For spherical or partially spherical shells as shown in Figure 2.30, we consider only the rotation of the tangent $\theta_o$ and the radial outward displacement $\delta_o$, both occurring at the rim of the sphere. The internal bending moment $M(\phi)$ and the internal shear force $V(\phi)$ are seldom

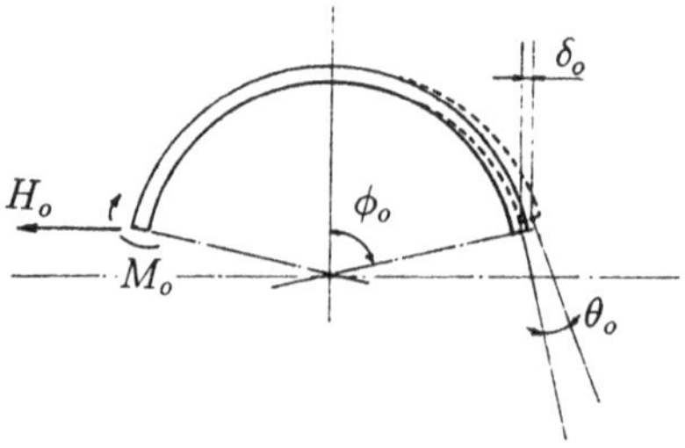


**FIGURE 2.30.** Spherical cap under symmetric loading.

needed in practical problems. Nor are we concerned about the exact functional form.

$R_m, t$—Mean radius and thickness (mm)

$E$—Young's modulus $(\mathrm{Kg/mm})^2$

$\nu$—Poisson's ratio

$$\beta = \frac{[3(1-\nu^2)]^{1/4}}{\sqrt{R_m t}} \ \mathrm{mm}^{-1} \tag{2.76}$$

$$\lambda = \beta R_m \tag{2.77}$$

For a full hemisphere the following formulas give the displacement and rotation at the rim.

$$\delta_o = H_o \frac{2R_m\lambda}{Et} + M_o \frac{2\lambda^2}{Et} \tag{2.78}$$

$$\theta_o = H_o \frac{2\lambda^2}{Et} + M_o \frac{4\lambda^3}{R_m Et} \tag{2.79}$$

For a more extensive set of formulas for cylinders and spheres, $(y, \theta, \delta, V(x), M(x)$ at any point, stress calculations, etc.) see ASME B & PV Code, [20] . For shells of other forms, such as cones, refer to Roark and Young, [8]. For a classic treatment of cylindrical shells, and other related topics—namely rectangular and circular plates on elastic support—refer to Hetenyi, [7].

## 2.9 STRUCTURAL DISCONTINUITY—THE CONCEPT

Structural discontinuity exists in pressure vessels where two dissimilar geometries are joined together. In the Figure 2.31(a), a cylindrical pipe and a spherical closure are shown. Although manufactured as one piece, for the purposes of structural calculations, we look at them as though the two are joined integrally. At the tan line the radial displacements of the cylinder and sphere due to an internal pressure of $p$ are, respectively,

$$\frac{pR^2(2-\nu)}{2Et} \quad \text{and} \quad \frac{pR^2(1-\nu)}{2Et}$$

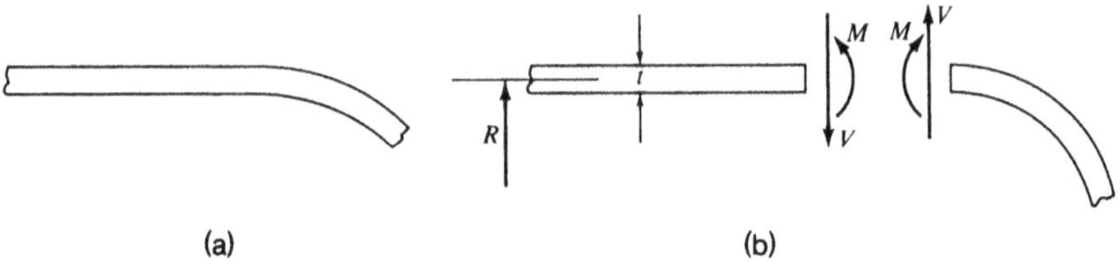


**FIGURE 2.31.** (a) Cylinder with hemispherical ends. (b) redundant forces and moments at the discontinuity.

The natural displacements are incompatible with each other. Rather, they are in the ratio $(2 - \nu)$:$(1 - \nu)$, that is about 2.5:1. Such free displacements cannot happen because the two geometries are structurally held together. Rather, the cylinder and sphere pull and push each other around the circumference at the section *yy* in such a way that they would have a common displacement and rotation at *yy*.

We can visualize this by considering their free bodies and introducing ring forces and ring moments as in Figure 2.31(b). They act in such a way that they are equal and opposite to each other, thus causing no effect on the overall equilibrium of other external forces.

The two unknowns $F$ and $M$ are determined from the following two conditions. These equations are nothing but the compatibility conditions.

- The total radial displacements of the cylinder and sphere due to pressure $p$, $F$, and $M$ must be equal.
- The total rotation of the meridion for the cylinder and sphere due to pressure $p$, $F$, and $M$ must be equal.

In other words,

$$(\delta_{pr} + \delta_M + \delta_F)_{CYL} = (\delta_{pr} + \delta_M + \delta_F)_{SPH} \tag{2.80}$$

$$(\theta_{pr} + \theta_M + \theta_F)_{CYL} = (\theta_{pr} + \theta_M + \theta_F)_{SPH} \tag{2.81}$$

Substituting for the $\delta$s and $\theta$s from the formulas for cylinder and sphere, we have the following for $\delta$,

$$\left[\frac{pR^2(2 - \nu)}{2Et} + \frac{2\beta_c^2R^2}{Et}M - \frac{2\beta_c R^2}{Et}F\right] = \left[\frac{pR^2(1 - \nu)}{2Et} + \frac{2\beta_s^2R^2}{Et}M - \frac{2\beta_s R^2}{Et}F\right] \tag{2.82}$$

and the following for the rotation $\theta$.

$$\left[0 + \frac{4\beta_c^3R^2}{Et}M - \frac{2\beta_c^2R^2}{Et}F\right] = \left[0 - \frac{4\beta_s^3R^2}{Et}M - \frac{2\beta_s^2R^2}{Et}F\right] \quad (2.83)$$

Assume that the cylinder and sphere have the same radius and thickness, $\beta_c = \beta_s = \beta$. For this condition, Equation 2.82 gives

$$F = \frac{p}{8\beta}$$

and the Equation 2.83 gives

$$M = 0$$

*Stress in the Cylinder*

The stresses in the cylinder and sphere are now easy to calculate. After converting $F$ and $M$ in terms of $p$, we get the following for the stress in the cylinder.

$$\text{Longitudinal } \sigma_L = \frac{pR}{2t} \pm \frac{3pB(\beta x)}{4\beta^2t^2} \quad (2.84)$$

$$\text{Hoop } \sigma_h = \frac{pR}{t} - \frac{pR}{4t}D(\beta x) \mp \frac{3\nu p}{4\beta^2t^2}B(\beta x) \quad (2.85)$$

*Stress in the Sphere*

$$\text{Longitudinal } \sigma_L = \frac{pR}{2t} \mp \frac{6M}{t^2} \quad (2.86)$$

$$\text{Hoop } \sigma_h = \frac{pR}{2t} + \frac{2\beta^2R}{t}M + \frac{\beta R}{t}F \quad (2.87)$$

However, for the purposes of categorization of stress as primary and secondary, it is better to keep $F$ and $M$ in the above formulas as they are, instead of expressing them in terms of $p$. The presence of $p$ tends to hide the secondary nature of $F$ and $M$.

### 2.9.1 The Value of Discontinuity Analysis in Design

Substituting for $F$ in Equation 2.87 we see that the hoop stress in the sphere may be expected to exceed the membrane stress in the sphere by 50%. No general statement can be made about the percentage excess, because real life problems may involve other geometries or thicknesses. However, it is safe to assume that *very close to the discontinuities, stresses can exceed the nominal value by a considerable amount*. The following points are worth bearing in mind about discontinuities.

- The mold engineer is better off avoiding a weld line at the structural discontinuity.
- In shapes such as cylinders or spheres, the direction of the greater stress is easy to identify. In many other cases, it is not as easy. In other cases, the weld line, if unavoidable, may be located perpendicular to the lower stress magnitude.
- It is better to locate the applied loads away from the discontinuities.
- Of all shapes of integral end closures, the semi-ellipsoidal closures are known to cause the minimum discontinuity stresses. Where possible, use them in preference to hemispherical closures.
- Discontinuities exist where the product of the two principal curvatures (called the *Gaussian curvature*) changes sign. In pipe bends, the plane normal to the plane of the bend represents the line of discontinuity. See Exercise 18.
- It may appear that by clever design the negative signs in the formulas for $F$ and $M$ can be taken advantage of, to obtain reduced stresses. In doing so, one may get into high compressive stresses very close to the discontinuity areas, and the full magnitude of tensile stresses just a short distance away. Such situations are candidates for buckling.
- For unreinforced thermoplastics, the apparent advantage arising from discontinuity stresses is applicable only for the short term. Since the discontinuity stresses are secondary, they relax. This means that the initial advantage in stress disappears with time. In the long term analysis, we may not take any credit from the discontinuity stresses.

**Example 2.13**

Pipes are almost always used to move fluids between two vessels. Quite often, such pipes are subjected to a temperature difference, $\Delta T$. See Figure 2.32. Assuming that the vessel is very rigid compared to the pipe, calculate the additional stresses caused by the discontinuity at the pipe-to-vessel junction.

*Solution* Since the vessel is infinitely rigid, all the expansion caused by

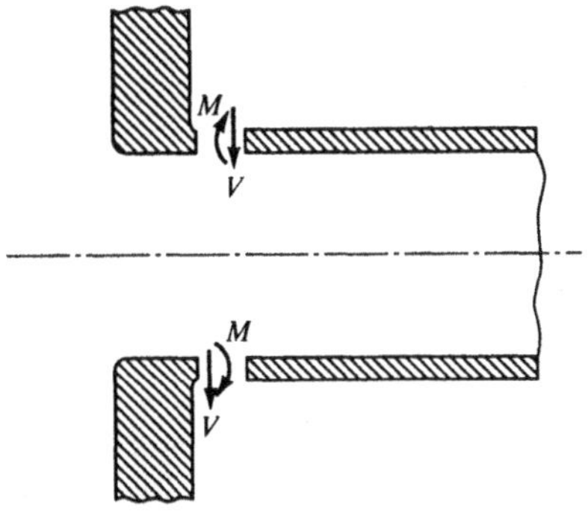


**FIGURE 2.32.** Thermal stresses in a nozzle.

$\Delta T$ has to be taken up by the pipe. The displacement $\delta = R\alpha\,\Delta T$, and the rotation $\theta = 0$. For the directions of $P_o$ and $M_o$ in the figure, we can write the compatibility conditions as follows.

$$\frac{2P_o\beta}{k}\left(\frac{t}{R}\right) - \frac{2M_o\beta^2}{k}\left(\frac{t}{R}\right) = R\alpha\,\Delta T$$

$$\frac{2P_o\beta^2}{k}\left(\frac{t}{R}\right) - \frac{4M_o\beta^3}{k}\left(\frac{t}{R}\right) = 0$$

Solving,

$$F = -\frac{E\alpha\,\Delta T}{\beta}\left(\frac{t}{R}\right)$$

$$M = -\frac{E\alpha\,\Delta T}{2\beta^2}\left(\frac{t}{R}\right)$$

Stress in the pipe caused by pressure has to be superimposed on the secondary stresses. Accordingly, the hoop stress in the pipe $\sigma_h$ is given by

$$\sigma_h = \frac{pR}{t} + \frac{P_o}{t} \pm \frac{6M_o}{t^2}$$

where the positive sign applies on the outer surface and negative sign applies on the inside.

## 2.10 APPLICATIONS IN THE THEORY OF PLATES

The theory of flat plates is very important to the plastics product design. We had earlier mentioned that because of constraints in manufacturability, and the low modulus of the plastics material, it appears that the plastics components design has a universal limitation on thickness, possibly under 20 mm. Hence ribs become a natural requirement. Waffles, triangular grids, and honeycombs are some of the common grid forms of ribs. The waffle is simply a grid of ribs laid out in two perpendicular directions. It renders the plate orthotropic in stiffness. The examples in this section are orthotropic plates. Since orthotropic plates are a generalization of isotropic plates, the discussions here are applicable also for the isotropic plates. The triangular and honeycomb shapes of rib layout produce a more complex effect on the overall stiffness of the plate and are not discussed in this book.

In its elementary form, the theory of isotropic plates is an extension of beam theory to two dimensions. The derivation of the governing equation for isotropic plates is shown in Figure 2.33. The basic assumptions of the plate theory are comparable to those of the beam theory which are as follows.

- There is a plane in the plate that has only zero strains in $x$ and $y$ directions, in response to applied moments; this plane is called the neutral plane.
- Plane sections which are normal to the neutral plane remain plane and normal.
- There are no normal stresses in the direction of the thickness of the plate.

As in the case of beams,

a plate is thin if its linear dimensions are all greater than 15 times the thickness.

In a manner similar to classical elasticity, several fundamental problems in plates are solved using the *inverse approach*. Typically, one finds a function or functions that satisfy the governing equation, and works back to find the problem for which they are solutions. An alternative procedure is to assume the solutions in the form of double Fourier series, and

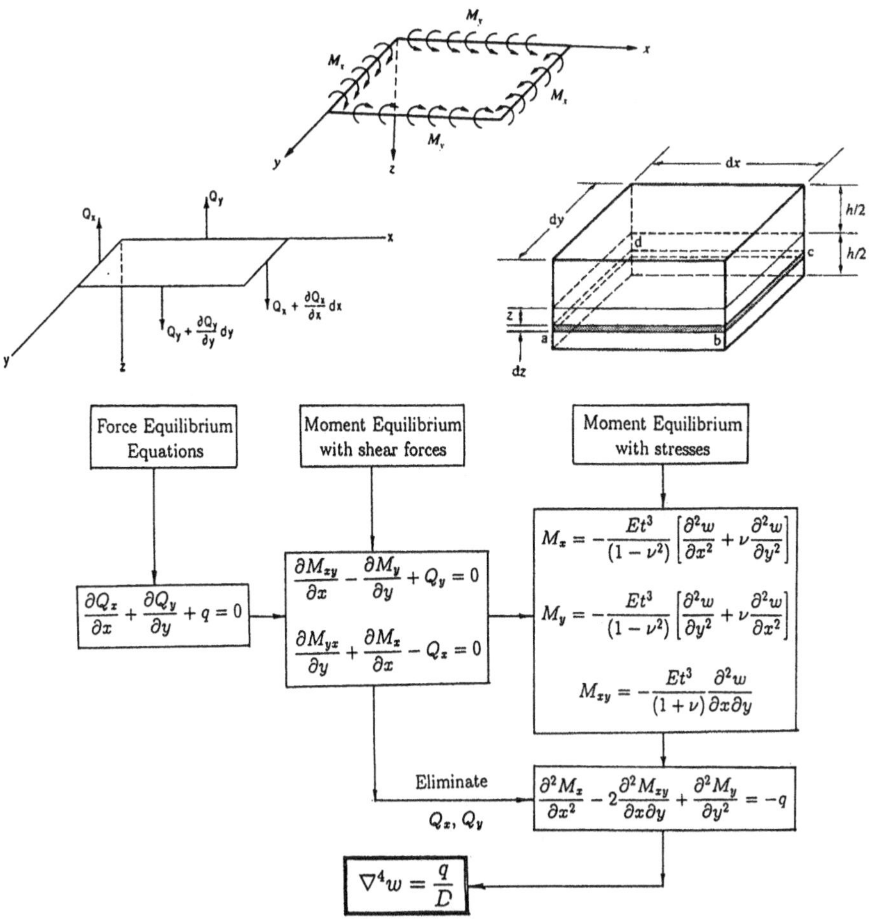


**FIGURE 2.33.** Derivation of isotropic plate equation.

determine the coefficients on the basis of boundary conditions. Subsequently, the superposition of several solutions is used to solve a real life problem, or to create a catalog of pre-engineered solutions.

Such solutions are few in number, and are far from sufficient for plastics design. Some reference books containing catalogs of solutions for the isotropic and orthotropic plates are listed at the end of this chapter.

### 2.10.1 Stiffness and Stress Analysis of Orthotropic Plates

For the purposes of deflection calculations, natural frequency calculations, or buckling analysis we know that the stiffness is important. Local features like fillets and small holes hardly matter in these analyses. For such calculations, a ribbed plate may be substituted by a flat plate of uniform thickness having modified elastic properties. Basically, the additional stiffness of the ribs is treated as an overall increased thickness. If the ribs are not identical in all respects in two perpendicular directions, then the equivalent model will also have orthotropic elastic moduli, that is two different moduli along the two directions. One may also imagine that the stiffness of the ribs is uniformly "smeared" over the plate width and length.

However, such an approach is not adequate for the analysis of stress. The stress distribution depends on further details of geometry, which have been ignored in the above approach. Two examples are presented for the stiffness analysis of orthotropic plates.

### 2.10.2 Governing Equation for Orthotropic Plates

The governing equation for orthotropic plates is different from the one in Figure 2.33 since it must reflect orthotropic moduli. The flexural rigidity term $D$ is to be replaced, and accordingly the governing equation is rewritten as follows:

$$D_x \frac{\partial^4 w}{\partial x^4} + 2B \frac{\partial^4 w}{\partial^2 x \partial^2 y} + D_y \frac{\partial^4 w}{\partial y^4} = q(x, y) \tag{2.88}$$

where $D_x$ is the flexural rigidity of the plate along the $x$-direction, $D_y$ along the $y$-direction, and $D_{xy}$ is the torsional rigidity. More specifically, $D_x$ is calculated based on the moment of inertia viewing the cross section along $x$. The torsional rigidity is usually approximated as the geometric mean of the other two flexural rigidities. Strictly, $D_{xy}$ is supposed to be obtained from a torsion test of the plate. *All the rigidities are expressed on "per unit length" basis. Hence their units are kg mm$^2$/mm.*

Obviously, here we have a more complex equation than the one for isotropic case. A few ready made solutions are available in the literature. The following two examples are representative of the techniques used for rectangular orthotropic plates. By virtue of their generality, they also can be used for isotropic plates.

**Example 2.14**

A translucent ribbed FRP plate (Figure 2.34) made of a material with $E = 700$ kg/mm$^2$ and $\nu = 0.38$ is used as a sun roof panel in a factory machine shop in a tropical climate. The plates are supported from below

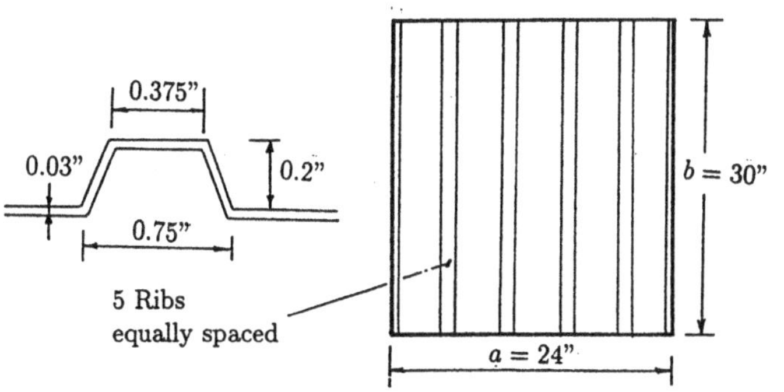


**FIGURE 2.34.** Stiffness of a ribbed plate is orthotropic.

by steel members, so only a unit size of 600 × 800 mm of a fully clamped plate may be considered. (Note carefully the orientations of length and width in relation to the ribs. This orientation is important in orthotropic plates.) Determine the short term deflection at the center for fixed edge conditions and for a normal equivalent of the slant wind pressure of 0.05 kg/cm$^2$.

*Solution* The following steps outline the overall procedure. Calculations of centroid positions, moments of inertia about the centroid, etc. are simply stated and not given in detail.

*Step 1.* Find the rigidity of the plate along the $x$ and $y$ directions. This is done by finding the second moment of area in the two directions.

*Step 2.* Find the torsional modulus $B$ from the approximate relation $B = \sqrt{D_x D_y}$, per unit length.

*Step 3.* Find the maximum deflection using the formula for the deflection at any point. [6].

$$w \approx \frac{49p_o}{8} \frac{\left(x^2 - \frac{a^2}{4}\right)^2 \left(y^2 - \frac{b^2}{4}\right)^2}{7D_x b^4 + 4Ba^2b^2 + 7D_y a^4}$$

The formula for $w$ in Step 3 is really an approximate summation of a double Fourier series.

*Step 1.* Viewed along the $x$-axis, the position of the neutral axis from the bottom,

$$\bar{y} = 1.16 \text{ mm}$$

The second moment of area of cross section about the neutral axis is found to be

$$I_x = 2.50 \text{ mm}^4/\text{mm}$$

Along the $y$-direction, the moment of inertia is simply that of the thickness of the unribbed panel per unit width

$$I_y = 0.083 \text{ mm}^4/\text{mm}.$$

*Step 2.* Based on the modulus of 700 kg/mm², and $\nu = 0.28$, the flexural rigidities $D_x$, $D_y$, and $D_{xy}$ are found to be

$$D_x = 1900 \text{ kg mm}^2/\text{mm}$$

$$D_y = 63.3 \text{ kg mm}^2/\text{mm}$$

$$D_{xy} = 346.8 \text{ kg mm}^2/\text{mm}$$

*Step 3.* Substituting the various values in the formula cited above, the maximum deflection at the center of the plate is computed to be 109 mm.

***Commentary***

*For the case of an isotropic plate, we have $D_x = D_y = D_{xy} = 63.3$ kg mm²/mm. These values give a maximum deflection at the center of 2130 mm! This value of deflection is absurdly high, and is an artifact of the "bending only" assumptions in the development of the theory. In reality, for very thin plates when the plate has curved enough, it begins to resist normal applied loads by in-plane stretch or membrane mode. The stiffness of plates in this mode is very high. This points to the need for using a refined theory of plates. In many practical situations with plastics, the plate (or shell) deflections are high enough to warrant the use of an approach that can consider membrane and bending capability. Many finite element codes currently available in the market have this capability. It is advisable to use them almost in all situations.*

**References**

1. Beer, F. P. and Johnston, E. R., *Mechanics of Materials*, 1981, McGraw Hill Book Company, New York.

2. Timoshenko, S. P. and Gere, J. M., *Mechanics of Materials*, 2nd Ed., 1984, PWS Engineering, Boston.
3. Den Hartog, J. P., *Advanced Strength of Materials*, 1952, McGraw Hill Book Company, Inc.
4. Faupel, J. H. and Fisher, F. E., *Engineering Design*, 1981, 2nd Ed., John Wiley & Sons, New York.
5. Timoshenko, S. and Woinowsky-Kreiger, *Theory of Plates and Shells*, 1959, McGraw Hill Book Co. New York, NY.
6. Szilard, R., *Theory and Analysis of Plates—Classical and Numerical Methods*, 1976, Prentice Hall Inc. Englewood, NJ.
7. Hetenyi, M. *Beams on Elastic Foundation*, 1961, University of Michigan Press, Ann Arbor, MI.
8. Roark, R. J. and Young, W. C., *Formulas for Stress and Strain*, (Table 26), 5th Ed., 1982, McGraw Hill, New York, NY.
9. Murray, A. D. and Rusch, K. C., "Ford Escort All-Thermoplastic Bumper," SAE publication SP-566, *Plastics in Passenger Cars, International Congress and Exposition*, Feb 27–Mar 2, 1984.
10. *Annual Book of ASTM Standards*, Vol. 8.01, 1990, Standard D-790, "Standard Test Methods for Flexural Properties of Unreinforced and Reinforced Plastics and Electrical Insulating Materials," ASTM, Philadelphia, PA.
11. Williams, J. G., *Stress Analysis of Polymers*, 2nd Ed, 1980, Halsted Press, New York, NY, USA.
12. Bulson, P. S., *Buried Structures—Static and Dynamic Strength*, 1985, Chapman and Hall, London, UK.
13. Moser, A. P., *Buried Pipe Design*, 1990, McGraw Hill, New York, NY.
14. Young, O. C. and Trott, J. J., *Buried Rigid Pipes, Structural Design of Pipelines*, 1984, Elsevier Applied Science Publishers Ltd, Essex, UK.
15. *AWWA Standard for Fiberglass Pressure Pipe*, 1989, American Water Works Association, Denver, CO.
16. Timoshenko, S. P. and Goodier, J. N., *Theory of Elasticity*, 3rd Ed., (Chapter 4), 1970, McGraw Hill Book Co., New York, NY.
17. Seely, F. B. and Smith J. O., *Advanced Mechanics of Materials*, 2nd Ed, John Wiley and Sons, New York, NY.
18. Adams, R. D. and Wake, W. C., *Structural Adhesive Joints in Engineering*, 1984, Elsevier Applied Science Publishers, London, UK.
19. Harvey, J. F., *Theory and Design of Modern Pressure Vessels*, 1974, 2nd Ed., Van Nostrand Reinhold, New York, NY.
20. *ASME Pressure Vessel and Piping Code*, Section VIII, Div. 2, Non-Mandatory Appendices, Latest Edition. (One edition every three years, latest one in 1986.)
21. *Annual Book of ASTM Standards*, Vol. 8.04, 1988, Standard F-845, *Standard Specification For Plastic Insert Fittings for Polybutylene (PB) Tubing*, ASTM, Philadelphia, PA.
22. Ford, D. C. and Reinhart, F. W. (editors) *Plastics Piping Manual*, 1976, Plastics Pipe Institute, New York, NY.

**Exercises**

**2.1.** Explain qualitatively why a rope has a lower bending and compressive stiffness than a rod of the same diameter, but would have about the same axial tensile stiffness.

**2.2.** A vegetable cutting board is placed over a kitchen sink, which supports it along its long side. The board is 300 mm × 500 mm, and is made of Nylon 66. Modulus of the material and Poisson's ratio of the material may to taken to be 1100 MPa and 0.3 respectively. A cutting force of 5 kg is applied on the board at its center. Calculate the thickness needed to limit the deflection to 0.02 mm.

**2.3.** A cantilevered steel chair is used as office furniture. The skeletal frame is built from a thin walled pipe of 25 mm × 0.75 mm. Find the vertical deflection of the point *a* with respect to *d*. Assume that the length of the pipe *cd* rests on the floor. Modulus of the material is $21.1 \times 10^3$ kg/mm$^2$.

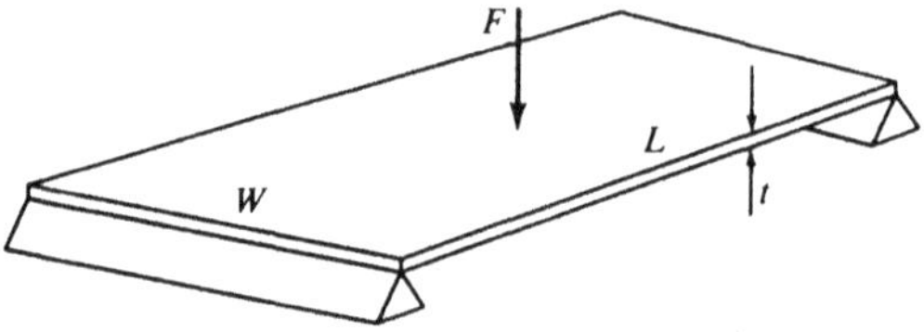


**EXERCISE 2.2.**

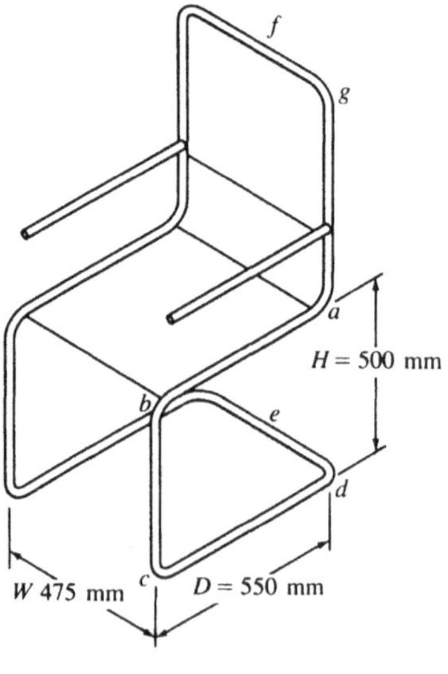


**EXERCISE 2.3.**

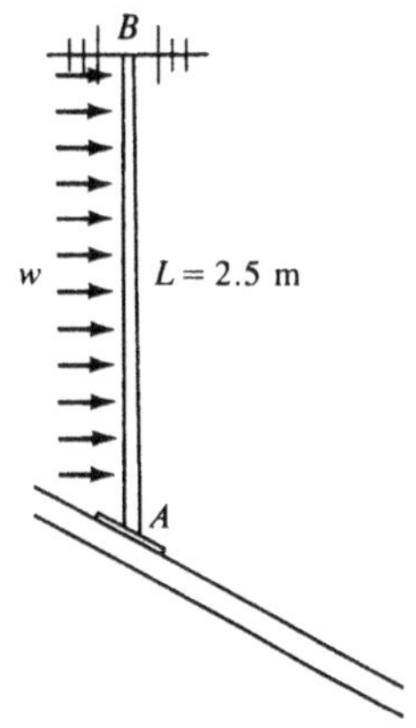


**EXERCISE 2.4.**

**2.4.** Suppose that the television antenna on the roof of a house is made of polycarbonate filled with 30% by volume of chopped glass. It is a hollow pipe of height 2.5 m, fixed at the bottom. The size of the pipe varies linearly from its bottom (60 mm × 5 mm) to its top (30 mm × 5 mm). The wind load under gale conditions is equivalent to 0.05 kg/mm. Calculate the tip deflection of the antenna, given $E = 900$ kg/mm$^2$, and $\nu = 0.28$.

**2.5.** The handle on a pressure cooker is attached to the pot by screws. The total weight of contents in the pot is about 6 kg, acting through a line 175 mm from the tip of the handle. The handle itself is 220 mm long, tapering from a larger section at $A$ to a smaller section at $B$, as shown. The pot is lifted by hand by using forces $F_1$ and $F_2$ as shown. Determine the maximum stress in the handle.

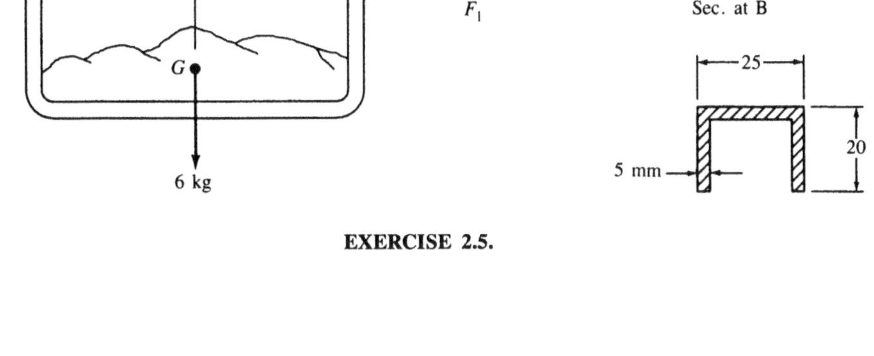


**EXERCISE 2.5.**

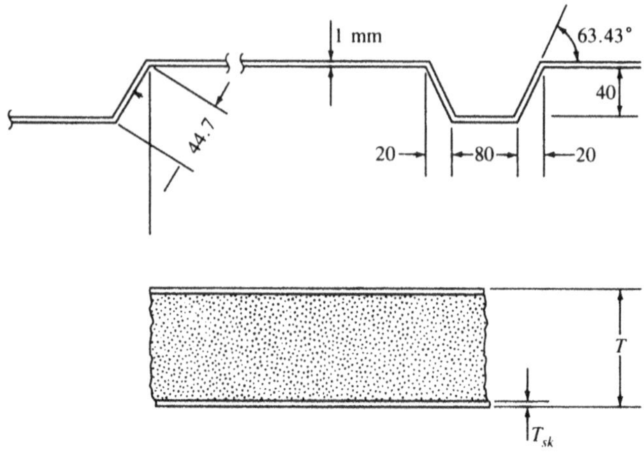


**EXERCISE 2.6.**

**2.6.** The current design of the side panels of a semi trailer is ribbed aluminum sheet with dimensions as shown in the figure. Replace it by a sandwich panel of the same flexural stiffness with a skin of thickness $T_{sk}$ mm and a total thickness not exceeding 50 mm. Estimate the weight savings as a function of $T_{sk}$.

**2.7.** Critically review the approach used in the piping flexibility of Example 2.7. Note that only bending stress was considered. What other stresses are possible? Develop a method leading to the calculation of stresses other than bending.

**2.8.** The addition of pipe clamps at judicious locations can reduce the pull-out forces on a tee-to-pipe joint. Assume an L-shaped layout of pipe *AbaB*, as shown, with tee fittings at *A* and *B*, and clamps at *a* and *b*. Discuss the factors involved in the placement of the clamps *a* and *b*, and evolve a logic diagram to identify the feasible solutions.

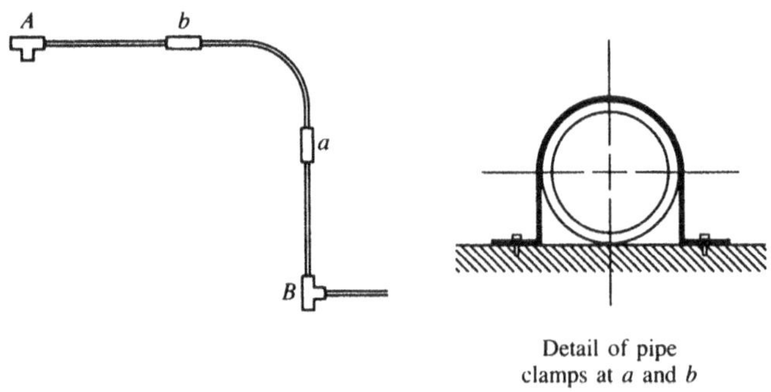


**EXERCISE 2.8.**

**2.9.** In the tee-fitting of Example 2.9, find out the manual force needed to initially pull the tube over the fitting. Note that the interference is minimal to make this easy. Assume a friction factor of 0.2.

**2.10.** With reference to the example of the rotating hub (Example 2.8), arrange the following variables in decreasing order of effectiveness in improving the torque taking capability.

(a) interference(s).
(b) substitute stronger plastic material for the polypropylene.
(c) substitute stiffer plastic material for the polypropylene.
(d) increase the radius at the interface.
(e) increase the friction factor, by knurling and other similar ideas.
(f) increase the dimension $L$—width of contact.

Give reasons.

**2.11.** For a semi-infinite cylinder find the end ring loads and moments needed to produce a deflection $\delta$ but no rotation, that is $\theta = 0$.

**2.12.** Based on the previous problem, find the deflection and rotation under a concentrated ring load on an infinite cylinder.

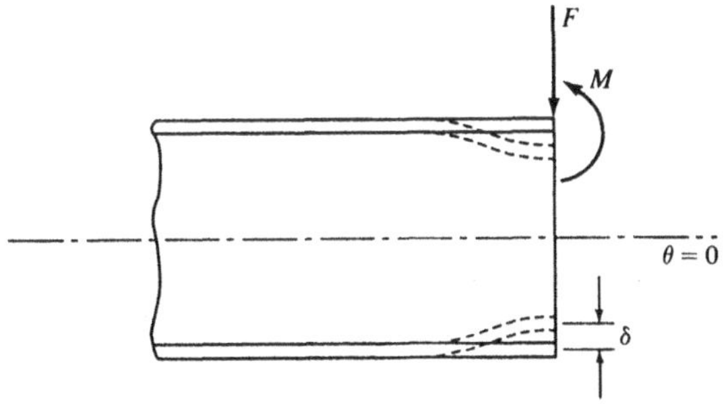


**EXERCISE 2.11.**

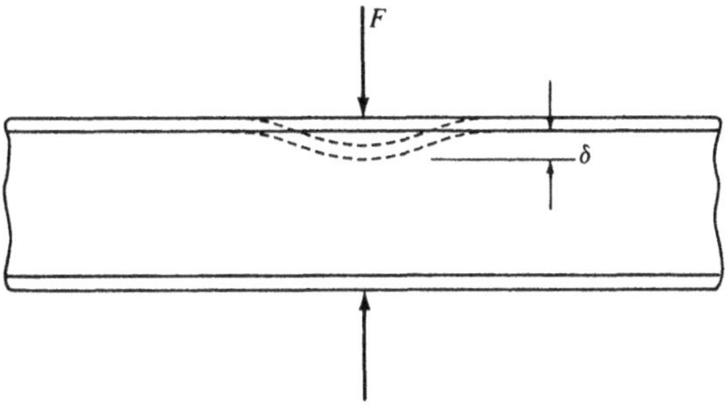


**EXERCISE 2.12.**

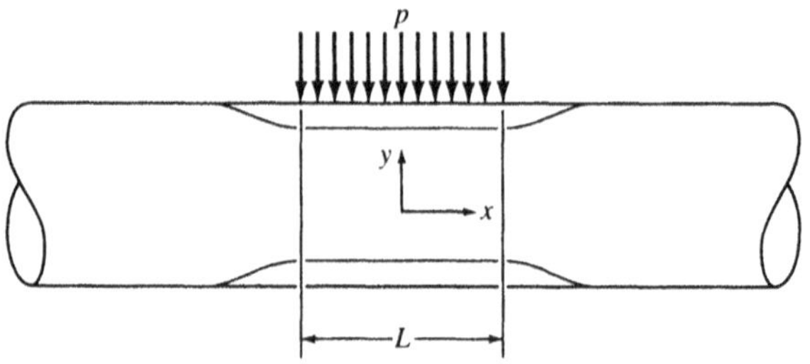


**EXERCISE 2.14.**

**2.13.** Prove all the formulas in Equations 2.72 through 2.75 for an infinite cylinder.

**2.14.** By using the superposition principle, the results for a concentrated ring load acting over an axial length *dx* can be integrated to obtain the results for an axially distributed load. Use this idea to calculate the $\delta(x)$ and $\theta(x)$ caused by a uniformly distributed load as shown.

**2.15.** *Term Paper:* Prove that the deflection at the center of a pipe of finite length $L$, due to radial load $P_o$ acting at the same location is given by

$$y_c = \frac{P_o \beta}{2k} \frac{2 + \cos \beta L + \cosh \beta L}{\sin \beta L + \sinh \beta L}$$

Also show that the deflection at the end of the pipe is given by

$$y_{end} = \frac{2P_o \beta}{k} \frac{\cos \beta L/2 \cosh \beta L/2}{\sin \beta L + \sinh \beta L}$$

Prove that for sufficiently long cylinders, such as $\beta L > 2\pi$, the deflection at the end is almost zero, even if the ends are left free. Thus provide a fully analytical proof for the finite element example on page 66. (Suggestion: Refer to Chapter 5 of Reference 3, for a treatment using beams on elastic foundations. You can establish the $k$ value for a pipe, and proceed.)

**2.16.** PRI (Pipe Research Institute) creep tests involve end-capping a cylindrical specimen, and keeping it under pressure until a creep failure occurs. Some of the results may have to be discarded because the failure occurs near the end cap. Can all failures be made to occur in the gage length by a suitable redesign of the specimen?

**2.17.** In a cylinder-sphere junction of a pressure vessel, which location has the maximum total stress?

(a) outside of cylinder
(b) outside of sphere
(c) inside of cylinder
(d) inside of sphere

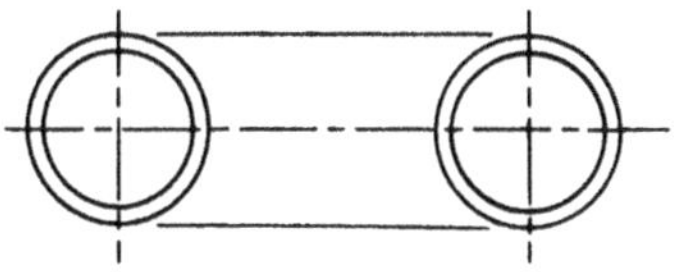

EXERCISE 2.18.

**2.18.** Mark the line on a torus where the Gaussian curvature changes sign. If you have access to finite element software, stress analyze a torus under internal pressure, and verify that the location change of Gaussian curvature is a discontinuity from the point of view of stress analysis.

**2.19.** In concentric conduits inner pipes are supported from the outer pipe by ribs. Such ribs are all the more necessary at pipe bends (elbows). In the light of the Gaussian curvatures, where would you locate such supports?

**2.20.** Standard calculation procedures for vessel buckling of shells are based on an $l/d$ ratio. For a known $r/t$ ratio, convert the $l/d$ ratio into a $\beta l$ ratio, and show that $\beta l$ also governs the buckling of shells.

**2.21.** Show that when the thickness of a cylinder is very small, the thick cylinder formula Equation 2.34 reduces to the well known formula

$$\sigma_h = \frac{pR_m}{t}$$

**2.22.** Derive the displacement formula (Equation 2.40) using Equation 2.34, Equation 2.37, and Hooke's law. In particular, which of the two conditions—plane stress or plane strain—does it represent?

**2.23.** For a semi-infinite pipe, 150 mm o.d. × 12.5 mm thick, made of rigid PVC for which $\nu = 0.35$. a. Find $\beta$ for the pipe. b. Find the axial distances from the end at which internal bending moments which arise from an applied ring force are *exactly* zero. c. Find the distances from the end at which the internal shear forces caused by an applied ring force at the end are exactly zero. d. Find the distances at which the internal moments caused by a combined application of ring force and ring moment at the ends vanish exactly.

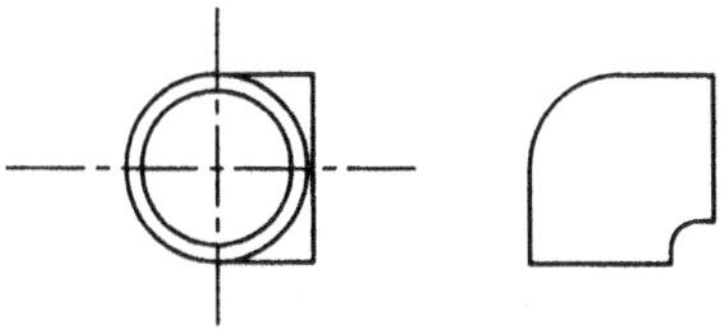

EXERCISE 2.19.

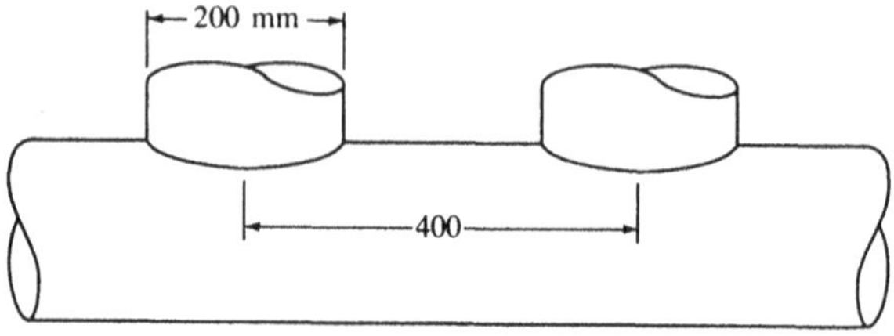


**EXERCISE 2.24.**

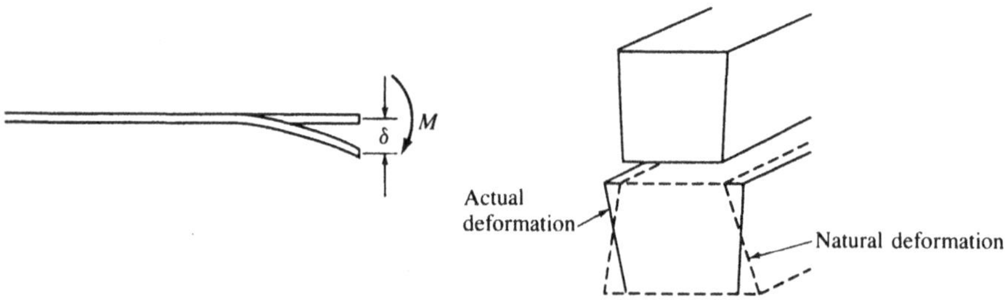


**EXERCISE 2.25.**

**2.24.** A large PVC of size 450 mm o.d. × 45 mm thickness has two openings each of diameter 200 mm and spaced 400 mm apart, center to center. These holes are located on the same longitudinal line. Would you think that the stress field in the vicinity of one opening would interfere with that at the other? Give reasons.

**2.25.** A moment in the circumferential direction arises because of the Poisson's effect, and proportional to $M(x)$. This moment is not discussed in the text. The sketch above illustrates the origin of the moment. Determine its magnitude. Also determine the error incurred in the hoop stress if this moment were to be completely neglected? (Note that Reference 20 considers this moment in the stress calculation on shell walls.)

**2.26.** Plot the axial stress on the outside and inside surfaces of a cylinder, which are caused by the rigid welding (or other integral joining) of a hemispherical cap of matching size.

**2.27.** For a spherical cap, the meridional variation of the stress can be expressed as function of the meridional angle $\phi$. Plot the stresses on the sphere because of the discontinuity described in the previous problem. Show that the stress fades with distance, and that the factor $\beta$ governs the pattern of decay.

**2.28.** Based on the previous two problems, determine how far away a weld line should be if the cylinder-sphere combination was injection molded. The factor of safety at the weld line relative to the local stress must be at least the same as that of the cylinder. Assume that the strength at a weld is only 70% of the standard value for the material.

# 3

# Beyond Elastic Behavior

## 3.1 INTRODUCTION

Polymers are normally ductile in nature and have the capability to *yield* under high enough stress. Yielding, without which the material would break in a brittle manner, is helpful in structures to smooth out peak stresses and convert them into strain concentrations. Also, yielding allows sizing of a part to be done on the basis of strength of materials formulas without worrying about stress concentrations, at least initially. In plastics the distinction between yield and ultimate strength was considered unnecessary. However, the trend has changed, especially after the thermo-forming process became commercially viable. Hence, there is a clear need to understand the stress-strain relationship while yielding is in progress. In this chapter, we discuss the phenomenological aspects of yielding and another similar mode of failure called "crazing."

### 3.1.1 True Stress and Considier Construction

Plastics can exhibit brittle or ductile behavior in a tension test. Brittle behavior is characterized by an abrupt failure of the test specimen. The fracture surface does not exhibit any stretch (or elongation). The stress strain curve abruptly ends, as shown in curve 1 of Figure 3.1. On the contrary, when a ductile polymer test specimen is loaded beyond a certain level, it begins to form a neck, and increase in length by a fairly large amount. While this takes place the load (not stress) on the specimen may barely increase, or stay constant or even decrease. This phase is associated with yielding. This behavior is illustrated by curve 2 of Figure 3.1. The

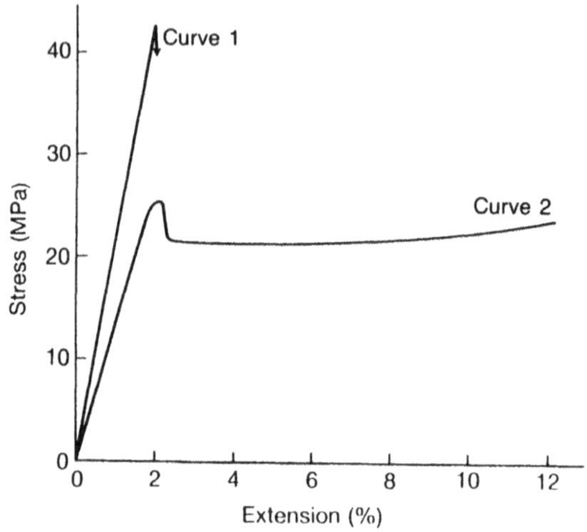


**FIGURE 3.1.** Typical stress strain curve for polymers. Curve 1 is for brittle plastics and curve 2 is for ductile plastics. Plastics may have very high ductility, with ultimate strains over 800%.

volume of the material is known to remain[1] constant while yielding takes place. In other words, the reduction in area is exactly made up for by the increase in length. Hence,

$$AL = A_0 L_0, \quad \text{or} \quad A = \frac{A_0 L_0}{L}$$

Since length-after-yield $L = L_0(1 + \epsilon)$, the area after yield can be written as $A = A_0/(1 + \epsilon)$. The factor $(1 + \epsilon)$ is sometimes called the *Extension Ratio*. The engineering definition of stress is "load divided by the original area." Using this, we can find the true stress, which is load divided by the true area.

$$\sigma_{tr} = \frac{P}{A} = \frac{P(1 + \epsilon)}{A_0} = \sigma_0(1 + \epsilon)$$

The load supported by the specimen, $P$, in terms of the true stress is given

[1] In reality, it is almost constant. For materials with different yield strengths in tension and compression, it is not exactly constant.

by

$$P = A_0 \sigma_0 = \frac{A_0 \sigma_{tr}}{1 + \epsilon} \tag{3.1}$$

Considier (1870) explained the load drop in the testing machine during yielding (of steel, of course) by expressing $P$ in terms of true stress $\sigma_{tr}$, as above. By differentiating $P$ with respect to $\epsilon$ we see that

$$\frac{dP}{d\epsilon} = \frac{A_0}{(1 + \epsilon)^2} \left[ (1 + \epsilon) \frac{d\sigma_{tr}}{d\epsilon} - \sigma_{tr} \right] \tag{3.2}$$

The condition for $P$ reaching a maximum is obviously

$$\frac{d\sigma_{tr}}{d\epsilon} = \frac{\sigma_{tr}}{1 + \epsilon} \tag{3.3}$$

Consider the points 1 and 2 on the (true) stress strain diagram, in Figure 3.2. The point 1 is just any point on the curve, whereas the point 2 is the point of tangency from the point $A(-1, 0)$.[2] It is seen that the condition for the load reaching a maximum (Equation 3.3) is satisfied exactly at the point 2, and only at the point 2. Thus the point 2 represents

[2] Note that this point denotes a physical impossibility. At this point, the material experiences extreme compression and vanishes completely!

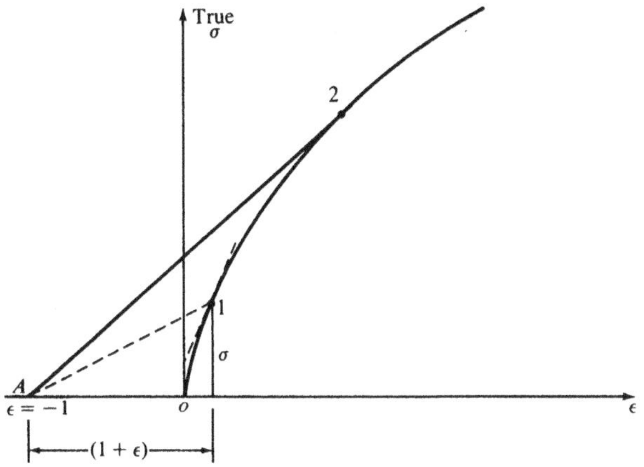


**FIGURE 3.2.** True stress strain diagram of a ductile polymer.

the point of maximum load (or the "instability" point). Considier used this geometric construction to calculate the instability load in a test, in the absence of sensitive recording instruments in his times.

This concept can be used to determine whether or not a plastic specimen would

- deform homogeneously without forming a neck,
- form a neck, or
- form a neck and cold draw.

These properties are important characteristics for the thermo-forming of polymers. Referring to Figure 3.3(a), we see that it is impossible to draw a tangent from $\epsilon = -1$. These specimens do not drop load and deform homogeneously throughout the test. This behavior is typical of rubber-like materials. In Figure 3.3(b), it is possible to draw just one tangent. Such specimens neck and drop load before breaking. In Figure 3.3(c), it is possible to draw two tangents, one point of tangency for load drop and the other for increase of load a second time. Such specimens, neck, draw, and usually have a very high ductility, with strains at failure exceeding 250%[3] for most plastics. Failure strains for certain high impact grades may reach as high as 1000%. The use of Considier construction is not only to determine the necking and drawing capabilities of the material, but also to

[3] DuPont's Delrin is an example of the case (b), and Zytel is an example of case (c), as can be seen from References 1 and 2.

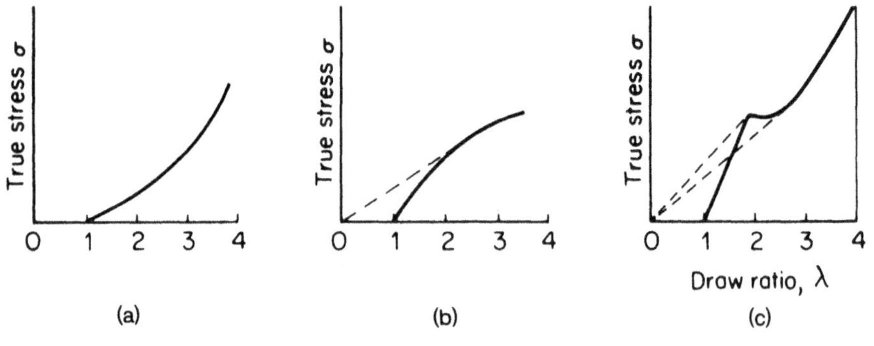


**FIGURE 3.3.** Different types of stress strain curves of polymers, and Considier construction.

determine the loads and stresses at which it takes place. This is vital for the calculation of power requirements in forming operations.

## 3.2 ONSET OF YIELD

Yielding computations must be done in two phases, to detect its onset and to calculate the plastic flow of the material under stress. Detecting its onset is important to determine when in a calculation should one switch from Hooke's laws to plastic flow rules.

For materials which fail in tensile tests in a brittle manner, there is no need to detect the onset. Instead, we may use a *maximum stress criterion* of failure. This criterion simply states that the material will break when any one principal stress component reaches the ultimate tensile strength of the material. The application of this criterion is therefore straightforward. In the context of finite element software, the principal stresses are obtained directly from the software itself. To the design engineer, brittle materials provide no latitude in design. Since they are completely intolerant to stress concentrations, components made of such materials are fragile at almost any size. Fragility exists at the peak stress areas.

Our interest is in the more ductile materials, which have strains to failure of the order of at least 15%, and preferably 200%. The onset criterion which has been found to work well for ductile materials is the *Von Mises criterion*. We need a few new definitions before discussing the Von Mises criterion.

### 3.2.1 Deviatoric and Hydrostatic Stress Components

Since the volume remains constant under yielding, we may develop the criterion by watching the elastic components of stress that produce volume changes. The change in volume $e$ per unit volume is given by

$$e = (1 + \epsilon_x)(1 + \epsilon_y)(1 + \epsilon_z) - 1 = (\epsilon_x + \epsilon_y + \epsilon_z).$$

This is the same as Equation 1.23. From Chapter 1, adding up Equations 1.24, we obtain

$$(\sigma_x + \sigma_y + \sigma_z) = (3\lambda + 2G)e = 3Ke \tag{3.4}$$

where $K$ is the bulk modulus. This can also be written as

$$\sigma_{ave} = Ke,$$

thereby suggesting that for stress states with zero average stress, there is no volume change.

Any state of stress can be split into two parts, one with equal principal stresses called the *hydrostatic stress component* and the other containing the rest called the *deviatoric stress component*. For example:

$$\begin{bmatrix} \sigma_x & \tau_{xy} & \tau_{xz} \\ \tau_{xy} & \sigma_y & \tau_{yz} \\ \tau_{xz} & \tau_{yz} & \sigma_z \end{bmatrix} = \begin{bmatrix} (\sigma_x - \sigma_{ave}) & \tau_{xy} & \tau_{xz} \\ \tau_{xy} & (\sigma_y - \sigma_{ave}) & \tau_{yz} \\ \tau_{xz} & \tau_{yz} & (\sigma_z - \sigma_{ave}) \end{bmatrix}$$

$$+ \begin{bmatrix} \sigma_{ave} & 0 & 0 \\ 0 & \sigma_{ave} & 0 \\ 0 & 0 & \sigma_{ave} \end{bmatrix}$$

The first matrix on the right side is the deviatoric stress, and the second gives the hydrostatic stress components. Note that the average of the deviatoric stress components is zero, and hence this stress component is a candidate for description of plasticity, both onset and flow. In subscript notation we write this briefly as

$$\sigma_{ij} = s_{ij} + \sigma_{ave}\,\delta_{ij}$$

**Example 3.1**

At a point in a structure the stress in a local $(x, y, z)$ coordinate system is calculated to be

$$[\sigma_{ij}] = \begin{bmatrix} 5.6 & 2.7 & -0.9 \\ 2.7 & 12.0 & 10.5 \\ -0.9 & 10.5 & -1.1 \end{bmatrix} \text{MPa}$$

Find the hydrostatic and deviatoric stress components:

*Solution* The average normal stress $\sigma_{\text{ave}} = \frac{1}{3}(5.6 + 12.0 - 1.1) = 5.5$ MPa. Hence the hydrostatic stress component is

$$\sigma_{\text{ave}}[\delta_{ij}] = \begin{bmatrix} 5.5 & 0 & 0 \\ 0 & 5.5 & 0 \\ 0 & 0 & 5.5 \end{bmatrix} \text{MPa}$$

The deviatoric stress component is

$$[s_{ij}] = [\sigma_{ij}] - \sigma_{ave}[\delta_{ij}] = \begin{bmatrix} 5.6 & 2.7 & -0.9 \\ 2.7 & 12.0 & 10.5 \\ -0.9 & 10.5 & -1.1 \end{bmatrix}$$

$$- \begin{bmatrix} 5.5 & 0 & 0 \\ 0 & 5.5 & 0 \\ 0 & 0 & 5.5 \end{bmatrix} \text{MPa}$$

$$= \begin{bmatrix} 0.1 & 2.7 & -0.9 \\ 2.7 & 6.5 & 10.5 \\ -0.9 & 10.5 & -6.6 \end{bmatrix} \text{MPa}$$

### 3.2.2 Stress Invariants

Returning to our consideration of the onset of plasticity, we are indeed looking for a criterion that is a function of $s_{ij}$. Furthermore, the criterion must be independent of directions, and hence must be expressible in terms of the so-called *invariants*. Invariants are functions of stress at a point that remain unchanged when referred to different types and orientations of coordinate systems. In particular, they remain constant with respect to rotation of coordinates. Yielding, being a coordinate-independent phenomenon, can be represented by invariants. The first, second, and third invariants of stress are:

$$J_1 = \sigma_1 + \sigma_2 + \sigma_3 \text{ in terms of principal stresses} \tag{3.5}$$

$$= \sigma_x + \sigma_y + \sigma_z \text{ in terms of } x, y, \text{ and } z \text{ components.} \tag{3.6}$$

$$J_2 = \sigma_1\sigma_2 + \sigma_2\sigma_3 + \sigma_3\sigma_1 \text{ in terms of principal stresses} \tag{3.7}$$

$$= (\sigma_x\sigma_y + \sigma_y\sigma_z + \sigma_z\sigma_x) - (\tau_{xy}^2 + \tau_{yz}^2 + \tau_{zx}^2) \text{ in terms of } x, y, z \text{ components.} \tag{3.8}$$

$$J_3 = \sigma_1\sigma_2\sigma_3 \text{ in terms of principal components.}$$

$$= \text{Determinant of the stress matrix, for other coordinate systems.} \tag{3.9}$$

If we apply these formulas to the deviatoric stress components $s_{ij}$, its invariants are given below. The primes are used for deviatoric stresses and its invariants.

$$J_1' = \sigma_1' + \sigma_2' + \sigma_3' = 0 \tag{3.10}$$

$$J_2' = (\sigma_1'\sigma_2' + \sigma_2'\sigma_3' + \sigma_3'\sigma_1') \tag{3.11}$$

$$= -\tfrac{1}{3}\left[(\sigma_1^2 + \sigma_2^2 + \sigma_3^3) - (\sigma_1\sigma_2 + \sigma_2\sigma_3 + \sigma_3\sigma_1)\right]$$

$$= -\tfrac{1}{6}\left[(\sigma_1 - \sigma_2)^2 + (\sigma_2 - \sigma_3)^2 + (\sigma_3 - \sigma_2)^2\right] \tag{3.12}$$

(note primes)

$$J_3' = \sigma_1'\sigma_2'\sigma_3' \tag{3.13}$$

### 3.2.3 Development of Von Mises Criterion

A possible form of the criterion for the onset of plasticity is $f(J_1', J_2', J_3') = 0$. However, since $J_1' = 0$, for deviatoric stress, this criterion reduces to $f(J_2', J_3') =$ a constant. Further, we make the assumption that the material does not exhibit the so-called *Bauschinger Effect*. This effect considers the increase in tensile yield strength due to strain-hardening and the decrease in the compressive yield strength due to strain-softening. In other words, if a stress state $\sigma_{ij}$ produces yielding in tension so will a stress $-\sigma_{ij}$ produce yielding in compression. Since $J_3'$ is an odd function of stress components, its presence in $f(\cdots)$ violates the no-Bauschinger assumption. Hence, the form of Von Mises criterion is reduced to $f(J_2') = -K^2$. A further simplification of the criterion is obtained by assuming $f(J_2')$ to be just equal to $J_2'$ itself. It has been shown that $J_2'$ is directly related to *distortion strain energy* of the body. That is, it is associated with shape changes *not* involving a volume change. Based on this, the Von Mises criterion is

$$J_2' = \sigma_1'\sigma_2' + \sigma_2'\sigma_3' + \sigma_3'\sigma_1' = -K^2, \tag{3.14}$$

for onset and during plasticity. $K$ is easily determined by requiring that for one dimension this criterion must agree with laboratory test results. Since a test specimen begins to yield at $\sigma_1 = \sigma_y$, $\sigma_2 = \sigma_3 = 0$, we have

$$\sigma_{\text{ave}} = \frac{1}{3}\sigma_y; \quad \sigma_1' = \frac{2}{3}\sigma_y; \quad \sigma_2' = \sigma_3' = -\frac{1}{3}\sigma_y$$

Substituting these in $J_2'$, we get $K = \sigma_y/\sqrt{3}$. With this value for $K$ we can write a few alternative forms of the Von Mises criterion.

In terms of principal stress components (not deviatoric), the criterion is

$$(\sigma_1 - \sigma_2)^2 + (\sigma_2 - \sigma_3)^2 + (\sigma_3 - \sigma_1)^2 = 2\sigma_y^2 \tag{3.15}$$

In terms of the $x$, $y$, and $z$ components, the most convenient form of the criterion is

$$(\sigma_x - \sigma_y)^2 + (\sigma_y - \sigma_z)^2 + (\sigma_z - \sigma_x)^2 + 6\left(\tau_{xy}^2 + \tau_{yz}^2 + \tau_{zx}^2\right) = 2\sigma_y^2 \tag{3.16}$$

A common but erroneous notion about yielding is that when any component of stress exceeds $\sigma_y$, no further increase in stress is possible. This is incorrect, since the Von Mises criterion does not place restrictions on the individual components of stress. A feature of this criterion is that it determines whether or not yielding occurred at a point, a go/no-go type of decision. It is not a stress-strain relationship, and hence is not capable of calculating the post-yield strains.

### 3.2.4 Geometric Representation of the Von Mises Criterion

For two dimensional plane stress problems, we have $\sigma_3 = 0$. The criterion for yield reduces to

$$J_2' = \sigma_1^2 + \sigma_2^2 - \sigma_1\sigma_2 = \sigma_y^2 \tag{3.17}$$

This equation represents an ellipse in the $\sigma_1\sigma_2$ plane. If the point $(\sigma_1, \sigma_2)$ lies inside the ellipse, then yielding does not occur. If the point is on the ellipse, then yielding just begins. For points outside the ellipse, yielding is in progress. The ellipse is shown in Figure 3.4. It intersects $x$ and $y$ axes at $(\pm\sigma_y, 0)$, and $(0, \pm\sigma_y)$ respectively. It is centered at the origin, because of the absence of the Bauschinger effect. Its axis is tilted at 45° to the $x$ axis, thus permitting stress combinations higher than $\sigma_y$ to exist without causing yielding. See the shaded area in Figure 3.4.

In three dimensions, the Von Mises criterion represents a cylindrical surface, with axis lying along the [111] direction. Many interesting material behaviors can be interpreted geometrically using the surface. For example, hydrostatic components lie along the [111] axis. Points $\sigma_i$, $(i = 1, 2, 3)$ as well as $\sigma_i - p$, $(i = 1, 2, 3)$ may be seen to map onto the same point on the Von Mises yield surface. This is equivalent to the statement that yielding is

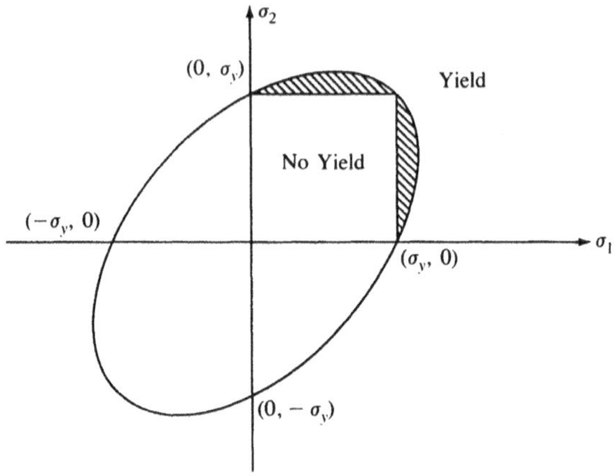


**FIGURE 3.4.** Von Mises criterion for yielding maps into an ellipse for two dimensional stress.

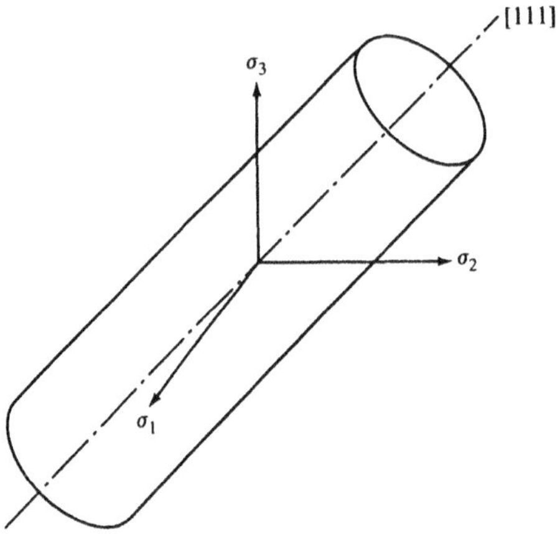


**FIGURE 3.5.** Von Mises yield surface for three dimensional stress state. Viewed along the [111] axis, this surface has a circular cross section. Any stress state and the same state superimposed with any hydrostatic stress component will map on the same point on this surface.

insensitive to the hydrostatic components of stress. Figure 3.5 represents this cylinder.

**Example 3.2**

The shaded area in Figure 3.4 denotes that there can exist combinations of stresses in two dimensions greater than $\sigma_y$, while yielding itself does not occur. Investigate this statement for three dimensions.

*Solution* Consider a state of stress in which the principal stresses are

$$\sigma_1 = 1.5\sigma_y; \sigma_2 = 1.1\sigma_y, \quad \text{and} \quad \sigma_3 = 1.2\sigma_y$$

For this state, $J_2$ is given by

$$-6J_2 \equiv (\sigma_1 - \sigma_2)^2 + (\sigma_2 - \sigma_3)^2 + (\sigma_3 - \sigma_1)^2 \leq 2\sigma_y^2$$

$$\sigma_y\sqrt{0.4^2 + 0.1^2 + 0.3^2} = 0.51\sigma_y < \sqrt{2}\,\sigma_y$$

While each of the components is greater than $\sigma_y$, yielding does not begin. It is instructive to vary one of the components, say $\sigma_2$. If this component is increased by an amount $0.1\sigma_y$, the expression for $J_2$ decreases.

$$\sqrt{-6J_2} \equiv \sigma_y\sqrt{0.3^2 + 0.2^2 + 0.3^2} = 0.469\sigma_y < \sqrt{2}\,\sigma_y$$

On the other hand, it it is decreased by $0.1\sigma_y$, then,

$$\sqrt{-6J_2} \equiv \sigma_y\sqrt{0.5^2 + 0^2 + 0.3^2} = 0.583\sigma_y$$

This state of stress is closer to yielding than the first two. If it is completely removed, then

$$\sqrt{-6J_2} \equiv \sigma_y\sqrt{0.4^2 + 1.1^2 + 1.5^2} = 1.902\sigma_y > \sqrt{2}\,\sigma_y$$

which means that yielding is in progress already. In general, combinations of tensile and compressive principal stresses at the same point are more conducive to yielding than having all components be of the same sign.

**Example 3.3**

A PVC pipe of size 150 mm o.d.$\times$ 5 mm thickness is kept under internal pressure at room temperature. The material has an initial tensile yield strength of 11 MPa (1.5 ksi). At what value of the internal pressure would the pipe just begin to yield?

*Solution* The internal pressure causes hoop and axial stresses in the body of the pipe which are in fact the principal stresses. Let us first consider that the stress is bi-axial

$$\sigma_1 = \sigma_a = \frac{pd_0}{2t} = p\frac{150}{2 \times 5} = 7.5p$$

$$\sigma_2 = \sigma_h = \frac{pd_0}{4t} = p\frac{150}{4 \times 5} = 15p$$

Check if

$$\sigma_1^2 - \sigma_1\sigma_2 + \sigma_2^2 = \sigma_y^2$$

$$(7.5^2 - 7.5 \times 15 + 15^2)p^2 = \sigma_y^2$$

From this,

$$p = \sqrt{\frac{121}{168.75}}\ p = 0.847 \text{ MPa} = 123 \text{ psi}$$

However, a tri-axial stress state exists on the inner pipe surface. Here in addition to $\sigma_a$ and $\sigma_h$, as above, there exists a compressive stress $-p$ normal to the wall. Since this point has a combination of negative and positive stresses, yielding is likely to occur at a lower pressure

$$\sqrt{[(8.5^2 + 16^2 + 7.5^2]} = \sqrt{2}\,\sigma_y$$

Or,

$$p = 0.793 \text{ MPa } (115 \text{ psi.})$$

***Commentary***

*For a similar reason, yielding of a pressurized spherical vessel can be studied only in three dimensions. Because in the curved surface of the sphere at the midthickness, the stresses are equal in all directions. However, on the inner surface, there is a normal compressive stress which promotes yielding at a lower pressure.*

**Example 3.4**

Since yielding is a tri-axial phenomenon, we can generalize the term "Factor of Safety" to apply to $\sqrt{-J_2}$, rather than to any individual stress component.[4] In this sense, calculate the FOS available in a PVC pipe of

[4] Later, in Chapter 7, we use this generalization for situations of different strengths in different directions.

250 mm × 40 mm subjected to an internal pressure of 1.378 MPa (200 psi). Assume that the yield strength of the material is 11 MPa.

*Solution* First of all, it is unlikely that a pipe as thick as 40 mm—presumably extruded—would behave in a ductile manner, because of the unavoidable orientation and lack of homogeneity. Setting aside this condition, we calculate the triaxial stress at a point on the inside surface of the pipe, using thick cylinder formulas. Assuming $p$ to be the pressure at incipience of yielding,

$$\sigma_h = p\left(\frac{r_0^2 + r_i^2}{r_0^2 - r_i^2}\right) = p\left(\frac{125^2 + 85^2}{125^2 - 85^2}\right) = 2.72p$$

$$\sigma_a = \frac{\text{Load}}{\text{Area}} = \frac{\pi r_i^2 p}{\pi(r_0^2 - r_i^2)} = p\frac{85^2}{125^2 - 85^2} = 0.86p$$

$$\sigma_r = -p$$

Note that the 2-to-1 ratio of hoop and axial stresses does not hold for thick pipes. Substituting,

$$\left[(2.72 + 1)^2 + (0.86 + 1)^2 + (2.72 - 0.86)^2\right]p^2 \leq 2\sigma_y^2$$

$$20.75p^2 \leq 2(11^2)$$

$$p = 3.4144 \text{ MPa } (496.6 \text{ psi})$$

Compared to this, the internal pressure is only 1.378 MPa. The available FOS is 2.478.

**Example 3.5**

A square plate is made from 25% glass filled Nylon with a yield strength of 4644.4 MPa (32,000 psi). The plate is of size 10 mm × 400 mm × 400 mm. It is clamped rigidly all around its edges and is loaded by a uniformly distributed load of $w$ Newtons per sq mm. At what value of $w$ does yielding begin?

*Solution* The plate has a variable stress field. However, due to symmetry, the stress is maximum at the center of the plate, and the two principal stresses are equal. Using Roark tables [3], we see that the maximum stress can be expressed as

$$\sigma \max = p\beta \frac{(\text{Long side})^2}{(\text{thickness})^2}$$

Since the plate is square,

$$\sigma_1 = \sigma_2 = \frac{\beta p 400^2}{15^2} = 0.3078 \frac{400^2}{10^2} p = 294.5p$$

$$\sigma_1^2 - \sigma_1 \sigma_2 + \sigma_2^2 = \sigma_y^2$$

Substituting $\sigma_1 = \sigma_2$ in this, we get

$$\sigma_1 = \sigma_y$$

Or, $492.5p = 4644$ MPa; $p = 9.43$ MPa (139 psi)

### 3.2.5 Sensitivity to Hydrostatic Stress

Materials with different yield strengths in tension and compression are sensitive to the hydrostatic stress components. Most polymers have a higher compressive yield strength than tensile strength (Table 3.1). The center of the yield ellipse for these materials is shifted more towards the

**Table 3.1. Typical Values of Yield Stresses (MPa) For Some Polymers**

| Polymer | $\sigma_{yc}$ | $\sigma_{yt}$ |
|---|---|---|
| PVC | 67.52 | 57.19 |
| PE | 14.47 | 11.02 |
| PP | 43.41 | 32.38 |
| PTFE | 14.47 | 11.71 |
| Nylon | 66.83 | 61.32 |
| ABS | 42.72 | 44.79 |

compressive side, that is into the third quadrant. This denotes that a larger range of combinations of stress is possible on the compressive side. The immediate consequence is that the volume change is not constant during plastic deformations. Hence the deviatoric stress components are no longer relevant.

To include the hydrostatic stress in the criterion in the simplest form, we modify the criterion to,

$$AJ_1 + BJ_2 = 1$$

where $A$ and $B$ are material dependent constants. Note that $J_1$ is simply equal to 3 × (hydrostatic stress). When expanded, this equation becomes,

$$A(\sigma_1 + \sigma_2 + \sigma_3) + B\left[(\sigma_1 - \sigma_2)^2 + (\sigma_2 - \sigma_3)^2 + (\sigma_3 - \sigma_1)^2\right] = 1$$

To determine the constants $A$ and $B$, we express the conditions that under uniaxial stress conditions tensile yield must occur when $\sigma_1 = \sigma_{yt}$ and compressive yield must occur when $\sigma_2 = \sigma_{yc}$. These two conditions give:

$$A\sigma_{yt} + 2B\sigma_{yt}^2 = 1$$

$$-A\sigma_{yc} + 2B\sigma_{yc}^2 = 1$$

Solving,

$$A = \frac{\sigma_{yc} - \sigma_{yt}}{\sigma_{yt}\sigma_{yc}} \quad \text{and,} \quad B = \frac{1}{2\sigma_{yt}\sigma_{yc}} \tag{3.18}$$

Substituting, the modified Von Mises criterion becomes

$$2(\sigma_{yc} - \sigma_{yt})(\sigma_1 + \sigma_2 + \sigma_3) + \left[(\sigma_1 - \sigma_2)^2 + (\sigma_2 - \sigma_3)^2 + (\sigma_3 - \sigma_1)^2\right] = 2\sigma_{yt}\sigma_{yc} \tag{3.19}$$

**Example 3.6**

An injection-molding grade of polycarbonate has yield strengths of 68.9 MPa (10000 psi) in tension and 79.2 MPa (11500 psi) in compression. Write the Von Mises criterion in three dimensions for this material.

*Solution* Different strengths in tension and compression indicate sensitivity of yielding to the hydrostatic stress component. Hence the Von Mises criterion is written as

$$AJ_1 + BJ_2 = 1,$$

where,
$$A = \frac{\sigma_{yc} - \sigma_{yt}}{\sigma_{yc}\sigma_{yt}} = \frac{79.2 - 68.9}{79.2 \times 68.9} = 1.887 \times 10^{-3}$$

$$B = \frac{1}{2 \times 79.2 \times 68.9} = 9.163 \times 10^{-5}$$

Using these results the yield criterion can be written in expanded form as

$$20.6(\sigma_1 + \sigma_2 + \sigma_3) + \left[(\sigma_1 - \sigma_2)^2 + (\sigma_2 - \sigma_3)^2 + (\sigma_3 - \sigma_1)^2\right] = 10914$$

For two dimensions, this equation reduces to

$$10.3(\sigma_1 + \sigma_2) + (\sigma_1^2 - \sigma_1\sigma_2 + \sigma_2^2) = 5456.9$$

It is easily verified that in one dimension, the resulting quadratic has the roots $\sigma_{yt}$ and $\sigma_{yc}$.

### 3.2.6 Tresca Yield Criterion

Another form of the yield criterion that is easier to use in calculations is known as the *Tresca yield criterion*. It is based on the argument that the maximum shear stresses, rather than the shear strain energy, are responsible for producing distortions (without change in volume). Hence any one of these shear components exceeding the yield stress of the material would produce yield. Since the maximum shear stress is equal to one half the difference between the principal stresses (recall Mohr's circle), the Tresca criterion can be written as

$$\max(|\sigma_1 - \sigma_2|, |\sigma_2 - \sigma_3|, |\sigma_3 - \sigma_1|) \geq \sigma_y$$

An advantage of the Tresca criterion is its simplicity and the ease with which it can be used in the post-yield stress analysis of *elastic-perfectly plastic* materials, a material model that describes most ductile polymers. The following examples illustrate the use of the Tresca criterion.

**Example 3.7**

Prove that the use of the Tresca criterion for pressured thin cylinders is equivalent to using the outer diameter in the calculation of hoop stresses.

*Solution* Yielding occurs first on the inner surface, where the criterion $|\sigma_h - (-p)|$ is the highest.

$$|\sigma_h - (-p)| = \frac{pd_i}{2t} + p = \frac{p(d_i + 2t)}{2t}$$

$$= \frac{pd_0}{2t}$$

***Commentary***
*Use of outer diameter for hoop stress calculation is considered as a "conservative" estimate of stresses. It is not quite so, because it is just the realistic estimate of the yield criterion.*

**Example 3.8**

A pipe with ends capped is subjected to a pressure $P$, which is steadily being increased. The pipe has an outer radius of $a$, and an inner radius of $b$. Assume that $P$ has reached a value at which the plasticity boundary has reached a radius $\rho$. Assume also that the material of the pipe is elastic perfectly-plastic, and has a yield strength of $\sigma_y$. Find the stresses in the pipe in the plastic, and in the elastic regions. Use the Tresca criterion of yielding.

*Solution* The equation of equilibrium of a cylindrical pipe in polar coordinates is

$$\frac{d\sigma_r}{dr} - \frac{\sigma_h - \sigma_r}{r} = 0 \tag{3.20}$$

Since inside plastic region and on its boundary the Tresca criterion is satisfied, we may write

$$\sigma_h - \sigma_r = \sigma_y$$

Using this in Equation 3.20, we get

$$\frac{d\sigma_r}{dr} - \frac{\sigma_y}{r} = 0,$$

which is solved as

$$\sigma_r = A + \sigma_y \ln r \tag{3.21}$$

$$\sigma_h = A + \sigma_y(1 + \ln r) \tag{3.22}$$

In order to determine the constant of integration $A$, we express the condition that at the boundary of plasticity, that is at radius $\rho$, elasticity solutions apply. Let us assume that the radial stress $\sigma_r$ at this radius is $\bar{P}$. Using thick cylinder solutions,

$$\text{Tresca Criterion} \equiv (\sigma_h - \sigma_r) = \sigma_y$$

$$\sigma_y = \left.\frac{\bar{P}\rho^2(a^2 + r^2)}{r^2(a^2 - b^2)}\right|_{r=\rho} + \left.\frac{\bar{P}\rho^2(a^2 - r^2)}{r^2(a^2 - \rho^2)}\right|_{r=\rho}$$

$$= \bar{P}\left(\frac{2a^2}{a^2 - \rho^2}\right)$$

$$\bar{P} = \frac{\sigma_y(a^2 - \rho^2)}{2a^2}$$

We use the condition that at $r = \rho$, the radial pressure is $\bar{P}$. That is,

$$A + \sigma_y \ln \rho = \bar{P} = \frac{\sigma_y(a^2 - \rho^2)}{2a^2}$$

Or,

$$A = \sigma_y\left[\frac{1}{2} - \frac{\rho^2}{2a^2} - \ln \rho\right]$$

Substituting in Equations 3.21 and 3.22, we obtain the stresses, for the region $b \leq r \leq \rho$.

$$\sigma_r = \frac{\sigma_y}{2}\left[1 - \frac{\rho^2}{a^2} + 2\ln\frac{r}{\rho}\right] \tag{3.23}$$

$$\sigma_h = \sigma_r + \sigma_y = \frac{\sigma_y}{2}\left[3 - \frac{\rho^2}{a^2} + 2\ln\frac{r}{\rho}\right] \tag{3.24}$$

$$\sigma_a = \frac{\bar{P}b^2}{a^2 - b^2} \tag{3.25}$$

The last equation above for axial stress is simply obtained from elasticity, since no help is available from the Tresca criterion. In the region $\rho \leq r \leq a$, elasticity solutions prevail. Therefore,

$$\sigma_r = -\sigma_y \frac{\rho^2}{r^2} \frac{a^2 - r^2}{2a^2} \tag{3.26}$$

$$\sigma_h = \sigma_y \frac{a^2 - \rho^2}{2a^2} \left[ 1 + \frac{\rho^2 (a^2 - r^2)}{r^2 (a^2 - \rho^2)} \right] \tag{3.27}$$

Axial stress is the same as in the plastic region. However, so far we have not related the applied internal pressure $P$ and the extent of penetration of plasticity, namely the radius $\rho$. This is simply accomplished by requiring that Equation 3.23 gives $-P$ at the inner radius. Hence,

$$P = \frac{\sigma_y}{2} \left[ \ln \frac{b}{\rho} + \frac{\rho^2}{a^2} - 1 \right] \tag{3.28}$$

This equation in $\rho$ can be solved by trial and error.

An actual calculation procedure must start by solving for $\rho$ from equation 3.28 and substituting for $\rho$ in the other equations for stress. A plot of $\sigma_r$ and $\sigma_h$ are shown qualitatively in Figure 3.6.

***Commentary***

*Note that because of the absence of the axial stress in the entire problem, the Tresca criterion is unable to distinguish between a pipe that is end-capped and an open ended one.*

## 3.3 POST-YIELD STRESS-STRAIN RELATIONSHIP

When yielding is under progress, the stress and strain do not have a one-to-one relationship as they do in elastic behavior. Rather, they are history dependent. Levy and Von Mises independently proposed that instead of the *total* strain, the *incremental* strain can be related to the deviatoric stress. Their analysis is restricted to materials with no Bauschinger effects and assumes that *the increment in strain and deviatoric stress have the same principal directions, and are proportional to each other.* That is,

$$\frac{de_1}{\sigma_1'} = \frac{de_2}{\sigma_2'} = \frac{de_3}{\sigma_3'} = d\lambda$$

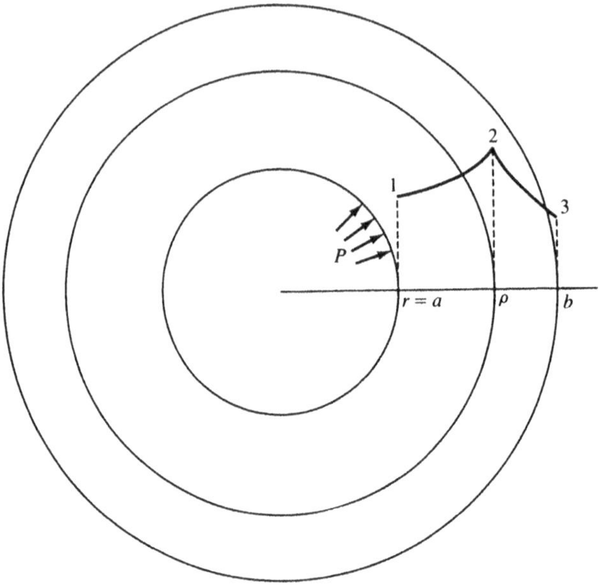


**FIGURE 3.6.** Hoop stress distribution in a partially yielded pipe. The curve 1-2 denotes the yielded portion of the wall, and the curve 2-3 the elastic part. Note that the stress in the plastic zone increases logarithmically with the radius. The radial stress has a similar distribution.

where $d\lambda$ is a parameter dependent on the state of plastic strain, and is not a material property. It can be shown that $d\lambda$ is directly related to the local slope of the one dimensional stress strain curve for the material. Being a rule of incremental computation, the Levy-Mises equation is not amenable to hand calculations.

## 3.4 CRAZING

Crazing is a time dependent failure mode of glassy thermoplastics and some semicrystalline polymers. Typically, crazes are wedge-shaped zones in the material where extremely oriented fibrils are formed. The fibrils connect two bulk surfaces, and hence the material is continuous unlike in a crack. Figure 3.7 is a schematic of a craze. Craze is a porous material, with volume fractions as low as 50%. Due to their orientation, the fibrils may sustain a stress as high as 200 MPa, (30,000 psi). Depending upon the conditions prevailing in the neighborhood of a craze, the fibrils may draw

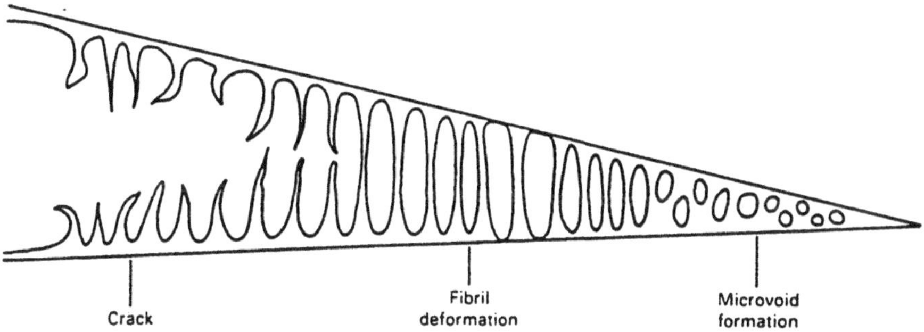


**FIGURE 3.7.** Formation of fibrils in a craze.

material from the bulk and extend, or they may fracture, which is perceived as cracking of the material. Fracture of fibrils makes the two bulk surfaces discontinuous and this is the mechanism of crack initiation in plastics. More of this is discussed later in Chapter 6 on fracture mechanics.

In structures, crazing can therefore be considered as the *beginning* of a failure just like yielding. Residual stresses, molded-in stresses, or thermal stresses can cause crazing. In transparent plastics, such as aircraft windows, crazing can reduce visibility. One can therefore expect crazing in peak stress areas (or stress concentrations). In this respect, quantitative identification of the occurrence of crazing is important. The factor of safety on crazing can, however, be low, theoretically unity. Unfortunately, a full three dimensional description of crazing is not available yet. In the plane normal to the plane of a craze, however, a description is available, mainly in the work of Sternstein et al. [5]. Plane stress conditions are assumed. If $\sigma_1$ and $\sigma_2$ are the principal stresses in that plane, then craze occurs if

$$|\sigma_1 - \sigma_2| = A - \frac{B}{\sigma_1 + \sigma_2} \tag{3.29}$$

where $A$ and $B$ are material properties which are functions of temperature and time.

In a two dimensional stress plot the craze envelope may be superimposed on the Von Mises envelope. See Figure 3.8. It can be shown that the $\sigma_1 + \sigma_2 = 0$ line is an asymptote to the craze curve. Thus, below this line crazes cannot occur.

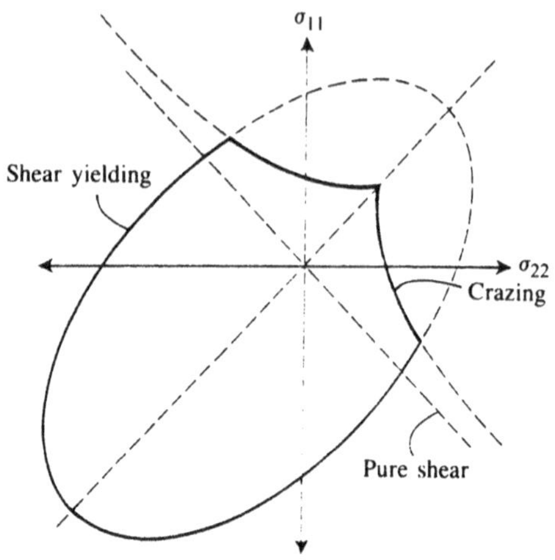


**FIGURE 3.8.** Effect of stress combinations on crazing and yielding.

**References**

1. *DuPont Delrin Acetal Resin, Design Handbook*, E. I. DuPont de Nemours and Co., Wilmington, DE.
2. *Zytel Nylon Resin, Design Handbook*, E. I. DuPont de Nemours and Co., Wilmington, DE.
3. Young, W. C., *Roark's Formulas for Stress and Strain*, 6th Ed. 1989, McGraw Hill Publishing Company, New York, NY.
4. Sternstein, S. S., and Ongching, L., American Chemical Society, Polymer Preprints, 10, 1117, 1969.

**Exercises**

**3.1.** Write the expression for the invariants $J_1$, $J_2$, and $J_3$ for the two dimensional stress state. By using the stress transformation rules of Chapter 1, verify explicitly that they are in fact constant for all directions.

**3.2.** Show that the two stress states

$$\begin{bmatrix} \sigma_1 & 0 \\ 0 & \sigma_2 \end{bmatrix} \quad \text{and} \quad \begin{bmatrix} \sigma_1 + c & 0 \\ 0 & \sigma_2 + c \end{bmatrix}$$

lead to the same $J_2'$. Write an analogous statement for three dimensions.

**3.3.** Substitute $\sigma_1 = \sigma_2 = \sigma_3 = \sigma_y$ in Equation 3.15. You see that yielding does not occur. But if you substitute $\sigma_1 = \sigma_2 = \sigma_y$ in Equation 3.17, yielding does occur. Explain. Prove, therefore, that a spherical vessel subjected to a pressure that causes the membrane stress to reach yield strength would in fact yield, whereas a chunk of plastic immersed in a liquid which has a pressure equal to the $\sigma_y$, would not yield.

**3.4.** Prove that the Tresca criterion of yielding, when applied to pressurized pipes, is effectively equivalent to using the outer diameter in the calculation of hoop stress as opposed to the inner diameter.

**3.5.** Example 3.3 can be redone in three dimensions with additional rigor. Use the thick shell formulas for stress and show that the yield pressure is less than 0.793 MPa (115 psi) as arrived at in this example.

**3.6.** An injection molding grade of polycarbonate has a yield strength of 69 MPa in tension and 79 MPa in compression. A thin section 1.5 mm thick, 10 mm wide, and 30 mm long, made of this material is used as a snap-on closure in a dot-matrix printer. What is the highest tip deflection this can take before yielding?

**3.7.** A rectangular cross section 5 mm wide and 15 mm deep is subjected to a bending moment of 110 Nm. If the material is PVC with tensile yield strength of 57 MPa and the compressive yield of 68 MPa, show that the cross section undergoes yielding. Calculate and sketch the distribution of bending stress in the cross section.

**3.8.** In the above problem, if the moment is removed, find the residual stress. Neglect time effects.

**3.9.** Outline a procedure for finding the maximum load per area that a rectangular plate can take without yielding if its strengths are different in tension and compression.

**3.10.** For materials listed in Table 3.1, calculate the constants $A$ and $B$ and write the modified Von Mises criteria.

**3.11.** A PVC pipe has an inner diameter of 125 mm and a thickness of 20 mm. The material has a yield strength of 57 MPa in tension and 68 MPa in compression. If the ends are closed and the pipe pressurized, what pressure would just cause yielding?

# 4

# Rationale of Stress Analysis

## 4.1 DESIGN BY ANALYSIS

One of the evolutionary changes in the plastics industry is the shift from the concept of "design-by-rules" to one of "design-by-analysis." Design-by-rules is the method of design based on prior experience, rather than by structural mechanics. In design-by-analysis, stress analysis plays a central role [1]. It is not a new concept in mechanical engineering, since such a concept arose in railroad engineering, and then was adopted in automotive, and aircraft industries, and later in the nuclear power industry with higher levels of sophistication. In its simplest form, the concept emphasizes the thorough understanding of the in-service behavior of a product well before its manufacture. Simply, the design-and-analysis technique is the most economical way to simulate the working of the product, before its blueprints are made. The concept of design-by-analysis is already in vogue in mature industries. With the increasing use of plastics for structural functions, the relevance of design-by-analysis is evident. Thanks to powerful personal computers, the engineer is able to perform several tasks of design-by-analysis at a reasonable cost. Typically, the geometry of a part is created with CAD software and is data-based in a format that can be ported to many different software modules. For example, it can be used as input for a mold flow analysis, a mold design program, a material estimation program, a drafting program, or a finite element analysis program. Until the late 1980s, such facilities were out of the reach of any but the very largest industries.

There is, at the present time, a wealth of accumulated design experience in the plastics industry, and mature computational techniques from metals are available. However, solutions to problems unique to the plastics, such as, for example, the mold design, remain substantially in their formative stages. Nonetheless, the motivating force in the industry at this time is the transition to the design-by-analysis concept. This concept enables engineers to analyze every known aspect of a product design.

## 4.2 OBJECTIVES OF STRESS ANALYSIS

The objectives of stress analysis are:

- to ensure satisfactory structural performance of the component under all known and foreseeable operating conditions, and
- to ensure, a priori, that the component carries a reasonable margin of safety commensurate with the frequency of loading, category of loading, and the severity of consequences of a failure.

The meaning of the term *structural performance*, of course, varies with context. It may also vary with the type of product, and the type of loading. In one case, for example, it may mean adequate long term strength and long term stiffness. In another case, it may mean accumulated long term strains. If repetitive loads are anticipated, then it may mean freedom from fatigue, crazing, excessive local yielding, catastrophic crack growth, etc. Likewise, under loads of higher frequency, vibration damping characteristics assume importance in the design. In fact, for a real product, it means a combination of such criteria.

Incidentally, we have also expanded on the significance of the word *failure*. In the early years of metals design, only one form of failure was recognized and used for design. That was yielding. However, in its modern use, the word simply means lack of structural performance of any type. In its extended sense, failure of a part may be attributed to:

- material behavior under stress, and
- geometric parameters of the product, such as areas of cross section, moment of inertia, radius of curvature, etc.

Examples of the former are yielding, fatigue, crazing, cracking, creep, creep rupture, etc. In these types of failures, the stress at a point tells the whole story. An engineer will only need to compare the stress at that point with an appropriate material property. The way stress is distributed away from that point is not relevant to the context.

Examples of the latter failures are buckling, excessive deflections of the component, and vibration. In these types of failures, a knowledge of the local stresses is not sufficient in the assessment of failure. Rather, the geometry of the structure, the methods used for joining, etc. become additional factors. In any case, the structural performance requirements and hence the failure modes of a component become clear, when its service objectives are well defined.

### 4.2.1 Uncertainties in Design and Analysis

Quite often, the question is asked, "Does the analysis correspond to the real life?" The honest answer is "It may." Uncertainties arise in the design and analysis process because of many reasons beyond the control of the engineer. Consequently, there is never an exact correspondence between analysis and real life. However, this does not diminish in any manner, the utility and relevance of stress analysis. Some of the uncertainties confronting the plastics engineer are given below.

*Variability of material properties:* There is a statistical variation of properties of "identical materials" tested under "identical conditions."

*Influence of testing methods on the material properties:* The test methods have a considerable effect on the measured properties. Unlike the situation in the metals, plastics test specimens are quite often molded, not machined out of a stock. This gives rise to orientation in the specimen, which is no longer representative of the material. Further, given the fact that the specimens may have oriented properties, the test results are influenced by the way specimen is gripped in the testing machine, the speed of testing, etc. Keeping track of the strain during a tension test can be tricky too. Suppose that a strain gage is used for this purpose. (This is hardly ever done!) The modulus of the plastic material being two orders of magnitude less than the metal of the strain gage foil, there is a local stiffening of the specimen because of the bonding of the gage, and strains measured by the gage are lower than elsewhere. The gage affects the measurement!

*Influence of environmental parameters:* Environmental variables such as temperature, humidity, etc., affect the material property data. In the first place, these parameters may go unrecognized as relevant variables during the test. Secondly, even if recognized, and documented, they cannot be factored easily into the stress calculations of an actual product.

*Operating loads not well defined:* Design conditions for components are not always well defined. Components whose design is governed by mandates and national codes tend to have clearly defined loading conditions. An automobile bumper, or a football helmet are examples of components covered by standards and codes. [2, 3, 4] Hence, their design objectives are very clear. However, more utilitarian plastic components often have no well defined set of operating conditions. Hence, in the latter case, the design conditions and real life loading may not match at all. A plastic picnic chair is an example of this.

*Exact material behavior may be unknown:* It is very likely that a material is used in a design on the basis of extrapolated long term properties. This happens because of the worldly priorities in a production environment. In real life, the component may behave differently than predicted.

*Material behavior known only qualitatively:* It is well known that, for all plastics, the creep strain is a function of stress, time, and temperature. There is practically no data available for any plastic giving creep strain *as a function of all three parameters*. In this light, consider a microwaveable container with a tight lid. Although this is not a perfect example, it is simple enough to drive home the point. Let us suppose that the objective is to determine the number of heating and cooling cycles that the container will withstand. The container sees thermal cycling, and the tightness depends on the temperature. The tightness of the lid is lowered at higher temperatures, because of the drop in modulus with *temperature as well as with time*. In the cooling cycle, two things happen with essentially opposite effects. First, because of the temperature drop, there is an increase of modulus, and the tightness increases. Second, because of the passage of time, there is a decrease of tightness because of relaxation. The combined effect of the two factors can be assessed only if the temperature is incorporated into the expressions for creep strain. Presently, creep strain is available for most materials *at a temperature*, and not for variable temperatures.

*Limitations in stress analysis techniques:* Stress analysis techniques have their own limitations too. In the example of the microwaveable container above, suppose that a description of the creep behavior was made available in the form

$$\epsilon_c = A\sigma^m t^n f(T).$$

It does *not* necessarily mean that the behavior of the container over a thermal cycle can therefore be calculated accurately. Computational techniques are behind the corresponding mathematical techniques, and are not yet commercially available for common isothermal viscoelastic problems, not to mention the variations in temperature.

A similar comment applies also for post-buckling analysis, post-yield fracture mechanics, etc., all of which require sophisticated programs and competent personnel.

Many plastics components, such as a part flexing and contacting another part, depend, for their function, on the low modulus. The complexity of this problem lies in the fact that the boundary condition, with which the problem was started, changes abruptly because of contact, and also because of the continuous increase in the contact zone. Such a problem can be solved only by a few sophisticated finite element software packages.

Another area where computational techniques are limited is the so-called "coupled-problem."

> A coupled problem is one that combines more than one discipline of engineering science for its formulation and solution.

Suppose that we perform a moldfill analysis for an injection molding process. We know that viscosity opposes flow, flow causes heat from friction work, thus raising the temperature. The heat originally in the melt, together with the heat generated by friction-work have two paths of flow, one to lose heat to the metallic mold material, and the other to distribute within the melt by conduction and convection. The local temperature of the melt in the mold will depend on both these factors. At the present time, there is no calculation tool that can go into details of such second order effects.

Consider another example of calculating the shrinkage stresses. Shrinkage stress builds up in a mold while cooling is taking place. But, while still in the liquid stage, there is practically no modulus for the material, and hence no stress can develop. The question of solving for the shrinkage stress involves, in the least, modulus as a function of temperature and a temperature at which we may treat the material to become a solid, and hence develop stress. Evidently, only approximate solutions are currently possible.

*Structural behavior may be unknown:* In general, in the metals, a design against yielding is also generally good against buckling and vibration. In the case of plastics, this is far from true. Table 4.1 gives a comparison of the ratio $E/\sigma_y$ for several metallic and plastic materials. This ratio is representative of the stiffness a part will have if it was designed to just resist yielding. The metals have a high ratio, and hence, in most cases, designing for stress is adequate for stiffness also. However, the low modulus of elasticity of plastics leads to low structural stiffness and consequently to various forms of remedial measures in the component design such as ribs, stiffeners, etc. In terms of their geometry there may be

**Table 4.1. Why Design for Strength in Plastics is Not Enough Protection Against Buckling**

| Material | E (psi) | $\sigma_y$ (psi) | $\frac{E}{\sigma_y}$ |
|---|---|---|---|
| METALS: | | | |
| Steel (structural) | 30,000,000 | 36,000 | 833 |
| HSLA Steel | 30,000,000 | 50,000 | 600 |
| Aluminum | 10,400,000 | 47,000 | 220 |
| UNREINFORCED PLASTICS: | | | |
| Polyvinyl Chloride | 65,000 | 4,200 | 16 |
| Polycarbonate | 1,200,000 | 8,000 | 150 |
| Polybutylene | 35,000 | 2,000 | 18 |
| ABS | 350,000 | 5,000 | 70 |
| Polyester | 350,000 | 8,500 | 41 |
| REINFORCED PLASTICS: | | | |
| Filled Polyester (35% glass, mica) | 1,200,000 | 12,500 | 96 |
| Filled Polyethylene (30% glass) | 850,000 | 7,500 | 113 |
| Carbon Epoxy T 300/5208 | 21,300,000 | 213,000 | 100 |
| S-Glass Epoxy | 6,260,000 | 185,000 | 34 |

Based on References 6, 7, and 8.

no precedents existing in the metals. Hence, the exact response of the structure to applied loads cannot be easily understood and analyzed.

*Secondary structural behavior:* By this term we mean that there are types of structural responses that are generally neglected in the metallics design but which need to be considered in the design plastics. To cite an example, we consider creep buckling.

Consider a plastic column under compressive axial load. Because of the unavoidable eccentricity of loading, there is initially a small bending stress. Over a period of time, the eccentricity grows because of creep. The primary effect is an increase in the bending stress. The secondary effect arises because of the combination of axial load and bending moment which may become critical for certain eccentricity, and may cause buckling over time. This time-dependent buckling, caused by creep, is called creep buckling. Note that the secondary behavior arises from the initial eccentricity that is small and is usually ignored.

*Limitations in modeling a problem:* Invariably, a problem is analyzed by isolating the problem domain from its surroundings, and substituting the surroundings by the forces or displacements they impose on the problem domain. For example, a small area of contact force is taken to be equivalent to a uniformly distributed pressure; a few points on the structure are arbitrarily taken to be held "fixed" or "simply supported." While such assumptions are needed for simplification and to limit the problem size, they do not represent the real conditions. Rather, they represent a tendency to err on the safe side.

*All possible failure modes may not be considered:* Such cases arise because of ignorance of what the possible modes of failure are for a given product and a given set of loadings. This is the most likely omission in the case of new products.

*Not associating the stress category with a failure mode:* Stresses belong in well defined categories, and each category is associated with a particular type of failure of the material. Also, the word "failure" denotes many different events. *Thus, it is not always true that the greatest stress is all that counts.* Readers with some experience in design will readily observe that there are numerous points in a part where the design stress is exceeded because of stress concentration effects, etc., yet such excessive stresses do not result in the failure of the part. Clearly, failure is related to the stress and its distributions in a more complex way. Such points are discussed later in this chapter.

## 4.3 FACTOR OF SAFETY (FOS)

The preceding discussions make it clear that the safe working of a product is a realistic objective in design, although absolute accuracy in analysis is not. Since the engineer cannot afford to consider and allow for each of the uncertainties as listed, an escape is provided in the concept of "factor of safety." The factor of safety (FOS) is supposed to scale down the working stress, and the terminology carries with it an expression of conservatism and fear. It is truly meant to be so. *Factor of safety is a factor of ignorance!* The objective of all analyses is to compare a calculated quantity with an allowable material property, and to ensure that the two quantities will be separated by a certain factor. This factor of separation is the factor of safety. It may be applied to any design quantity, such as the tensile strength of the material, yield stress, number of stress cycles, the required natural frequency of a structure, the critical buckling load for a structure, an allowable initial defect size, etc.

### 4.3.1 What Size Should the FOS Be?

In the context of "design-by-analysis," there is no unique answer to the question, "What size should the FOS be?" Design rules and material manufacturers' guidelines usually give a single number, one as large as six to ten, to be applied on the short term ultimate strength of the material. Such an approach is not satisfactory, because it gives no indication about how much margin is available on other modes of failure. Many modes of failure depend on the knowledge of other independent material properties, and hence a single factor of safety simply cannot provide an omnibus protection against all modes of failure. *Hence, the design must require different FOSs for different modes of failure.*[1] A list of recommended factors of safety for plastics are given in Table 4.2, and their rationale are discussed in the following.

***Short Term Strength***

Consider, for example, the selection of a short term design stress for a material for which the yield stress and the tensile strength are given. (It must be noted that the yield point for thermoplastics is not clearly identifiable in simple tests. There is, however, a point on the stress-strain curve at which the reversibility of strain ceases. One may use the offset method described in ASTM D-790, to determine the yield point of the material.) Since yield is the earliest mode of failure to occur in any case, we may apply an FOS of 1.5 on $\sigma_y$. The ultimate tensile strength represents a catastrophic failure, and we may therefore require a higher FOS than is applicable for yield strength. A value of 3.0 on $\sigma_U$ is typical. Obviously, the design stress is the lower of the two. The numbers 3.0 and 1.5 are based on the time-tested steel structural design standards [9].

$$\text{Design Stress} = \frac{\sigma_y}{1.5} \text{ or } \frac{\sigma_U}{3.0} \text{ whichever is lower.}$$

[1] The concept of graded FOS is reflected in the current standards for underground glass reinforced plastic pipelines AWWA C-950-81. [5].

**Table 4.2. Recommended Factors of Safety**

| | Criterion | Recommended FOS |
|---|---|---|
| a. | Short term strength | 1.5 on $\sigma_y$ *and* 3 to 4 on $\sigma_u$. |
| b. | Stiffness for vibration | 2.5 to 3 on forcing frequency. |
| c. | Thermal stress (sustained) | 1.5 on $\sigma_y$. |
| d. | Thermal stress (intermittent) | Accumulated strain not to exceed 0.5% or viscoelastic limit. |
| e. | Fatigue | 10 on number of cycles or, 2 on the stress amplitude. |
| f. | Creep | Design stress is the smallest of $\frac{\sigma_\epsilon}{FOS_\epsilon}, \frac{\sigma_{cr}}{FOS_{cr}}, \frac{\sigma_R}{FOS_R}$. (See also text.) |
| g. | Nonlinear elastic behavior | Same as (a) above, however, scaled up by the ratio $\frac{E_t}{E_{sec}}$. |
| h. | Buckling | 2 to 2.5 to avoid buckling. Perform post-buckling analysis to design for functional buckling. |

A vast majority of thermoplastics have $\sigma_y$ and $\sigma_U$ so close to one another that ASTM D-790 allows $\sigma_y = \sigma_U$ numerically. However, from the material point of view, these two are different.

***Stiffness***

Imagine the case of a design for stiffness, where a certain component is assembled to be in contact with a vibrating mechanism. The natural frequency of the component should be higher than the forcing frequency. How much higher? In order to avoid resonance, it is an established practice, to require a factor of separation of 2.5 or even 3 from the forcing frequency. This ensures that the start-up and shutdown of the forcing mechanism will not lead to resonance. Higher frequency components of vibration do exist both in the force and in the structure. Experience with structural dynamic analysis shows that several higher modes of natural vibrations have frequencies that are quite close to each other, or may even be coincident. Some may be close to the forcing frequency. They do not, in general, pose a serious problem from the stress point of view, because of the inherently high damping properties of plastics. *Note that if long term stiffness is a criterion, then the long term modulus must be used in the calculation of stiffness.*

***Thermal Stress***

Thermal loads on a structure may be sustained or intermittent. For sustained thermal loads, the response of the structure is one of continuous relaxation. Hence, the initial stress at time $t = 0$ is the highest. Hence, *as far as stress is concerned*, a low FOS on short term yield should be adequate. Ideally, an FOS = 1 on $\sigma_y$ is enough. However, since thermal stress may never occur in isolation, an FOS = 1.5 on $\sigma_y$ appears pragmatic.

For intermittent thermal loads, the picture is quite different. Local strains may also have to be considered as an additional control parameter, besides stress, for which the above FOS applies anyway.

The response of the structure to intermittent thermal loads is represented by a number of cycles of stress, relaxation, and recovery, as shown in Figure 4.1. The thermal cycle may act at intervals long enough to allow complete recovery of the initial conditions. Or, the intervals may not be long enough, with the result that at the end of each load cycle, the component has a residual stress and strain, which accumulate over cycles.

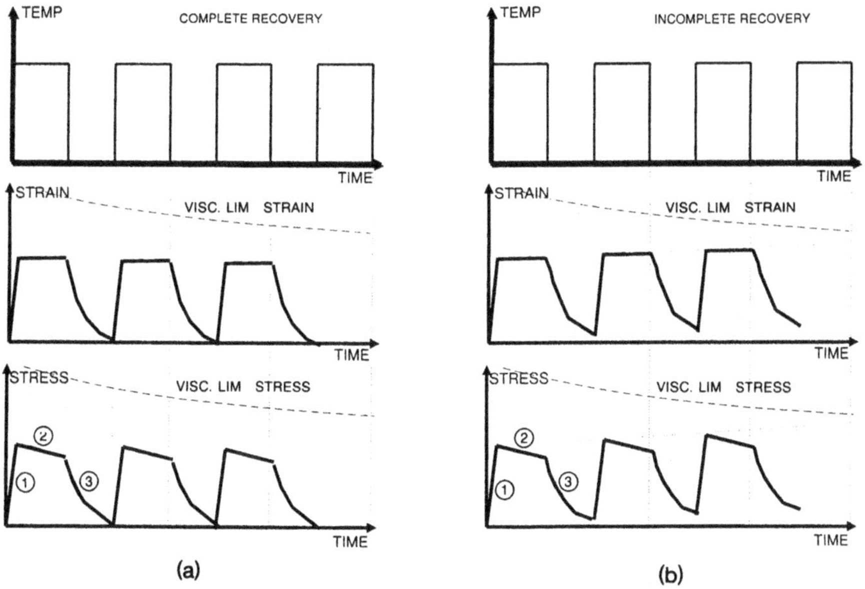


**FIGURE 4.1.** Viscoelastic response to intermittent thermal stress. (a) Long intervals. (b) Short intervals. (1) Creep. (2) Relaxation. (3) Recovery.

It is appropriate to take the strain at yield $\epsilon_y$, as obtained from simple static tension tests, as the limit for accumulated thermal strain. In any case, the engineer has to refer to the material manufacturer's recommendations for an appropriate strain limit. Given enough time, all materials begin to exhibit some nonlinearity of viscoelastic behavior. That is, the stress-strain relation ceases to be exactly linear. One may use this knowledge to set up a limit on the strain. Such a limit, called the "viscoelastic strain limit," is the state of stress and strain at which the linear viscoelastic calculations lose their accuracy, and are no longer reliable. Note that the viscoelastic strain limit is not a material failure, it is only a breakdown of the accuracy of the calculations. The viscoelastic limit decreases slowly with time. See Figure 4.1.

Thus, for intermittent thermal loads, an arbitrary strain of $\epsilon_y$, or the viscoelastic strain limit, whichever is lower, applies.

***Fatigue Loading***

Once again, fatigue loading presents a complex situation, and there is no single criterion to assign a factor of safety, since it could be applied on the number of cycles ($N$) or on the allowable stress amplitude ($\sigma_a$) for a given number of cycles. The important point, however, is that fatigue data tend to have a high spread, and it is advisable to take a factor of safety of 10 on $N$, or 2 on $\sigma_a$. Additional consideration must be given to the conditions under which the component is operating, such as mean stress, environment, etc.

***A Situation of Creep***

In situations of creep, three events must be considered in the design, namely (1) excessive strain, (2) crazing, and (3) creep rupture. (Note that these occurrences are local to a point, and can be related to the stress at that point.) Protection against each of the three events may be provided either by limiting the stress or the strain.

First, one may prescribe a limiting strain value of 0.5 to 5.0%, or $\epsilon_y$ at the end of the life of the product, depending upon the material used. The stress $\sigma_\epsilon$ corresponding to the limiting strain is selected, and an FOS is applied on that stress. Denote this FOS as $\text{FOS}_\epsilon$.

Second, one may find the stress $\sigma_{cr}$ that would initiate crazing at the end of the life of the product, and apply an FOS on the craze-stress. Denote this FOS as $\text{FOS}_{cr}$.

Third, one may find the stress $\sigma_R$ that would produce rupture at the end of the life of the product, and apply a FOS on the rupture-stress. Denote this FOS as $\text{FOS}_R$.

Obviously, the design stress is the smallest of the three. Hence,

$$\sigma_{des} = \text{The smallest of } \frac{\sigma_\epsilon}{\text{FOS}_\epsilon}, \frac{\sigma_{cr}}{\text{FOS}_{cr}}, \frac{\sigma_R}{\text{FOS}_R}.$$

We note that the FOS on rupture stress must be high, say 3.0, because it is associated with a catastrophic failure. The other two FOSs are associated with beginnings of failure, and hence *may* be as low as 1. *Choosing a design stress on the above basis does not protect the design from stress concentration effects*. In other words, the design stress prevails at most portions; however, at the SCF areas, some or all of the above criteria may be violated. The sequence of events in a stress concentration area is one of the following two possibilities: (1) the material fails in a ductile manner, in this case, gross deformation occurs first, followed by crazing, crack initiation and crack propagation; (2) the material fails in a brittle manner. An analysis must be performed to check such locations, to determine which event can occur. Comprehensive long term material data must be available for such analysis, in addition to analysis software.

***Nonlinear Elastic Materials***

The question of assigning an FOS for nonlinear elastic materials is once again tricky. Primarily, an FOS must be applied on the allowable strain, because, in the microscopic scale, strain is the fundamental mechanism producing failure.[2] But that requires a strain analysis rather than a stress analysis. Despite the fact that both analyses use the same techniques of computation, the two analyses are quite different in their objective. Strain analysis with a nonlinear stress-strain relation can be expensive. The alternative to this is to use the secant modulus, (the ratio of yield stress to yield strain) and perform a regular linear stress analysis. Under such circumstances, the FOS may be applied to the stress. The FOSs discussed above for stresses are applicable in these analyses too. However we need to recognize some important differences arising from nonlinearity.

We need to consider four different cases. (See Figure 4.2).

1. Load controlled problem using tangent modulus
2. Load controlled problem using secant modulus
3. Displacement controlled problem using tangent modulus
4. Displacement controlled problem using secant modulus

[2] Strain is the basic parameter in the failure mechanisms in the microscopic scale for all materials, metals and nonmetals. With metals, however, this is equivalent to describing failure in terms of stress, because of the linear stress-strain relationship.

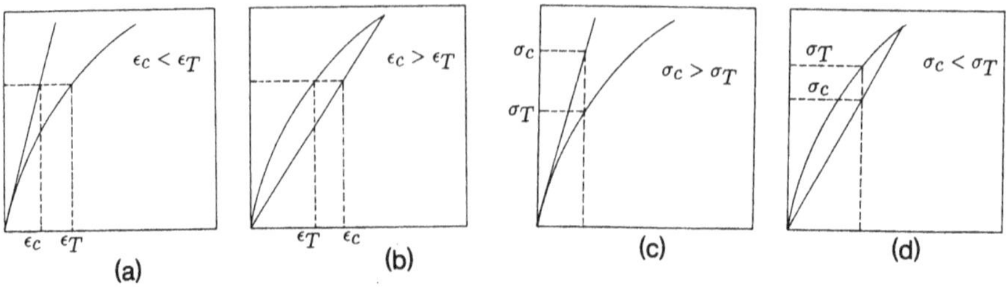


**FIGURE 4.2.** Nonlinear elastic material. (a) Load controlled problem using tangent modulus. (b) Load controlled problem using secant modulus. (c) Displacement controlled problem using tangent modulus. (d) Displacement controlled problem using secant modulus.

In a load controlled problem, case 1, the strain is underestimated by the analysis, as can be seen from Figure 4.2(a). If the secant modulus is used, as in case 2, then the strain is overestimated. See Figure 4.2(b).

Correspondingly, in a displacement controlled problem using tangent modulus, the stress is overestimated, Figure 4.2(c). Lastly, if secant modulus is used then stress is underestimated, as shown in Figure 4.2(d).

If the design criterion (be it stress or strain) is overestimated by the chosen procedure, then the analysis errs on the conservative side. This is acceptable because the calculation works in the same direction of conservatism as the factor of safety itself. If the design criterion is underestimated, then it is prudent to use a modified FOS defined by

$$\text{Corrected FOS} = \left(\frac{\text{Tangent Modulus}}{\text{Secant Modulus}}\right) \text{FOS}$$

The modified FOS is an over-correction of the FOS! The actual stress strain curve lies between the tangent modulus line and the secant modulus line. *However, if the true stress strain curve was used to perform a nonlinear elastic analysis, then there is no need to modify the FOS.*

***Buckling***

This is another tricky situation. Let us begin by explaining some basics of buckling. Buckling is a condition of instability. That is, when buckling is imminent, the curve relating the load to the displacement of the structure is not well behaved at certain loads. There are two kinds of buckling, and there is some kind of similarity between them and the neutral equilibrium and unstable equilibrium of a particle. Figure 4.3 suggests the similarity.

In the case of neutral equilibrium, any new position is a stable position. Euler column buckling, also referred to in the literature as eigenvalue buckling, presents a similar picture. At the critical buckling load, any

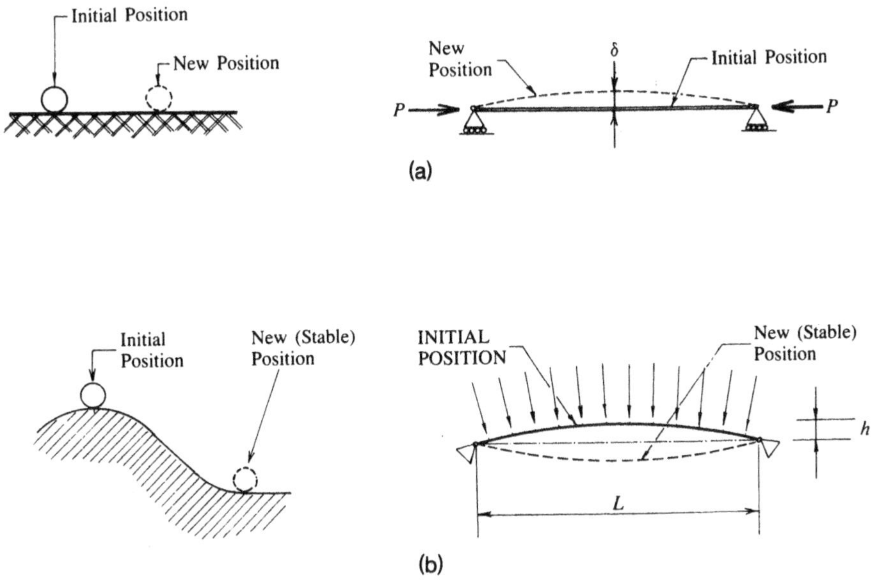


**FIGURE 4.3.** Types of structural instabilities. (a) Eigenvalue buckling; any new position is a stable one. (b) Snap-through buckling, in which stress reverses sign with monotonic increase in $h$. There is a unique new stable position for the applied load. This buckling occurs typically in shallow shells ($h/L \ll 1$).

lateral displacement $\delta$ corresponds to an acceptable configuration of the column.

In the case of unstable equilibrium, the particle may be thought of as being on top of a hill, and any small displacement reduces the potential energy and the final stable configuration is at the foot of the hill. This corresponds to the snapping through of a shallow dish such as in an oil can. There are no intermediate stable configurations. In the process of snapping from the initial to the final configuration, the load-deflection curve for the component reverses trend. That is, a *decrease* in load accompanies an increase in $\delta$, which is equivalent to the shedding of potential energy. Figure 4.4 illustrates the load-deflection curve for a simple two member truss. See References [10] and [11] for a detailed discussion of this aspect of instability.

There are functional designs which utilize the snap-through dish made of reinforced plastic. For example, there are low pressure safety valve designs, in which excess pressure, if any, causes a dish to snap-through and move a small needle valve which, in turn, releases the pressure. It is important that after the pressure is reduced, the dish *must spring back* to

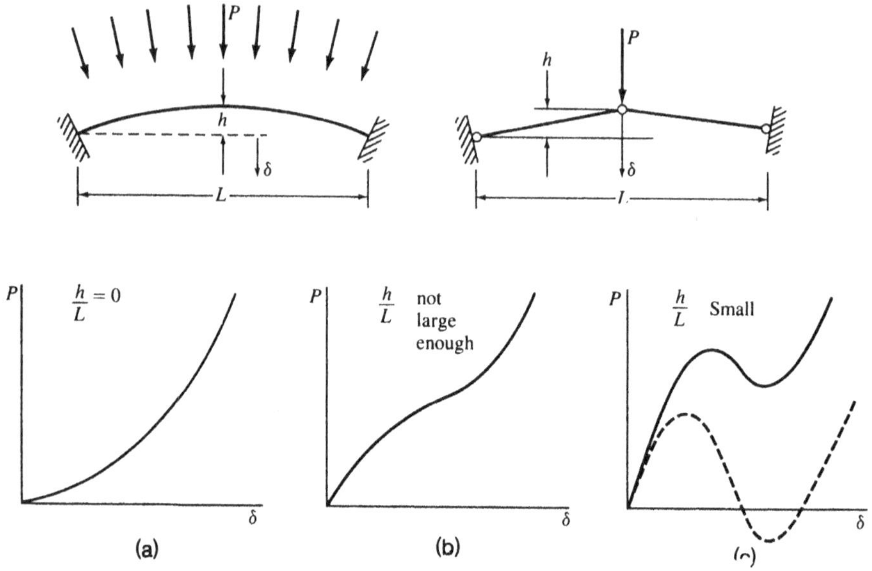


**FIGURE 4.4.** Example of snap-through buckling. Types of behavior of shallow structures. In (a) for flat plates, stress stiffening would cause the structure to exhibit a large increase in load for a small increase in deflection. For structures which are not so deep (b) an inclextion may be observed in the load-deflection curve. For shallow curved structures (c) an actual reversal of trend takes place in the load-deflection curve. Capturing this trend of the curve in a finite element analysis requires a high degree of skill.

its original position. Bottle caps used for vacuum-sealed jelly and jam bottles in grocery stores are another example.

Returning to the discussion of the FOS for buckling, we note that in the case of a snap-through dish or similar components, buckling is a *mode of operation, and not a mode of failure*. Hence, an FOS on these is inappropriate. Rather, a detailed study is needed to ensure that precisely at the designated loads, buckling will take place, and upon removal of loads, the component will return to its original configuration.

However, for the "Euler column" type of buckling (or eigenvalue buckling, as it is called) we need to consider an FOS. Again, eigenvalue buckling has, like vibration, several "modes," the lowest being the overall buckling of the entire structure. An FOS of 2 or 2.5 is appropriate on the overall buckling load. However, this is not all. Local buckling of stiffeners and ribs also needs to be considered. Unfortunately, computational facilities for local buckling are very limited, at least at the personal computer level. Assuming that one is available, an FOS of 2 or 2.5 on local buckling load is once again appropriate. Note that the remedy for local buckling is to reconfigure the area *locally*, rather than to attempt an overall resizing of the part. It is recognized that the FOS recommended here against eigenvalue buckling is rather high. The reason for this is that the eccentricity of loading cannot be factored into the eigenvalue buckling calculations. Once again, if long term modulus is relevant, then the critical buckling load must be calculated using that modulus.

## 4.4 BASIS FOR FACTOR OF SAFETY

From the foregoing, it is clear that the rational assignment of factor of safety should take into account the following.

*Severity of consequences of failure:* The severity levels of the various modes of failure are different. Some are catastrophic, while some others are the *beginnings* of failure. Those that are the beginnings of failure permit scheduled replacement of parts, and therefore, may be assigned a low factor of safety. Those that represent catastrophic failures, must be kept at a higher factor of separation, and hence should be assigned a higher factor of safety. Furthermore, there can be loads which are too severe to design for economically, such as in the case of acts of God. Nevertheless, they are to be listed and their probability of occurrence is to be evaluated. Generally, the very severe combination of loads will occur with very low probability. Hence, it is only rational to scale down the FOS progressively with higher and higher loads, on the grounds of their lower and lower probability levels of occurrence.

Loads of higher severity, that have been properly identified and analyzed, can be associated with lower FOS. Examples of this are product testing. A pressurized container, for instance, is tested at or above 150% of its rated working pressure. Obviously, the test is an overload on the product. The FOS carried by the product during the test is low. The lower FOS must be justified by the low frequency of the test loads.

Exceptions do exist, however. Where human lives may be involved, catastrophic failures must be avoided at any cost. Prior experience with a product may help reduce the FOS in certain cases.

*Accuracy of material property characterization:* It was mentioned that fatigue data are subjected to wide variations, because fatigue is sensitive to microscopic irregularities which are not controllable in an experiment. Since high variance corresponds to low confidence level, the FOS must be high, notwithstanding the fact that fatigue data might represent only the *beginning* of failure.

The engineer has to examine how the data for fatigue properties are presented. A material manufacturer's recommendation is seldom likely to present the raw data points; rather, they represent the lower limit of the data and thus may represent, typically, a 99% confidence level. In such a case, the engineer may deem that the spread of the data has already been corrected for. Hence the factor of safety can be lowered. Or the engineer may have a curve of in-house tests, which might have all of the raw data points marked upon the curve. Sometimes, only raw data points are shown with no curve fitted through them. In such a case, the mean curve carries a low confidence level, and accordingly the factor of safety has to be high, as discussed previously.

*The stress category:* The FOS depends upon whether the stress is one that causes creep, or fatigue, or is one that relaxes with time, whether the problem is load controlled, or displacement controlled. The apparent long term effects of stress will depend on its origin.

## 4.5 INTEGRATION OF STRESS ANALYSIS WITH DESIGN

We have seen that, entirely apart from the techniques of analysis, there is enough to learn about dealing with the results of an analysis. We may say that a rational stress analysis is one that meshes with (1) solid mechanics, for computation techniques, (2) service loads, to determine the origins of stress and categorize them, (3) failure modes, to assign a rational FOS, and (4) the material behavior and material property data, for providing input for stress analysis, and also to provide the design with adequate conservatism. The interrelation is illustrated in Figure 4.5. A comprehensive design and analysis procedure involves:

- Listing all independent service loads,
- Listing all possible combinations on concurrent occurrences of such loads,
- Categorizing the load combinations on the basis of their severity and probability,
- Defining broad guidelines for the extent of failure that can be tolerated under the various types of loading, and
- Sizing, detailing, and stress-analyzing the product to meet the above criteria.

### 4.5.1 Severity Categories of Loads

Components which are part of a more complex system may be loaded in a complicated way because of the operating characteristics of the system

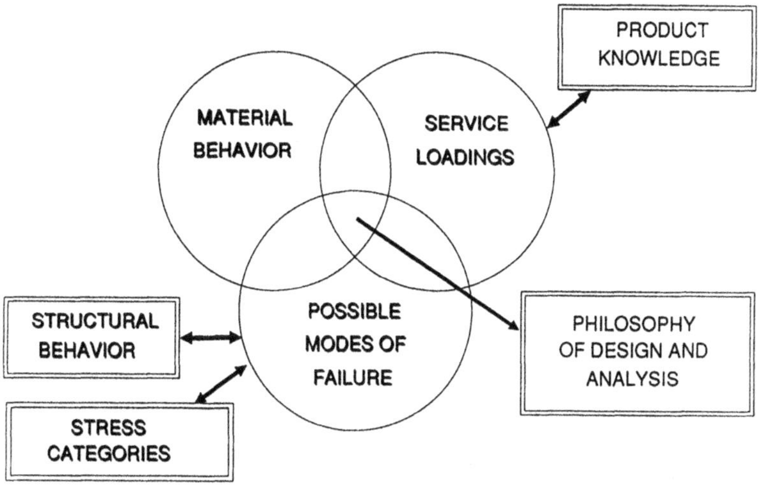


**FIGURE 4.5.** Stress analysis in relation to other aspects of design.

itself. The various independent loads and their combinations can correspond to different levels of severity. An example of an underground pipeline is given in Table 4.3. In this example, the normal loading conditions are very clear to understand. The test condition is an intentional overloading for a short duration. For the test/overload/emergency conditions, the part cannot carry the same level of FOS as for normal operating conditions. However, the FOS for the test/overload/emergency conditions cannot be below unity. Furthermore, these conditions prevail only for short durations, and hence there is no need to consider all modes of failure. Occurrences of such conditions must be few relative to the normal operating conditions. In other words, the designer must ensure that the probability of occurrence of the overload conditions is acceptably small, and will not entail catastrophic failure. If the probability is in fact high, then such loads are better classified under normal operating conditions!

It is not the intent to discuss here how the FOS must be scaled with the severity of loads, nor to classify loads according to their severity. Suffice it to say, that there are conditions under which a limited amount of damage can be, and will be sustained because of low FOS, by the component without adverse effect on the operation of the major system of which it is a part. There are, likewise, conditions under which the component will sustain severe damage, but will permit safe withdrawal from service and replacement. It is entirely up to the system designer to exercise the system knowledge and expertise available, and to classify which loading conditions

**Table 4.3. Severity Levels for a Plastic Pipe for Underground Gas Service.**

| Independent Load Cases | Load Case Name | Possible Combinations | Severity Category |
|---|---|---|---|
| Internal pressure | A | B | Normal |
| Soil overburden | B | A+B | Normal |
| Hydro testing | C | C | Test |
| Installation, handling | D | D | Normal |
| Thermal gradients | E | A+B+E | Normal |
| Flotation | F | A+B+E+F | Emergency |
| Vehicles passing over | G | A+B+E+G | Normal |
| Earthquake | H | A+B+E+F+G+H | Fault |
| Safety valve malfunction | I | A+B+I | Emergency |

*Columns 1 and 2 are simply listings of loads that can be analyzed as independent load cases. However, in reality, loads combine as shown in column 3. Column 4 is a possible way to categorize the severity level of the consequence of failure. For example, there can be no internal pressure, until line is installed, and hence A cannot occur by itself, except with at least B.*

belong to which level of severity. Thus, the component design specifications resulting from the system design must state what operating conditions must be classified under what category. We recall once again that this is easily possible with components covered by standards and codes. With other simpler components with no stringent demands, there may be no need at all to classify normal, test, overload, and other conditions.

## 4.6 STRESS CATEGORIES

The significance of any stress depends on the origin and its spatial distribution. Based on their origin, they can be classified as *primary* and *secondary*. Based on the spatial distribution, one can have an additional category, viz., *peak*.

> Primary stresses are those that are required for equilibrium with externally applied loads.

This definition implies that primary stresses are associated with load controlled problems. Primary stresses have the characteristic that, in each direction, the stresses induced by them add up to the external applied forces. It was stated in Chapter 1 that primary stress distribution is substantially independent of the material properties (although they are utilized in the derivation of the solution). Because of the required equilibrium with external forces, primary stresses last as long as do the applied loads. Because of the enduring presence of the stresses, the strains should increase viscoelastically.

Stresses which arise as a result of having to satisfy a strain or displacement adjustment in the structure are called the secondary stresses.

Thermal stress, shrinkage stress, stresses at interference fitted joints, wedges, etc. are examples of secondary stress. Because, they arise as a result of internal adjustments in strain, and are not caused by external loads. An important characteristic of secondary stress is that the stresses add up to zero in all directions, since they arise *without* the application of forces. Hence there are no forces with which they are in equilibrium. The origin of stress, in these cases, is traceable to the imposed strain. In such cases the imposed strain remains constant over time, and by the nature of viscoelasticity, the stress distribution, as a whole, must decay. *Thus, the secondary stresses possess a self-limiting characteristic*. Therefore, in viscoelastic materials, the initial magnitude of stress is also the maximum. Another important characteristic of secondary stress is that its magnitude, *in general* cannot be reduced by increasing the size of the component. Rather, the opposite is true.

Figures 4.6 (a) and (b) show qualitatively the response of a viscoelastic material to primary and secondary stresses.

Peak stresses are simply stress concentrations.

A stress distribution qualifies to be called a stress concentration if, and only if, very high stresses are distributed in a very small volume of the material.[3] A stress concentration has no effect on the displacements and strains at a short distance. For ductile materials, stress concentrations are *never* the cause of failure in one cycle of application. This is true whether

[3] The question arises, "How small is small?" Reference 1 has some criteria to determine this.

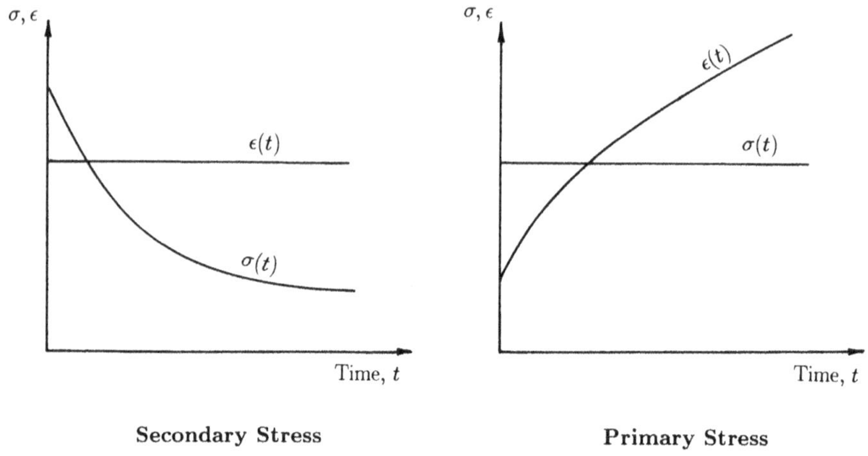


**FIGURE 4.6.** Viscoelastic responses under primary and secondary stresses.

the stress is caused by applied loads or by applied displacements. Peak stresses are potential sources of fatigue failure, or crazing, or crack initiation, all of which require repeated application, elapsed time, or both.

Primary stresses are further classified, on the basis of their spatial distribution, as *primary membrane stress and primary bending stress*. Membrane stress is a stress that is constant over the thickness. Bending stress is one that varies linearly over the thickness of the part. See Figure 4.7. (Such a subclassification is not necessary for secondary stresses, because of their self limiting nature.) In terms of severity of failure, the membrane stress is more serious than is bending stress. It is well known that when the maximum bending stress reaches yield stress (plasticity), further loads spread the plastic zone, until the entire cross section becomes plastic. This is illustrated in Figure 4.7. Several text books on the strength of materials such as References [12] and [13] discuss this aspect of yielding. Thus, the bending stress belongs to a lower severity category than does membrane stress.

There is a close relationship between the category of stress and the failure mode for which it is responsible. Table 4.4 is a summary of failure modes and the stress categories that are responsible for the various modes of failure. Both the stress-dependent and structure-dependent modes of failure are considered in this table. Consider the following statements:

- For thermoplastics, which exhibit viseoelasticity, primary stress causes creep, and secondary stresses should relax.

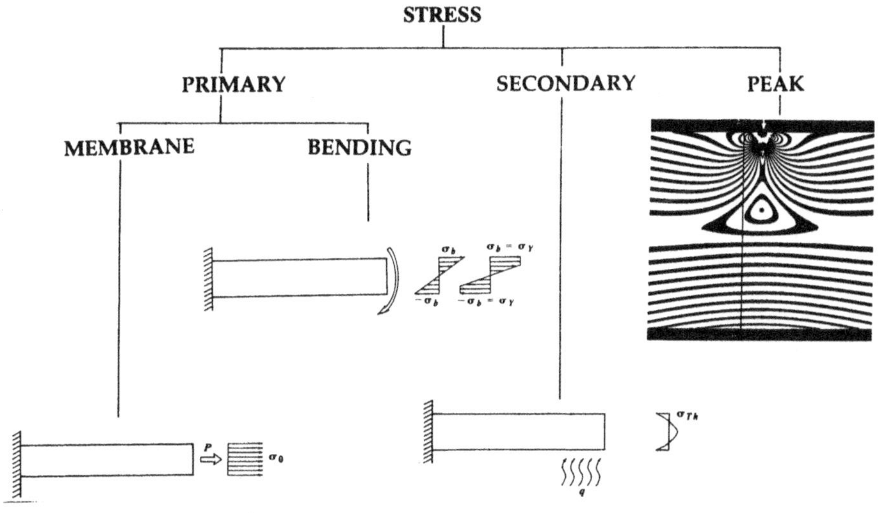


**FIGURE 4.7.** Categories of stress.

- A membrane stress reaching the yield strength value of the material, is at the end of failure, whereas a bending stress reaching the yield strength is at the beginning of failure.
- The stress in the "eye" of a stress concentration reaching the yield strength value is of no concern for a one-time application. A similar statement is difficult to make if the stress at the "eye" exceeds the yield stress. It needs detailed analysis to make categorical statements.
- Thermoplastic materials creep at any stress and almost at any temperature. Also, given enough time, they craze at almost any stress level. However, crazing depends on the local stress at a point, regardless of the *distribution*. Thus, an area of *sustained, static* stress concentration is a potential source of crazing.
- Under repeated loading, it is well known that a crack initiates at the stress concentration. The crack propagates to a critical length and the part fails catastrophically. Thus, an area of *dynamic, cyclic* stress concentration is a site of potential fatigue failure.
- Stress concentration has a negligible effect on the overall stiffness of the structure and hence on the natural frequencies of vibration. Therefore, in a finite element analysis of a part for calculating its natural frequencies, there is no need to model fine details like fillets, since such geometric details have no effects on the overall stiffness.
- A craze is the beginning of a failure. So is the occurrence of a predesignated creep strain. In contrast, creep rupture is the end of a failure.

**Table 4.4. Relationship Between Failure Modes and Stress Categories.**

| Failure Mode | What Governs | PM[1] | PB[1] | S[1] | Pk[1] |
|---|---|---|---|---|---|
| Ultimate failure in a single application | Distribution | ■ | ■ | ■ | |
| Gross yielding in a single application | Distribution | ■ | ■ | ■ | |
| Excessive static deflection | Distribution | ■ | ■ | ■ | |
| Excessive short term strains (elasto plastic) | Local stress | | | | ■ |
| Excessive long term deflection (creep) | Distribution | ■ | ■ | | |
| Excessive long term deflection (creep) | Distribution | ■ | ■ | | |
| Crazing | Local stress | | | | ■ |
| Creep rupture | Local stress | | | | ■ |
| Crack initiation | Local stress | | | | ■ |
| Crack propagation | Local stress intensity $K_I$ | | | | ■ |
| Fatigue | Local stress | | | | ■ |
| Excessive relaxation | Distribution | | | ■ | |
| Buckling (overall) | Structure | | | | Note 3 |
| Buckling (local) | Structure | | | | Note 3 |
| Low natural frequency | Structure | | | | Note 3 |
| Resonance | Structure | | | | Note 3 |

1. PM: Primary Membrane Stress. PB: Primary Bending Stress. S: Secondary Stress. Pk: Peak Stress.

2. When the symbol ■ appears more than once in a row, then it means that the sum of the stress categories is responsible for the mode of failure indicated. For example, "Ultimate failure in a single application" occurs because of the sum total of (PM + PB + S). Note that the secondary stress, acting alone, is never a cause of a failure in a single application; however, when acting in combination with primary stresses, it can add to their effects on failure.

3. Structure dependent failures are immune to the category of stress. For example, buckling can be caused by thermal stress or by primary stress.

- A fatigue crack initiation is the beginning of a failure, since catastrophic failure occurs only after the crack propagates to a critical size.
- A case of coexisting primary and secondary stresses involves creep and relaxation simultaneously. The net effect is the superposition of the two (subject to the linearity of viscoelasticity not being violated). Essentially, after long enough times, the secondary stresses disappear; the displacements which originally caused it, remain; primary stresses remain; and strains caused by the primary stresses creep with time.

## 4.7 HOW TO IDENTIFY STRESS CATEGORIES

In the preceding section we discussed the significance of stress categories with respect to the failure modes. No mention was made about how to recognize a stress category, when one is encountered. Historically, the technology of stress analysis started with beam theory, framed structures, plate theory, and then the shells. These theories use forces and moments as the basic unknowns, rather than the stresses themselves. Forces cause membrane stresses and moments cause bending stresses. Hence the categorization is natural in these calculation procedures. However, when the material behavior under stress concentration came to be understood better, there arose a need to obtain the stress concentration factors for many different shapes. This gave rise to a number of tabulated elastic solutions and experimental solutions (or both) for stress concentration factors.

Thus, until the advent of FEA as the major stress analysis tool, the procedures had a built-in scheme for categorization. However, there is a major problem with the finite element technique. Unlike the case with the classical theories, stress categorization does not occur naturally in the finite element method. It simply gives the sum total of all the stress categories. Most users of finite elements tend to look at the global maximum magnitude of stress as the only criterion in the analysis, which is not quite correct. Furthermore, one analysis cannot provide all categories of stress. It may be necessary to perform different finite element analyses at the same location to assess different categories of stress.

The importance of this problem has been considered by some large FEA programs. ANSYS, for example, has a capability to categorize stresses. The reader is referred to classic papers on this topic. [14, 15]

A few broad guidelines are provided in the following for recognizing the category of stresses.

- Any calculation which uses the formula

$$\text{Stress} = \frac{\text{Applied Load}}{\text{Area}}$$

  gives primary membrane stress.
- A plane truss analysis leads to primary membrane stress.
- Euler's beam theory leads to primary bending stress.
- As a consequence, three-dimensional frame analysis leads to primary membrane and primary bending stresses separately.
- A completely analytical solution to a beam, a plate, or a shell problem leads to membrane forces and bending moments. Membrane stresses and bending stresses are just one step beyond that.
- Any numerical scheme (not FEA) for a classical structural theory (typically plates and shells) gives membrane forces and bending moments. Hence stresses calculated using such procedures are primary membrane and bending stresses.
- A finite element analysis using only plate/shell elements gives membrane forces and bending moments. Consequently, the stresses obtained from such an analysis belong to the primary membrane and bending stress categories.
- A finite element analysis using plane elements or three-dimensional elements gives peak stresses, but not the other categories of stresses.
- An analytical solution resulting from the theory of elasticity gives the entire stress field, and hence gives peak stresses, but not the other categories.
- There is, in general, no foolproof way of identifying primary and secondary categories of stresses, except a thorough understanding of the physical origin of stress. A tentative norm is that a stress is primary if it is independent of material properties,[4] and secondary if it is proportional to the Young's modulus.
- Tabulated formulas for stresses in a handbook may be difficult to categorize into primary and secondary. Dependence of stress on material properties may be obscured by rewriting the formula in a variety of different ways. All handbooks give a brief description of the theory used in the derivation of the formulas. It behooves the engineer to refer to the theory used in the light of the stress categories, before applying the results.

---

[4] Certain distributions of body forces lead to a dependence on Poisson's ratio in a weak manner.

- Thermal stresses, by whatever calculation method they are arrived at, are secondary stresses. Likewise, stresses near a structural shape change (discontinuity) are secondary stresses.
- Primary stresses are recognized by the fact that they sum up to the external applied load. Secondary stresses add up to zero, since they are not in equilibrium with any applied load.
- Application of stress concentration factors from handbooks like References [16] and [17] leads to peak stresses.

In plastics components, the shapes are unconventional and hence developing the the development of the appropriate intuition needs a lot of practice.

**Example 4.1**

Use the formulas for tangential stresses in a rotating annular disk and show that the distribution exactly balances the centrifugal force applied on the disk.

*Solution* The tangential stress distribution is given by

$$\sigma_\theta = \left(\frac{3+\nu}{8}\right)\rho\omega^2\left[b^2 + a^2 + \frac{a^2b^2}{r^2} - \frac{1+3\nu}{3+\nu}r^2\right]$$

The summation of the stresses across a radius can be computed as a force in the manner shown below. The summation is simply the integral across the radial line from $a$ to $b$.

$$\int_a^b \sigma_\theta\,dr = \left(\frac{3+\nu}{8}\right)\rho\omega^2\left[(b^2+a^2)(b-a) + a^2b^2\left(\frac{1}{b} - \frac{1}{c}\right)\right.$$
$$\left. - \frac{(1+3\nu)}{3(3+\nu)}(b^3 - a^3)\right]$$
$$= \rho\omega^2(b^3 - a^3)$$

The total centrifugal force acting across the same radius is obtained as one half of the force on a half-disk. It is gotten by integrating the components of elemental centrifugal forces normal to the plane of section, $x - x$.

$$F_c = \int r\omega^2 dm$$

$$= \int_r \int_\theta r\omega^2 (\rho r\, dr\, d\theta) \sin\theta$$

$$= \rho\omega^2 \int_a^b \int_0^\pi r^2 \sin\theta \, dr\, d\theta$$

$$= \rho\omega^2 \int_a^b r^2 dr \int_0^\pi \sin\theta \, d\theta$$

$$= 2\rho\omega^2 (b^3 - a^3)$$

The force on just one of the radii is one-half of this force, and it is seen that the stresses balance the applied force, in the overall.

**Example 4.2**

Consider a polybutylene pipe carrying hot water. The inner surface temperature is $T_i$, and the outer wall temperature is taken to be the reference temperature, that is, zero. The temperature distribution across the wall thickness can be shown to be

$$T = T_i \frac{\log(b/r)}{\log(b/a)},$$

where $a$ and $b$ are the inner and outer radii of the pipe. Further, it can be shown that the thermal stress across the wall thickness in the hoop direction is given by

$$\sigma_\theta = \frac{\alpha E T_i}{2(1 - \nu)\log(b/a)} \left[ 1 - \log\frac{b}{r} - \frac{a^2}{(b^2 - a^2)} \left( 1 + \frac{b^2}{r^2} \right) \log\frac{b}{a} \right]$$

Show that this distribution of hoop stress is equivalent to a net zero force.

*Solution* As in the last example, the net force can be obtained by integrating the stress from $a$ to $b$. Also recall that

$$\int \log r\,dr = (r \log r - r) \quad \text{and} \quad \int \frac{dr}{r^2} = -\frac{1}{r}$$

$$\begin{aligned}\int_a^b \sigma_\theta\,dr &= \frac{\alpha E T_i}{2(1-\nu)\log(b/a)}\left[r - r\log b + r\log r - r \right.\\ &\quad \left. - \frac{a^2}{(b^2-a^2)}\log\frac{b}{a}\left\{r - \frac{b}{r}\right\}\right]\Big|_a^b \\ &= \cdots\left[-(b-a)\log b + (b\log b - a\log a)\right. \\ &\quad \left. - \frac{a^2}{(b^2-a^2)}\frac{(b-a)(b+a)}{a}\log\frac{b}{a}\right] \\ &= \cdots\left\{a\log\frac{b}{a} - a\log\frac{b}{a}\right\} \\ &= 0\end{aligned}$$

This shows that at any angular position, at any point along the length, the net summation of thermal stresses across the thickness is zero.

**Example 4.3**

Here we discuss a particularly important variation in the category of temperature induced stresses. Consider a long bar held fixed between its ends. Its temperature is raised by an amount $T$ above the ambient temperature. Discuss the consequences.

*Solution* The stress induced is axial, uniform all over the bar cross section and length. Its magnitude is given by

$$\sigma = E\alpha T,$$

$\alpha$ being the thermal expansion coefficient. The total force across a cross section is easily seen to be $E\alpha TA$, where $A$ is the area of the cross section.

In this problem, we have a situation in which temperature stresses are induced by holding the ends fixed. Such a condition requires the use of an external force. The application of a uniform temperature alone cannot create such a state of stress.

*The stress in the foregoing example is not "thermal" stress, according to our definition.* It is referred to in the literature as "expansion stress." Expansion stress is also self-limiting, will relax with time, and is induced by temperature. But it is *not* equivalent to a zero total force. Therefore, the FOS for expansion stress has to be high; it can cause buckling under temperature rise, and can cause every kind of failure that a primary stress can. The only salient difference is its time-based behavior. Expansion stresses are typical of pipelines.

**References**

1. Bernstein, M. D., *Design Criteria for Boilers and Pressure Vessels in the U.S.A.*, in ASME PVP-Vol. 162, *1988 International Design Criteria of Boilers and Pressure Vessels*, 1988, ASME.
2. ANSI Z-90.1-1971, *Am. Natl. Std., Specifications for Protective Headgear for Vehicular Users (High Protection).*
3. ANSI Z-89.1-1981, *Am. Natl. Std., Requirements for Protective Headwear for Industrial Workers.*
4. SAE Handbook, Vol. 4, Specification No. SAE-J980a, *Bumper Evaluation Test Procedure—Passenger Cars*, published by Soc. of Automotive Engineers, Warrendale, PA.
5. AWWA Standard No. C950-88, 1988, *Standard for Fiberglass Pressure Pipe*, published by American Water Works Association, Denver, CO.
6. *Modern Plastics Encyclopedia*, 1984-85, Vol. 61, Number 10A, McGraw Hill, New York, NY.
7. *Laminate Analysis Program*, In-house software, L. J. Broutman and Associates, Chicago, IL.
8. Metals Handbook, Vol. 1, *Properties and Selection, Irons and Steels*, and Vol. 2, *Properties and Selection, Nonferrous Alloys and Pure Metals*, American Society for Metals, Metals Park, OH.
9. *Manual of Steel Construction—Allowable Stress Design*, 9th Edn, Published by Amer. Inst of Steel Construction, Chicago, IL.
10. *Bathe, K. J. Finite Element Procedures in Engineering Analysis*, 1982, Prentice-Hall, Englewood Cliffs, NJ.
11. Cook, R. D., Malkus, D. S., and Plesha, M. E., *Concepts and Applications of Finite Element Analysis*, 1989, 3rd Edn., John Wiley & Sons, New York, NY.
12. Beer, F. P. and Johnston, E. R. Jr., *Mechanics of Materials*, 1981, McGraw Hill, New York, N.Y.
13. Gere, J. M. and Timoshenko, S. P., *Mechanics of Materials*, 1984, PWS Publishers, Boston, MA.
14. Kroencke, W. C., *Classification of Finite Element Stresses According to ASME Section III Stress Categories*, Pressure Vessel and Piping, Analysis and Computer, ASME June 1974.

15. Kroencke, W. C., Addicott, G. W., and Hinton, B. M., *Interpretation of Finite Element Stresses According to ASME Section III* ASME Paper No. 75-pvp-63, ASME Second National Congress on Pressure Vessels and Piping, June 1975.
16. Roark, R. J. and Young, W. C., *Formulas for Stress and Strain*, 5th Edn., 1975, McGraw Hill, New York, N.Y.
17. Peterson, R. E., *Stress Concentration Design Factors*, 1953, John Wiley & Sons, Inc., New York, N.Y.
18. Timoshenko, S. P. and Goodier, J. N., *Theory of Elasticity*, 3rd Edn., 1970, McGraw Hill, New York, N.Y.
19. Harvey, J. F., *Theory and Design of Modern Pressure Vessels*, 2nd Edn., 1974, Van Nostrand Reinhold, New York NY.

**Exercises**

**4.1.** A rubber band is stretched around a pack of papers. Categorize the stress in the band.

**4.2.** In a plastic clamp as shown in the figure, serrations are used to prevent the clamp from slipping. In what category does the stress at point $A$ belong? Also identify a few points which are likely to fail early, and state what mode of failure can be expected there.

**4.3.** $PQ$ in the figure is the cantilevered portion in the battery cover of a pocket radio. What category of stress does the opening/closing operation create at the point $R$? Explain.

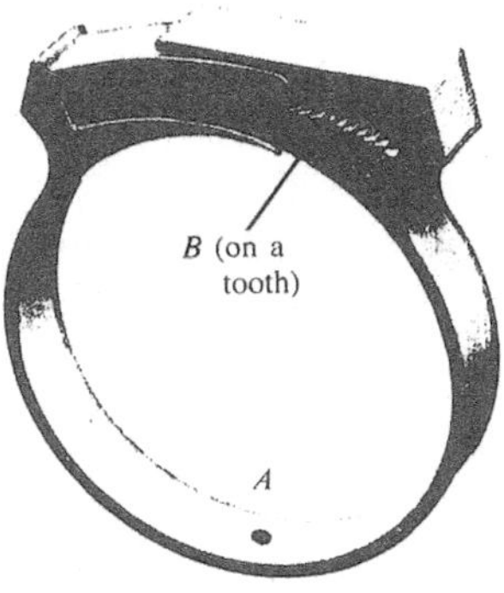


**EXERCISE 4.2.** Serrated clamp.

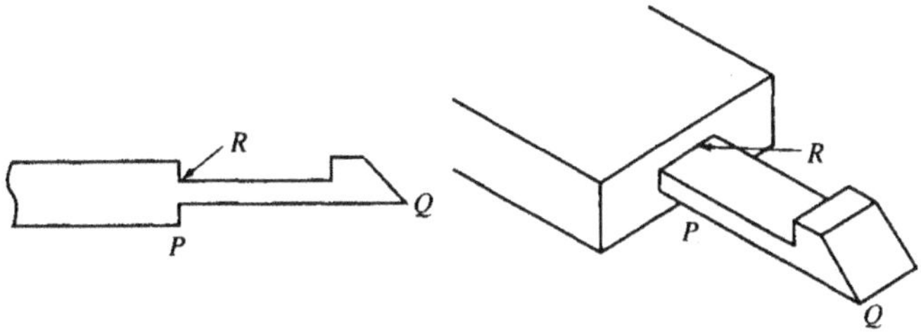


**EXERCISE 4.3.** The snap-fit element in a pocket radio.

**4.4.** In the previous problem, draw a qualitative diagram of the stress history at the point $R$, caused by repeated opening and closing of the battery chamber. Assume linear viscoelastic behavior of the material.

**4.5.** Three kinds of loading a 3-point bend specimen are shown in the figures (a), (b) and (c). Draw the qualitative diagram of stress and strain histories of the beam specimen in each of these cases.

**4.6.** Two specimens of the same net cross section (see figure) are loaded in such a way that the tensile stress at the point $P$ on the specimens are both equal to the yield point of the material. Which one is likely to survive the loading better? Explain.

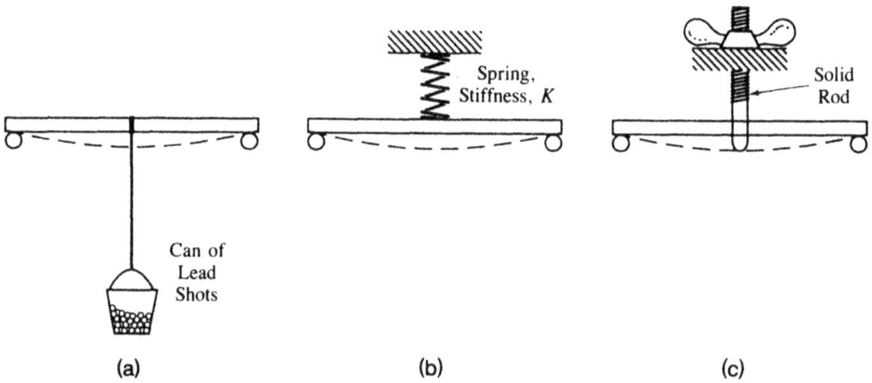


**EXERCISE 4.5.** Three ways to load a three-point bend specimen.

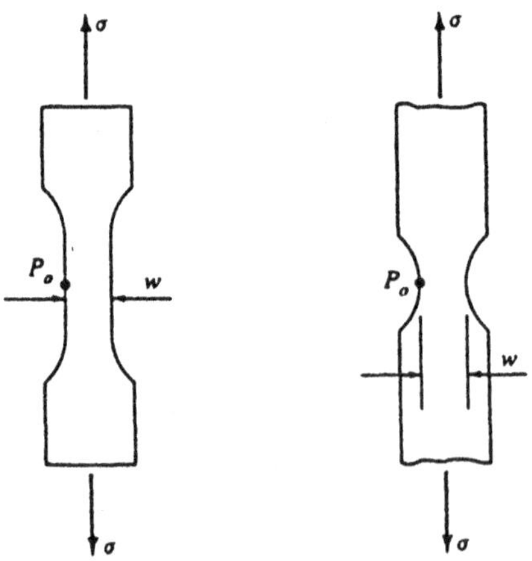


**EXERCISE 4.6.** Two tensile specimens of the same cross-sectional area.

**4.7.** Figure 4.12 (see color insert) is a contour plot of the $\sigma_{yy}$ stress distribution in a plate with a round hole. Only a quadrant is shown. The material property, the hole size, the plate size, and the applied load are also marked on the diagram. Assume that the yield point of the material is 5 kg/mm$^2$. Is the component suitable for repeated application of the load? Why?

**4.8.** Set up a design of a circular nylon shaft press-fitted to another circular drive shaft, transmitting power in a hand drill.

**4.9.** Redo the above example, without a press fit, but with the drive and driven shafts provided with square shaped male and female assemblies, as shown in the figure.

**4.10.** Which design would you prefer for the closure of a container containing water at 1.5 kg/cm$^2$—threaded closure or bayonet type closure? Give reasons.

**4.11.** Compare press-fits and snap-fits with respect to the stress histories on the mating components.

**4.12.** *Suggested Term Paper:* In this chapter the definition of stress concentration is loose. This is true in almost all texts. Much is left to interpretation; when a number of stress contours crowd around a particular point, then it is considered a stress concentration. Since stresses concentrate or rarefy in all circumstances, what would be a rigorous way of defining stress concentration?

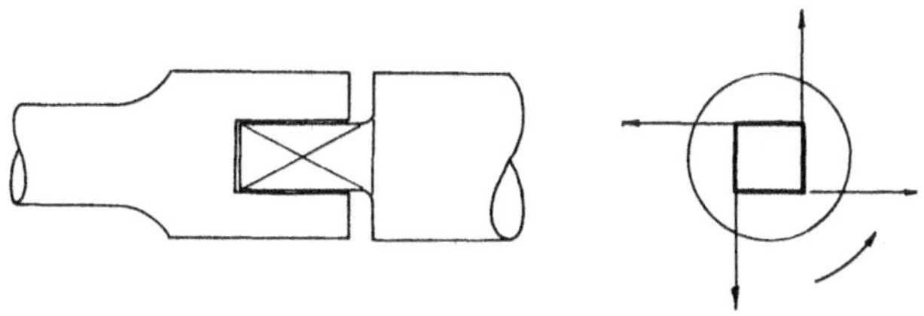

**EXERCISE 4.9.** Square ends in square sockets can be used to transmit power, but tend to do it by local contact stresses at corners.

# 5

# Applied Viscoelasticity

## LIST OF SYMBOLS USED

$\sigma_{**}$ —A stress. Subscripts 1, 2, 3 denote principal stresses.
$\sigma(t)$ —Stress as a function of time, $t$
$\sigma 1, \sigma 2$ —Successive stress amplitudes of dynamic stress.
$\epsilon_{**}$ —Strain. Subscripts 1, 2, 3 denote principal strains.
$\epsilon(t)$ —Strain as a function of time, $t$
$\epsilon 1, \epsilon 2$ —Successive strain amplitudes of dynamic strain.
$E_1, E_2$ —Moduli of spring elements in the standard model.
$E(t), E_R, E_U$ —Relaxation Moduli and its extreme values.
$E^*, E', E''$ —Complex, Storage and Loss Moduli respectively.
$J(t), J_R, J_U$ —Creep Compliance and its extreme values.
$J^*, J', J''$ —Complex, Storage and Loss Compliances, respectively.
$G_1, G_2$ —Shear Moduli represented by spring elements in a standard model.
$G(t), G_R, G_U$—Same significance as the corresponding $E_{**}$'s, except that these apply in shear mode.
$G^*, G', G''$ —Complex, Storage and Loss Shear Moduli, respectively.
$\tau_\tau, \tau_\sigma$ —Retardation and Relaxation times of single element viscoelastic models.
$\omega$ —Frequency of cyclic loading.
$\delta$ —Loss angle.
$H(\tau), L(\tau)$ —Relaxation and Retardation Spectra, continuous functions

of time, represent the distribution of $\tau_\sigma$ and $\tau_\epsilon$ for models involving many elements.

$e$ —The base of natural logarithms, or, a strain.

## 5.1 INTRODUCTION

The term *viscoelasticity* stands for all the aspects of time dependent response of stress to strain and vice versa. It is the characteristic that contrasts the behavior of plastics most from that of metals. It manifests itself in the form of many different time dependent phenomena. In the description of such phenomena we may consider applied stress, loading types, loading rate, strain, strain rate, time elapsed, and lastly the temperature, as independent variables. Temperature is perhaps the strongest variable of all. Because of the large number of variables involved, the description of viscoelasticity has to be made necessarily through its aspects, namely *creep*, *relaxation* and *recovery*. Creep is the strain response to a stress that is constant with time. Relaxation is the stress response to an applied constant strain. Recovery is the strain response to a stress that has been removed. Besides these aspects, a whole range of special behaviors and properties are revealed when periodic loading is applied. In a manner very similar to voltage and currents in alternating current electric circuits, there is time difference between applied stress and resulting strain. Strain lags behind applied stress. A description of viscoelasticity in three dimensions is very complex. Nonlinearities in viscoelastic behavior only add to the complexity. Furthermore, there are no simple failure criteria in the context of viscoelastic behavior. Depending upon the level of stress, the time elapsed, and the type of material, the material can "fail" due to crazing, ductile creep rupture, or brittle creep rupture. The different modes of failures correspond to different and independent mechanisms.

There has been a reasonable amount of success in characterizing and using the linear viscoelastic behavior in the design of plastic components. Of course, a considerable amount of experimental work and data processing has been needed to achieve this end. This chapter deals with certain aspects of calculation techniques for linear viscoelastic solids and of material data acquisition.

## 5.2 ASPECTS OF VISCOELASTICITY

### 5.2.1 Creep

Creep is the increase in strain with time in response to a constant applied stress, at a constant temperature. One can imagine a series of tensile tests, each one subjected to various stress levels. For a given specimen, the stress remains constant throughout the test. The strains are retrieved at various times, and plotted as a family of curves as shown in Figures 5.1(a). Higher levels of applied stress cause higher strains for any time, and the strains grow with time. By linearity we mean a linear relationship between stress and strain. Creep strains are usually nonlinear in time. In short, we look for a relation of the form

$$\epsilon(t) = \sigma \times (\text{A function of time}).$$

This can be observed by cross-plotting strain as a function of stress, as shown in Figure 5.1(b). This figure can also be considered to have been obtained from a batch of creep experiments. Each specimen is loaded at different stress levels. The instant of initial loading is taken as the starting time of the test. Strains are observed on each of the specimens at times $t_1, t_2, t_3, \cdots$, etc. The stress-strain relation is said to be linear if they are proportional to each other at any time. That is,

$$\frac{\epsilon_1}{\sigma_1} = \frac{\epsilon_2}{\sigma_2} = \frac{\epsilon_3}{\sigma_3}$$

The lines in Figure 5.1(b) are known as *isochronous* creep curves. The linearity applies only within a certain envelope of stress and strain combi-

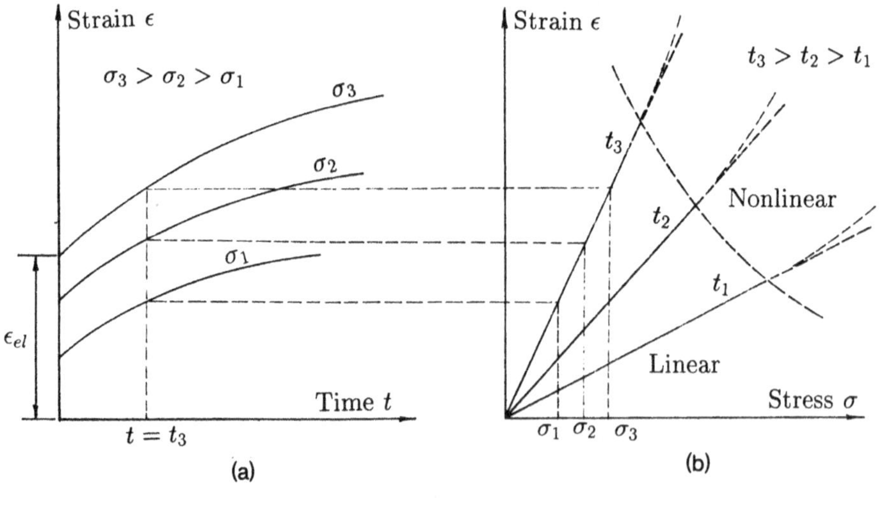


**FIGURE 5.1.** Creep.

nations. Outside this envelope, strains grow faster than linearly. In the linear region, strain can be written as

$$\epsilon(t) = \sigma J(t)$$

We can also define the property *creep compliance* $J(t)$ as the ratio of strain to stress at any time. Stress is constant while strain increases with time. Thus the creep compliance is a function of time, $t$, and is defined as

$$J(t) = \frac{\epsilon(t)}{\sigma} \tag{5.1}$$

Creep compliance is a very important factor in the time dependent design of plastics.

***Characteristics of Creep in Plastics***

The creep strains in plastics do not increase indefinitely, and approach a final value asymptotically. Given enough time, they also recover completely upon the removal of the load. Figure 5.2 shows the trend in the value of creep compliance with time. Initially at very short times, the compliance is low, and is almost equal to the reciprocal of Young's modulus. This is denoted by the *unrelaxed compliance*, $J_U$. The terminal value of creep strain (that is, strain at the end of infinite time) represents the maximum compliance the material can have. This value is denoted by $J_R$, the subscript $R$ being used for *fully relaxed compliance*. The time scale on the $x$-axis may be of the order of 10 to 20 years, which represents the reasonable expected life of a product.

The strain denoted by $\epsilon(t)$ may be assumed to include the elastic component (the time independent part). Elastic components are also linear and thus its addition ensures linearity of the total strain.

### 5.2.2 Relaxation

Relaxation is the counterpart of creep. It is the decline in str^ss with time, in response to a constant applied strain, at constant temperature. Once again, one can imagine a series of tests in which the test specimens are given initial strains of $\epsilon_1, \epsilon_2, \epsilon_3, \cdots$, etc. In such tests, the stresses are typically computed from the load cell readings, taken at different times. The stresses are plotted as a family of curves against time, as shown in Figure 5.3(a). Higher initial stains correspond to higher curves and vice

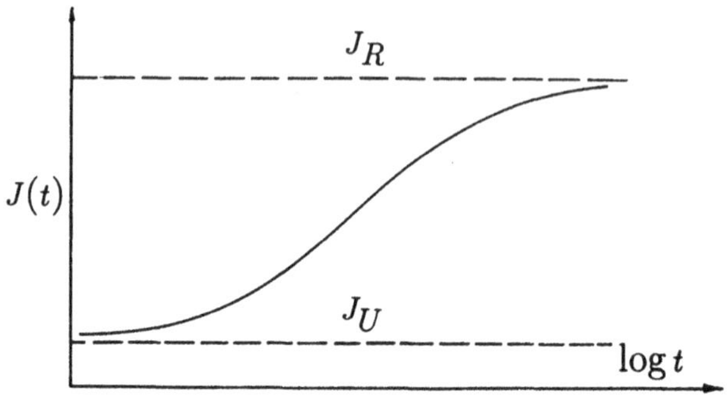


**FIGURE 5.2.** Creep compliance as a function of time.

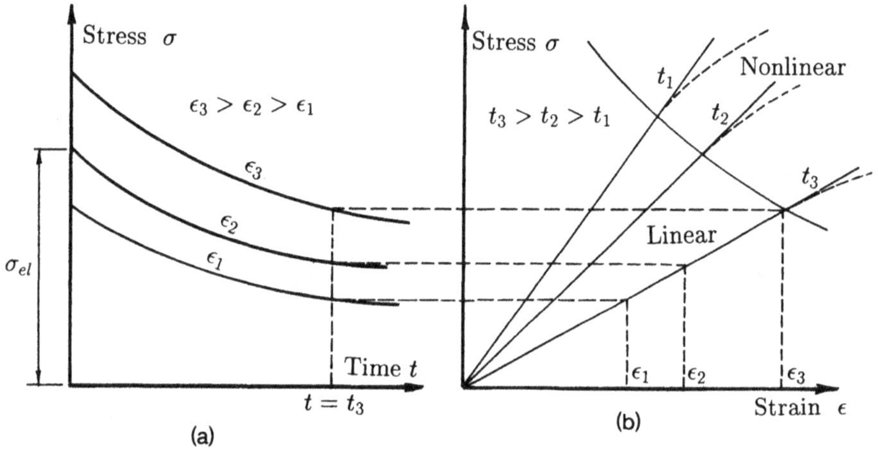


**FIGURE 5.3.** Relaxation.

versa. Linear viscoelasticity implies a relationship of the form,

$$\sigma(t) = \epsilon \times (\text{A function of time}).$$

The linearity can also be visualized by cross plotting the relaxation curves, as shown in Figure 5.3(b). As with the isochronous curves, the stress strain curves are linear within an envelope of stress and strain. Outside this envelope, the stress relaxes *slower* than linearly. In the linear range,

$$\frac{\sigma_1}{\epsilon_1} = \frac{\sigma_2}{\epsilon_2} = \frac{\sigma_3}{\epsilon_3}$$

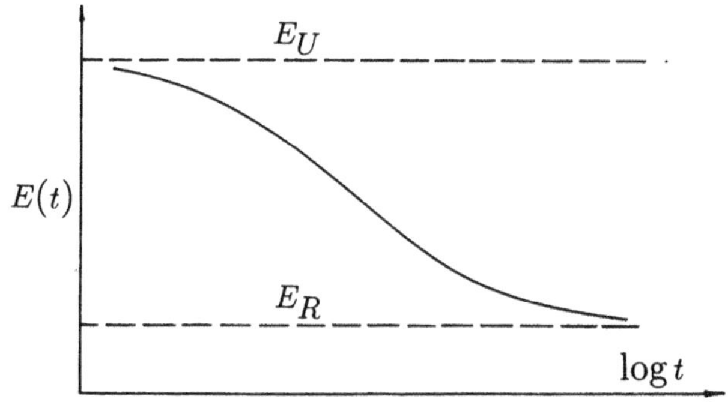


**FIGURE 5.4.** Relaxation modulus as a function of time.

We can define the property called *relaxation modulus* $E(t)$, as the ratio of the instantaneous stress to the initial applied strain. Thus $E(t)$ is a function of time.

$$E(t) = \frac{\sigma(t)}{\epsilon} \tag{5.2}$$

In general, the relaxation modulus is not *exactly* the inverse of creep compliance. We will show later (Example 5.9) that it is so, only to a degree of approximation.

Initially, the relaxation modulus has its maximum value $E_U$, or the *unrelaxed modulus* which is the same as the Young's modulus. At time $t \to \infty$, it attains a minimum value $E_R$, or the *fully relaxed modulus*. The trend is indicated in Figure 5.4.

### 5.2.3 Recovery

Recovery is a tendency of the plastic to return to the stress-free state upon removal of an applied load. The recovery is not instantaneous but takes place over time. This can be demonstrated by subjecting a specimen to a stress $\sigma_o$, holding the stress level for some time, say till $t = \tau$, and removing the load. Because of creep, strains grow during the interval $0 < t < \tau$. Removal of the load results in the instantaneous reduction of the *elastic strain*, followed by a slow reduction in creep strain. Theoretically, all the strains can be recovered, providing enough time is allowed.

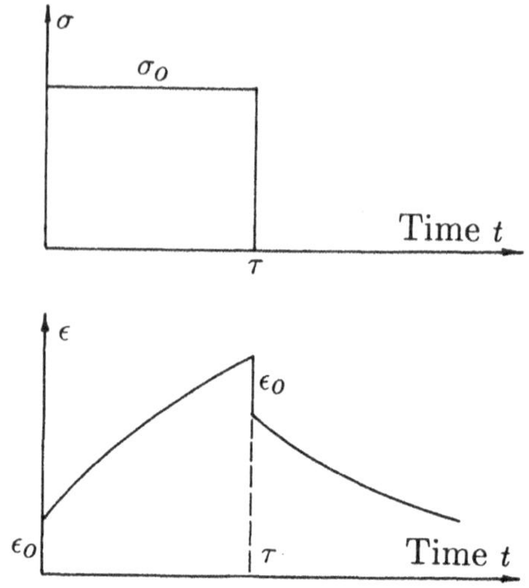


**FIGURE 5.5.** Recovery.

Since strains exist during recovery while stress is zero, no simple stress strain relationship can be written. Figure 5.5 illustrates this behavior.

## 5.3 VISCOELASTIC MODELS

It is apparent that a simple rule like Hooke's law cannot describe the different time dependent behaviors. One needs a description that takes into account the history-dependent behavior of the material. There are two approaches to accomplish this.

1. Spring Dashpot Models: An arrangement of springs and dashpots in parallel and series can be used as the analog of viscoelastic behavior. The load and displacement in the arrangement is time dependent, and the algebraic expressions are *formally similar* to the stress-strain relation. There are a few arrangements that are helpful in visualizing the plastics behavior, at least qualitatively. This model is also called the *differential representation* of viscoelasticity.
2. Boltzmann Superposition Principle: With its origins in electrical circuit theory, this principle is a special way of adding stresses and strains on the time scale, in a history-dependent manner. This principle simulates the creep, relaxation, recovery, and other dynamic behavior in a manner that is true to reality. This approach is also called the *integral representation* of viscoelasticity.

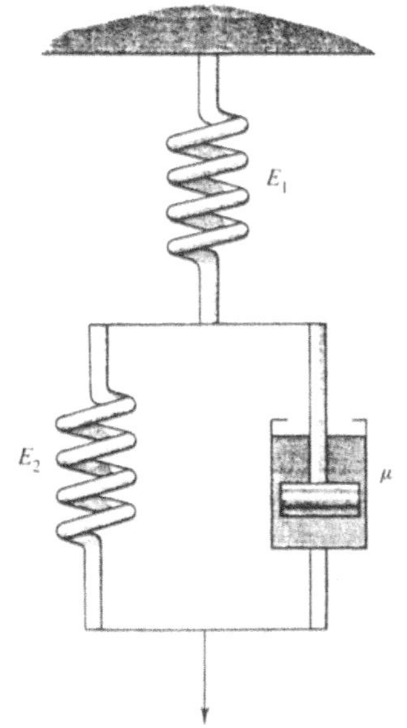


**FIGURE 5.6.** Standard viscoelastic model.

It can be shown that the integral and differential representations are equivalent to each other.

## 5.4 SPRING DASHPOT MODELS

The spring dashpot models are aids to visualize and write the equations between stress, strain, and their time derivatives. Such models are able to explain the creep, relaxation, recovery, dynamic behavior and several rate processes.[1] This section will elucidate such behaviors, and procedures to account for them in the stress analysis. We begin with an elementary model called the *standard viscoelastic model*, (also called the *Zener model*), shown in Figure 5.6. It consists of a primary spring of stiffness $E_1$ in series with another element which itself is a parallel arrangement of a secondary spring of stiffness $E_2$ and a dashpot of viscosity $\mu$. These springs and dashpots are to be construed as representations of the material stress and

---

[1] However, it is found that in general an infinite number of springs and dashpots of suitable properties are needed to exactly fit the experimental results. See Example 5.1, and Section 5.5.

strain behavior. In these configurations the following correspondence holds.

$$\text{Tensile or Shear Stress} \rightarrow \text{Load}$$

$$\text{Tensile or Shear Strain} \rightarrow \text{Displacement}$$

The term *viscoelasticity* is derived from this similarity. We recall that springs act in a time-independent fashion, giving immediate responses to stress and strain. The dashpots behave in quite an opposite manner. When in isolation, they are completely time-dependent, and never give an immediate response. Their responses are represented by the relation

$$\text{Stress} = \text{Viscosity} \times \text{Strain Rate}$$

In text books [1], two other basic models are discussed, namely the *Maxwell* and *Voigt* models. It is sufficient for our purposes to state that the Maxwell model, consisting of a spring and dashpot in series, represents relaxation well, but not creep or recovery. The Voigt model, consisting of a spring and dashpot in parallel, represents creep and recovery very well, but not relaxation. Both models represent recovery well enough. The standard model has the minimum number of ingredients to represent all behaviors well, at least qualitatively.

### 5.4.1 Technique of Writing the Differential Equation

We want to write the differential equation connecting the (strain) total displacement—that is, the displacements between end points—and the (stress) total load acting between the ends. Details of load sharing between the various elements are unnecessary. The order of the differential equation depends on the number of elements present in the system. In the case of the standard model, it is of the second order. More elements lead to higher order of the differential equation. Since formation of higher order equations involves more manipulation of the equations, we can use a shortcut known as the *operator method* of writing the differential equation.[2]

Table 5.1 shows a systematic procedure for developing the equation. Simple single elements are the basic units. The Maxwell and the Voigt models use the *operator* concept for forming the equation. Although the table shows only normal stresses and strains, it is nonetheless also valid for

[2] Reference 2 discusses this method in some detail. However, this approach is more elaborately discussed in books about the dynamics of mechanical systems, or oscillations of electrical circuits.

**Table 5.1. Technique of Setting Up Differential Equation for a Given Material Model.**

| | | |
|---|---|---|
| **Spring** | $\sigma = E\epsilon$ |  |
| **Dashpot** | $\sigma = \mu\dot{\epsilon}$ | 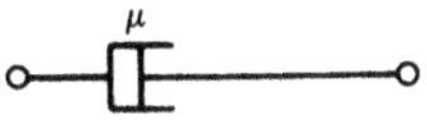 |
| **Maxwell** | $\sigma = E\epsilon_1 = \mu\dot{\epsilon}_2$<br>$\epsilon = \epsilon_1 + \epsilon_2$ or $\dot{\epsilon} = \dot{\epsilon}_1 + \dot{\epsilon}_2 = \frac{\dot{\sigma}}{E} + \frac{\sigma}{\mu}$<br>In operator notation,<br>$\dot{\epsilon} = \left(\frac{1}{\mu} + \frac{1}{E}\frac{d}{dt}\right)\sigma$ |  |
| **Voigt** | $\epsilon = \frac{\sigma_1}{E}$, and $\dot{\epsilon} = \frac{\sigma_2}{\mu}$<br>$\sigma = \sigma_1 + \sigma_2 = E\epsilon + \mu\dot{\epsilon}$<br>In Operator Notation, $\sigma = \left(E + \mu\frac{d}{dt}\right)\epsilon$ | 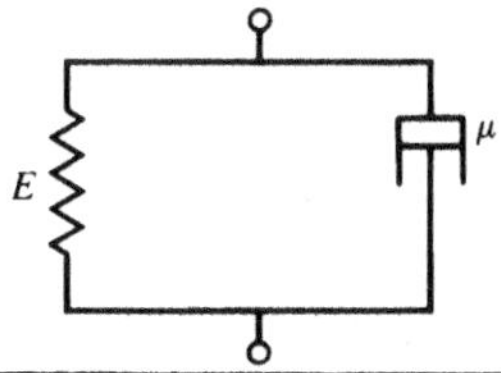 |

shear stresses and strains. $G$ would replace $E$, and $\mu$ would denote the shear viscosity. Rheological testing machines are more conveniently set up for shear mode because less power is needed to run the machine in shear. In the following we use normal stresses and strains unless otherwise stated. For example, for the Voigt and the Maxwell models, we can write respectively,

$$\sigma = \left(E + \mu\frac{d}{dt}\right)\epsilon \quad \text{(Voigt)}$$

$$\sigma = \frac{\dot{\epsilon}}{\left(\frac{1}{\mu} + \frac{1}{E}\frac{d}{dt}\right)} \quad \text{(Maxwell)}$$

**Table 5.1. (Continued)**

| | | |
|---|---|---|
| Standard Model | $\sigma = E_1\epsilon_1 = \left(E_2 + \mu\frac{d}{dt}\right)\epsilon$<br>In Operator Notation,<br>$\epsilon = \epsilon_1 + \epsilon_2 = \frac{\sigma}{E_1} + \frac{\sigma}{E_2 + \mu\frac{d}{dt}}$<br>Text uses the expanded form,<br>$\frac{d\epsilon}{dt} + \frac{E_2}{\mu}\epsilon = \left(\frac{E_1 + E_2}{E_1\mu}\right)\sigma + \frac{1}{E_1}\frac{d\sigma}{dt}$ | 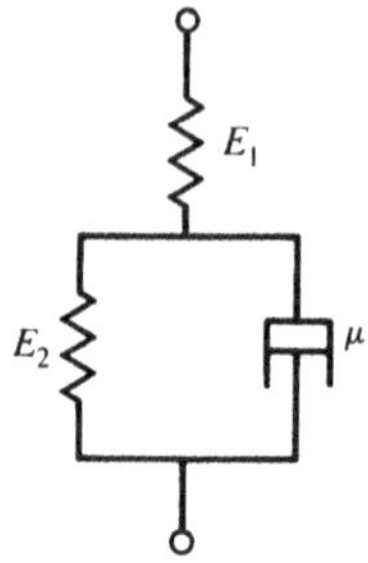 |
| Another Form of Standard Model | $\epsilon = \frac{\sigma_1}{E_1}$<br>In Operator Notation,<br>$\sigma = \sigma_1 + \sigma_2 = E_1\epsilon + \frac{\dot{\epsilon}}{\left(\frac{1}{\mu} + \frac{1}{E_2}\frac{d}{dt}\right)}$<br>Or, $(E_1 + E_2)\frac{d\epsilon}{dt} + \frac{E_1E_2}{\mu}\epsilon = \frac{d\sigma}{dt} + \frac{E_2}{\mu}\sigma$ | 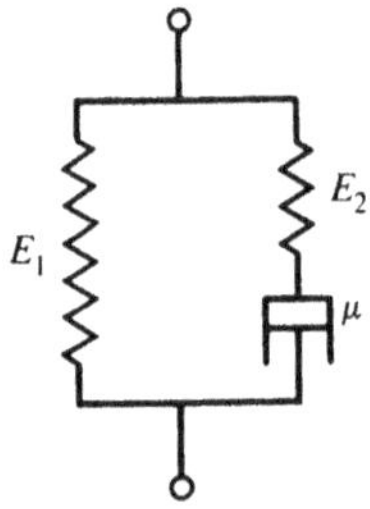 |
| Several Voigts in series | Strains are additive.<br>$\epsilon = \sigma\sum_i \frac{1}{E_i + \mu_i\frac{d}{dt}}$ | 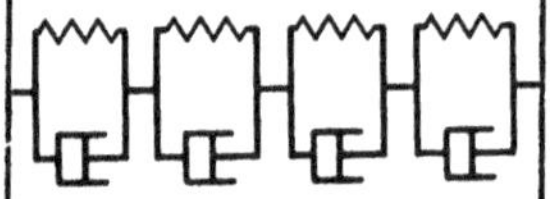 |
| Several Maxwells in parallel | Stresses are additive.<br>$\sigma = \dot{\epsilon}\sum_i \frac{1}{\frac{1}{\mu_i} + \frac{1}{E_i}\frac{d}{dt}}$ | 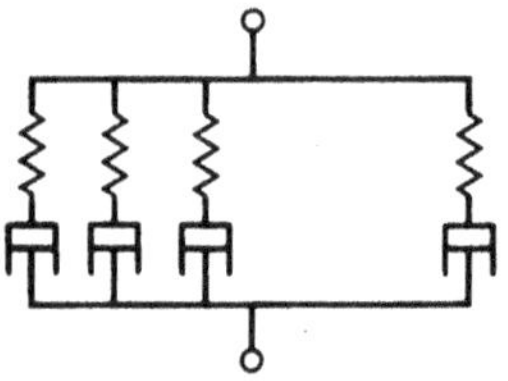 |

Basically, the symbol $d/dt$ can be treated as though it is an algebraic quantity. Thus, for the standard model, since strains are additive, we can write

$$\epsilon = \frac{\sigma}{E_1} + \frac{\sigma}{\left(E_2 + \mu \frac{d}{dt}\right)}$$

Although the expression $d/dt$ is incomplete, it is treated as a symbol in all the algebraic manipulations. One can extend this concept further to several Voigt elements in series, or to several Maxwell elements in parallel. The expressions are shown in Table 5.1. It is easily seen why higher order configurations lead to higher order equations.

### 5.4.2 Creep Behavior of the Standard Model

From Table 5.1 we obtain the stress-strain equation of the standard model

$$\frac{d\epsilon}{dt} + \frac{E_2}{\mu}\epsilon = \left(\frac{E_1 + E_2}{E_1\,\mu}\right)\sigma + \frac{1}{E_1}\frac{d\sigma}{dt}. \tag{5.3}$$

Since the stress is constant in creep, $d\sigma/dt$ is set to zero. Equation 5.3 may be specialized to describe the creep behavior.

$$\dot{\epsilon} + \frac{1}{\tau_\sigma}\epsilon = \left(\frac{E_1 + E_2}{E_1\,\mu}\right)\sigma; \quad \text{where, } \tau_\sigma = \frac{\mu}{E_2} \text{ sec} \tag{5.4}$$

The time $\tau_\sigma$ is called the *relaxation time*. The subscript $\sigma$ is used to denote that stress is held constant. Its meaning is explained below. Since the initial strain is equal to the elastic strain,

$$\epsilon(0) = \frac{\sigma}{E_1}.$$

The solution for this first order differential equation is given by

$$\epsilon = \frac{\sigma}{E_1} + \frac{\sigma}{E_2}\left[1 - \exp\left(-\frac{t}{\tau_\sigma}\right)\right] \tag{5.5}$$

The solution agrees with observation qualitatively in that it predicts the asymptotic increase of creep strain from the initial elastic strain. It is seen that the initial or the so called *unrelaxed modulus* is given by setting $t = 0$. Since the strain is simply the initial strain, the unrelaxed modulus,

$$E_U = \frac{1}{J_U} = E_1. \tag{5.6}$$

The fully relaxed modulus is the modulus of the material after a long time has passed. This is obtained by letting $t \to \infty$. We obtain the fully relaxed modulus,

$$E_R = \frac{1}{J_R} = \frac{E_1 E_2}{E_1 + E_2} \tag{5.7}$$

At any time $t$, the compliance of the material is given by

$$J(t) = \frac{1}{E_1} + \frac{1}{E_2}\left[1 - \exp\left(-\frac{t}{\tau_\sigma}\right)\right] \tag{5.8}$$

The trend of the variation of $[J(t)]^{-1}$ is shown in Figure 5.7.

The physical meaning of the relaxation time $\tau_\sigma$ becomes evident if we consider the time dependent part of the creep strain at times $t_0$ and $(t_0 + \tau_\sigma)$. They are marked as $\epsilon 1$ and $\epsilon 2$ in Figure 5.8(a). It is easy to see that

$$\frac{\epsilon 2}{\epsilon 1} = \frac{1}{e}$$

Hence, $\tau_\sigma$ can be defined as the time interval needed to reduce the time dependent part of the strain by a factor of $e$.

### 5.4.3 The Relaxation Behavior Predicted by the Standard Model

As before, we start with the general equation of the Standard Model from Table 5.1, and set the strain constant. That is $\dot{\epsilon} = 0$. Therefore the equation representing relaxation is

$$\dot{\sigma} + \frac{\sigma}{\tau_\epsilon} = \frac{E_1 E_2}{\mu}\epsilon,$$

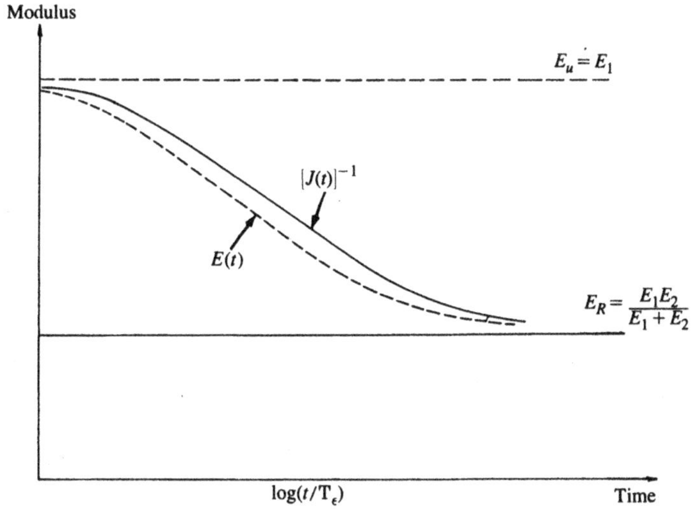


**FIGURE 5.7.** Variation of unrelaxed and relaxed moduli as functions of time.

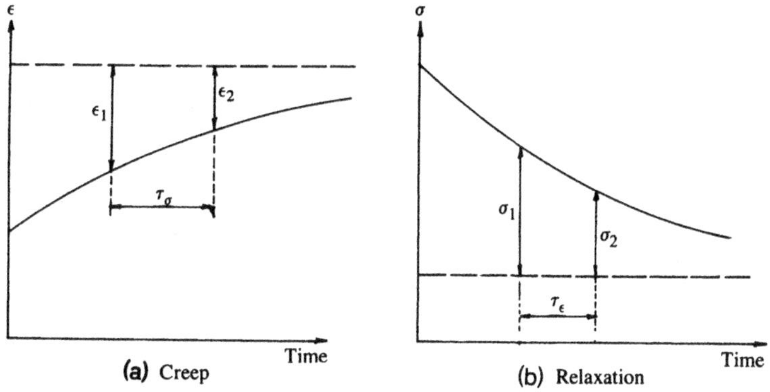


**FIGURE 5.8.** The physical meaning of $\tau_\sigma$ and $\tau_\epsilon$.

where

$$\tau_\epsilon = \frac{\mu}{E_1 + E_2} \tag{5.9}$$

$\tau_\epsilon$ is called the *retardation time*. The subscript $\epsilon$ denotes that the strain is held constant. It is observed that

$$\frac{\tau_\sigma}{\tau_\epsilon} = \frac{J_U}{J_R}$$

The initial condition for this problem is somewhat awkward. The strict definition of relaxation requires that an initial strain be applied instantaneously, but is is impossible to do this in practice because of the presence of viscous element in the model. However, we set the initial condition as

$$\sigma_o = \epsilon E_1.$$

Using such a condition, the solution for the stress is given by

$$\sigma = \epsilon \frac{E_1 E_2}{E_1 + E_2}\left[1 + \frac{E_1}{E_2}\exp\left(-\frac{t}{\tau_\epsilon}\right)\right] \tag{5.10}$$

This equation represents a stress that decays with time. Its similarity to the discharge of a capacitor is worth noting. At time $t = 0$, the initial conditions give the initial unrelaxed modulus, $E_U$. Likewise, at time $t \to \infty$, we obtain the fully relaxed modulus $E_R$. It is easy to see that the moduli *are the same as for creep*. The modulus at any time $t$ is rearranged as

$$\left.\begin{matrix}\text{Relaxation}\\ \text{Modulus}\end{matrix}\right\} E(t) = E_1 - \frac{E_1^2}{E_1 + E_2}\left[1 - \exp\left(-\frac{t}{\tau_\epsilon}\right)\right] \tag{5.11}$$

The trend of the variation of $E(t)$ with time is also plotted in Figure 5.7. The relaxation modulus is less than the inverse of the creep compliance for all values of $t$, except that at very small and very large times, they are equal.

The meaning of retardation time $\tau_\epsilon$ is similar too. It is the time interval needed for the drop in the time dependent part of the stress by a factor of

$e$. Referring to Figure 5.8(b), this means

$$\frac{\sigma 2}{\sigma_1} = \frac{1}{e}$$

### 5.4.4 Recovery Behavior Predicted by the Standard Model

The recovery behavior assumes that the creep precedes it and that at the instant of removal of load, the stress and strain are independent quantities as far as what follows the removal of the load. The recovery is simply a *negative creep*, starting concurrently at the instant of load removal. This fact can be expressed by changing $\sigma$ in Equation 5.4 to $\sigma[1 - U(t)]$ where $U(t)$ is the unit step function, defined by

$$U(t - t') = \begin{cases} 0, & \text{if } t < t' \\ 1, & \text{otherwise} \end{cases}$$

It simply states that just before time $t = t'$ (in this case $t' = 0$) there is stress and no stress is present thereafter. The differential equation for recovery is

$$\dot{\epsilon} + \frac{1}{\tau_\sigma}\epsilon = \frac{E_1 + E_2}{E_1 \mu}\sigma[1 - U(t)] \tag{5.12}$$

The initial conditions are

$$\text{at } t = 0, \epsilon = \epsilon_0 \text{ and } \sigma = \sigma_0$$

$\epsilon_0$ and $\sigma_0$ cannot be assumed to be related because the initial state was arrived at in a history dependent manner. With these initial conditions, the solution for recovery is given by

$$\epsilon = \left[\epsilon_0 - \left(\frac{E_1 + E_2}{E_1 E_2}\right)\sigma_0\right]\exp\left(-\frac{t}{\tau_\sigma}\right) \tag{5.13}$$

As $t$ becomes larger, the strain tends to zero, denoting complete recovery.

**Example 5.1**

Polyethylene has a short term Young's modulus of 148.9 MPa (21600 psi). Its apparent modulus at the end of 10 years is known to be about 78.3 MPa

(11350 psi), and this value may be assumed to be its fully relaxed modulus. $\mu$ may be taken as $1.2 \times 10^{12}$ Pa.s. Find the spring constants $E_1$ and $E_2$ assuming that the material is represented by the standard model. Calculate the relaxation time, retardation time, relaxation modulus, and creep compliance.

*Solution*

$$\text{Young's Modulus} = E_U = E_1 = 148.9 \text{ MPa}$$

$$\text{Relaxed Modulus} = E_R = \frac{E_1 E_2}{E_1 + E_2} = 78.3 \text{ MPa}$$

From the second relation, $E_2$ is obtained as 165.14 MPa.

$$\text{Relaxation Time, } \tau_\sigma = \frac{\mu}{E_2} = \frac{1.2 \times 10^{12} \text{ Pa.s}}{165.14 \times 10^6 \text{ Pa}} = 7266s = 2.018h$$

$$\text{Retardation Time, } \tau_\epsilon = \frac{\mu}{E_1 + E_2} = \frac{1.2 \times 10^{12} \text{ Pa.s}}{314.04 \times 10^6 \text{ Pa}} = 3821s = 1.061h$$

$$\text{Relaxation Modulus, } E(t) = E_1 - \frac{E_1^2}{E_1 + E_2}\left[1 - \exp\left(-\frac{t}{\tau_\epsilon}\right)\right]$$

$$= 148.9 - \frac{148.9^2}{314.04}\left[1 - \exp\left(-\frac{t}{1.061}\right)\right]$$

$$= 78.3 + 70.6 \exp - \left(\frac{t}{1.061}\right)$$

$$\text{Creep Compliance, } J(t) = \frac{1}{E_1} + \frac{1}{E_2}\left[1 - \exp\left(-\frac{t}{\tau_\sigma}\right)\right]$$

$$= \left[1.278 - 0.606 \exp\left(-\frac{t}{2.018}\right)\right] \times 10^{-2} \text{ MPa}^{-1}$$

***Commentary***

*The expression for $E(t)$ reveals the inadequacy of just one element in the standard model. For example, according to the values of the constants $E_1$, $E_2$, and $\mu$, the fully relaxed modulus is reached in just about 10 hours, whereas* *the experiments show that it is reached only after 10 years! This explains the need for a continuous distribution of $\tau_{**}$'s.*

**Example 5.2**

For PVC an approximate single element standard model can be constructed using $E_1 = 3840$ MPa, $E_2 = 10110$ MPa, and $\mu = 5 \times 10^{14}$ Pa.s. Discuss (a) the strain response of this material for a constant stress rate and (b) the stress response for a constant strain rate.

*Solution*

***Constant Stress Rate $\dot{\sigma}$***
We may substitute the condition $\sigma = \dot{\sigma} t$ in Equation 5.3 and get

$$\frac{d\epsilon}{dt} + \frac{\epsilon}{\tau_\sigma} = \left(\frac{E_1 + E_2}{\mu E_1}\right)\dot{\sigma} t + \frac{\dot{\sigma}}{E_1}$$

Solving this differential equation we get,

$$\epsilon = A \exp\left(-\frac{t}{\tau_\sigma}\right) + \dot{\sigma} t \frac{E_1 + E_2}{E_1 E_2} - \frac{\dot{\sigma}\mu}{E_2^2}$$

At time $t = 0$, strain is zero, $\epsilon = 0$. Hence, $A = \dot{\sigma}\mu / E_2^2$. Thus the complete solution is

$$\epsilon = \frac{\sigma}{E_R} - \frac{\dot{\sigma}\mu}{E_2^2}\left[1 - \exp\left(-\frac{t}{\tau_\sigma}\right)\right]$$

This can be also written as

$$\epsilon = \sigma J(t) - \frac{\sigma}{E_2}\left[\frac{\tau_\sigma}{t} + \exp\left(-\frac{t}{\tau_\sigma}\right)\left(1 - \frac{\tau_\sigma}{t}\right)\right]$$

The first term is the isochronous strain, and the second term is always positive. Thus, the strain is always less than the isochronous strain, that is strain lags behind stress. Substituting numerical values,

$$\tau_\sigma = \frac{\mu}{E_2} = \frac{5 \times 10^{14}}{10110 \times 10^6} = 49456s = 13.7h$$

$$E_R = \frac{E_1 E_2}{E_1 + E_2} = 2783 \text{ MPa}$$

With these, we get

$$\epsilon = \dot{\sigma}[13.55e^{-0.00136t} + 3.59t - 13.55].10^{-4} \text{ mm/mm}$$

Note that $\dot{\sigma}$ must be in MPa/hr, and $t$ in hours.

***Constant Strain Rate, $\dot{\epsilon}$***

Substituting the condition $\epsilon = \dot{\epsilon}t$ in Equation 5.3, we get

$$\dot{\sigma} + \frac{\sigma}{\tau_\epsilon} = E_1\dot{\epsilon} + \frac{E_1\dot{\epsilon}t}{\tau_\sigma}$$

With the boundary condition that at $t = 0$, $\sigma = 0$, this equation can be solved as

$$\sigma = E_1\dot{\epsilon}t + \frac{E_1}{E_2}E_R\dot{\epsilon}\tau_\epsilon\left[1 - \exp\left(-\frac{t}{\tau_\epsilon}\right)\right]$$

The first term is the isochronous stress, and the second term is always positive. This once again shows that strain lags behind the stress. Substituting the numerical values,

$$\tau_\epsilon = \frac{\mu}{E_1 + E_2} = 35842s = 9.96h$$

$$\sigma = \dot{\epsilon}[3840t + 10528(1 - e^{-0.1004t})] \text{ MPa}$$

Note that $\dot{\epsilon}$ must be in mm/mm/hour, and $t$ in hours.

***Commentary***

*In all rate processes, the strain always lags behind the isochronous strain. This is important especially in tensile testing plastics materials that creep easily. If the machine is one that applies displacement (rather than load), then the rate of displacement is directly responsible for the additional stress, leading the strain, as seen from the second term. This can result in fictitiously higher apparent modulus for the material.*

**Example 5.3**

By using a spring of high stiffness in parallel with a standard model, show that metal-insert molded components mitigate creep, and improve relaxation and recovery.

*Solution* First of all, there is a certain lack of rigor in the way the proof is required. This is because the viscoelastic model is for a material point in the plastic, and not for a structural configuration. Metal inserts help reduce creep by the load-sharing pattern; metal takes almost 90% of the stress and the plastic only 10%. The creep applies only to this small value of stress, and that appears in the overall as the dimensional stability of the structure. Adding a stiff spring in parallel to the standard model is therefore a crude approximation. Nonetheless we may proceed to do so, because it demonstrates the order of magnitudes involved.

From Figure 5.9, we can set up the differential equation of the model as

$$\sigma = \epsilon E_0 + \frac{\epsilon}{\dfrac{1}{E_1} + \dfrac{1}{E_2\,\mu\dfrac{d}{dt}}}$$

We note that the metal modulus $E_0 \gg E_1$, $E_2$, and may be up to 100 times

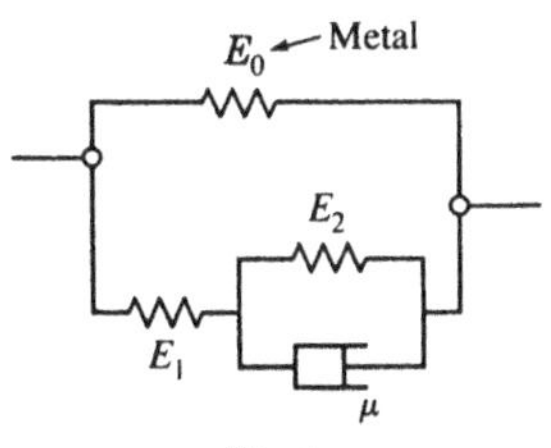


**FIGURE 5.9.** A composite structure consisting of a metallic element and a viscoelastic element sharing load in parallel can be represented as above. $E_0$ represents the stiffness of the metal and the standard model represents the viscoelastic material.

larger. We also note that the symbol $d/dt$ is an "operator." Simplifying,

$$\mu\frac{d\sigma}{dt} + (E_1 + E_2)\sigma = (E_0E_1 + E_1E_2 + E_2E_0)\epsilon + \mu(E_0 + E_1)\frac{d\epsilon}{dt}.$$

***Creep***

Consider a constant stress $\sigma_0$, and $\dot{\sigma} = 0$. The equation for the model becomes,

$$\mu(E_0 + E_1)\frac{d\epsilon}{dt} + (E_0E_1 + E_1E_2 + E_2E_0)\epsilon = (E_1 + E_2)\sigma_0$$

We may define the relaxation time of this model as

$$\tau_\epsilon = \mu\frac{E_0 + E_1}{E_0E_1 + E_1E_2 + E_2E_0} \approx \frac{\mu}{E_1 + E_2},$$

which is the same as for the plain plastic material. Thus,

$$\frac{d\epsilon}{dt} + \frac{\epsilon}{\tau_\epsilon} = \frac{(E_1 + E_2)}{\mu(E_0 + E_1)}\sigma_0$$

The boundary condition is that at $t = 0$, $\epsilon = \sigma_0/(E_1 + E_0)$, being the elastic strain.

$$\epsilon(t) = A\exp\left(-\frac{t}{\tau_\epsilon}\right) + \frac{\sigma_0}{E_1 + E_0}$$

The use of the boundary condition renders $A = 0$, which means that the model is creepless!. However, if we did not use the approximation that $E_0 \gg E_1$, and $E_2$, then $A$ would not vanish exactly, and would leave some small amount of creep taking place.

***Relaxation***

We modify the general equation with the condition that a constant strain $\epsilon_0$ is applied starting from time $t = 0$.

$$\frac{d\sigma}{dt} + \frac{\sigma}{\tau_\epsilon} = \frac{1}{\mu}(E_0E_1 + E_1E_2 + E_2E_0)\epsilon_0$$

Using the initial condition that at $t = 0$, $\sigma = \epsilon_0(E_1 + E_0)$, this equation can be solved to give

$$\sigma = \epsilon_0(E_0 + E_1) - \frac{E_1^2}{E_1 + E_2}\epsilon_0\left[1 - \exp\left(-\frac{t}{\tau_\epsilon}\right)\right]$$

The first term is the elastic component of stress, and is time independent. The second term represents a quantity that is of the order of 1/50th to 1/100th of the magnitude of the first term, and also decays with time. This shows that the portion of stress that undergoes relaxation is limited to about 2% of the initial stress. The creep components of strain being only a small part of the total, recovery is limited to this part. Once again, in the overall, recovery appears to be very fast.

## 5.5 THE TIME SPECTRA CONCEPT

Despite its merits in describing creep, relaxation, and recovery, the standard model does not predict the useful structural life of plastic very well. For example, the reduction in the modulus over an interval of $4\tau_\epsilon$ is equal to 1.83% of its initial value. Thus, the useful life of a plastic predicted by the standard model is only about $4\tau_{**}$. This does not conform with observation. Most plastics have a lifetime that exceeds this by about three or four decades. This difficulty is circumvented by assuming several elements in the viscoelastic model, with various relaxation times. We know that the primary spring gives the material its Young's modulus, $E_1$, and that the secondary spring and dashpot affect the values of $\tau_\epsilon$ and $\tau_\sigma$. Hence, in order to fit the model with real life, we need just one primary spring and several secondary elements. Consider the series of Voigt and Maxwell models shown in Table 5.1 in combination with one primary spring. Looking at the operators in the stress strain equations, we see that,

- each secondary element contributes a characteristic time $\tau$,
- there are as many $\tau$'s as the number of secondary elements available in the model, and
- the time dependent part of the modulus (or compliance) results from the complementary function of the differential equation, and hence is the sum of the contributions of each element.

A larger number of Maxwell models in parallel leads to a relaxation modulus of the form

$$E(t) = \sum_{i=1}^{n} E_i \exp\left(-\frac{t}{\tau_i}\right)$$

An extension of this concept is to consider that there are *infinite* elements in the model. The elements are such that their characteristic times are continuously distributed. In this case,

$$E(t) = E_R + \int_0^\infty H(\tau)\exp\left(-\frac{t}{\tau_i}\right) d\tau \tag{5.14}$$

The term $E_R$ denotes the contribution of the primary spring. In the same manner, an infinite number of Voigt models in series gives a creep compliance of the form

$$J(t) = J_U + \int_0^\infty L(\tau)\left[1 - \exp\left(-\frac{t}{\tau}\right)\right] d\tau \tag{5.15}$$

The functions $H(\tau)$ and $L(\tau)$ are respectively called the *relaxation time spectrum* and the *retardation time spectrum*[3] for the material. These functions are important in the understanding of chemical mechanisms of relaxation, rather than for stress analysis.

## 5.6 DYNAMIC BEHAVIOR OF LINEAR VISCOELASTIC MATERIALS

Under strain cycling and stress cycling, viscoelastic materials display more aspects, involving new properties. Additional apparent moduli are exhibited by the material under stress and strain cycling. At first, the Standard Model can be used to elucidate the behavior, and later an infinite string of Maxwell and Voigt models can be considered. Basically, we investigate the dependence of the responses upon the frequency of cycling. The motivation for this study is as follows. The ultimate aim is to be able to determine the modulus of the material at any time—usually long times. Since real time tests for long durations are expensive to perform, it is expedient to obtain the desired property as a function of frequency and subsequently to

[3] Some authors prefer to use the more expressive terminology "strengths of relaxation and retardation."

convert it to a function of time. Such a conversion is obtained by a mathematical process known as Fourier Transforms.

$$\begin{array}{c}\text{Fourier Transform}\\ \left.\begin{array}{c}\text{Frequency Dependent}\\ \text{Function}\end{array}\right\} \xrightarrow{\downarrow} \left\{\begin{array}{c}\text{Time Dependent}\\ \text{Function}\end{array}\right.\end{array}$$

### 5.6.1 Stress Cycling of the Standard Model

We consider a sinusoidal variation in the applied stress of the form

$$\sigma = \sigma_0 \cos \omega t$$

We once again begin with the differential equation of the Standard Model, which takes the form

$$\frac{d\epsilon}{dt} + \frac{\epsilon}{\tau_\sigma} = \sigma_0 \left( \frac{E_1 + E_2}{\tau_\sigma E_1 E_2} \right) \cos \omega t - \sigma_0 \frac{\omega}{E_1} \sin \omega t \tag{5.16}$$

No initial conditions are needed since we are interested only in the steady state solution, which is given by the particular solution of this equation. Solving, we get

$$\epsilon = A \cos \omega t + B \sin \omega t \tag{5.17}$$

where,

$$A = \sigma_0 \left[ \frac{1}{E_1} + \frac{1}{E_2(1 + \omega^2 \tau_\sigma^2)} \right]$$

$$B = \sigma_0 \frac{\omega \tau_\sigma}{E_2(1 + \omega^2 \tau_\sigma^2)}$$

Since $B$ is always positive, the strain can be expressed as,

$$\epsilon = \sqrt{A^2 + B^2} \cos(\omega t - \delta) \tag{5.18}$$

where,

$$\tan \delta = \frac{B}{A} = \frac{E_1 \omega \tau_\sigma}{E_1 + E_2(1 + \omega^2 \tau_\sigma^2)} \tag{5.19}$$

showing that

the strain lags behind the applied stress.

Also, the term $A \cos \omega t$ represents the in-phase component of strain, and $B \sin \omega t$ the out-of-phase component. Therefore, we define an *in-phase* or *storage modulus* $E'_\sigma$ by the following (subscripts "$\sigma$" denote stress cycling).

$$\frac{1}{E'_\sigma} = \frac{1}{E_1} + \frac{1}{E_2(1 + \omega^2 \tau_\sigma^2)} \tag{5.20}$$

The *out-of-phase* or *loss modulus* $E''_\sigma$ is likewise defined as

$$\frac{1}{E''_\sigma} = \frac{\omega \tau_\sigma}{E_2(1 + \omega^2 \tau_\sigma^2)} \tag{5.21}$$

The out-of-phase modulus is called loss modulus because the energy per unit volume dissipated from the specimen is directly proportional to this modulus. (See Example 5.6.) It is also the basis for calculating the power required to drive the stress cycling. The total energy needed is obtained by integrating the strain energy density over the volume of the material, for a given time.

Note that the $E'_\sigma$ and $E''_\sigma$ obtained above are valid only for stress cycling, and are different from the $E(t)$ obtained from static loading tests.

### 5.6.2 Strain Cycling of the Standard Model

Strain cycling is analyzed by starting with the boundary condition

$$\epsilon = \epsilon_0 \cos \omega t$$

The equation for the standard model is modified to

$$\dot{\sigma} + \frac{\sigma}{\tau_\epsilon} = \epsilon_0 \frac{E_1}{\tau_\sigma} \cos \omega t + \epsilon_0 (E_1 \omega) \sin \omega t \tag{5.22}$$

The steady state solution for this equation is given by

$$\sigma = A' \cos \omega t + B' \sin \omega t \tag{5.23}$$

where,

$$A' = \epsilon_0 \left( \frac{E_1 E_2}{E_1 + E_2} \right) \left[ 1 + \frac{E_1}{E_2} \frac{\omega^2 \tau_\epsilon^2}{1 + \omega^2 \tau_\epsilon^2} \right]$$

$$B' = \epsilon_0 \left( \frac{E_1^2}{E_1 + E_2} \right) \left[ \frac{\omega \tau_\epsilon}{1 + \omega^2 \tau_\epsilon^2} \right]$$

We see that the storage and loss moduli exhibited by the material in strain cycling are different from those obtained in stress cycling. They may be defined as follows:

$$E'_\epsilon = \left(\frac{E_1 E_2}{E_1 + E_2}\right)\left[1 + \frac{E_1}{E_2}\frac{\omega^2\tau_\epsilon^2}{1 + \omega^2\tau_\epsilon^2}\right] \quad \text{(In-Phase)} \tag{5.24}$$

$$E''_\epsilon = \left(\frac{E_1^2}{E_1 + E_2}\right)\left[\frac{\omega\tau_\epsilon}{1 + \omega^2\tau_\epsilon^2}\right] \quad \text{(Out-of-Phase)} \tag{5.25}$$

$$\tan\delta = \frac{\omega\tau_\epsilon}{\left[\omega^2\tau_\epsilon^2 + \dfrac{E_1}{E_2}(1 + \omega^2\tau_\epsilon^2)\right]} \tag{5.26}$$

The subscripts $\epsilon$ denote strain cycling. As before, we can express the stress as the sum of in-phase and out-of-phase components and observe that

the stress leads the strain, also in strain cycling.

It can also be shown that the phase angle $\delta$ is the same for both types of cycling.

### 5.6.3 Complex Modulus and Compliance

Since stress and strain have in-phase and out-of-phase components, we can draw an analogy from the theory of complex numbers. By doing this, it is possible to see the relationship between the various moduli. Let us consider the stress cycling in the first place. We may treat the stress $\sigma_0 \cos\omega t$ as a component of the vector $\sigma_0$. So also, the strain can be considered to be the components of another vector, at an angle to the stress vector.

$$\sigma = \sigma_0 \exp i\omega t$$

$$\epsilon = \epsilon_0 \exp i(\omega t - \delta)$$

$$E^* = E'_\sigma + iE''_\sigma = \frac{\sigma_0}{\epsilon_0}\exp i\delta = |E^*|\exp i\delta$$

Thus,
$$E'_\sigma = |E^*|\cos\delta$$
$$E''_\sigma = |E^*|\sin\delta$$
$$\tan\delta = \frac{E'_\sigma}{E''_\sigma}$$

If we consider the strain cycling, then again we can derive a similar set of equations, giving the complex compliance.

$$\epsilon = \epsilon_0 \exp i(\omega t - \delta)$$
$$\sigma = \sigma_0 \exp i\omega t$$
$$J^* = J' - iJ'' = \frac{\epsilon_0}{\sigma_0}\exp(-i\delta) = |J'|\exp(-i\delta)$$

Thus,
$$j' = |J^*|\cos\delta$$
$$J'' = |J^*|\sin\delta$$
$$\tan\delta = \frac{J'}{J''}$$

It can be shown that the complex compliance and complex modulus are inverses of each other.

$$J^* = \frac{1}{E^*} \tag{5.27}$$

The above equation relates the individual components of the compliance and modulus.

**Example 5.4**

A rheological testing machine is a specialized machine that can retrieve in-phase and out-of-phase moduli under cyclic loading conditions. Such a machine was used to test a blend of polypropylene in a cyclic bending stress program. The same specimen is tested in two different frequencies of loading at the same temperature. The following results were obtained.

At $\omega_1 = 500$ rad/s, $E' = 3.962 \times 10^3$ MPa, and
$$E'' = 2.084 \times 10^2 \text{ MPa}$$
At $\omega_1 = 6.28$ rad/s, $E' = 3.330 \times 10^3$ MPa, and
$$E'' = 1.440 \times 10^2 \text{ MPa}$$

Assume that the material can be treated as a single element standard model. Find $E_1$, $E_2$, and $\mu$.

*Solution* We need to use Equations 5.20 and 5.21 repeatedly. Simply by substitution we obtain the following four equations.

$$\frac{1}{3.962 \times 10^3} = 0.254 \times 10^{-3} = \frac{1}{E_1} + \frac{E_2}{E_2^2 + \mu^2\omega_1^2} \tag{5.28}$$

$$\frac{1}{3.330 \times 10^3} = 0.3003 \times 10^{-3} = \frac{1}{E_1} + \frac{E_2}{E_2^2 + \mu^2\omega_2^2} \tag{5.29}$$

$$\frac{1}{2.084 \times 10^2} = 0.47985 \times 10^{-2} = \frac{\mu\omega_1}{E_2^2 + \mu^2\omega_1^2} \tag{5.30}$$

$$\frac{1}{1.44 \times 10^2} = 0.69444 \times 10^{-2} = \frac{\mu\omega_2}{E_2^2 + \mu^2\omega_2^2} \tag{5.31}$$

Subtract Equation 5.28 from Equation 5.29

$$0.0479 \times 10^{-3} = E_2\left[\frac{1}{E_2^2 + \mu^2\omega_1^2} - \frac{1}{E_2^2 + \mu^2\omega_2^2}\right] \tag{5.32}$$

Divide Equation 5.30 by Equation 5.31 to get

$$(E_2^2 + \mu^2\omega_1^2) = \frac{\omega_1}{\omega_2}(E_2^2 + \mu^2\omega_2^2) \tag{5.33}$$

Substitute Equation 5.33 in Equation 5.32

$$\frac{E_2}{E_2^2 + \mu^2\omega_2^2} = 4.1937 \times 10^{-7} \tag{5.34}$$

Use Equation 5.34 in Equation 5.28

$$E_1 = 3968 \text{ MPa}$$

From Equation 5.33,

$$\mu^2(\omega_1^2 - 115.22\,\omega_2^2) = 114.22 E_2^2$$

which gives,

$$\mu = 0.02157\, E_2$$

Using this in Equation 5.30, we can extract

$$E_2 = 19.157 \quad \text{MPa}$$

$$\mu = 0.4132 \quad \text{MPa.s}$$

**Example 5.5**

Show that only the out-of-phase compliance (or modulus) is responsible for the power dissipation in a sinusoidal loading of a viscoelastic material.

*Solution*

Let

$$\sigma = \sigma_0 \cos \omega t$$

$$\epsilon = \sigma_0[J' \cos \omega t \; - J'' \sin \omega t]$$

Power dissipated in a small volume of the material per cycle is given by

$$\Delta P = \int_0^{2\pi/\omega} \sigma \frac{d\epsilon}{dt}\, dt$$

$$= -\sigma_0^2 \omega \int_0^{2\pi/\omega} \cos \omega t [J' \sin \omega t + J'' \cos \omega t]\, dt$$

$$= -\sigma_0^2 \omega \int_0^{2\pi/\omega} [J' \cos \omega t \sin \omega t + J'' \cos^2 \omega t]\, dt$$

$$= -\sigma_0^2 \omega \left[0 - J''.\frac{\pi}{\omega}\right]$$

$$= \pi \sigma_0^2 J''$$

If the stresses and strains throughout the material volume are within the linear viscoelastic range, then the summation over the volume of the element will depend only on $J''$.

For strain cycling also, a similar calculation can be performed leading to the same conclusion.

**Example 5.6**

The standard viscoelastic model of a plastic has in-phase and out-of-phase compliances of $J' = 2.518 \times 10^{-4}$ $(\text{MPa})^{-1}$, and $J'' = 0.1323 \times 10^{-4}$ $(\text{MPa})^{-1}$ at a sub zero temperature. A 300 mm long bending specimen of width 30 mm and depth 10 mm (Figure 5.10) is loaded as a simply supported beam with a central load of 25 N at a frequency 80 Hz. Find the energy dissipated by the specimen.

*Solution* The maximum magnitude of stress $\sigma_0$ is given by

$$\sigma_0 = \frac{3}{2}\frac{PL}{BD^2} = 3.75\ \text{N/mm}^2$$

At any point in the specimen, distance $x$ from a support, and $y$ down below the neutral axis, the stress is proportional to $x$ as well $y$. It can be seen to be

$$\sigma = \sigma_0 \frac{x}{150}\frac{2y}{D} = 0.05\frac{xy}{D}$$

Due to the sinusoidal variation, we may write it as

$$\sigma = 0.05\frac{xy}{D}\cos\omega t$$

$$\epsilon = \sigma_0[J'\cos\omega t - J''\sin\omega t]$$

$$\dot{\epsilon} = -\omega\sigma_0[J'\sin\omega t + J''\cos\omega t]$$

$$\frac{d\mathscr{E}}{dV} = \int_0^{2\pi/\omega} \sigma\dot{\epsilon}\,dt \quad \text{per cycle, per unit volume.}$$

$$= \pi J''\left(\frac{0.05xy}{D}\right)^2.$$

Compare this with the results of Example 5.5. The total energy dissipated for the whole specimen is obtained by integrating this over the volume of

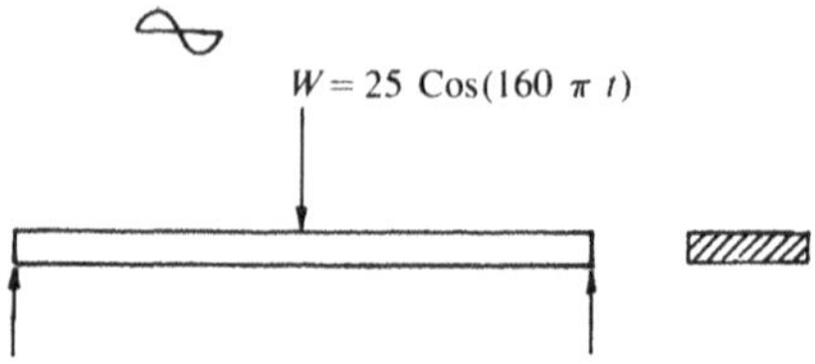


**FIGURE 5.10.** Viscoelastic beam subjected to cyclic loading.

the specimen. Since $x$ is taken for the left half only,

$$\mathscr{E} = 2 \times 2 \times \frac{\pi J'' 0.0025}{D^2} \int_x \int_y x^2 y^2 B \, dx \, dy \quad \text{per cycle}$$

$$= \frac{4\pi J''(0.0025)}{D^2} B . \left( \frac{x^3}{3} \right) \Big|_0^{L/2} \left( \frac{y^3}{3} \right) \Big|_0^{D/2}$$

$$\times 0.01 \frac{\pi J'' B D L^3}{576}$$

$$= 5.845 \text{ N.mm per cycle}$$

$$= 5.845 \times 80 = 467.6 \text{ N.mm/sec}$$

$$= 0.4676 \text{ Watts}$$

### 5.6.4 Closing Remarks on the Spring Dashpot Models

The spring dashpot models are useful in developing qualitative thinking, in the first place. The time spectra concept helps, in principle, in retro-fitting the time based data for a real life material, without having to determine the individual spring and dashpot constants. The time spectra concept is a technique for the determination of the combined effect.

However, this approach is seldom taken for material characterization. Alternative approaches are available to retrieve long term properties from short term tests. One such approach is to take rheological measurements from a dynamic test as a function of frequency, and convert them to functions of time by Fourier transforms. Another approach is to perform several short term tests at various temperatures. By using a principle known as the *Time-Temperature Transformation* or the *Williams-Landel-Ferry* (WLF) equations, it is possible to transform higher temperature data to equivalent long time data. A cursory discussion of these two methods is given in Section 5.15.2.

For actual structural calculations, Boltzmann's Superposition Principle appears to be more powerful.

## 5.7 BOLTZMANN'S SUPERPOSITION PRINCIPLE

Earlier, it was stated that the Boltzmann's Superposition Principle (BSP) is a way of describing the various aspects of viscoelasticity. Also, this principle assumes only the linear additivity of strains. We can develop the BSP equation for creep in three steps, as shown in Figure 5.11.

In the first step (Figure 5.11(a)), consider the removal of a stress $\sigma_1$ at time $t = t_1$. BSP states that the strains due to the removal of stress can be treated as the addition of a negative creep strain, starting from time $t_1$. Thus, the strain after stress removal is given by

$$\epsilon(t) = \sigma_1 J(t) - \sigma_1 J(t - t_1)$$

Incidentally, this also can be treated as the description of recovery. Note carefully that the term "recovered strain" means the difference between the strain if stress were not removed, and the strain resulting from the removal of stress, both strains being calculated for the same time, $t$.

In the second step (Figure 5.11(b)), consider an increment from stress level $\sigma_1$ to $\sigma_2$, at time $\tau = t_1$. For this purpose, we consider that $\sigma_1$ is removed, and $\sigma_2$ added. Because of the removal of $\sigma_1$, there is a strain

$$\epsilon_1 = \sigma_1[J(t) - J(t - t_1)]$$

The addition of $\sigma_2$ causes another strain $\epsilon_2$ which begins at time $t_1$.

$$\epsilon_2 = \sigma_2 J(t - t_1)$$

The actual strain is the sum of these two strains, and is given by

$$\epsilon(t) = \sigma_1 J(t) + (\sigma_2 - \sigma_1) J(t - t_1)$$

In the third step (Figure 5.11(c)), we consider an infinitesimal increment $d\sigma$ of the stress, taking place at time $\tau$. The increment in strain caused by this is $d\epsilon$, given by

$$d\epsilon = d\sigma J(t - \tau)$$

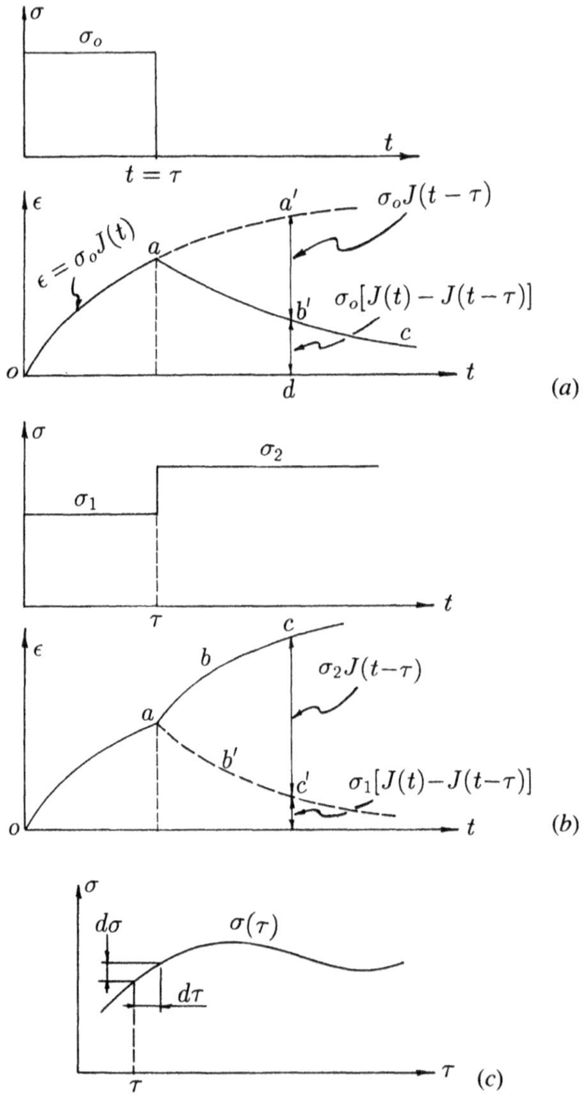


**FIGURE 5.11.** Development of the Boltzmann's superposition principle from physical reasoning.

For an entire history of the variation of the stress, we simply need to integrate the strain increments.

$$\epsilon(t) = \int_{-\infty}^{t} J(t-\tau)d\sigma$$

$$= \int_{-\infty}^{t} J(t-\tau)\frac{\partial\sigma}{\partial\tau}d\tau$$

The above equation translates as follows.

> The strain at any time $t$ is the result of all changes in the stress that took place until $t$. Also, in the beginning the material must be in a stress- and strain-free state.

For practical problems, it is informative to separate the time-independent portion of strain and also the compliance function and write it explicitly in the above equation. Thus,

$$\epsilon(t) = \frac{\sigma_0}{E} + \int_{-\infty}^{t} J(t-\tau)\frac{\partial\sigma}{\partial\tau}d\tau \tag{5.35}$$

and,

$$\sigma(t) = \epsilon_0 E_R + \int_{-\infty}^{t} E(t-\tau)\frac{\partial\epsilon}{\partial\tau}d\tau \tag{5.36}$$

Note that $J(t)$ and $E(t)$ in the above denote the history-dependent components *only*. Although not a standard notation, we will follow this convention in solving example problems.

### 5.7.1 Creep Predicted by BSP

The usefulness of BSP over the spring dashpot models is that in the latter, the compliance (or modulus) must always be an exponential function of time, whereas BSP allows the use of any reasonable curve fit. For constant temperature applications this is very useful, since tabular data commonly available from data-books can be curve-fitted and used directly in BSP. See examples in Section 5.11.

We assume that the stress $\sigma_0$ is applied at time $t = 0$. Therefore,

$$\sigma = \sigma_0 U(\tau)$$

$$\frac{\partial\sigma}{\partial\tau} = \sigma_0 \delta(\tau)$$

where $U(\tau)$ is the unit step function defined earlier, and $\delta(**)$ is Dirac's delta function or the impulse function, defined as

$$\delta(x - x_o) = \begin{cases} \infty, & \text{for } x = x_o \\ 0, & \text{elsewhere} \end{cases}$$

and, $$\int_{-\infty}^{\infty} f(x)\delta(x - x_o)\, dx = f(x_o).$$

$$\begin{aligned} \epsilon(t) &= \frac{\sigma_o U(0)}{E} + \int_{-\infty}^{t} J(t - \tau)\frac{\partial \sigma}{\partial \tau}\, d\tau \\ &= \frac{\sigma_0}{E} + \int_0^t J(t - \tau)\sigma_0 \delta(\tau)\, d\tau \\ &= \frac{\sigma_0}{E} + \sigma_0 J(t) \end{aligned} \tag{5.37}$$

The last equation is exactly the way phenomenological creep is described, since $J(t)$ is an observed material characteristic and is not derived from a material model. For example, it can be seen that BSP gives the same results for the standard model as does Equation 5.5. Because the time-dependent and time-independent components of $J(t)$ for this model are respectively

$$\frac{1}{E_2} \text{ and } \frac{E_1 + E_2}{E_1 E_2}$$

### 5.7.2 Relaxation Predicted by BSP

We start with the conditions,

$$\epsilon = \epsilon_0 U(\tau)$$

$$\frac{\partial \epsilon}{\partial \tau} = \epsilon_0 \delta(t)$$

$$\begin{aligned} \sigma(t) &= E_R \epsilon_0 + \int_{-\infty}^{t} E(t - \tau)\epsilon_0 \delta(\tau)\, d\tau \\ &= E_R \epsilon_0 + E(t)\epsilon_0 \end{aligned}$$

Once again, BSP is able to retrieve the phenomenological description of relaxation, without having to use any material model, and it is seen that BSP reproduces the results of the standard model as time-independent and time-dependent components of moduli, which are respectively

$$\frac{E_1 E_2}{E_1 + E_2} \quad \text{and} \quad \frac{E_1^2}{E_1 + E_2}$$

### 5.7.3 Recovery Predicted By BSP

Consider a stress applied for a duration $0 < t < t_1$ and removed suddenly at time $t_1$. This can be described by the function

$$\sigma = \sigma_0[U(t) - U(t - t_1)]$$

Substituting in BSP,

$$\epsilon(t) = \frac{\sigma_0}{E} + \int_0^{t_1} J(t - \tau)\sigma_0 \delta(\tau)\, d\tau - \left\{\frac{\sigma_0}{E} + \int_{t_1}^{t} \sigma_0 J(t - \tau)\delta(\tau - t_1)\, d\tau\right\}$$

$$= \sigma_0[J(t) - J(t - t_1)]$$

Note that in the above equation the elastic strains were subtracted out. This is the consequence of supposing that the removal of a stress is the equivalent of the addition of a new negative stress, which comes with its own negative *elastic* strain. This way, elastic strains are recovered instantaneously, and the creep strains are recovered over time. In fact, this equation is simply a step in the derivation of BSP.

### 5.7.4 Response to Sinusoidal Stresses Using BSP

Consider an input stress of

$$\sigma(\tau) = \sigma_0 \cos \omega\tau$$

$$\frac{\partial \sigma}{\partial \tau} = -\sigma_0 \omega \sin \omega\tau$$

$$\epsilon(t) = -\sigma_0 \omega \int_{-\infty}^{\infty} J(t - \tau) \sin \omega\tau\, d\tau$$

The upper limit of $\infty$ in the integration represents the response after sufficient time, or the steady state response. We may transform this equation by setting $(t - \tau) = s$, and obtain

$$\epsilon(t) = \sigma_0 \omega \int_{-\infty}^{\infty} J(s) \sin \omega(s - t)\, ds$$

$$= \sigma_0 \omega \int_{-\infty}^{\infty} J(s) J(s)[\sin \omega s \cos \omega t - \cos \omega s \sin \omega t]\, ds$$

$$= \sigma_0 \left\{ \omega \int_{-\infty}^{\infty} J(s) \sin \omega s\, ds \right\} \cos \omega t - \sigma_0 \left\{ \omega \int_{-\infty}^{\infty} J(s) \cos \omega s\, ds \right\} \sin \omega t$$

$$= \sigma_0 [f_1(\omega) \cos \omega t - f_2(\omega) \sin \omega t]$$

$$= \sigma_0 [J' \cos \omega t - J'' \sin \omega t]$$

Thus, a sinusoidal stress input causes a sinusoidal strain output, only lagging by a certain angle. This result is identical to what was obtained for spring dashpot models.

A similar conclusion is obtained also for strain cycling. Thus, BSP is able to explain every phenomenon of viscoelasticity as do the spring dashpot models.

**Example 5.7**

A polypropylene test specimen of size 2 mm × 20 mm × 250 mm is pulled along its length on a "hard" testing machine which applies a displacement of ends of 0.50 mm in one minute, and holds the ends in that position thereafter. Trace the history of stress in the specimen for 200 hours. Take the Young's modulus of the material as 825 MPa, and the time dependent modulus as $2000t^{-0.09}$ MPa, where $t$ is measured in seconds.

*Solution* The strain rate is given by

$$\dot{\epsilon} = \frac{0.50 \text{ mm}}{200 \text{ mm} \times 60 \text{ sec}} = 3.333 \times 10^{-5} \, (\text{sec})^{-1} \text{ in } 0 \leq t \leq 60$$

$$= 0, \text{ thereafter.}$$

The input strain history is shown inset in Figure 5.12. The stress history is

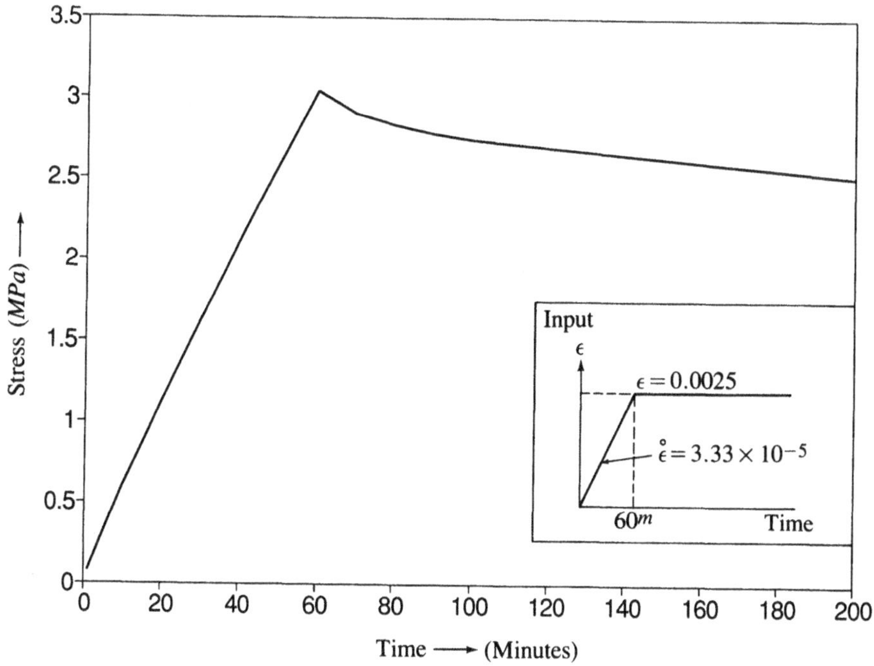


**FIGURE 5.12.** Response of a viscoelastic tension test specimen to a strain ramp input.

given by

$$\sigma(t) = \int_{-\infty}^{t} E(t-\tau)\dot{\epsilon}\, d\tau$$

$$= \int_{0}^{t} E(t-\tau)\dot{\epsilon}\, d\tau$$

$$= 2000 \times 3.333 \times 10^{-5} \int_{0}^{t} (t-\tau)^{-0.09}\, d\tau$$

The integral is split into two phases, first when the strain is being applied $(0 \le t \le 60)$, and the second when the strain is held constant $(t \ge 60)$.

***The Range*** $(0 \le t \le 60)$

$$\sigma(t) = 0.06666 \int_{0}^{t} (t-\tau)^{-0.09}\, d\tau$$

$$= 0.06666 \left[ \frac{(t-\tau)^{0.91}}{0.91} \right]_{t}^{0}$$

$$= 0.07325 t^{0.91}$$

***The Range (t ≥ 60)***

$$\sigma(t) = \int_0^{60} E(t-\tau)\dot{\epsilon}\,d\tau + \int_{60}^{\infty} E(t-\tau)\dot{\epsilon}\,d\tau$$

$$= \int_0^{60} (\cdots) + 0, \quad \text{since} \cdot \epsilon = 0\text{, for the second range.}$$

$$= -0.07325\,[(t-\tau)]|_0^{60}$$

$$= 0.07325\left[t^{0.91} - (t-60)^{0.91}\right]$$

Note that the calculations consider only the history dependent portion of the stress. The stress history is shown in Figure 5.12.

**Example 5.8**

A PVC bolt is subjected to an axial tensile stress history as shown in the inset in Figure 5.13. The relaxation modulus $E(t) = 78.73 \times 10^3(20.5 +$

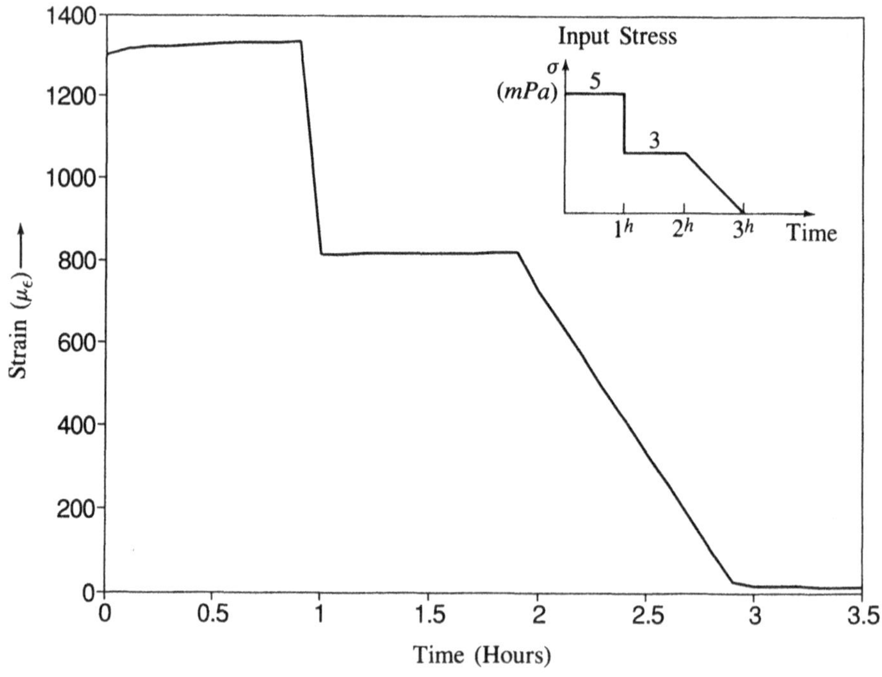


**FIGURE 5.13.** Use of BSP for several jump loads, involving creep and recovery.

$0.557t^{0.305})^{-1}$ MPa, ($t$ in hours). Using the approximation $E(t)J(t) \approx 1$, calculate the strain history in the bolt.

*Solution*

$$J(t) = \frac{1}{E(t)} = 12.7(20.5 + 0.557t^{0.305}) \text{ microstrains}$$

$$= (260.35 + 7.0739t^{0.305}) \text{ microstrains}$$

$$= (a + bt^n) \text{ for brevity.}$$

From the figure, we have the following for the stress rate.

$$\frac{\partial\sigma}{\partial\tau} = 5\delta(\tau) \text{ in } 0 \le t \le 1$$

$$= -2\delta(\tau) \text{ in } 1 \le t \le 2$$

$$= -3 \text{ in } 2 \le t \le 3$$

$$= 0 \text{ elsewhere.}$$

In the Range $0 \le t \le 1$

$$\epsilon(t) = \int_0^t 5\delta(\tau)J(t-\tau)\,d\tau = 5J(t)$$

$$= 5(a + bt^n)$$

$$= 1301.75 + 35.37t^{0.305}$$

In the Range $1 \le t \le 2$

$$\epsilon(t) = \int_0^1 (\cdots) - \int_1^t 2\delta(\tau - 1)J(t-\tau)\,d\tau$$

$$= 5(a + bt^n) - 2[a + b(t-1)^n]$$

$$= 781 + 35.37t^{0.305} - 14.15(t-1)^{0.305}$$

In the Range $2 \leq t \leq 3$

$$\epsilon(t) = \int_0^1 (\cdots) + \int_1^2 (\cdots) - \int_2^t 3J(t - \tau)\, d\tau$$

$$= 3a + 5bt^n - 2b(t-1)^n - 3a(t-2) - \frac{3b}{1+n}(t-2)^{1+n}$$

$$= 781 + 35.37t^{0.305} - 14.15(t-1)^{0.305}$$

$$- 781(t-2) - 16.26(t-2)^{1.305}$$

In the Range $t \geq 3$

$$\epsilon(t) = \int_0^1 (\cdots) + \int_1^2 (\cdots) + \int_2^3 (\cdots)$$

$$= 5bt^n - 2b(t-1)^n - \frac{3b}{1+n}\left[(t-2)^{1+n} - (t-2)^{1+n}\right]$$

$$= 35.37t^{0.305} - 14.15(t-1)^{0.305} - 16.26\left[(t-2)^{1+n} - (t-3)^{1+n}\right]$$

It may be verified that the jump in the value of strain, if any, is exactly equal to the *elastic* component. The history-dependent component of strain, as well as its time-derivative are continuous in the entire range of the problem. Also, after 3 hrs, the calculation automatically shows that strain is independent of $a$, that is, the instantaneous responses no longer occur, and only the recovery is taking place. Thus, BSP can track creep or recovery behavior automatically, since by the BSP one is the negative of the other. The strain response is shown in Figure 5.13.

## 5.8 USE OF LAPLACE TRANSFORMS IN BSP

A mathematical technique known as *Laplace Transform*[4] is particularly useful in the applications of BSP. The reader is referred to a brief treatment and catalog of results in Appendix 3. Standard undergraduate mathematics text books such as Reference 3 are recommended for a fuller

---

[4] If the reader is not familiar with Laplace Transforms, it may be advisable to read the material in Appendix 3 and Reference 3.

understanding of the technique. Basically, the Laplace Transform is a technique of transforming a differential equation—especially in time, $t$—into an algebraic equation containing a dummy variable $s$. A solution is obtained in the transformed variable $s$, which is inverse-transformed to obtain what is really the solution of the original differential equation.

The use of Laplace Transform in BSP is intuitive. The formal similarity of BSP with the *convolution* integral in Appendix 3, suggests that such a relationship must exist between the creep compliance and the relaxation modulus of the material. We can find the exact relationship as follows. Let $\bar{J}(s)$ and $\bar{E}(s)$ be the Laplace transforms of $J(t)$ and $E(t)$ respectively.

$$\int_0^{\infty} \exp(-st)J(t)\,dt = \bar{J}(s)$$

$$\int_0^{\infty} \exp(-st)E(t)\,dt = \bar{E}(s)$$

Using the idea of the convolution integral we see that the $L$-transform of the BSP equation itself is given by

$$\bar{J}(s)s\bar{\sigma} = \bar{\epsilon}$$

$$\bar{E}(s)s\bar{\epsilon} = \bar{\sigma}$$

Hence,
$$\bar{J}(s)\bar{E}(s) = \frac{1}{s^2} \tag{5.38}$$

The inverse of Equation 5.38 clearly involves the convolution of $J$ and $E$, on the left side. It is expressed as

$$\int_0^t J(t-\tau)E(\tau)\,d\tau = t$$

or,
$$\int_0^t E(t-\tau)J(\tau)\,d\tau = t \tag{5.39}$$

Thus, if one of the functions is known, then the other can be computed.

**Example 5.9**

The time dependent part of relaxation modulus $E(t)$ of PVC is given by $22800t^{-0.305}$ MPa. Calculate its creep compliance.

*Solution* In the following $\Gamma(\cdots)$ denotes the Euler Gamma or the factorial function. Tabulated values of Gamma functions are available in Reference 4. We use just one of its properties, namely

$$\Gamma(1-n)\Gamma(1+n) = \frac{\sin n\pi}{n\pi}$$

$$E(t) = 22800t^{-0.305}\ \text{MPa}.$$

$$\mathscr{L}[E(t)] = \bar{E}(s) = 22800\frac{\Gamma(1-0.305)}{s^{(1-0.305)}}$$

However,
$$\bar{J}(s) = \frac{1}{s^2\bar{E}(s)}$$

$$= \frac{1}{22800\Gamma(1-0.305).s^{(1+0.305)}}$$

$$\mathscr{L}^{-1}[\bar{J}(s)] = J(t) = \mathscr{L}^{-1}\left[\frac{1}{22800\Gamma(1-0.305).s^{(1+0.305)}}\right]$$

$$= 43.860\times 10^{-6}\frac{t^{0.305}}{\Gamma(1-0.305)\Gamma(1+0.305)}$$

$$= 43.860\times 10^{-6}\frac{\sin(0.305\pi)}{0.305\pi}t^{0.305}$$

$$= 37.450\times 10^{-6}t^{0.305}$$

***Commentary***

*This result is about 15% lower than the arithmetic inverse of the modulus. However, when the elastic and viscous effects are combined, the error appears to reduce to less than 0.5%. This demonstrates the practicality of the approximation*

$$J(t) = \frac{1}{E(t)},$$

*when the total effects are included in E and J.*

## 5.9 THE CORRESPONDENCE PRINCIPLE

The Laplace Transform can be used to derive the so-called *Correspondence Principle*. It is based on the similarity of the elasticity equations and the

Laplace transforms of the viscoelastic equations. Table 5.2 lists the family of equations in linear elasticity, and the transforms of the viscoelastic equations. The formal similarity in equilibrium, boundary conditions, stress strain equations, etc. is very striking. Note the operator expressions in the last row for stress-strain relations. The operators $\mathscr{P}$ and $\mathscr{Q}$ are the result of the differential operators for spring dashpot models in Table 5.1. $\overline{\mathscr{P}}$ and $\overline{\mathscr{Q}}$ are not the exact Laplace transforms of the operators, but are such that

$$L[\mathscr{P}\sigma] = \overline{\mathscr{P}}(s)\bar{\sigma}$$

$$L[\mathscr{Q}\epsilon] = \overline{\mathscr{Q}}(s)\bar{\epsilon}.$$

Note that the bulk modulus $K$, is *not transformed*, and is treated as just a constant. Also note that the *deviatoric stress* components $s_{ij}$ have been used. These are explained further below. The correspondence principle can be stated as follows.

> Due to the similarity of the equations in Table 5.2, the closed-form elastic solutions are also the Laplace transforms of the viscoelastic solution of the same problem, provided that
>
> - the modulus $G$ is replaced by $\overline{\mathscr{Q}}/\overline{\mathscr{P}}$,
> - the bulk modulus $K$ is left unchanged, and other elastic constants $E$ and $\nu$ are expressed in terms of $G$ and $K$.
>
> The inverse transform after the changes, gives the viscoelastic solution.

Figure 5.14 explains the sequence of steps involved in the use of correspondence principle. A clear but short treatment is available in Reference 2.

***Reexamination of Elasticity in Three Dimensions***

Let us take a different look at the linear elasticity equations in three dimensions. In particular, we split the stress strain equations into two parts, one involving volume change and the other involving zero volume change. In Chapter 1, Equations 1.23, it was shown that the stress can be expressed as the sum of two components—one that produces volume change, and another that does not. Using subscript notations for brevity,

$$\begin{aligned}\sigma_{ij} &= s_{ij} + \frac{1}{3}(\sigma_{xx} + \sigma_{yy} + \sigma_{zz}) \\ &= s_{ij} + \sigma_{ave},\end{aligned}$$

Table 5.2. Correspondence Principle

| | Elastic Eqns | Transformed Viscoelasticity |
|---|---|---|
| Equilibrium | 1. $\sigma_{ij,j} + b_i = 0$ | 1. $\overline{\sigma}_{ij,j} + \overline{b_i} = 0$ |
| Strain-Displacement | 2. $\epsilon_{ij} = \frac{1}{2}(u_{i,j} + u_{j,i})$ | 2. $\overline{\epsilon}_{ij} = \frac{1}{2}(\overline{u}_{i,j} + \overline{u}_{j,i})$ |
| Force and Displacement Boundary Conditions | 3. $\sigma_{ij} n_j = T_i^n$ on $S_1$<br>$u_i = u_i^*$ on $S_2$ | 3. $\overline{\sigma}_{ij} n_j = \overline{T_i^n}$ on $S_1$<br>$\overline{u_i} = \overline{u_i^*}$ on $S_2$ |
| Stress-Strain Relations | 4. $\sigma_{ij} = 2G\epsilon_{ij}$<br>$\sigma_{ii} = 3K\epsilon_{ii}$ | 4. $\overline{\mathcal{P}}(s)\overline{\sigma}_{ij} = 2\overline{\mathcal{Q}}(s)\overline{\epsilon}_{ij}$<br>$\overline{\sigma}_{ii} = 3K\overline{\epsilon}_{ii}$ |
| Alternate Stress-Strain Relations | 4. $\sigma_{ij} = (2G + \delta_{ik}\delta_{kj})\epsilon_{ij}$<br>In one dimension $\sigma = E\epsilon$ | 4. $\overline{\mathcal{P}}(s)\overline{\sigma_{ij}} = 2\overline{\mathcal{Q}}(s)\overline{\epsilon_{ij}}$<br>(Gives Pseudoelasticity) |

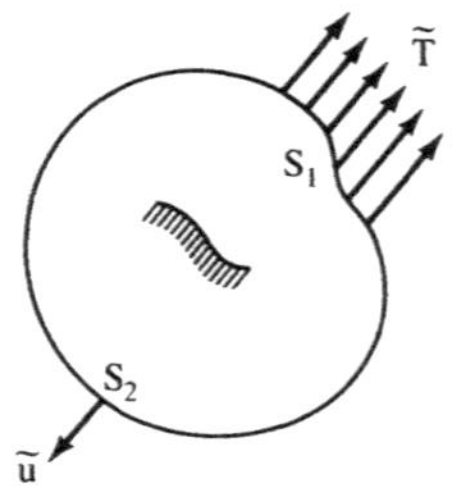

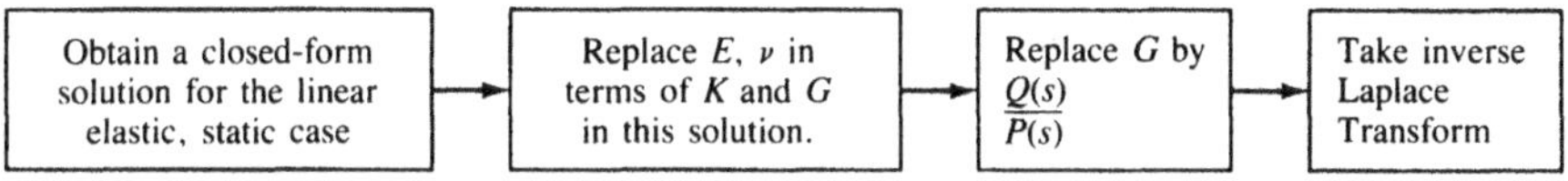


**FIGURE 5.14.** Steps involved in using correspondence principle for viscoelastic solution.

where, $\sigma_{ave}$ is the average normal stress.

$$\epsilon_{ij} = e_{ij} + \frac{1}{3}(\epsilon_{xx} + \epsilon_{yy} + \epsilon_{zz})$$
$$= e_{ij} + \epsilon_{ave},$$

where, $\epsilon_{ave}$ is the average normal strain. Equations 1.23 and 1.24 can be written in a slightly different way as follows.

$$\sigma_{ii} = s_{ii} + \sigma_{ave} = 3\lambda\epsilon_{ave} + 2G\epsilon_{xx}$$
$$= 3\lambda\epsilon_{ave} + 2G(e_{xx} + \epsilon_{ave})$$
$$= 2Ge_{xx} + (3\lambda + 2G)\epsilon_{ave} = 2Ge_{xx} + K\epsilon_{ave} \tag{5.40}$$

By the definition of $K$, the terms $\sigma_{ave}$ and $\epsilon_{ave}$ are connected by

$$\sigma_{ave} = 3K\epsilon_{ave}.$$

Thus, the stress-strain equations get split up into two parts as shown in Equation 5.41, one containing all of the volume expansion, and the other without it.

$$s_{ij} = 2Ge_{ij} \text{ and } \sigma_{ave} = 3K\epsilon_{ave} \tag{5.41}$$

In the above equations, the shear stresses are also included, by virtue of the symbolism used.

## 5.10 CORRESPONDENCE PRINCIPLE FOR 3-D VISCOELASTICITY

### 5.10.1 Practical Assumptions for 3-D Problems in Viscoelasticity

All elastic properties of thermoplastics vary to some extent with time. Of all of those properties, the variation of $K$ is the least, and hence is assumed to be a constant. Replacing $G$ by $\bar{\mathcal{Q}}/\bar{\mathcal{P}}$ introduces the dummy variable $s$ in the solution. Since only two elastic constants are needed for the description of the material, the other properties can be expressed in terms of these two. We may use

$$\nu(t) = \frac{3K - 2G(t)}{2(3K + G(t))} \quad \text{and} \quad E(t) = \frac{9KG(t)}{3K + G(t)} \tag{5.42}$$

Several other relations connecting elastic constants are needed in the course of problem solving. As an aid in such situations, Table 5.3 lists the various relations between any two given elastic constants.

Another approach is to treat the Poisson's ratio as a constant in time, and obtain the Young's modulus $E(t)$ from the models or from experiments. This approach is very simple because of its complete similarity to linear elasticity. For this reason it is called *pseudoelasticity*. This approach is based on the observation the $\nu$, although subject to a variation of up to 35 to 40%, is still a weak variable in the expressions for stress and strain, and can be treated as constant in engineering calculations.

The correspondence principle uses the "constant $K$" approach and the pseudoelasticity uses the "constant $\nu$" approach. For a viscoelastic model, the stress and strain are related via suitable differential operators $\mathscr{P}$ and $\mathscr{Q}$. See Table 5.1 for how such operators are set up. For example, for a

**Table 5.3. Relationship Between Elastic Constants**

| *Given* | To Obtain | | | |
|---|---|---|---|---|
| | $E$ | $G$ | $\nu$ | $K$ |
| $E, G$ | – | – | $\left(\frac{E}{2G} - 1\right)$ | $\frac{EG}{3(3G - E)}$ |
| $E, \nu$ | – | $\frac{E}{2(1+\nu)}$ | – | $\frac{E}{3(1 - 2\nu)}$ |
| $E, K$ | – | $\frac{3EK}{(9K + E)}$ | $\frac{1}{2}\left(1 - \frac{E}{3K}\right)$ | – |
| $G, \nu$ | $2G(1+\nu)$ | – | – | $\frac{2G(1+\nu)}{3(1-2\nu)}$ |
| $G, K$ | $\left(\frac{9KG}{3K+G}\right)$ | – | $\frac{3K - 2G}{2(G + 3K)}$ | – |
| $\nu, K$ | $3K(1-2\nu)$ | $\frac{3K(1-2\nu)}{2(1+\nu)}$ | – | – |

Maxwell model,

$$\mathscr{P}\sigma = \mathscr{Q}\epsilon,$$

where,
$$\mathscr{P} = \left(\frac{1}{\mu} + \frac{1}{E}\frac{d}{dt}\right)$$

$$\mathscr{Q} = \frac{d}{dt}$$

Similarly for a Voigt model,

$$\mathscr{P} = 1$$

$$\mathscr{Q} = E + \mu\frac{d}{dt}$$

Taking Laplace transforms of the stress strain equations for a Maxwell equation we get,

$$\left(\frac{1}{\mu} + \frac{s}{E}\right)\bar{\sigma} = (s)\bar{\epsilon}$$

Thus,
$$\bar{\mathscr{P}} = \left(\frac{1}{\mu} + \frac{s}{E}\right) \quad \text{and} \quad \bar{\mathscr{Q}} = s$$

By a similar procedure we may prove that for the Voigt model,

$$\bar{\mathscr{P}} = 1 \text{ and } \bar{\mathscr{Q}} = (E + \mu s)$$

For a standard model,

$$\bar{\mathscr{P}} = \left(\frac{E_1 + E_2}{E_1\,\mu} + \frac{s}{E_1}\right)$$

$$\bar{\mathscr{Q}} = \left(\frac{E_2}{\mu} + s\right)$$

It is now clear that in the transformed solution, $G$ is a function of $s$.

Let us consider a few illustrations of the correspondence principle.

**Example 5.10**

Find the deflection of an end-loaded cantilever as a function of time. Assume the material of the beam can be represented as a standard model, with constant $G$, $G_2$, and $\mu$.

*Solution*

***STEP 1***

Write the solution in terms of $K$ and $G$

$$\delta = \frac{WL^3}{3EI} = \frac{WL^3}{3I}\left(\frac{1}{9K} + \frac{1}{3G}\right)$$

***STEP 2***

Replace $G$ by $\overline{\mathscr{Q}}/\overline{\mathscr{P}}$.

$$\bar{\delta} = \frac{L^3}{3I}\left(\frac{1}{9K} + \frac{\overline{\mathscr{P}}}{3\overline{\mathscr{Q}}}\right)\overline{W}$$

$$= \frac{L^3}{3I}\left(\frac{1}{9K} + \frac{G + G_2 + \mu s}{3G(G_2 + \mu s)}\right)\overline{W}$$

$$= \frac{L^3}{3I}\left(\frac{1}{9K} + \frac{1}{3G} + \frac{1}{3(G_2 + \mu s)}\right)\overline{W}$$

***STEP 3***

Take the inverse Laplace transform. We note that $\overline{W}/(G_2 + \mu s)$ is the product of two functions and hence the inverse is the convolution of $W$ and $[\exp(-G_2 t/\mu)]$.

$$\delta(t) = \frac{WL^3}{3EI} + \frac{L^3}{9I\mu}\int_0^t W(t-\tau)\exp\left(-\frac{G_2\tau}{\mu}\right)d\tau$$

(Note $W$ is a constant).

$$= \frac{WL^3}{3EI} + \frac{WL^3}{3GI}\left[1 - \exp\left(-\frac{G_2 t}{\mu}\right)\right]$$

***Commentary***

*The power of the correspondence principle is evident, since it demonstrates the use of the viscoelastic stress strain relations for structural problems. At time* $t = 0$, *the elastic solution is retrieved, and the deflection as* $t \to \infty$ *is bounded. However, the long term modulus reverse-calculated from this solution, obtained by letting* $t \to \infty$ *is not equal to* $E_R$*! This anamoly is a consequence of treating K as constant, and does not arise in pseudoelasticity.*

**Example 5.11**

A viscoelastic plastic plug snugly fits the inside of a rigid metallic round tube. The plug is axially loaded with a compressive stress of $\sigma_0$. Assume the material to be a Voigt model material. Find the transverse stress developed in the plug as a function of time $t$.

*Solution*

***STEP 1***

Elasticity Solution

From elasticity, since the transverse strains are zero, the transverse stresses are given by

$$\sigma_{xx} = -\frac{\nu}{1-\nu}\sigma_0 = \frac{3K - 2G}{3K + 4G}(-\sigma_0)$$

For a Voigt material, $\bar{\mathscr{P}} = 1$, and $\bar{\mathscr{Q}} = (G + \mu s)$.

***STEP 2***

Replace $G$

$$\begin{aligned}\bar{\sigma}_{xx} &= \frac{3K\bar{\mathscr{P}} - 2\bar{\mathscr{Q}}}{3K\bar{\mathscr{P}} + 4\bar{\mathscr{Q}}}(\bar{\sigma}_0) \\ &= \frac{3K - 2G - 2\mu s}{3K + 4G + 4\mu s}(-\bar{\sigma}_0) \\ &= \frac{\bar{\sigma}_0}{2} - \frac{9K}{6K + 8G + 8\mu s}\bar{\sigma}_0\end{aligned}$$

***STEP 3***

Take the inverse Laplace transform. We note that the second terms is the convolution of $\sigma_0$ and $\exp(-4G + 3K/4\mu)\tau$.

$$\begin{aligned}\sigma_{xx} &= \frac{\sigma_0}{2} - \frac{9K\sigma_0}{8\mu}\left[\exp\left(-\frac{4G + 3K}{4\mu}\right)t - 1\right] \\ &= \frac{\sigma_0}{2} - \frac{\sigma_0(1+\nu)}{2(1-\nu)}\left[\exp\left(-\frac{4G + 3K}{4\mu}\right)t - 1\right] \\ &= -\frac{\nu}{1-\nu}\sigma_0 - \frac{\sigma_0}{2}\exp\left(-\frac{4G + 3K}{4\mu}\right)t\end{aligned}$$

**Example 5.12**

A viscoelastic body is acted upon by a load $W$, in response to which its elastic deflection $u = Wf(x_i)/E$, where $E$ is its Young's modulus, and $f(x_i)$ is a function of geometry. Assume the standard viscoelastic model for the material and find the time dependent displacement $u(t)$.

*Solution* This problem is similar to the previous two, except in generality.

$$u = \frac{W}{E} f(x_i) = Wf(x_i)\left(\frac{1}{9K} + \frac{1}{3G}\right)$$

$$\bar{u} = Wf(x_i)\left(\frac{1}{9K} + \frac{\bar{\mathscr{P}}}{3\bar{\mathscr{Q}}}\right)$$

$$= Wf(x_i)\left(\frac{1}{9K} + \frac{G_1 + G_2 + \mu s}{3G_1(G_2 + \mu s)}\right)$$

$\bar{\mathscr{P}}$ and $\bar{\mathscr{Q}}$ for the standard model was obtained from Table 5.1. Taking the inverse transform, we see that the viscoelastic solution is given by

$$u(t) = \frac{Wf(x_i)}{E} + \frac{Wf(x_i)}{G_2}\left[1 - \exp\left(\frac{G_2 t}{\mu}\right)\right]$$

The characteristics of the solution of the three examples are similar, because they are all load-controlled problems and hence creep is the associated phenomenon. We consider a displacement-controlled problem in the following example.

**Example 5.13**

Consider the press-fit of one cylinder over another, as discussed in Chapter 1. Let us consider the interfacial pressure $P$ as functions of the dimensions of the pipe, their Young's modulus, and the interference $\delta$. Therefore,

$$P = E\,\delta f(a, b, c)$$

where $f(a, b, c)$ is a function of the sizes of the cylinders. If the materials of the pipe were the same, and can be represented by the standard viscoelastic model, obtain a relaxation solution for $P$ as a function of time.

*Solution*

$$P = E\,\delta f(a,b,c) \quad \text{(Elastic Solution)}$$

$$= \frac{9KG}{3K+G}\delta f(a,b,c)$$

$$\bar{P} = 9Kf(a,b,c)\frac{(G_2+\mu s)G_1}{3K(G+G_2+\mu s)+G(G_2+\mu s)}\bar{\delta}$$

After some manipulations of the fraction in $s$, and taking the inverse transform, we can see that

$$P(t) = E\,\delta f(a,b,c) - E\delta\frac{1}{1+\dfrac{G_2}{3K}+\dfrac{G}{G_2}}\left[1-\exp\left(-\frac{3G_2+E}{3\mu}\right)t\right]$$

The second term in the solution is a decaying part, thus denoting a relaxation of pressure.

It is to be noted that the relaxation or creep behavior due to a certain loading cannot *automatically* be detected by the correspondence principle, because use of the correspondence principle starts with an elastic solution. The load- or displacement-controlled nature of an elastic *solution* is a consequence of the way the problem is idealized and is not a feature of the mathematics used to solve the problem. Thus, the elastic solution chosen must be in the right format, explicitly showing the load-controlled or displacement-controlled nature of the problem, so the use of correspondence principle will exhibit creep or relaxation behavior.

## 5.11 PSEUDOELASTICITY

Pseudoelasticity is a simplification of the correspondence principle. Consider the equation

$$\sigma_{ij} = (2G + \lambda\delta_{ik}\delta_{kj})\epsilon_{ij},$$

which is also a statement of Hooke's law in three dimensions. $\lambda$ in this equation is the same as the one used in Equations 1.23. The term in the parenthesis reduces to simple $E$ in one dimension, and the equation becomes

$$\sigma = E\epsilon$$

Since viscoelastic models discussed in this chapter are also basically one dimensional, we may use a guideline, only slightly different from the correspondence principle. We replace $E$ (instead of $G$) by the ratio of the transformed operators, and also treat Poisson's ratio as a constant.

$$E \rightarrow \frac{\bar{\mathcal{Q}}}{\bar{\mathcal{P}}}$$

$$\nu \rightarrow \nu \text{ no change}$$

The remarkable feature of the pseudoelastic concept is that it can go with any general $E(t)$—possibly determined from experiments—and construction of an actual spring-dashpot model is unnecessary.

**Example 5.14**

Prove that in static, load-controlled problems, pseudoelasticity is equivalent to replacing $1/E$ by the creep compliance $J(t)$.

*Solution* The stress field in load-controlled problems (Figure 5.15) is either (1) entirely independent of the modulus $E$ and $\nu$, or (2) involves Poisson's ratio, $\nu$. Since in pseudoelasticity, $\nu$ is taken as a constant, the stress field for either one of these cases is the same as it is in the case of linear elasticity. However, the displacements are proportional to $1/E$. We may write the general solution as

$$u = \frac{W}{E} f(x_i),$$

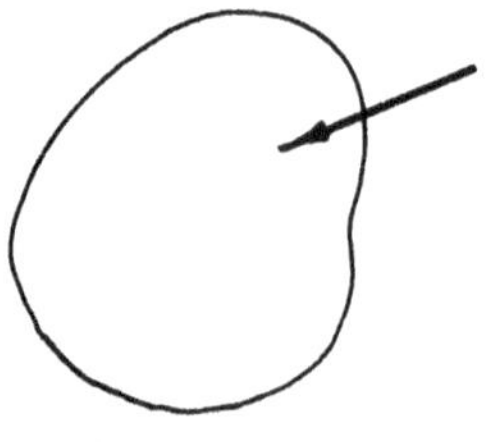

**FIGURE 5.15.** Load controlled problem for viscoelastic materials.

where $W$ is the applied load, and $f(x_i)$ is a function of the coordinates (or geometry). Assume a standard viscoelastic material, for which

$$\bar{\mathscr{Q}} = \left(s + \frac{1}{\tau_\sigma}\right) \text{ and } \bar{\mathscr{P}} = \frac{1}{E_1}\left(\frac{1}{\tau_\epsilon} + s\right)$$

Taking the Laplace transform of $u$, and replacing for $1/E$, we obtain

$$\bar{u} = \frac{\bar{W}f(x_i)}{E_1}\left[1 + \left(\frac{1}{\tau_\epsilon} - \frac{1}{\tau_\sigma}\right)\left(s + \frac{1}{\tau_\sigma}\right)^{-1}\right]$$

$$\begin{aligned} u(t) &= \frac{Wf(x_i)}{E_1} + \frac{Wf(x_i)}{E_1}\frac{E_1}{\mu}\int_0^t \exp\left(-\frac{\tau}{\tau_\sigma}\right) d\tau \\ &= \frac{Wf(x_i)}{E_1} + \frac{Wf(x_i)}{E_2}\left[1 - \exp\left(-\frac{t}{\tau_\sigma}\right)\right] \\ &= Wf(x_i)J(t) \end{aligned}$$

Thus, in load controlled problems, displacements, and hence strains, are obtained by replacing

$$\frac{1}{E} \to J(t).$$

The conclusion is however, applicable to any material model.

**Example 5.15**

Prove that in static, displacement-controlled problems, pseudoelasticity is equivalent to replacing $E$ by the relaxation modulus $E(t)$.

*Solution* In displacement-controlled problems (Figure 5.16), the displacements and strains are given and are independent of material properties.

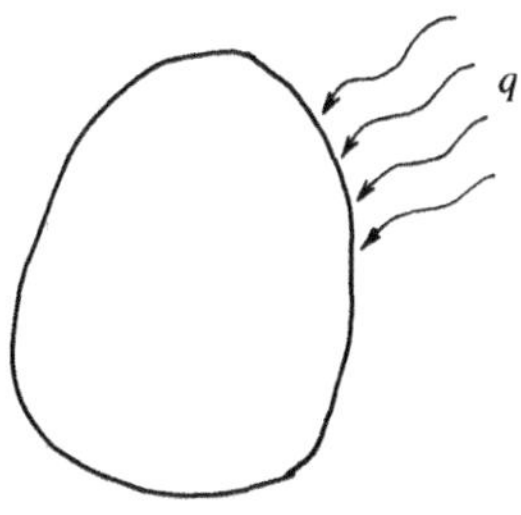


**FIGURE 5.16.** Displacement controlled problem for viscoelastic materials.

The stresses are directly proportional to modulus, $E$, and are expressible as

$$\sigma = Euf(x_i),$$

where $u$ is the applied displacement, and $E$ is the Young's modulus. As before, we may assume the standard model, take the Laplace transform and obtain,

$$\bar{\sigma} = \bar{E}\bar{u}f(x_i) = \frac{\bar{\mathscr{Q}}}{\bar{\mathscr{P}}}\bar{u}f(x_i)$$

$$= E_1\bar{u}f(x_i)\frac{\left(s + \dfrac{1}{\tau_\sigma}\right)}{\left(s + \dfrac{1}{\tau_\epsilon}\right)}$$

$$\sigma(t) = E_1uf(x_i)\left[1 - \frac{E_1}{\mu}\int_0^t \exp\left(-\frac{\tau}{\tau_\epsilon}\right)d\tau\right]$$

$$= E_1uf(x_i)\left[1 - \frac{E_1}{E_1 + E_2}\left\{1 - \exp\left(-\frac{t}{\tau_\epsilon}\right)\right\}\right]$$

$$= uf(x_i)\left[E_1 - \frac{E_1^2}{E_1 + E_2}\left\{1 - \exp\left(-\frac{t}{\tau_\epsilon}\right)\right\}\right]$$

$$= uf(x_i)E(t)$$

Thus, the stresses in displacement controlled problems are obtained by replacing

$$E \to E(t).$$

This observation is of course applicable to any other material model.

## 5.12 AN INTERIM SUMMARY

We have learned that the history dependent stress strain relation needs to be described not only by an instantaneous modulus (or compliance), but also by the use of a "hereditary" integral—BSP. The viscoelastic models using spring and dashpot elements help produce qualitative understanding, but do not sufficiently advance our knowledge and capability in *structural* and *stress analysis* of actual plastic components. However, the BSP accepts curve-fitted equations and is more powerful. Also, the spring dashpot model becomes unwieldy when the number of elements increases, since the order of differential equation also increases. BSP, on the other hand, is always a single integral in time, $\tau$. The description of the material behavior is contained in the function $C(t)$ or $J(t)$. The intuitive connection between BSP and Laplace transforms leads to a one-to-one "correspondence" between the transformed viscoelastic equations, and ordinary elastic equations. This helps to establish two alternative statements of the correspondence principle, which enables the long term material properties to be used for structural analysis much in the same manner as are the short term properties. The first statement, the exact one, supposes the bulk modulus $K$, treated as constant, and the shear modulus $G(t)$, obtained from rheological tests, as the two basic elastic properties. The second approach is called pseudoelasticity. This approach is more convenient but less formal. It supposes the Poisson's ratio $\nu$, treated as constant, and the tensile relaxation modulus $E(t)$, obtained from rheological or other tests, to be the basic material properties. The last two examples above may be taken to illustrate the way to use $E(t)$ and $J(t)$ in structural analysis.

## 5.13 COMMENTS ON THE USE OF PSEUDOELASTICITY

The ease of using pseudoelasticity is the motivation for obtaining $E(t)$ or $J(t)$ as a material property for plastics. In structural problems involving a single material, the engineer only needs to determine whether a given problem is a load-controlled one, or a displacement-controlled one. A linear elastic analysis can be used as a reference solution, from which the stresses or displacements can be prorated for the long term, using $E(t)$ or $J(t)$ as appropriate. If more than one material is involved, a finite element analysis using the respective long term properties is in order. *This statement applies by and large to rigid polymers.*

Important consequences of pseudoelasticity are as follows. They have been stated earlier informally.

- In load-controlled problems, stress remains constant. Strain is proportional to $J(t)$ which is an increasing function. Hence creep strain increases.
- In displacement-controlled problems, strain remains constant, and stress is proportional to $E(t)$, indicating relaxation. Molded-in stresses belong to this category.

- Combination problems do exist, such as thermal stress in combination with mechanical stresses. When the sum of the two stresses is within the linear viscoelastic range, it is sufficiently accurate to separate them out and deal with each one according to its category.

Pseudoelasticity can be applied to static or slow-transient problems only. Further, just as in the case of linear elasticity, results of pseudoelastic calculations in regions of stress concentrations need to be interpreted with the possible nonlinearity in mind. Typically in these regions, the long term strains can be greater than predicted by pseudoelasticity, and stress can relax. Thus failures such as crazing or stress-whitening related to the local strains can occur earlier than predicted by pseudoelasticity. Where geometric nonlinearity is involved, it may not be appropriate to prorate the elastic solutions, since the "correspondence" in the sense of Table 5.2 does not exist.

**Example 5.16**

Using pseudoelasticity, discuss the creep buckling of an axially loaded column of length $L$, which is pin-jointed at both ends.

*Solution* Consider the beam to be laterally deflected by an amount $w(x)$ shown in Figure 5.17. The Euler beam theory gives the relation

$$\frac{d^2w}{dx^2} + \frac{P_o w}{EI} = 0$$

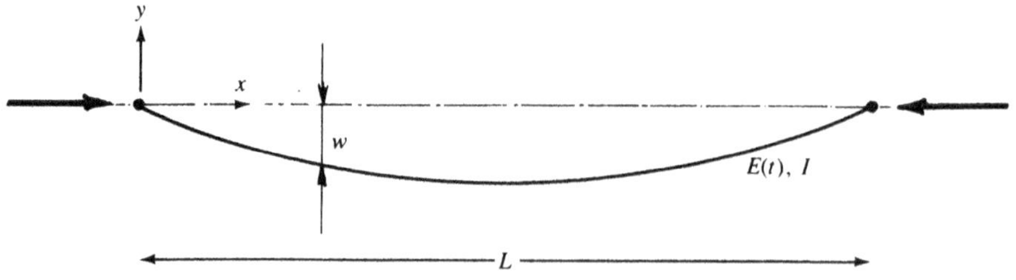


**FIGURE 5.17.** Simple model of creep buckling.

Since this is a load controlled problem we may rewrite this as

$$\frac{d^2w}{dx^2} + \frac{P_o w J(t)}{I} = 0$$

The solution for this differential equation is

$$w(x) = C \sin \sqrt{\frac{P_o J(t)}{I}}\, x$$

Since at $x = L$, the beam is pin-jointed, $w$ has to be zero. For a nontrivial solution, we must impose the condition

$$P_o = P_{cr} = \frac{n^2 \pi^2 I}{J(t) L^2},$$

where $n$ is any integer. The lowest value of $P_{cr}$ is obtained for $n = 1$. For other loads, the column will remain straight. At the above value(s), the column accepts any lateral deflection, which is a condition of instability in the sense that a small disturbance can lead to arbitrarily large deflections.

***Commentary***

*Conversely, if a load $P^*$ is given, then the above condition can be used to find the $J(t)$ that will cause buckling, and hence the time $t$ at which it will occur.*

*For more complex structures, the form of the equation for buckling load is still the same, if the support conditions remain constant over time. The equation can be expressed as*

$$P_{cr} J(t) = \text{Constant} = \frac{P_{cr,o}}{E_o}$$

*where $P_{cr,o}$ is the critical buckling load for linear elastic modulus $E_o$, and $P_{cr}$ is the critical buckling load after a time $t$. This relation shows that the computation of the long term buckling load is also obtained by prorating the critical load.*

**Example 5.17**

Reconsider the tee fitting discussed in Chapter 2, Example 2.9. Let us assume that the moduli $E(t)$ of the Acetal Copolymer tee fitting and the pipe material, at time $t = 60{,}000$ hours (about 6.5 years), reduce to 95 $kg/mm^2$ and 6 $kg/mm^2$, respectively. The aluminum pipe clamp does not reduce in modulus and maintains its modulus of $E = 7035$ $kg/mm^2$. The interferences $\delta_1$ and $\delta_2$ are still the same as their initial values.

Determine the interfacial pressures $P_1$ and $P_2$, at the end of $t = 60{,}000$ $h$ by using pseudoelasticity.

*Solution* Figure 5.18 shows the assembly of the tee fitting with the pipe. Note that this figure and the table below are reproduced from Example 2.9.

| | Fitting | Pipe | Clamp |
|---|---|---|---|
| Inner dia (mm) | 50.0 | 55.0 | 58.0 |
| Outer dia (mm) | 55.0 | 58.0 | 60.0 |
| Modulus ($kg/mm^2$) | 95 | 6 | 7035 |
| Poisson's ratio | 0.35 | 0.35 | 0.28 |

$$\delta_1 = \text{Interference between fitting and pipe} = 0.05 \text{ mm}$$

$$\delta_2 = \text{Interference between pipe and clamp} = 0.13 \text{ mm}$$

The solution procedure is similar to elastic analysis, except that we need to replace $1/E_{**} \rightarrow J(t)$. However, since $J(t)$ is not available readily, we may use the respective $1/E(t)$ in the place of $J(t)$. (Note that for computing the expression for $J(t)$ from $E(t)$, we need to have it as a function of time, and not as a number.) Substituting in the formulas of

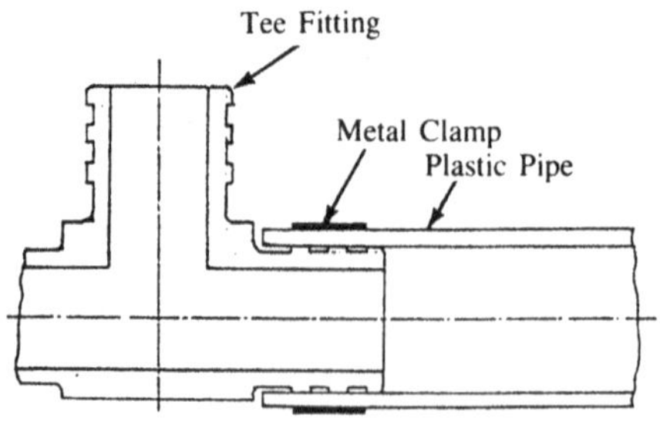


**FIGURE 5.18.** Press-fit of a tee and a pipe with a pipe clamp.

Example 2.9,

$$S_{11} = 90.93 \qquad S_{12} = -90.96$$

$$S_{21} = -86.26 \qquad S_{22} = 89.53$$

$$\Delta = S_{11}S_{22} - S_{12}S_{21} = 294.16$$

$$P_1 = \frac{1}{\Delta}[\delta_1 S_{22} - \delta_2 S_{12}] = 0.0554 \text{ kg/mm}^2 \text{ (79 psi)}$$

$$P_2 = \frac{1}{\Delta}[\delta_1 S_{11} - \delta_2 S_{21}] = 0.0548 \text{ kg/mm}^2 \text{ (78 psi)}$$

***Commentary***

*Since this problem is a displacement-controlled one, the initial values of pressures (obtained in Example 2.9) are the highest over time. So are the stresses. Considering the normal water pressures in domestic plumbing lines, (about 30 psi) the seal pressures are sufficient to hold the water pressure.*

*We may draw the comparison with the hose pipe going from the engine block of an automobile to the radiator, in which the "fitting" is steel, the "pipe" is the rubber hose, and the "clamp" is aluminum (or stainless steel). However only intermittently does this joint see higher pressure and temperature than does domestic plumbing. From the point of view of seal pressures only, if the steel clamp were replaced by a plastic clamp, the joint would need retightening about once every week! This does not happen simply because the metal clamps do not relax. This observation is important in designs in which plastic and metal work in tandem, and the plastic material is subject to creep and/or relaxation.*

## 5.14 FINDLEY'S CONSTANTS

One of the ways of describing the time-dependent behavior of plastics in linear and moderately nonlinear regions is by the use of Findley's constants. Findley[5] employed an equation of the type

$$\epsilon(t) = \epsilon_o \sinh \frac{\sigma}{\sigma_o} + \epsilon_t t^n \sinh \frac{\sigma}{\sigma_o}. \tag{5.43}$$

[5] In reality, the use of hyperbolic sine equations was suggested by Prandtl, and Findley investigated this equation in detail. In plastics literature, this has come to be called Findley's equations. Much of his work on the investigation of this equation for nonlinear viscoelasticity can be found in Reference 5. This is a book for the serious reader.

This equation has a time-independent and a time-dependent part, and involves five material constants. Findley's constants are hard to find. References 5 and 6 provide some information. Table 5.4 presents a partial list adapted from Reference 6. For small enough stresses, the $\sinh(\sigma/\sigma_o)$ term simply becomes $\sigma/\sigma_o$, retrieving linear viscoelasticity. Furthermore, the slope of the stress strain curve at the origin can be seen to be finite, unlike the power law. The most important use of Findley's constants arises because the $\sinh(\cdots)$ term is nonlinear. A stress-strain state at which the secant modulus differs from the initial modulus by a prescribed amount (arbitrarily 15%) is taken as the break down of the linearity assumption. This state, called the *viscoelastic limit*, provides an analog to the yielding behavior in elasticity, where the elastic assumption breaks down, although the structure does not.

The time $t$ in Equation 5.43 is nondimensional time. Although there are five constants in the equation, there really are only four independent constants, since $\epsilon_o$ and $\sigma_o$ are related by

$$\frac{\sigma_o}{\epsilon_o} = E_o, \text{ the Young's modulus.}$$

For small stresses, Findley's equation reduces to

$$\epsilon(t) \approx e_o \frac{\sigma}{\sigma_o} + \epsilon_t t^n \frac{\sigma}{\sigma_o}.$$

$$\approx \sigma \left[ \frac{1}{E_o} + \frac{t^n}{E_t} \right] \tag{5.44}$$

Findley's equations are particularly useful in problems of *sustained* applied loads. The complexity of using them for *applied displacement* problems is obvious from the form of Equation 5.43. Perhaps the best use of Findley's equations is to determine a limit state of stress and strain beyond which linearity is violated, and calculations are no longer accurate. It is very likely that the linearity violation occurs long before the fully relaxed conditions ($E_R$ or $J_R$) are reached. Thus, it hardly matters in engineering calculations that Findley's equations do not predict the asymptotic behavior of creep strains.

### 5.14.1 Viscoelastic Limit Stress and Strain

It is easily seen that the slope of the isochronous stress strain curves is $[J(t)]^{-1}$. When $\sigma/\sigma_o \ll 1$ or $\sigma/\sigma_t \ll 1$ then $\sinh(\sigma/\sigma_o) \approx \sigma/\sigma_o$ and

**Table 5.4.** **Findley's Constants For A Few Selected Materials. (Adapted From Reference 6.) All stresses and moduli in kg/mm² Relative Humidity = 50°; Temperature = 77°F.**

| Material | $n$ | $\epsilon_0$ | $\epsilon_t$ | $\sigma_0$ kg/mm² | $\sigma_t$ kg/mm² | $E_0$ kg/mm² | $E_t$ kg/ mm² |
|---|---|---|---|---|---|---|---|
| Polyethylene | 0.154 | 0.027 | 0.0021 | 0.4114 | .1617 | 15.2 | 78.1 |
| Polymonochloro-TriflouroEthylene | 0.0872 | 0.00810 | 0.00099 | 1.8284 | 1.0373 | 225.7 | 1047.8 |
| Polyvinylchloride | 0.305 | 0.00833 | 0.00079 | 3.263 | 1.1463 | 391.7 | 14416.3 |
| Polystyrene | 0.525 | 0.048 | 0.0000041 | 14.0646 | 0.4571 | 293.2 | 111462.7 |
| Glass Fabric/PE | 0.090 | 0.0034 | 0.000445 | 10.5485 | 9.8453 | 3101.3 | 22151.9 |

$\sinh(\sigma/\sigma_t) \approx \sigma/\sigma_t$. When $\sigma = \sigma_o$ or $\sigma = \sigma_t$ then the above approximation has an error of about 17%. *We may utilize this observation to arbitrarily select a stress at which the linearization of Findley's equation will carry an error of 15% as the viscoelastic limit stress. The corresponding strain is the viscoelastic limit strain.* The viscoelastic limits vary with the specified life expectancy of the product. Without going into many details, the limit stress and strain $\sigma_v$ and $\epsilon_v$ for a given life $T$ are the solution of the following equation(s).

$$\epsilon_0 \sinh\left(\frac{\sigma_v}{\sigma_0}\right) + \epsilon_t T^n \sinh\left(\frac{\sigma_v}{\sigma_t}\right) = \frac{1}{0.85}\sigma_v J(T) = \epsilon_v \tag{5.45}$$

Figures 5.19 and 5.20 are curves obtained by solving these equations for the materials listed in Table 5.4.

$J(T)$ is preferably obtained from the isochronous data. Alternatively,

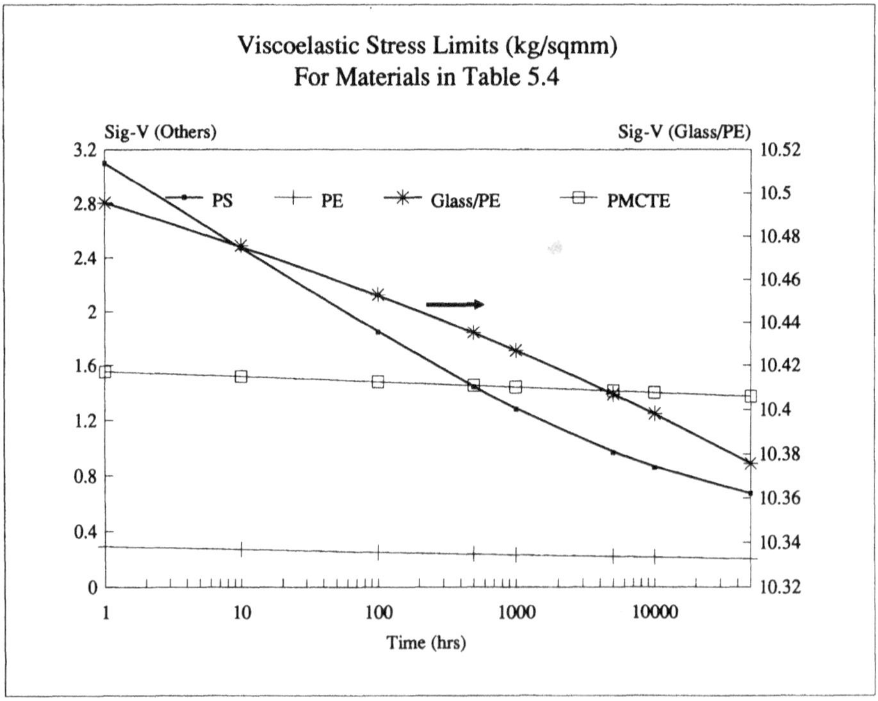


**FIGURE 5.19.** Viscoelastic stress limits (kg/mm$^2$) for materials in Table 5.4.

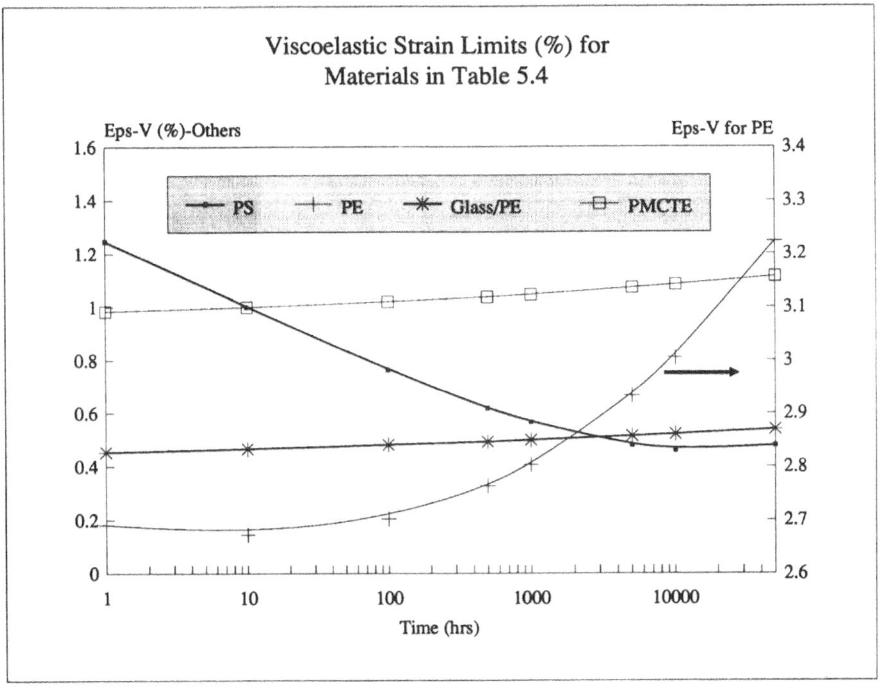


**FIGURE 5.20.** Viscoelastic strain limits for materials in Table 5.4.

Findley's constants can be used in the form

$$J(T) = \left( \frac{1}{E_0} + \frac{T^n}{E_t} \right)$$

**Example 5.18**

Calculate the relaxation of the interfacial pressure in the press-fit assembly of a PVC pipe over a steel spout, as shown in Figure 5.21. Take $r_o = 16$ mm, $r_i = 13$ mm, interference $\delta = 0.25$ mm, and $\nu = 0.35$.

*Solution* We may assume that the spout made of steel is infinitely stiff, and hence all the displacement is taken up by the plastic pipe. Thus, the problem is one of finding the pressure for which the dilatation of the inner

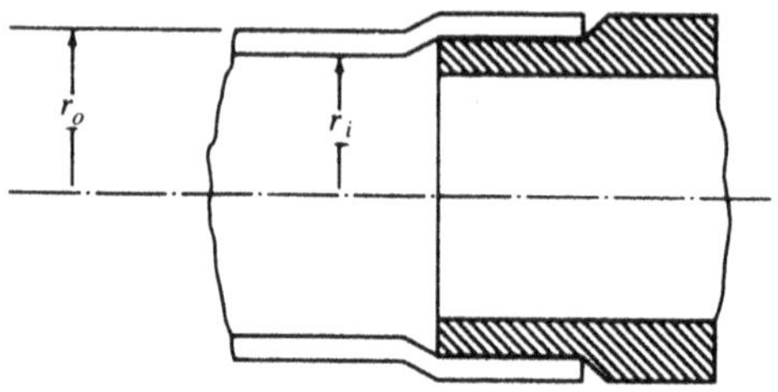


**FIGURE 5.21.** Relaxation of joint tightness in a simple spout-pipe assembly.

surface is equal to $\delta$. This is given by

$$p = \delta E(t)\left\{\frac{(r_o^2 - r_i^2)}{(r_o^2 + r_i^2) + \nu(r_o^2 - r_i^2)}\right\}$$

$$= 0.04775E(t) \text{ kg/mm}^2$$

However, the calculation of $E(t)$ needs some consideration. Since only strains are additive, we must proceed from $J(t)$. From Findley's equation we see that

$$J(t) = \left[\frac{1}{E_0} + \frac{t^n}{E_t}\right]$$

The Laplace transform of the right side is very complex. To make things easy, we may consider only the second term for the Laplace transformation. Let us denote by $j(t)$ the time dependent part of the compliance.

$$j(t) = J(t) - \frac{1}{E_0} = \frac{t^n}{E_t}$$

We use the symbol $e(t)$ to denote the time dependent modulus of the material, without considering the elastic component of strain.

$$\bar{j}(s) = \frac{\Gamma(1+n)}{E_t s^{1+n}}$$

Since $$\bar{j}(s)\bar{e}(s) = \frac{1}{s^2}$$

we have, $$\bar{e}(s) = \frac{E_t}{\Gamma(1+n)} \frac{1}{s^{(1-n)}}$$

$$e(t) = \left(\frac{\sin n\pi}{n\pi}\right) E_t t^{-n}$$

From Table 5.3, $E_0 = 390\ \text{kg/mm}^2$, $E_t = 20.5 \times 10^6$ psi $= 14415\ \text{kg/mm}^2$ and $n = 0.305$ for PVC. Based on this,

$$e(t) = 12300 \times t^{-0.305}$$

Furthermore, since the time dependent strain is additive to the elastic strain,

$$\frac{1}{E(t)} = \frac{1}{E_0} + \frac{t^n}{e(t)}$$

$$E(t) = \frac{E_0 e(t)}{E_0 t^n + e(t)}$$

The following table traces the interference pressure $P1$.

| Hrs | $E(t)$ | Press, $P1$ |
|---|---|---|
| 0 | 390 | 18.6 |
| 100 | 256 | 12.2 |
| 200 | 216 | 10.3 |
| 300 | 192 | 9.2 |
| 400 | 175 | 8.4 |
| 500 | 162 | 7.8 |
| 1000 | 124 | 5.9 |
| 2000 | 91 | 4.3 |
| 5000 | 58 | 2.8 |
| 10000 | 40 | 1.9 |
| 20000 | 27 | 1.3 |
| 50000 | 16 | 0.8 |
| 100000 | 11 | 0.5 |

***Commentary***

*The important point in this example is the procedure used for calculating $E(t)$. Instead of Laplace transforms we could have used an algebraic reciprocal.*

*This procedure would have led to about 15% error in E(t). The error is on the unsafe side, because it would have led to the incorrect conclusion that the joint is tighter than it actually is.*

**Example 5.19**

Suppose that polycarbonate material is being evaluated for the design of a vacuum flask, which is cylindrical in shape with end closures. The flask is to be 75 mm in diameter, about 375 mm long, and is to be closed at both ends with flat closures. The yield stress of the material is 6.3 $kg/mm^2$. The vacuum in the flask is $-1$ $kg/cm^2$. Other conditions are that:

- the stress in the part should carry a minimum FOS of 2 against crazing at 40000 hours.
- the design should have a minimum FOS of 1.5 on the stress needed for isochronous strain of 1% at time $t = 0$.
- the design should have a minimum FOS of 1.0 on the stress needed for an isochronous strain of 2% at time $t = 40000$ hours.
- the stress should have a minimum FOS of 1.5 against yielding.

*Solution* The material property curves shown in Figure 5.22 are reproduced here from Reference 7. All modes of failure except buckling depend

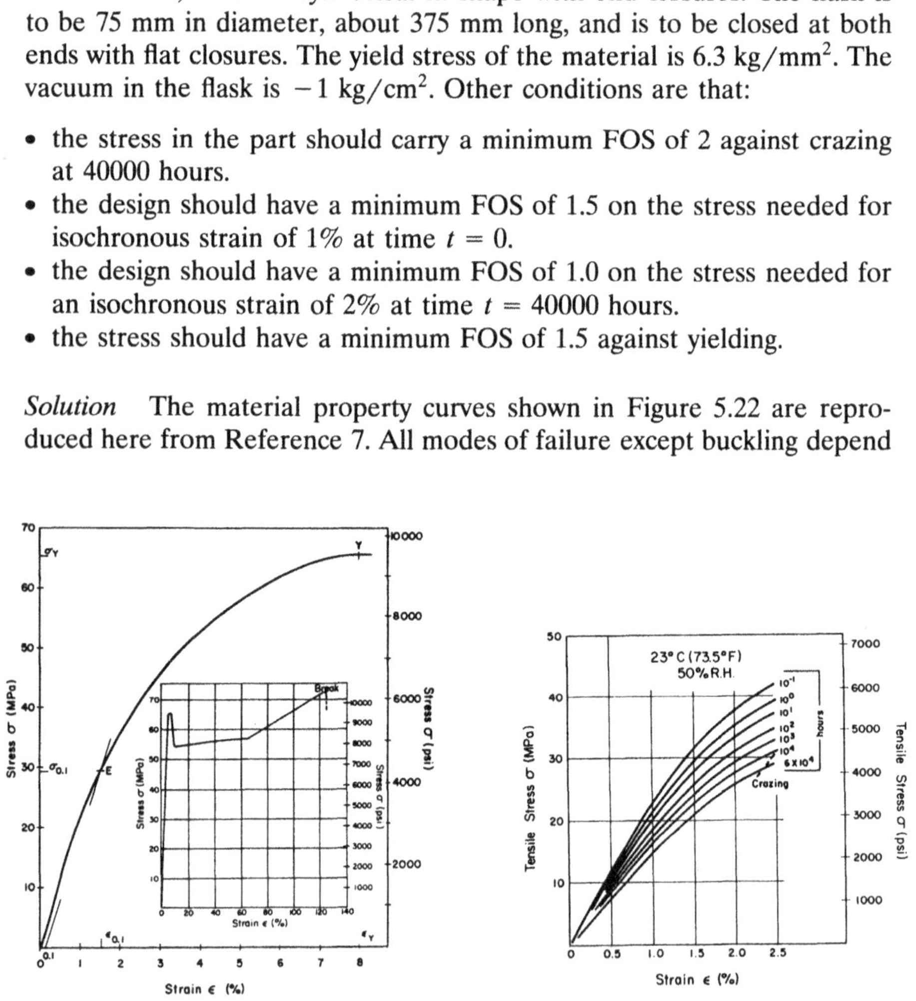


**FIGURE 5.22.** Short and long term properties of polypropylene. Inset in (a) is the elongation characteristics obtained from short term tests. Shown in (b) are the isochronous stress-strain curves. Note the beginning of crazing is marked in dotted lines. (*Reproduced with kind permission from Marcel Dekker Publishing Company.*)

on the stress induced, and hence the sizing of the flask can be based on a stress that meets all the above criteria.

1. Allowable stress for crazing 4150 (psi)/2.0 = 2.92/2.0 = 1.46 kg/mm$^2$. See Figure 5.22(b).
2. Allowable stress for 1% strain at $t = 0$ is equal to 3000 (psi) = 2.11/1.5 = 1.41 kg/mm$^2$. Refer to Figure 5.22(a).
3. Allowable stress for 2% strain at $t = 40000$ hours is 3900 psi = 2.74 kg/mm$^2$. Refer to Figure 5.22(b).
4. Allowable stress for yielding = 9000 (psi)/1.5 = 6.33/1.5 = 4.22 kg/mm$^2$. See Figure 5.22(a) inset.

The least of all these stresses answers all the criteria, and thus the allowable stress is chosen as 1.41 kg/mm$^2$.

The thickness needed for the flask is simply

$$t = \frac{pd}{2\sigma_{des}} = \frac{(.01\ \text{kg/mm}^2)(75\ \text{mm})}{2(1.41\ \text{kg/mm}^2)} = 0.266\ \text{mm}$$

However, this thickness is so low that the buckling mode of failure must be checked. Since the cylinder is of finite length, the ring buckling formulas cannot be used. The formula for the buckling pressure of finite length cylinders with rigid ends is obtained from the paper cited in Reference 8:

$$P_{cr} = \frac{2.42E}{(1-\nu^2)^{3/4}} \frac{(t/D)^{5/2}}{\left[(L/D) - 0.45(t/D)^{1/2}\right]}$$

It is appropriate to take $E$ as $E(t)$ at $t = 40000$ hours. From Fig. 5.22 we obtain $E(t)$ as 210,000 psi or 147.7 kg/mm$^2$; $\nu = 0.28$. This equation is solved by trial and error to obtain a value of $t = 2.1$ mm. This is 7.8 times the thickness obtained by stress criteria.

***Commentary***

*Recall that in Chapters 1 and 4 it was stated that the stress-based steel designs almost always have the necessary stiffness in most situations because of the high E of steel, and that quite the opposite is true for plastics. This example is an illustration of the point. For example, if we had proceeded with the more conventional approach of taking a large factor of safety of 8 on the yield strength, it would hardly have made any difference in the overall picture. The flask would still have buckled.*

*In an actual design, checking by a buckling formula may not be sufficient, since near the flat ends of the flask, there may be secondary stresses which would increase the compressive stresses and may lead to buckling near the ends. A detailed finite element analysis would be needed to determine the actual thickness required.*

### 5.14.2 Design Rules for Long Term Failure Modes

Associated with the long term behavior of plastics are the following modes of failure, listed in ascending order of seriousness. The recommended factors of safety (FOS) indicated in parenthesis are representative of the seriousness. The factors of safety recommended in the following are higher than indicated in Chapter 4, for two reasons: (1) There are bound to be stress concentration areas in a design based on simple sizing formulas. In such areas, long term effects can occur much earlier than the predicted life of the product. (2) Creep rupture data are subject to a wide variation.

1. Reaching viscoelastic limits (FOS = 1 or less subject to other information)
2. Excessive creep strain (FOS = 2)
3. Crazing, stress whitening (FOS = 2)
4. Ductile creep rupture (FOS = 6 on stress, or FOS = 2 on time) and
5. Brittle creep rupture (FOS = 6 on stress, or FOS = 2 on time).

In an ideal situation, the following long term properties of the material must be available for the design.

- The estimated life of the product, $T$.
- Short term properties.
- The criteria that the component must satisfy at the end of time $T$.
- Data on crazing, time taken to show crazing at a particular stress level, temperature, and of course the environment appropriate to the problem.
- Data on time taken for creep rupture at a particular stress level and temperature, and the nature of the rupture.

Stress analysis of the product must identify conservatively,

- Short term stress and strain, for comparison with short term allowables.
- Long term stress and strain at time $T$, to determine whether an adequate margin of safety is available in terms of stress, strain, or time against excessive creep, crazing, and rupture. As mentioned earlier, the accumulation of residual stress (strain) in the case of cyclic loading must be considered.

### 5.14.3 Ranking of Materials for Long Term Performance

As a part of material selection, materials are ranked in the order of certain prescribed criteria. For example, materials can be ranked in the order of stiffness, strength, weight, or price. In the context of viscoelasticity, a criterion is the ability of the material to sustain high stress for a long time without failure of any kind.

Let

$\sigma_v$ —Viscoelastic limit,
$\sigma_{crz}$—Stress needed to start crazing at time $T$,
$\sigma_e$ —Stress needed to produce the allowable strain at time $T$,
$\sigma_R$ —Stress needed to produce stress rupture at time $T$,
$\sigma_{2T}$—Stress needed to produce stress rupture at time $2T$.

Since each of the above carries a different FOS with it, we choose the design stress $\sigma_D$ to be

$$\sigma_D = \text{Min}\left\{\sigma_v, \frac{\sigma_{crz}}{2}, \frac{\sigma_e}{2}, \frac{\sigma_R}{6}, \sigma_{2T}\right\}$$

From the material data sheets of the candidate materials, we may compute $\sigma_D$ for all of them. The rank of a particular material is its position in a list of descending order to $\sigma_D$.

The above is just one of the several possible criteria for ranking. This is suitable for load controlled design problems. Setting up ranking criteria for cyclic loading conditions can be difficult. For example, a highly creep resistant material is also relaxation-resistant. See, for example, the $E(t)$ and $1/J(t)$ for the standard model. Placing demands for high creep resistance simultaneously with fast relaxation is unrealistic, although desirable. Should such conflicting criteria be unavoidable, we probably have a situation of metal insert-molded components. See Example 5.9 above in this light.

## 5.15 METHODS OF DETERMINING $E(T)$

The life expectancy of structural products is of the order of 10 to 20 years, over which time the modulus $E(t)$ (or $J(t)$) is not available for many materials. Whatever the status until now, there is a growing need to produce such data for plastics over a long period of time. There are two procedures for producing long term data for $E(t)$ and $J(t)$. They are:

1. Testing the material for short durations at various temperatures ranging from low to high relative to the glass temperature $T_g$ of the material, and using a time-temperature equivalence principle called the *WLF Principle*, it is possible to treat the high temperature data as equivalent to long range data and vice versa.
2. Testing the material at dynamic loads at various frequencies $\omega$, and by using the mathematical technique known as Fourier transforms, the $G(\omega)$ or $J(\omega)$ can be *numerically* transformed to functions of time, $G(t)$ and $J(t)$.

In either of these two methods, there is no need to visualize the material behavior via springs and dashpots. Concepts of retardation and relaxation times and spectra are dispensed with, which is good since they are not needed for mechanical engineering work. Furthermore, a rheological testing machine can be used for either of these methods. Reference 9 is an excellent source describing the experimental methods in this area.

### 5.15.1 Obtaining $E(t)$, $J(t)$ by the WLF Principle

Since only short term testing is involved, this is easy enough to perform. As mentioned before, results (typically moduli and compliance) obtained from the high temperature short term tests are "shifted" laterally on the time scale relative to the results obtained for short term, room temperature conditions. The amount of shift $a_T$ is as defined below. Figure 5.23 illustrates this procedure. The shift factor $a_T$ is given by the WLF equation

$$\log a_T = \frac{C_1(T - T_s)}{C_2 + (T - T_s)} \quad \text{(Temps in}\ ^\circ\text{K)} \tag{5.46}$$

where $C_1$ and $C_2$ are material constants, $T$ is the test temperature and $T_s$ is a characteristic reference temperature. It has been shown that when $T_s$ is selected to be the glass transition temperature $T_g$ of the plastic, $C_1$ and $C_2$ are 17.44 and 51.6 respectively, in appropriate units. $a_T$ represents the "shift" on the the horizontal (time) axis. This equation is valid in the temperature range of $T_g$ to $(T_g + 100)^\circ$ K.

Utilizing this equation, it is possible to obtain the $E(t)$ or $J(t)$ over a very large range of time, by shifting and patching a number of high temperature test results. Rheological testing machines provided with environmental chambers are used to perform the tests and construct the

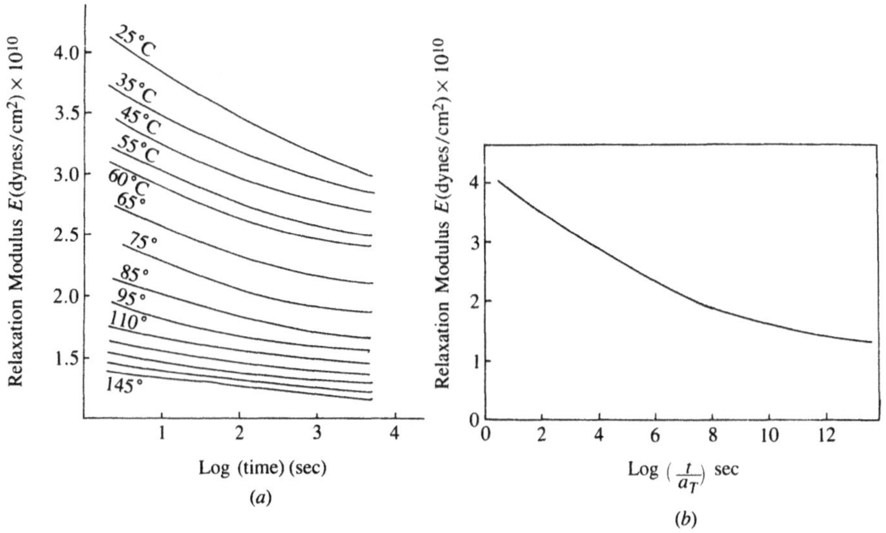


**FIGURE 5.23.** The concept of shifting as implied by the WLF equations. Measurements of relaxation modulus are taken at several temperatures for a manageable time range, as in (a). By treating the 25° C curve as the base, the other curves are shifted laterally by an amount $\log(a^T)$. The resulting master curve shown in (b) applies for 25° C. The same information can be reprocessed for other temperatures. Curves shown are for Nylon-66.

so-called "master curve," which is nothing but the shifted-and-patched curve of $E(t)$ (or a similar property) on the time scale. Obviously, when $T_s = T_g$, and the test temperature is $T_g$, the shift factor $\log a_T$ is zero, and no shifting is involved. Based on the sign of $\log a_T$ we see that curves obtained at lower temperatures would be shifted to the left, and those at higher temperatures would be shifted to the right. References 12, 13, and 14 are recommended for an in-depth understanding of this principle, and its formal proof.

The WLF equation can also be applied to tests involving combinations of frequency variation and temperature variation. Tests can be performed over a small range of frequency variation, and such tests may be repeated over several temperatures. By shifting it is possible to get a curve based on a broader range of frequencies. Since frequency $\omega$ is proportional to $[1/(\text{time})]$, the curves of modulus and compliance appear inverted left to right, when compared with the time based curves.

### 5.15.2 Obtaining $E(t)$, $J(t)$ by Fourier Transforms

Rheological testing machines are generally capable of controlling a wide range of input parameters, such as frequency of sinusoidal load, temperature, strain application, stress application, stress ramp, strain ramp, and also some combinations of these. These machines are extensively instrumented and provided with their own computation capability. It is quite possible to have the Fourier transform capability optionally built into the testing machine software.

Typically, a torsion experiment is chosen for this situation. Sinusoidal torque is applied at a preselected frequency and the storage and loss moduli are calculated by measuring the applied load and displacements. A wide range of frequency variation can be produced at relatively short times; one set of data points at a given $\omega$ (consisting of $E(t)$, $J(t)$, and $\omega$) may take about 15 minutes. Effectively, in about a week's time, it is possible to produce the data for about a 30 year period. The following integral transformation equations are used to numerically calculate the moduli as a function of time.

$$E(t) = E_0 + \frac{2}{\pi}\int_0^{\infty}\left[\frac{E' - E_o}{\omega}\right]\sin \omega t \, d\omega \tag{5.47}$$

$$E(t) = E_0 + \frac{2}{\pi}\int_0^{\infty}\frac{E''}{\omega}\cos \omega t \, d\omega \tag{5.48}$$

$$J(t) = \frac{2}{\pi}\int_0^{\infty}\frac{J'}{\omega}\sin \omega t \, d\omega \tag{5.49}$$

$$J(t) = \frac{2}{\pi}\int J''\omega(1 - \cos \omega t)\, d\omega \tag{5.50}$$

$E_0$—Young's Modulus

$E'$, $E''$—Components of Complex Moduli

Fourier transforms are so ubiquitous in several branches of engineering, that the algorithms are available in several forms—both software and hardware—and they are an essential adjunct in modern rheological testing machines.

## 5.16 CONCLUDING REMARKS

The peculiarity of viscoelasticity is that understanding the material behavior by itself involves a considerable amount of mathematical skills. Mathematical viscoelastic theory admits more than 30 different material properties. Some are time dependent, some are frequency dependent, some are for shear mode, and some for tensile mode. See for example Chapter 1 of Reference 10 and Reference 11. Although these properties are independent there are more general underlying connections. Many can be integral-transformed to one another.

The use of material properties for structural problems is quite another area. For static and slow transient problems, pseudoelasticity is the most suitable technique. Time dependent modulus $E(t)$ is a basic need in the solution. By using pseudoelasticity, a regular static finite element analysis can be used for long term analysis also. For intermittent loads, recovery calculations are needed, which in turn requires the BSP. There appear to be no computational tools available in the market as of the early 1990s, which incorporate any general creep (or relaxation) curve together with BSP.

In general, dynamic analysis of plastics structues is *very* complex. Random loads have their own spectrum and so does the material property. It is not quite accurate to use the current analysis tools for steel strucutres, since they invariably are based on the assumption that material properties are independent of the applied frequency.

The use of pseudoelasticity is best for *constant temperature problems*. For prescribed temperature fluctuations, no sufficiently reliable technique is available. One usually uses the worst-case scenario, by applying the data for the highest temperature encountered.

**References**


1. Williams, J. G., *Stress Analysis of Polymers*, 2nd Edn. 1980, Ellis Horwood Ltd., Chichester, U.K.
2. Mase, G. E., *Theory and Problems of Continuum Mechanics*, Schaum's Outline Series, 1970, McGraw Hill, New York, NY.
3. Kreyszig, E., *Advanced Engineering Mathematics*, 4th Edn. 1979, John Wiley & Sons, New York, NY.
4. Abramowitz, M. and Stegun, I. A., Editors, *Handbook of Mathematical Functions*, 1972, Dover Publications Inc., New York, NY.
5. Findley, W. N., Lai, J. S., and Onaran, K., *Creep and Relaxation of Nonlinear Viscoelastic Materials*, 1989, Republished by Dover Publications, New York, NY.
6. ASCE Manuals and Reports on Engineering Practice #63, *Structural Plastics Design Manual*, 1984, Published by Amer. Soc. of Civil Engineers, New York, NY.
7. Kahl, R. J., *Designing Polycarbonate Parts*, from *Plastics Products Design Handbook, Part A, Materials and Components*, Edited by Miller, E., pp 149–234, 1981, Marcel Dekker, New York, NY.
8. Windenburg, D. F., and Trilling, C., *Collapse by Instability of Thin Cylindrical Shell Under External Pressure*, Trans ASME, 1956, pp 819–825.

9. Murayama, T., *Dynamic Mechanical Analysis of Polymeric Material*, 1978, Elsevier Scientific Publishing Co, Amsterdam.
10. Ferry, J. D., *Viscoelastic Properties of Polymers*, 2nd Edn. 1970, John Wiley & Sons, New York, NY.
11. Trade Brochure entitled *Understanding Rheological Testing—Thermoplastics*, Published by Rheometrics, Inc. One Possumtown Road, Piscataway, NJ.
12. Ward, I. M., *Mechanical Properties of Solid Polymers*, 2nd Edn. 1983, John Wiley & Sons, Ltd, Chichester, UK.
13. McCrum, N. G., Buckley, C. P., and Bucknall, C. B., *Principles of Polymer Engineering*, 1988, Oxford University Press, New York, NY.
14. Nielsen, L. E., *Mechanical Properties of Polymers and Composites*, Vol 1, 1974, Marcel Dekker, Inc., New York, NY.

**Exercises**

**5.1.** You may have observed that the caps for jelly and jam bottles are made of metal, and marked with a cautionary note about the button at their center. Why are these caps made only of metal?

**5.2.** A balloon is inflated with air. Assuming no leak and no temperature change, which of the following scenarios is likely to happen?

Scenario 1: The balloon should increase in size, simply as a direct consequence of creep of the balloon material.

Scenario 2: Creep occurs. When it does, the pressure of the air falls, so does the stress, and hence creep rate is reduced. Eventually an equilibrium state is reached, when the increase in size would be exactly balanced by the tendency to reduce the pressure.

Scenario 3: This being an applied displacement problem, relaxation should occur. Beacuse of this, the stress is reduced. To achieve equilibrium with internal pressure, the size should shrink. However, because relaxation occurs at constant strain, the shrinking balloon presents a crinkled appearance. Thus, crinkling of the balloon which is normally observed is not caused by escape of air, but due to relaxation.

**5.3.** In creep rupture tests pressurized pipes are used. Water is used as the fluid under pressure. Constant pressure is maintained by pumping in water to make up for the swelling in volume. Discuss qualitatively the material behavior just before pipe failure, (1) if the pipe were sealed with pressurized water, and no pump was used, and (2) if the pipe were sealed with pressurized air and no pump was used.

**5.4.** A built-in beam of moment of area of cross-section $I$, and length $L$ is dipped at one end by an amount $\delta$ due to a snag. Use pseudoelasticity to determine the stress in the beam at time $t$.

**5.5.** The detail of an original design and a revised design of a snap fit are shown below. In the revised design the fillet prevents the snap from returning to its stress-free position. By assuming realistic dimensions, discuss the possible

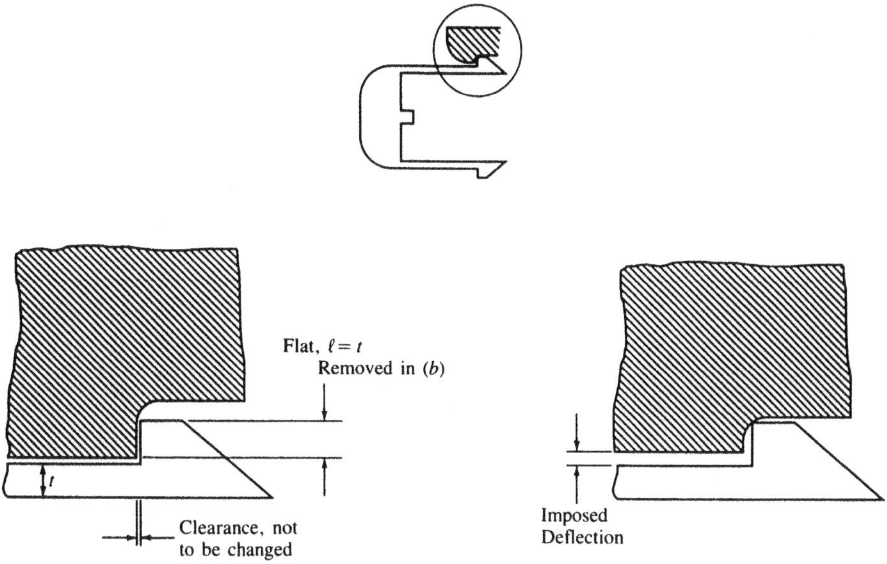


(*a*) Origiral Design (*b*) Revised design

**EXERCISE 5.5.**

effects of the revised design. Give reasons based on viscoelastic behavior to show that the snapping action may be lost.

**5.6.** Re-do the tee fitting example of Chapter 2 (Example 2.9) and determine the seal pressures at the end of 10,000 hours. Modulus for the tee fitting at 10,000 hours is 50 $kg/mm^2$, that of the pipe is 6.4 $kg/mm^2$. Their Poisson's ratios remain constant at 0.3 and 0.35, respectively. The metal ring does not creep and hence its modulus remains the same as at the beginning, that is, at 7,000 $kg/mm^2$. Poisson's ratio may be assumed to be 0.3. Show that it is the metal keeping the contact pressure.

**5.7.** Catheter tubes in hospitals are squeezed using plastic clips that have living hinges at one end as shown. Sometimes, these clips have to remain for several days in the clamped position. If a metal clip is used in parallel, indicate a strategic location for the metal clip, to minimize the stress in the living hinge.

**5.8.** A lawn chair is kept loaded on its seat area with 100 kg. Assume that this is equivalent to an axial load and a moment acting on each leg. The leg's cross section is shown below. The material of the chair is polyethylene. Find the time taken for the leg to buckle. The distance of the center of pressure to the centroid of the leg cross section may be assumed to be 250 mm.

**5.9.** A cantilevered beam is made of polypropylene and has a cross section of 25

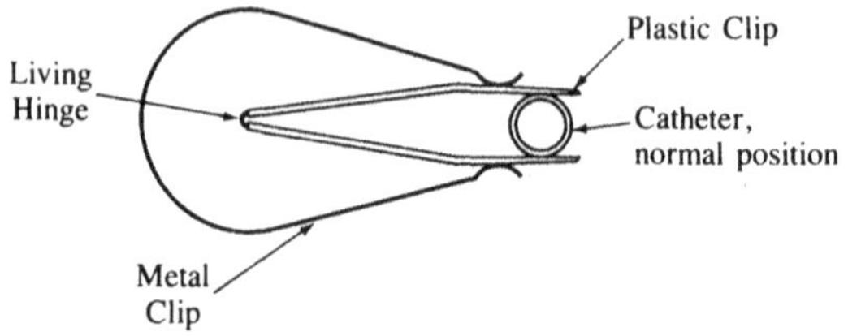


**EXERCISE 5.7.**

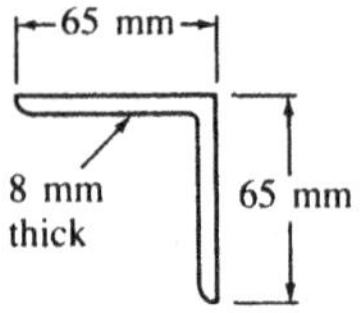


**EXERCISE 5.8.**

mm × 25 mm, and is 800 mm long. It is acted upon by a cyclic load of $30 \sin 500t$ kgs at its end, where $t$ is in seconds. At this frequency the in-phase and out-of-phase components of moduli are $3.962 \times 10^3$ MPa and $2.084 \times 10^2$ MPa, respectively. Find the maximum deflection at the tip.

**5.10.** Re-do Example 5.11 for standard viscoelastic material.

# 6

# Fracture Mechanics

## 6.1 INTRODUCTION

Fracture mechanics is a relatively new branch of stress analysis. It evolved from the need for a rational explanation of the several brittle failures-without-warning of ship hull structures, railroad structures, etc. which are made of normally ductile metals [1]. The development of fracture mechanics was initiated from several viewpoints. One such viewpoint is the strain energy consideration in a body in which a crack extends by just a small amount. Another approach is the application of linear elasticity for the determination of stress fields at the tip of a crack. Other approaches are equivalent to each other for the elastic behavior of the material in the close vicinity of the crack tip and diverge for plastic flow. Each approach defines its own version of the resistance of the material to fracture, which may be termed fracture toughness. It is used much in the same way as yield strength or ultimate strength is used. Basically, all approaches seek on the one hand, to determine the response of a cracked structure to applied loads, and on the other hand, to provide methods to measure the toughness of the material. The extension of fracture mechanics to plastics involves certain additional considerations. The high plastic flow capability and time dependence are the special characteristics of fracture mechanics as applied to plastics.

The basic premise in fracture mechanics is the existence of cracks or flaws in all structures. Such flaws can be microscopic in size. For this purpose local variations in density, microvoids, and orientation can be treated as flaws. These defects are potential sites for the initiation of a crack. This premise is more than just an assumption. It has been established that whatever the quality control system, a small percentage of products with crack-like defects "pass" all stages of inspection.

The growth of a crack violates the continuity of the material (compatibility conditions), and hence the theory of elasticity cannot be used as it is. In fact, the methods of stress analysis are good only for a static crack, and not for crack growth. Hence studies of crack growth caused by service loads is done by empirical laws.

## 6.2 AN OUTLINE

In this chapter we will discuss (1) the criterion for catastrophic propagation of a crack and (2) the methods for analyzing the slow growth of a crack under fatigue loading conditions. Item (1) is of basic importance to the study of fracture mechanics because it provides the means for the development and measurement of the so-called fracture toughness.

Catastrophic failure is characterized by the following four criteria.

1. A criterion based on the *Strain Energy Release Rate* ($G_c$) of the body during crack extension.
2. A criterion based on the stress field just ahead of the crack tip, also called the *Stress Intensity Criterion* ($K_{Ic}$).
3. A criterion based on a path-independent integral around the crack tip, which represents the loss of strain energy in the body. This criterion is called the *J-Integral Criterion* ($J_c$).
4. A criterion based on the amount of opening experienced at the crack tip (*CTOD*) before it extends. This is called the *Crack Tip Opening Displacement Criterion* or the *CTOD Criterion*.

As mentioned before, all these criteria are equivalent to each other for elastic behavior. However, they represent independent approaches when energy dissipation due to plastic flow is involved. In the following, we discuss the effect of fatigue loading, and the extension per load cycle of the crack. These considerations enable one to calculate the number of cycles that an initially cracked body can withstand.

Here, the stress intensity approach is given a preferential treatment simply because it utilizes the equations of elasticity directly, and thus it permits numerical calculations for a cracked structure.

### 6.2.1 Modes of Fracture

Fracture can occur in three different modes, as shown in Figure 6.1. These modes are derived directly from the three possible directions of stress that

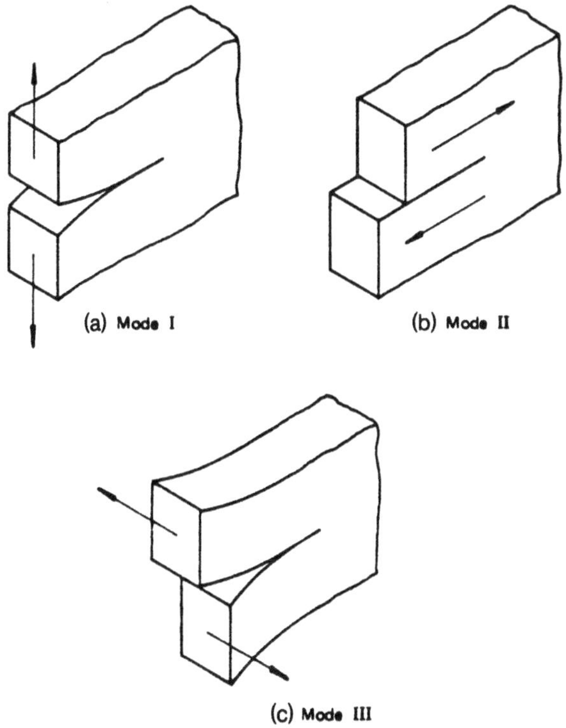


**FIGURE 6.1.** The three possible modes of fracture.

can exist on the plane of the crack. The first mode of fracture, designated by the subscript *I*, is the tensile fracture or the "opening mode" of fracture. This is by far the most predominant mode occurring in practice. The second mode, designated by *II*, is the "sliding mode" of fracture and involves pure shear forces only. The third mode, designated by *III*, is called the "tearing mode."

Here, we deal only with mode *I* of fracture.

## 6.3 STRAIN ENERGY RELEASE RATE CRITERION

This is the earliest criterion of fracture, and was proposed by Griffith [2] for fracture of brittle materials. Consider a two dimensional body of thickness $B$, with a crack embedded in it (Figure 6.2(a)). Suppose a load is applied on the body, and is gradually increased until a small growth in the crack length occurs. Let us assume elastic behavior of the body. The load

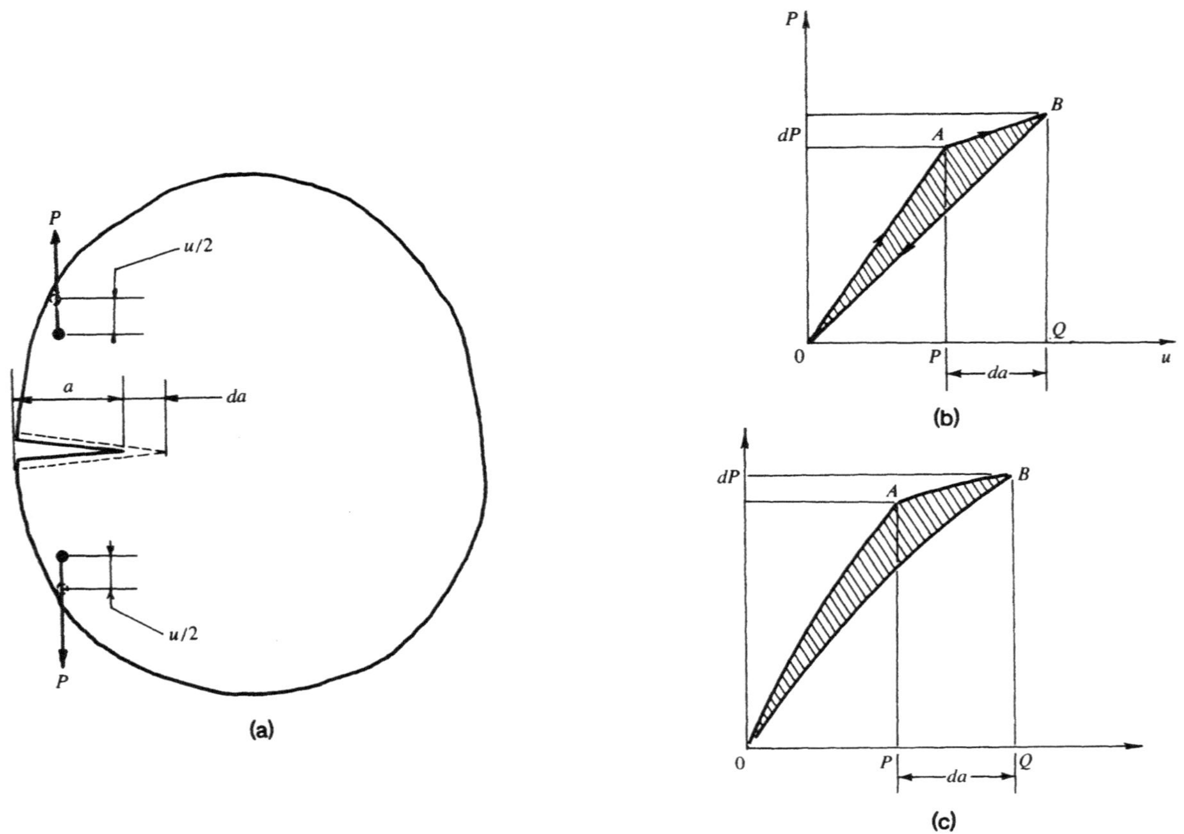


**FIGURE 6.2.** Energy released by extension of the crack. (a) Loading of an arbitrary shaped cracked body. (b) Linear elastic behavior. (c) Nonlinear elastic behavior. Shaded area *OAB* represents the strain energy released.

displacement curve is shown in Figure 6.2(b). We see that the curve is essentially linear up to the point $A$, where the crack extension occurs, during which time the point moves to $B$, along a different line $AB$. We are interested in the *change in the total energy* of the system due to crack extension. This change is really a release of energy by the system. Griffith's original proposal considered that this release of energy is used up exactly in producing new surfaces exposed by crack extension. Because of crack extension, the total elastic strain energy stored in the cracked body changes by an amount $\delta U_1$. This happens because of the change in stress distribution, some stressed parts becoming free surfaces, and also because of changes in external forces and displacements.

The total change in the energy of the system $dU$ is given by

$$dU = dU_1 - dU_2 \tag{6.1}$$

where

$dU_1$ = Change in strain energy stored in the body by crack extension

$dU_2$ = Incremental work done by force $P$ moving from $A$ to $B$.

From Figure 6.2 it is seen that the strain energy changes from the area $OAP$ to $OBQ$. This is so because, if unloaded, the load deflection curve will follow the line $BO$. Therefore, the triangle $OBQ$ is the strain energy available in the body after crack extension.

$$dU_1 = \frac{1}{2}(P + dP)(u + du) - \frac{1}{2}Pu$$

The increment in work done by the external forces, namely $dU_2$ occurs at a constant load $P$.

$$dU_2 = \left(P + \frac{dP}{2}\right) du$$

The change in the total energy of the system is therefore

$$\begin{aligned} dU &= dU_1 - dU_2 \\ &= \frac{1}{2}(u\,dP - P\,du), \end{aligned}$$

and it may be seen to be the wedge-shaped area $OAB$. The strain energy release rate $G$ is defined as simply the negative of $dU$ per unit crack length extension, per unit width of the body. It is negative because it is a release or shedding away of energy from the system.[1]

$$G = -\frac{1}{B}\frac{dU}{da}$$

$$= \frac{1}{2B}\frac{d}{da}(P\,du - u\,dP)$$

The strain energy release rate criterion states that fracture occurs when $G$ exceeds a critical parameter $G_c$. The physical interpretation of this criterion is easily seen by considering the equation $G = (\text{A const}) \times (dU_2/da - dU_1/da)$. In this equation, $dU$ is used for increasing the strain energy $dU_1$ of the body, and also to produce new surfaces. Thus, $(dU_2 - dU_1)$ represents the surplus energy available for producing the new surfaces. The excess, if any, is available to propagate the crack. Thus, the strain energy release criterion is equivalent to the statement that the crack propagates if the energy put into the system equals or exceeds the increase in strain energy plus the energy needed to produce new surfaces. We will discuss the methods of determining $G_c$.

## 6.4 STRESS ANALYSIS OF CRACKS

Before stating the $K_{Ic}$ criterion of fracture, we need to discuss the stress field just ahead of the crack tip in mode $I$ fracture.

The elastic stress analysis of sharp cracks leads to infinite stresses at the crack tip. This can be seen by first considering the stresses at the boundary of an elliptical opening in an infinite plate loaded with constant uniaxial stress $\sigma_0$. See Figure 6.3. The case of a sharp crack is obtained by letting the ellipse flatten continuously, until its minor axis is zero.

A more informative approach to crack tip stress analysis is to consider the admissible solutions (eigensolutions) for the so-called "stress functions"[2] in a polar coordinate system situated at the crack tip. See Figure 6.4. The exact procedures for calculating the stresses are outside the scope

[1] The symbol $G$ is used in this chapter to denote the strain energy release rate, and is not to be confused with the shear modulus of the material.

[2] Stress functions are a set of scalar functions $\Theta = f(r, \theta)$ which satisfy the bi-harmonic equation $\nabla^4\Theta = 0$. For a detailed discussion of the properties and use of such functions see References [3] and [4].

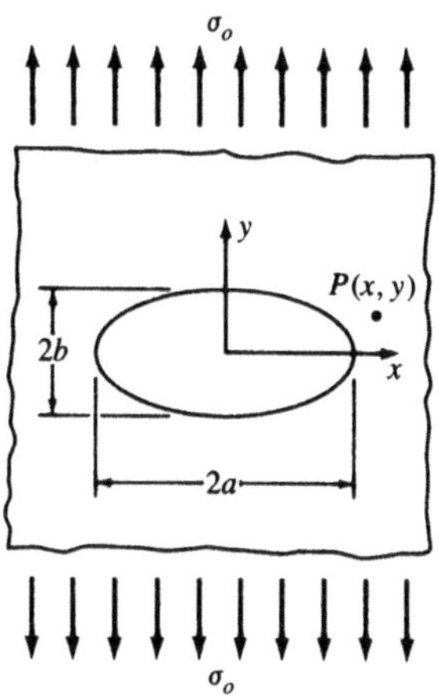


$$2B = (a + b) \qquad m = \left(\frac{a-b}{a+b}\right)$$

$$\alpha = \frac{x}{2B} + \sqrt{\left(\frac{x}{2B}\right)^2 - m}$$

$$z_2 = \sigma_o\left[1 + \frac{2(1+m)}{(\alpha^2 - m)}\right]$$

$$z_1 = \frac{\sigma_o}{2}\left[1 + \left(\frac{m^2-1}{\alpha^2-m}\right)\left\{1 + \left(\frac{m-1}{\alpha^2-1}\right)\left(\frac{3\alpha^2-m}{\alpha^2-m}\right)\right\}\right]$$

$$\sigma_x = \frac{z_2}{2} - z_1 \quad \text{along } x\text{-axis}$$

$$\sigma_y = \frac{z_2}{2} + z_1 \quad \text{along } y\text{-axis}$$

**FIGURE 6.3.** Stresses at points along the major axis of an elliptical hole in an infinite plate. Note that when $a = b$, the results coincide with that of a circle.

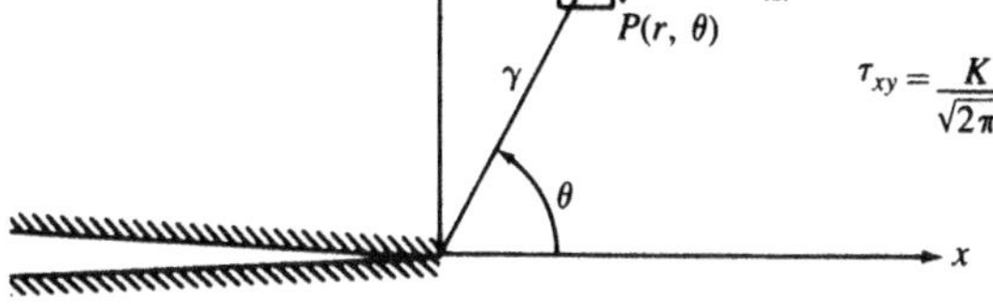


$$K = \sigma_0\sqrt{\pi a}$$

$$\sigma_{xx} = \frac{K}{\sqrt{2\pi r}} \text{ Cos } \theta/2\left[1 - \text{Sin }\frac{\theta}{2}\text{ Sin}\frac{3\theta}{2}\right]$$

$$\sigma_{yy} = \frac{K}{\sqrt{2\pi r}} \text{ Cos } \theta/2\left[1 + \text{Sin }\frac{\theta}{2}\text{ Sin}\frac{3\theta}{2}\right] + \theta_o$$

$$\tau_{xy} = \frac{K}{\sqrt{2\pi r}} \text{ Cos}\frac{\theta}{2}\text{ Sin}\frac{\theta}{2}\text{ Cos}\frac{3\theta}{2}$$

**FIGURE 6.4.** Stress field at a crack tip. Note that the point $P$ is identified by its polar coordinates, but the stress components are represented in the Cartesian system.

of this text. Rather, we need to know a few characteristics of the stress field at the crack tip.

1. For an infinite plate under uniaxial load as in Figure 6.4, the solutions for stress are of the form,

$$\sigma_{xx} = \frac{K}{\sqrt{2\pi r}} \cos\frac{\theta}{2}\left[1 - \sin\frac{\theta}{2}\sin\frac{3\theta}{2}\right] \tag{6.2}$$

$$\sigma_{yy} = \frac{K}{\sqrt{2\pi r}} \cos\frac{\theta}{2}\left[1 + \sin\frac{\theta}{2}\sin\frac{3\theta}{2}\right] + \sigma_o \tag{6.3}$$

$$\begin{aligned}\sigma_{zz} &= \nu(\sigma_{xx} + \sigma_{yy}), \text{ for plane strain}\\ &= 0, \text{ for plane stress}\end{aligned} \tag{6.4}$$

$$\tau_{xy} = \frac{K}{\sqrt{2\pi r}} \cos\frac{\theta}{2}\sin\frac{\theta}{2}\cos\frac{3\theta}{2} \tag{6.5}$$

$$\tau_{yz} = \tau_{xz} = 0 \tag{6.6}$$

where, $K = \sigma_o\sqrt{\pi\alpha}$.

2. Although in Figure 6.4 an infinite plate is considered, the solution at a crack tip is *formally the same for all shapes and crack orientations*. This is a direct consequence of the generality with which the eigensolutions are obtained. Any component of stress $\sigma_{**}$ is expressible in the form

$$\sigma_{**} = \frac{K}{\sqrt{2\pi r}} \times (\text{a function of } \theta).$$

where $K$ is a constant that depends on the applied stress, crack size, and its position and orientation relative to the geometry and size of the component. $K$ is always expressible in the form $K = C\sigma\sqrt{a}$, where $a$ is the crack length, $\sigma$ is the applied stress, and $C$ is a geometry dependent constant. This fact is true for other modes of fractures also, and hence for the combination.

3. From the above, the parameter $K$ can be seen to be the only variable that is problem-specific, and not the value of the local stress. This point of view dispenses with the infinite stresses. In other words, the *manner of approach* of the stress values to infinity is important. The parameter $K$ describes exactly that. $K$ is called the "stress intensity factor." The idea of using this factor to characterize the crack behavior was pro-

posed by Irwin [5]. Calculation of $K$ for even the simplest case, such as an infinite plate, uses the advanced mathematical theory of elasticity and is outside the scope of our discussion. However, a few pre-engineered solutions are available in References 6, 7, and 8. These solutions are used much in the same way as are stress analysis handbooks. A practical problem is simply modeled as closely as possible to one of the cases listed.

## 6.5 THE $K_{Ic}$ OR THE STRESS INTENSITY CRITERION

A critical value of $K$, denoted by $K_{Ic}$ determines the upper bound for $K$. $K_{Ic}$ is a material property, and represents the resistance of the material to cracking. It is a function of temperature and environment. For thermoplastics $K_{Ic}$ depends, in addition, on time because of their viscoelastic nature. This means that a crack can grow in length with time at a constant stress level. Since $K$ increases as $\sqrt{a}$, the effect of time is to produce an accelerated increase of the stress intensity factor. Of more importance is the fact that $K_{Ic}$ depends on whether the stress field is one of plane stress or plane strain. It has a maximum value for plane stress conditions and a minimum value for plane strain. Later, this point will be discussed in a more quantitative way. The minimum value is of importance in the design, and hence $K_{Ic}$ is called *plane strain fracture toughness.*

As of 1991 ASTM Subcommittee D-20 has released a Draft Standard for measurement of $K_{Ic}$ and $G_c$ for plastic materials.

In a practical design problem, one typically determines the $K_I$ for a known configuration of the crack. The crack size is usually assumed. The problem-specific constants are read from handbooks. The fracture toughness data $K_{Ic}$ is compared with $K_I$, and we require that

$$K_{Ic} \geq K_I \times \text{(a factor of safety, say, 2)}$$

See Figure 6.5 for a graphic explanation.

## 6.6 THE $J$-INTEGRAL CRITERION

The $J$-integral criterion, proposed by Rice [9] is based on the so-called $J$-integral defined by

$$J = \int_{\Gamma} W\,dy - \int_{\Gamma}\left(T_x \frac{\partial u_x}{\partial x} + T_y \frac{\partial u_y}{\partial x}\right) ds \tag{6.7}$$

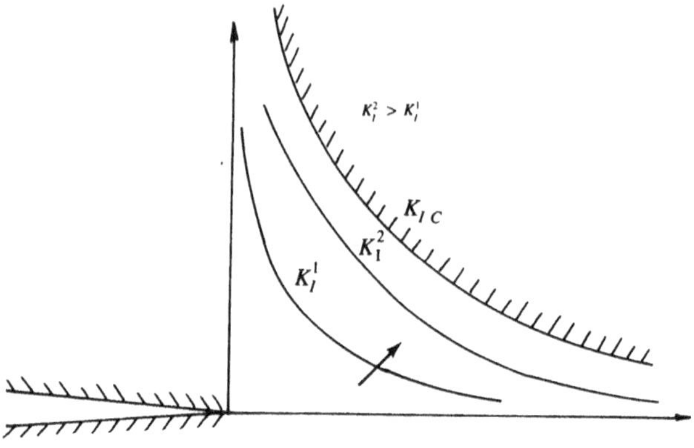


**FIGURE 6.5.** Stress intensity factors $K_I$, lower than fracture toughness $K_{Ic}$ represent contours below the shaded curve. With an increase of crack size or stress, the curve shifts upward.

where Γ is a contour starting on the lower face of the crack and ending on the upper face of the crack, and enclosing the crack tip. $W$ is the strain energy density. $T_x$ and $T_y$ are the components of the traction vector acting at the point on Γ and $u_i$ are the corresponding displacement components. See Figure 6.6. This integral, has the following properties.

- For elastic behavior, the integral represents the strain energy release rate, and is identical to $G$.

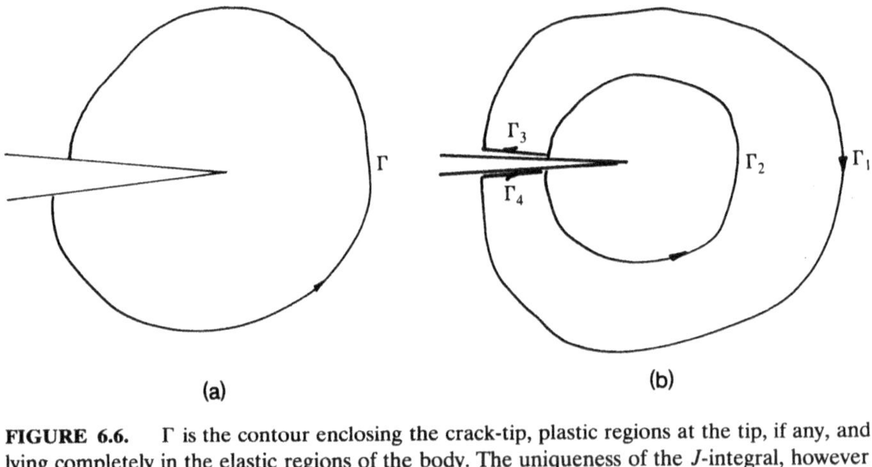


**FIGURE 6.6.** Γ is the contour enclosing the crack-tip, plastic regions at the tip, if any, and lying completely in the elastic regions of the body. The uniqueness of the $J$-integral, however is proved by considering two contours $\Gamma_1$ and $\Gamma_2$ bridged by $\Gamma_3$ and $\Gamma_4$ as shown, and by the fact that the $J$-integral around such a contour is exactly zero.

- *As long as the body is loaded in a monotonically increasing manner*, it includes the effects of plastic flow also.
- It is path independent, so any convenient path on which loads and displacements are known can be chosen. A good choice is a contour that encloses the crack tip and regions of plastic flow around the tip, and lies completely on elastic regions of the body. Performing the integration on such a contour is easy to program for computers.
- A critical parameter designated $J_c$ can be defined as a material property, and is used as a fracture criterion.

## 6.7 THE CTOD CRITERION

The CTOD criterion, proposed by Wells [10], is based on the argument that the stresses at the crack tip involve plastic flow of material, and that the critical parameters of fracture are related to the plastic strain capability of the material, which controls the fracture. The displacement normal to the plane of the crack, measured at the crack tip is a measure of the plastic strain. Thus, the crack tip opening displacement may be taken as a criterion of fracture.

For an infinite plate with a crack, the crack opening displacement $\delta$ can be shown to be

$$\delta = \frac{8\sigma_y a}{\pi E} \ln \sec\left(\frac{\pi\sigma_o}{2\sigma_y}\right) \tag{6.8}$$

where $\sigma_o$ is the applied stress perpendicular to the crack, $\sigma_y$ is the yield stress of the material, and $E$ is its modulus. The physical meaning of $\delta$ is shown in Figure 6.7.

## 6.8 REMARKS ON THE FRACTURE CRITERIA

The strain energy release criterion is simple to understand and also circumvents infinite stresses. There are a only few simple cases for which $G$ can be calculated easily. Such cases are the basis for the material testing for $G_c$, and generally not for the analysis of cracks. This criterion does not provide an analytical method for calculating the actual release rate for a general problem.

The stress intensity criterion is also equally simple to understand. The infinite stress at a crack tip is not a major obstacle to the understanding,

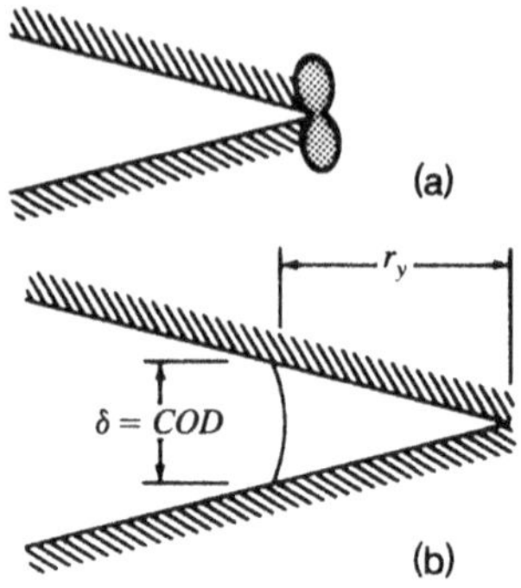


**FIGURE 6.7.** The physical meaning of COD. Plastic flow of material occurs in the shaded area, due to which there is a finite displacement $\delta$ at the crack tip. When $\delta$ reaches a critical value, the crack propagates catastrophically. $r_y$ is the plastic zone size. This is the region over which the fibrils pull the crack faces together. This pull has the same effect as a negative stress intensity factor.

since we deal only with $K$ at a crack tip, and not with the stress itself. Being rooted in the theory of elasticity, methods are available, however complex, to calculate $K$ for many cases. Moreover, the computation of $K$ for a general problem is possible in numerical methods, especially finite elements. Most finite element codes of the late 1980s provide a "crack tip element" as a standard feature. Such element formulations are based on the determination of the crack tip displacement field (not stress). The displacement is not singular, and is well behaved. The value of $K$ can be obtained from the displacement information. Because of its computability, the $K_{Ic}$ criterion is by far the most widely used criterion, both for metals and plastics.

It should be noted that the $J$-integral concept is abstract and is the most difficult of all the criteria to understand. Like the other criteria, it is also applicable only until the beginning of crack growth, and not for crack growth itself. This is because, although external loads are monotonic, there is an unloading of the stresses at the crack tip because of its extension and movement to a new location. The $J$-integral is derived only in two dimensions. This means that it is applicable strictly only for plane strain conditions, whereas any real life situation is a combination of plane strain and plane stress conditions. Furthermore, analytical solutions of the $J$-integral involving plasticity are rare. Numerical solutions such as finite element analysis are well suited for the $J$-integral approach.

In the elastic regime, the CTOD criterion has a one-to-one relationship with the $K$, $G$, and $J$ criteria. However, in the plastic regime, no solutions similar to Equation 6.8 for finite geometries are available. In general, the

CTOD criterion is considered more applicable to metals than to plastics. This is because the crack propagation in plastics takes place via a mechanism called "crazing" (see Section 6.20), which is a material behavior fundamentally different from yielding.

## 6.9 MORE ABOUT $G$

### 6.9.1 Determination of $G_c$

The double cantilever beam shown in Figure 6.8 is the simplest means of determining $G_c$. Each arm of the beam acts as a cantilever beam of length $a$. In order to calculate $G_c$, we have to calculate $G$ of the specimen. Suppose that one of the arms deflects vertically by an amount $u/2$, then

$$\frac{u}{2} = \frac{Pa^3}{3EI} = \frac{4Pa^3}{EBD^3}, \tag{6.9}$$

where $E$ is the Young's modulus of the material. We may rewrite this equation as $u = CP$, where $C$ is the compliance of the specimen, given by

$$C = \frac{8a^3}{EBD^3}$$

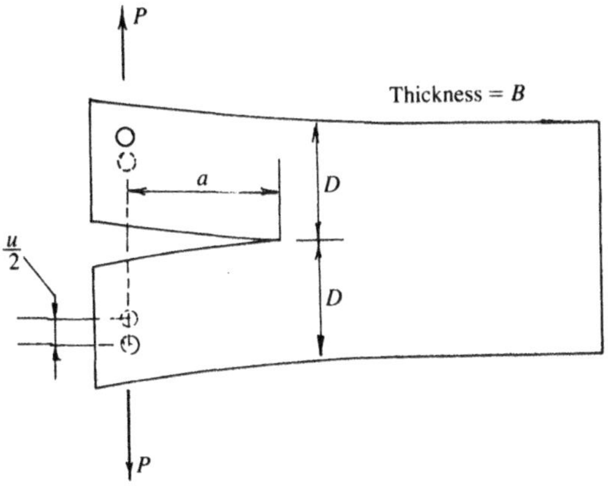


**FIGURE 6.8.** The double cantilever beam is the simplest configuration for measuring $G_c$. It is also a recommended shape for the specimen in the Draft ASTM Standard for fracture toughness testing.

Also,

$$\frac{du}{da} = C\frac{dP}{da} + P\frac{dC}{da}$$

$$G = \frac{1}{2B}\left[P\frac{du}{da} + u\frac{dP}{da}\right]$$

$$= \frac{1}{2B}\left[P\left(C\frac{dP}{da} + P\frac{dC}{da}\right) - CP\frac{dP}{da}\right]$$

$$= \frac{P^2}{2B}\frac{dC}{da} \tag{6.10}$$

$$= \left(\frac{12P^2a^2}{EBD^3}\right) \quad \text{or,} \tag{6.11}$$

$$= \frac{3}{16}\frac{EBD^3u^2}{a^4} \tag{6.12}$$

It can be seen that with $E$, $B$, $D$ and the crack length $a$ (or equivalently $u$) known, it is possible calculate $G$. It is only necessary to load a pre-cracked double cantilever specimen on a tension testing machine, and record $P$ and $u$, all the way up to failure of the specimen. Instantaneous values of $a$ can be calculated from Equation 6.9. $G$ can be calculated using Equation 6.11 or 6.12. The value of $G$ calculated at the point of fracture is the *critical strain energy release rate* $G_c$.

It is seen from Equations 6.11 and 6.12 that $G$ increases with $a$. It is also possible to resize the depth $D$ of the beam and $a$ to keep $G$ (almost) a constant. Such a shape is a tapered double cantilever beam with taper angle of about 11° and is useful as a test specimen configuration for experimental work. Another candidate shape with almost constant $G$ is a double torsion specimen. Both the shapes are shown in Figure 6.9. As was pointed out before, it is not as easy to calculate the load-displacement relationship for a general problem. This factor limits the utility of the strain energy release criterion for routine design and analysis, notwithstanding its ability to circumvent infinite stress fields.

### 6.9.2 *G* As a Contour Integral

The strain energy release rate can be calculated by means of an integral carried out over the boundary of the body. This procedure of calculation is

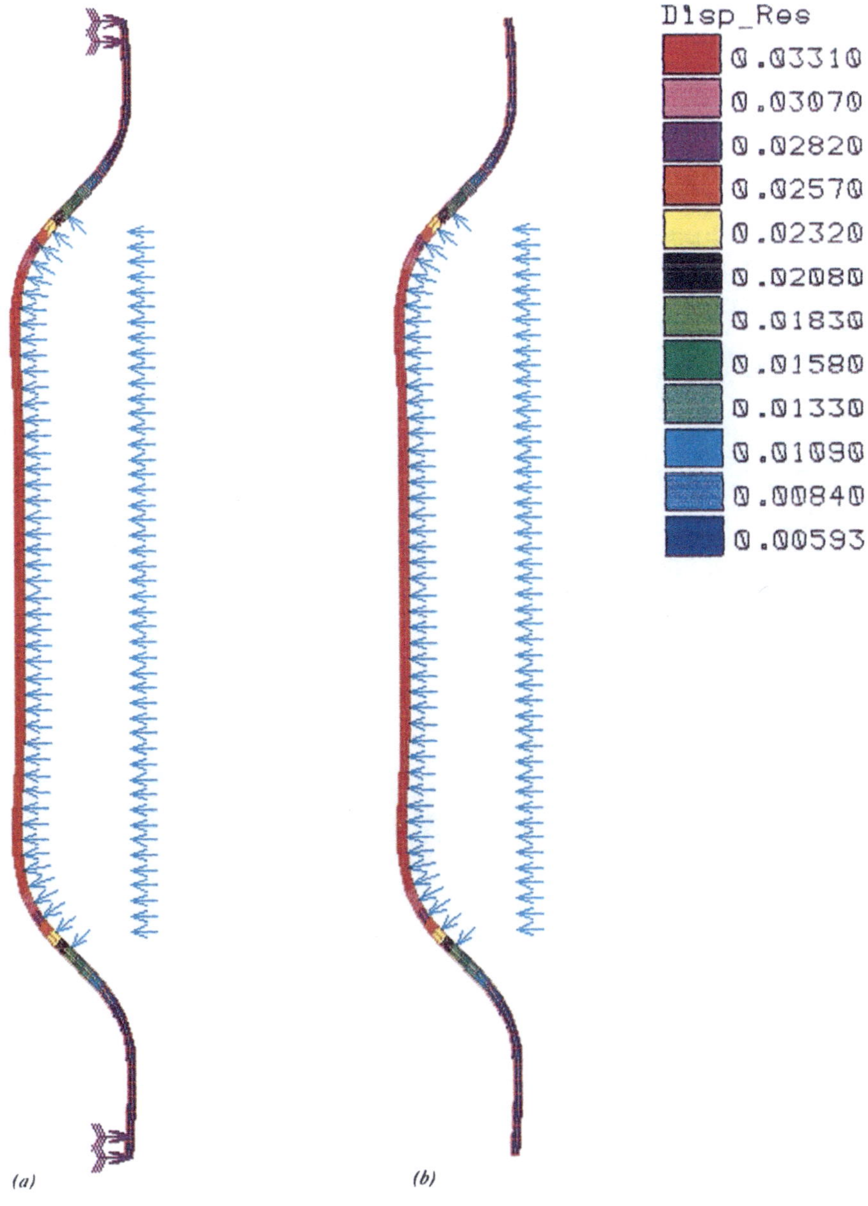


Figure 2.28. *Finite element analysis of ribbed and unribbed vessels.*

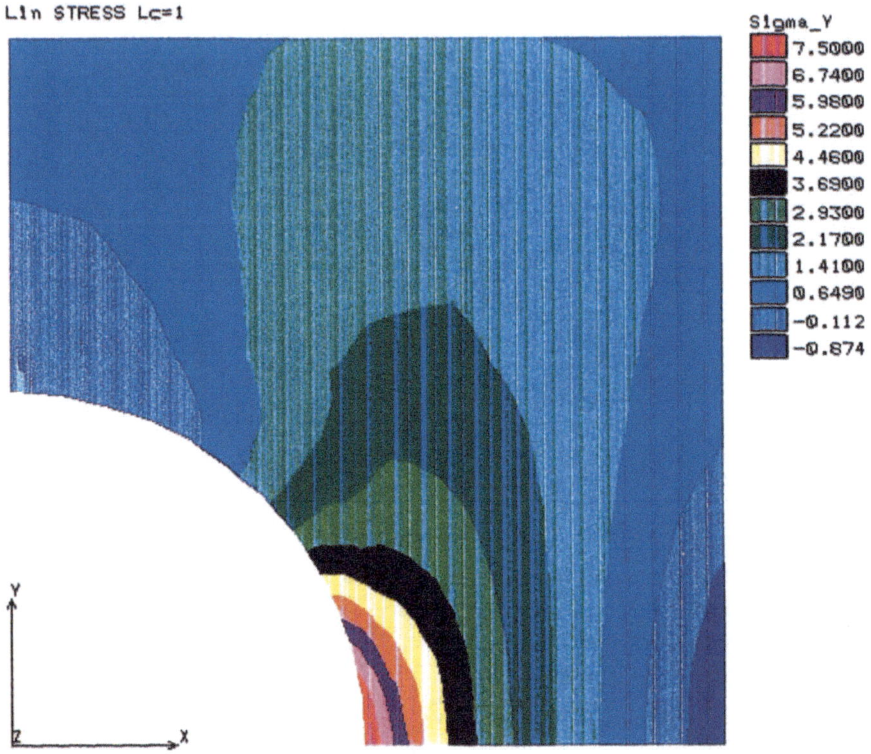


Figure for Exercise 4.7. *A plate with a large hole, in which stress at the hole boundary begins to feel the presence of the square plate boundary.*

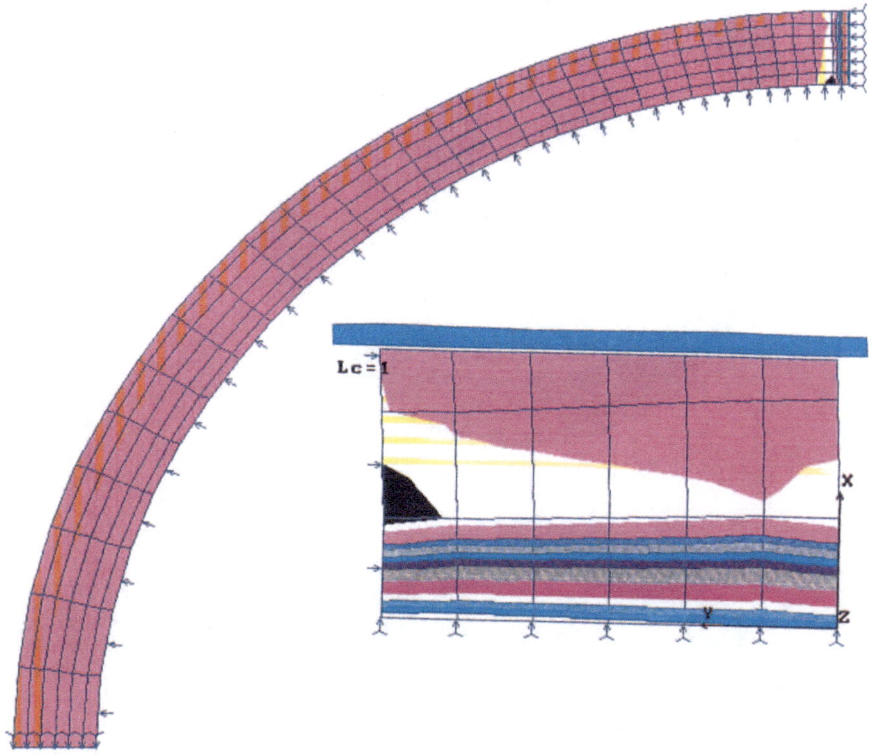


Figure 9.7. *A node on the axis of symmetry experiences a strain that is proportional to 1/r, r being the radial coordinate, which is zero. This causes some problems in the calculation of displacements, which shows spuriously in the stresses also.*

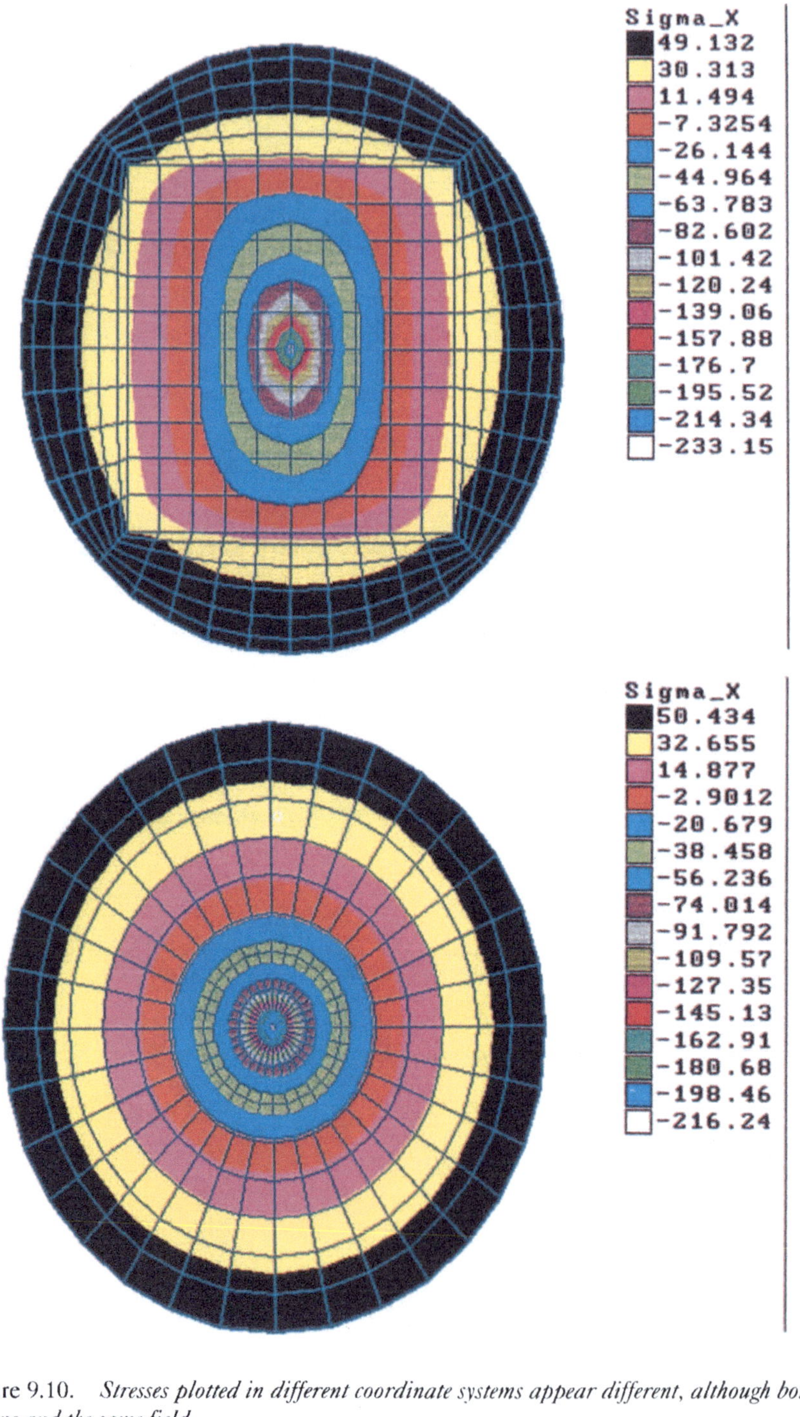


Figure 9.10. *Stresses plotted in different coordinate systems appear different, although both are one and the same field.*

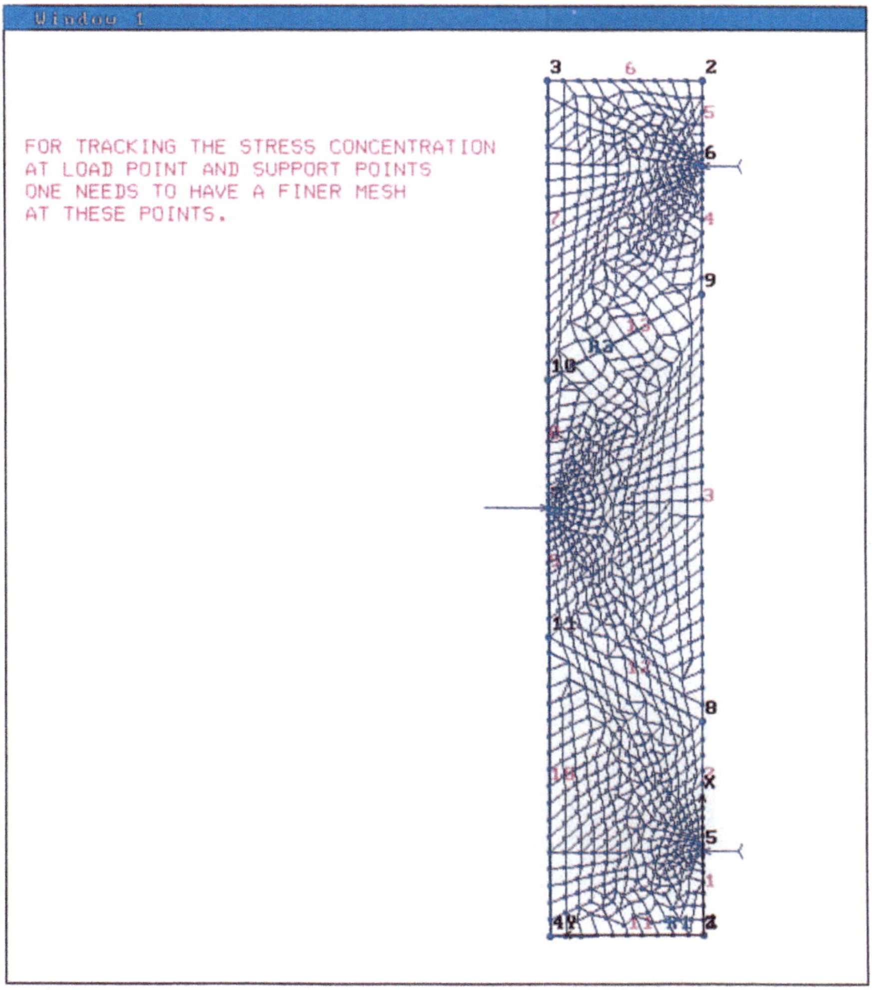


Figure 9.13. *A three-point bending specimen (a) showing the initial geometry, and (b) the final geometry created automatically by the computer by the h-refinement. The final geometry results in several iterations, and is guided by the error norm.*

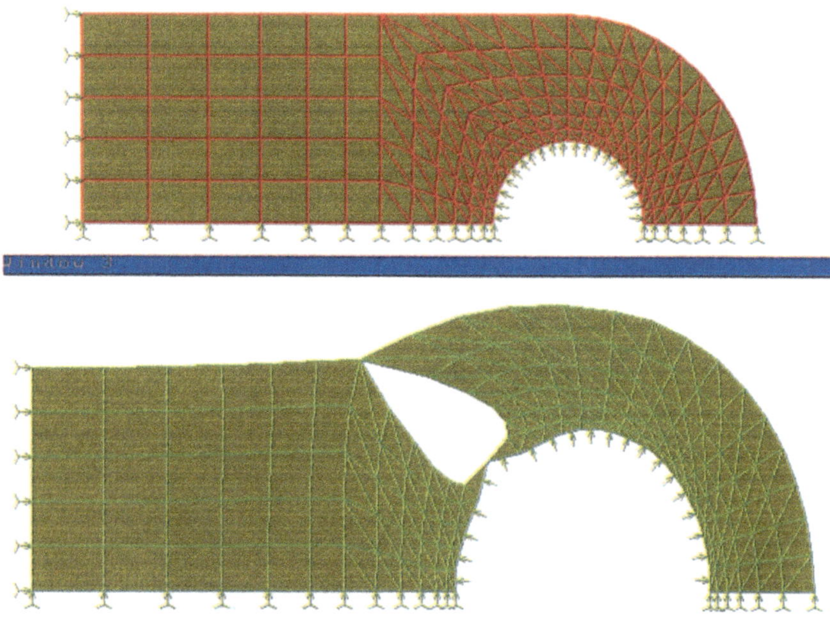

Figure for Exercise 9.3.

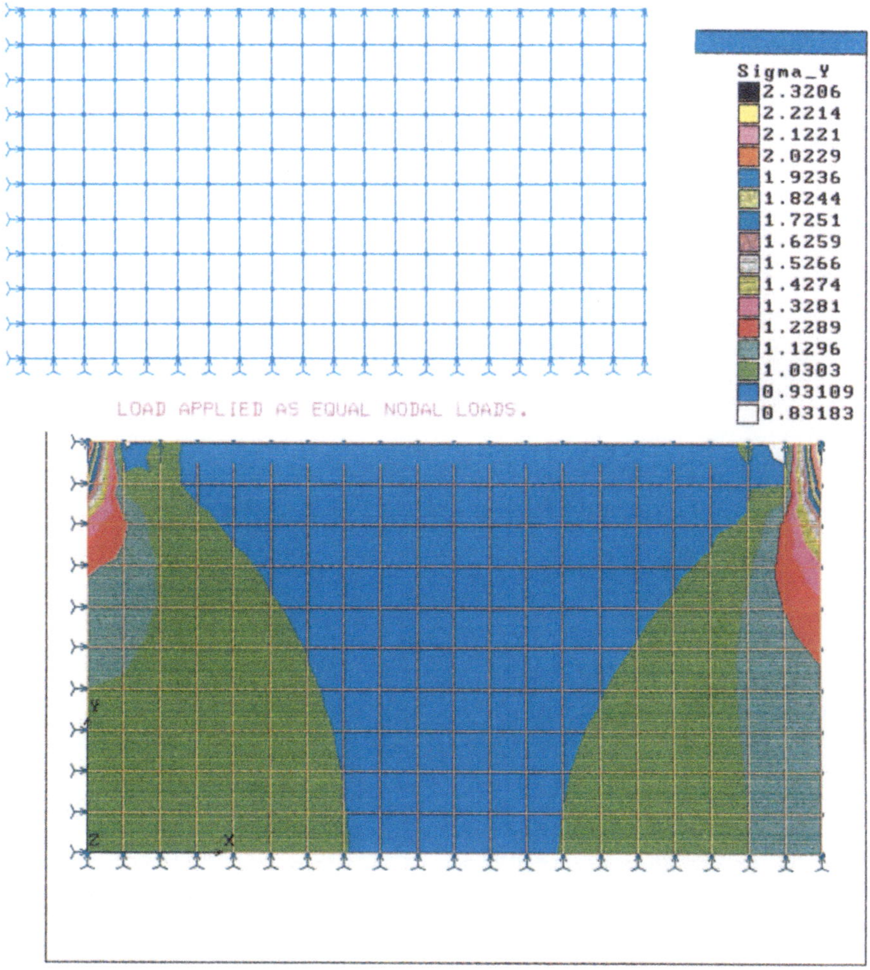


Figure for Exercise 9.18.

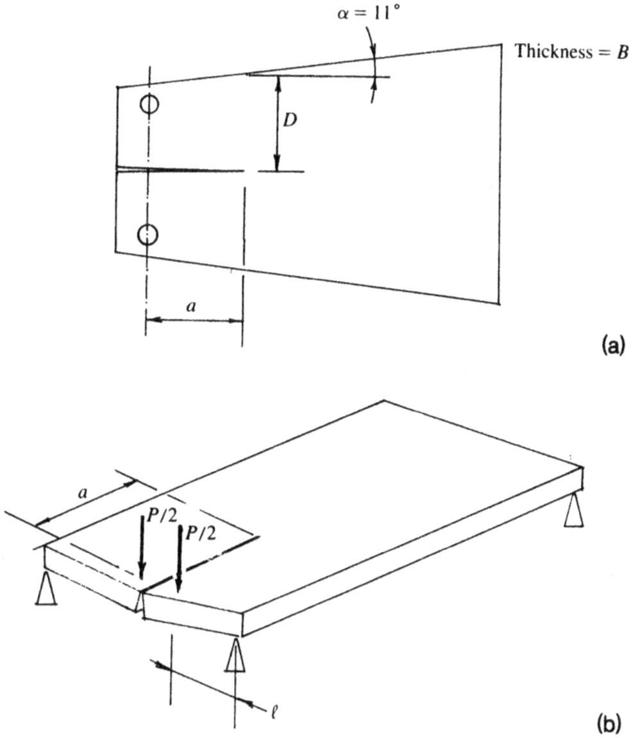


**FIGURE 6.9.** Two shapes of constant $G$. (a) The tapered double cantilever beam. (b) The double torsion specimen.

particularly useful to establish the connections among $G$, $K$, and $J$ in the elastic regime. It was proved earlier that

$$G = \frac{1}{2B}\left[P\frac{du}{da} - u\frac{dP}{da}\right]$$

When several concentrated forces $P_i$ are acting on the surface of the body, and $u_i$ are the corresponding displacements at the points of application of the forces, the energy terms are simply summations of the contributions from the individual forces. Thus, for this case,

$$G = \frac{1}{2B}\left[\sum P_i\frac{du_i}{da} - \sum u_i\frac{dP_i}{da}\right] \tag{6.13}$$

When the forces are distributed continuously rather than concentrated, the summation is replaced by an integral. The concentrated forces $P_i$ are simply replaced by $pB\,dL$, where $p$ is the pressure applied on the boundary and $dL$ is the elemental length of the boundary. Thus for the case of distributed forces,

$$G = \frac{1}{2}\left[\int p\frac{du}{da}\,dS - \int u\frac{dp}{da}\,dS\right] \tag{6.14}$$

Note that in this equation $p$ and $u$ are the surface forces and surface displacements. They both can be separated into normal and tangential components, namely $p_n$, $p_t$, $u_n$ and $u_t$ respectively. Because only products of like subscripts produce work terms, Equation 6.14 is expanded to

$$G = \frac{1}{2}\int\left[\left(p_n\frac{du_n}{da} + p_t\frac{du_t}{da}\right) - \left(u_n\frac{dp_n}{da} + u_t\frac{dp_t}{da}\right)\right]dS \tag{6.15}$$

This form of the $G$ is important for establishing its connections with $K$, $J$, and $COD$. A straightforward procedure is to take the boundary of the problem itself as the contour of integration. Alternatively, any simple interior contour enclosing the crack tip can be used. A circular contour with its center at the crack tip is a convenient choice. In the latter case, it is only necessary to recognize that on such contours $p_n$ represents the normal stress. (If the physical boundary of the problem is taken as the contour of integration then $p_n$ represents the actual applied pressure.) Likewise, $p_t$ represents the tangential (shear) stress. $u_n$ stands for the normal displacements and $u_t$ stands for the tangential displacements.

### 6.9.3 Relationship Between *G* and *K*

There is a unique relationship between $G$ and $K$ in the elastic regime. In order to calculate the relationship, we consider a circular contour around the crack, starting on the lower face and ending on the upper face, as shown in Figure 6.10. The actual calculations are too long to repeat here.

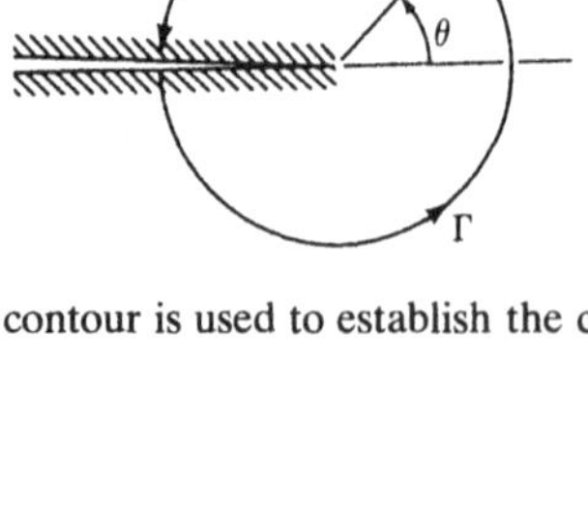


**FIGURE 6.10.** The circular contour is used to establish the connection between $G$ and $E$.

However, the important steps are listed. The reader can follow the calculations as an exercise.

*Step 1.* Note that at the crack tip the stress fields are described by Cartesian components, whereas their location is described by polar coordinates. The circular contour suggests that we express all quantities and derivatives in a polar coordinate system. Hence, convert the Cartesian stress components to polar coordinates, using the transformation rule of Chapter 1, or Mohr's circle.

$$\begin{Bmatrix} \sigma_r \\ \sigma_\theta \\ \tau_{r\theta} \end{Bmatrix} \rightarrow \begin{Bmatrix} \sigma_x \\ \sigma_y \\ \tau_{xy} \end{Bmatrix}$$

Also transform the displacements $u_x$ and $u_y$ to polar coordinates. This is simply done in a manner similar to a vector projection.

*Step 2.* This is a subtle step in the derivation. An advance of the crack tip by an amount *da also shifts the coordinate system*. Hence all points $(x, y)$ are referred to as $(x - da, y)$ after the crack extension. Thus $da = -dx$ and $dy = 0$, since $y$-coordinates remain unchanged. Conversion of the derivatives from Cartesian to polar coordinates is somewhat more involved. Start with the conditions,

$$\frac{d}{da} = \frac{\partial}{\partial a} + \frac{\partial}{\partial r}\frac{dr}{da} + \frac{\partial}{\partial \theta}\frac{d\theta}{da}$$

$$x = r\cos\theta \quad \text{and} \quad y = r\sin\theta$$

$$-da = dx = \cos\theta\, dr - r\sin\theta\, d\theta$$

$$dy = \sin\theta\, dr + r\cos\theta\, d\theta$$

*Step 3.* Derive the relationships for $d\sigma_r/da$, $d\sigma_\theta/da$, $d\tau_{r\theta}/da$, and similar derivatives for displacements $u_r$ and $u_\theta$. The final results for this step are

$$\frac{d\sigma_r}{da} = \frac{\partial\sigma_r}{\partial a} + \frac{\partial\sigma_r}{\partial\theta}\frac{\sin\theta}{r} - \frac{\partial\sigma_r}{\partial r}\cos\theta - 2\tau_{r\theta}\frac{\sin\theta}{r}$$

$$\frac{d\sigma_{r\theta}}{da} = \frac{\partial\sigma_{r\theta}}{\partial a} + \frac{\partial\tau_{r\theta}}{\partial\theta}\frac{\sin\theta}{r} - \frac{\partial\tau_{r\theta}}{\partial r}\cos\theta + (\sigma_r - \sigma_\theta)\frac{\sin\theta}{r}$$

$$\frac{du_r}{da} = \frac{\partial u_r}{\partial a} + \frac{\partial u_r}{\partial\theta}\frac{\sin\theta}{r} - \frac{\partial u_r}{\partial r}\cos\theta - u_\theta\frac{\sin\theta}{r}$$

$$\frac{du_\theta}{da} = \frac{\partial u_\theta}{\partial a} + \frac{\partial u_\theta}{\partial\theta}\frac{\sin\theta}{r} - \frac{\partial u_\theta}{\partial r}\cos\theta + u_r\frac{\sin\theta}{r}$$

*Step 4.* Lastly, write $G$ in its polar coordinate form (this does not need any transformation at all!)

$$G = \frac{1}{2}\int_{-\pi}^{\pi}\left[\left(\sigma_r\frac{du_r}{da} + \tau_{r\theta}\frac{du_\theta}{da}\right) - \left(u_r\frac{d\sigma_r}{da} + u_\theta\frac{d\tau_{r\theta}}{da}\right)\right] r\,d\theta$$

Substitute for the $\sigma$'s and $u$'s from the crack tip stress fields involving $K$, to obtain the expression

$$G = \frac{K^2}{E}, \quad \text{for plane stress} \tag{6.16}$$

$$= \frac{K^2(1-\nu^2)}{E}, \quad \text{for plane strain} \tag{6.17}$$

### 6.9.4 Relationship Between $G$ and $J$

Recall that $G$ was defined as the rate of energy release. That is

$$G = -\frac{1}{B}\frac{d(U_1 - U_2)}{da}$$

where,

$U_1$ = The strain energy of the body

$U_2$ = Work performed by the external forces

The strain energy of the system is simply the integration of the strain energy density, and hence

$$U_1 = \int WB\,dx\,dy$$

The rate of increase of $U_1$ with respect to an increase in $a$, is given by

$$\frac{dU_1}{da} = \frac{d}{da}\int WB\,dx\,dy$$

$$= -\frac{d}{dx}\int WB\,dx\,dy \text{ (since } da = -dx)$$

$$= -\int WB\,dy$$

External work performed by forces is simply the summation of (force × displacement). The tractions are simply the external forces. Hence,

$$U_2 = \int T_i u_i B\, ds$$

where the $T_i$ are distributed traction vectors acting on the surface. Therefore,

$$\frac{dU_2}{da} = \frac{d}{da}\int T_i u_i B\, ds$$

$$= \int T_i \frac{du_i}{da} B\, ds$$

$$= -\int T_i \frac{du_i}{dx} B\, dx$$

From the definition of $G$,

$$G = -\frac{1}{B}\frac{dU}{da}$$

$$= -\frac{1}{B}\left[-\int WB\, dy + \int T_i \frac{du_i}{dx} B\, dx\right]$$

$$= \left[\int W\, dy - \int T_i \frac{du_i}{dx}\, dx\right]$$

$$= J$$

The relationship between $G$ and $COD$ cannot be discussed until later in this chapter.

## 6.10 CALCULATION OF *K*

As stated before, the advantage of the $K_{Ic}$ criterion is that $K$ can be calculated for a number of cases. Problem-specific values of $K$ are usually obtained by identifying the problem geometry and the crack disposition with one of the pre-engineered solutions. References 6, 7, and 8 are well-known compendia of stress intensity factors. However, there are no simple procedures for calculating them for a given problem. For example, for the crack in an infinite plate, which is the simplest of all cases, we need to use the applications of conformal mapping to elasticity. For other cases, more advanced mathematics need to be used. A few solutions of practical utility are listed in Appendix 4. Quite often, a literature search is the best recourse for obtaining a solution that matches the problem on hand. In this respect, the *International Journal of Fracture* and *Journal of Engineering Fracture Mechanics* are excellent sources of solutions. Other methods available to the practicing engineer are boundary integral equations, or the finite element methods. In any case, a strong background in mathematics is essential for using these methods.

## 6.11 A FEW USEFUL RESULTS

### 6.11.1 Infinite Plate, Stress and Displacement Components

For the infinite plate with a crack, the stresses in the vicinity of the cracks were given in Equations 6.2, 6.3, and 6.5. The polar components, principal components and the displacements are also useful for further discussions.

$$\sigma_1 = \frac{K}{\sqrt{2\pi r}} \cos\frac{\theta}{2}\left(1 + \sin\frac{\theta}{2}\right)$$

$$\sigma_2 = \frac{K}{\sqrt{2\pi r}} \cos\frac{\theta}{2}\left(1 - \sin\frac{\theta}{2}\right)$$

$$\sigma_3 = 0 \quad \text{(Plane stress)}$$

$$= \frac{2\nu K}{2\pi r} \cos\frac{\theta}{2} \quad \text{(Plane strain)} \tag{6.18}$$

$$\sigma_r = \frac{K}{\sqrt{2\pi r}} \cos\frac{\theta}{2}\left(1 + \sin^2\frac{\theta}{2}\right)$$

$$\sigma_\theta = \frac{K}{\sqrt{2\pi r}} \cos\frac{\theta}{2}\left(1 - \sin^2\frac{\theta}{2}\right)$$

$$\tau_{r\theta} = \frac{K}{\sqrt{2\pi r}} \sin\frac{\theta}{2} \cos^2\frac{\theta}{2} \tag{6.19}$$

Also of importance to many derivations and calculations are the expressions for displacements.

$$u = 2(1+\nu)\frac{K}{E}\sqrt{\frac{r}{2\pi}}\cos\frac{\theta}{2}\left[1 - 2\nu + \sin^2\frac{\theta}{2}\right] \tag{6.20}$$

$$v = 2(1+\nu)\frac{K}{E}\sqrt{\frac{r}{2\pi}}\sin\frac{\theta}{2}\left[2 - 2\nu + \cos^2\frac{\theta}{2}\right] \tag{6.21}$$

(Plane strain)

$$u = \frac{2K}{E}\sqrt{\frac{r}{2\pi}}\cos\frac{\theta}{2}\left[1 + \sin^2\frac{\theta}{2} - \nu\cos^2\frac{\theta}{2}\right] \tag{6.22}$$

$$v = \frac{2K}{E}\sqrt{\frac{r}{2\pi}}\sin\frac{\theta}{2}\left[1 + \sin^2\frac{\theta}{2} - \nu\cos^2\frac{\theta}{2}\right] \tag{6.23}$$

(Plane stress)

These expressions are valid only in the very close vicinity of the crack tip. Note that $u$ and $v$ are obtained from the complex variables techniques for elasticity, and not from the traditional ones (that is, writing strains from stresses using Hooke's law and integrating strains). At a point on the crack flank, a distance $x$ away from the crack center, the vertical displacement $u$ is important in several calculations.

$$v = \frac{2\sigma(1-\nu^2)}{E}\sqrt{a^2 - x^2} \quad \text{Plane strain} \tag{6.24}$$

$$= \frac{2\sigma}{E}\sqrt{a^2 - x^2} \quad \text{Plane stress} \tag{6.25}$$

These expressions are valid far away from the crack tip, since points on the flank close to crack tip undergo plastic deformation.

### 6.11.2 Line Load on a Crack Embedded in an Infinite Plate

Line loads acting on the crack faces as shown in Figure 6.11 constitute another important case. Because of the asymmetry with respect to the crack tips, the stress intensity caused at ends $A$ and $B$ are different. For a

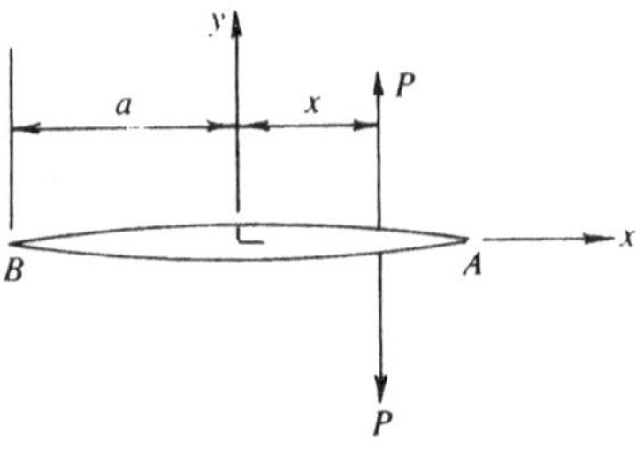


**FIGURE 6.11.** Line load acting on a crack. Note that $P$ is the force per unit thickness of the plate.

force $P$ per unit thickness of the plate (lbs/in) or (N/m), we have,

$$K_A = \frac{P}{\sqrt{\pi a}} \sqrt{\frac{a + x}{a - x}}\,; \qquad K_B = \frac{P}{\sqrt{\pi a}} \sqrt{\frac{a - x}{a + x}}$$

For a centrally loaded crack, $x = 0$, and hence

$$K_A = K_B = \frac{P}{\sqrt{\pi a}}$$

Unlike the case of stress applied far away, the stress intensity in this case *reduces* with increasing crack length at a constant load.

## 6.12 PRINCIPLE OF SUPERPOSITION FOR CALCULATING *K*

Since $K$ is directly proportional to the applied load, it is clear that different load cases cause their own $K$ at the tip of a crack, all of which are additive. If the loading can be expressed as a suitable combination of load cases, then the corresponding $K$'s can be added to obtain the total $K$. Its use is demonstrated in the following example.

### Example 6.1

A large plate is used to contain water under pressure $p_o$. Assume that there is a crack in the middle of the plate, and water is oozing out of the crack. Calculate the stress intensity factor $K$ caused by the water pressure, acting on the crack faces, as it flows out. Assume the plate to be infinite in size, for the purposes of this problem.

*Solution* Since the plate is considered infinitely large, we need to consider only the crack, without relation to the boundaries. Since superposition permits addition of $K$'s, we may extend it to integration of $dK$'s, since integration is after all only addition. This is illustrated in Figure 6.12. We consider the $dK$ due to a small load $p_o\, d\xi$ acting at a distance $\xi$ from the center of the crack.

$$dK_A = \frac{p_o}{\sqrt{\pi a}} \sqrt{\frac{a+\xi}{a-\xi}}\, d\xi$$

$$dK_B = \frac{p_o}{\sqrt{\pi a}} \sqrt{\frac{a-\xi}{a+\xi}}\, d\xi$$

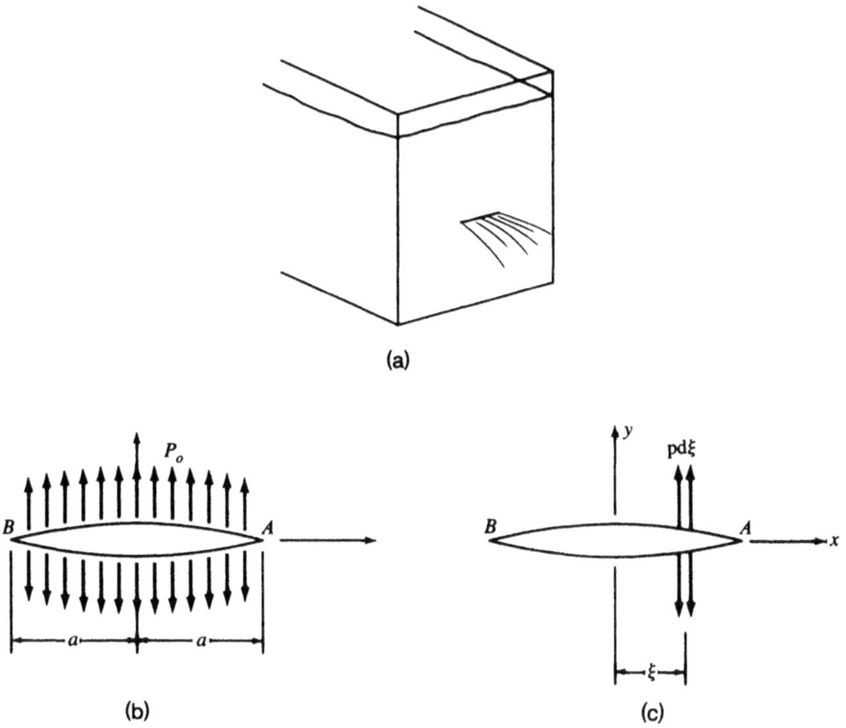


**FIGURE 6.12.** A crack in an infinite plate, acted upon by pressure on the crack faces. (a) The physical picture. (b) The mathematical idealization, and (c) The total effect of the pressure can be calculated by integrating the effect of an elemental force $p\,d\xi$, acting at a distance $\xi$.

We may treat the integral for $K_A$ as arising from two halves of the crack. Accordingly,

$$K = \int dK_{1A} + \int dK_{2A} = K_{1A} + K_{2A}$$

where, $$K_{1A} = \int_0^a \frac{p_o}{\sqrt{\pi a}} \sqrt{\frac{a - \xi}{a + \xi}}\, d\xi$$

$$K_{2A} = \int_0^a \frac{p_o}{\sqrt{\pi a}} \sqrt{\frac{a + \xi}{a - \xi}}\, d\xi$$

$$K = K_{1A} + K_{2A} = \frac{p_o}{\sqrt{\pi a}} \int_0^a \left( \sqrt{\frac{a + \xi}{a - \xi}} + \sqrt{\frac{a - \xi}{a + \xi}} \right) d\xi$$

$$= \frac{2p_o\sqrt{a}}{\sqrt{\pi}} \int_0^a \frac{d\xi}{\sqrt{a^2 - \xi^2}}$$

$$= \frac{2p_o\sqrt{a}}{\sqrt{\pi}} \arcsin \frac{\xi}{a} \bigg|_0^a$$

$$= p_o\sqrt{\pi a}$$

Because of symmetry, the $K$ at $A$ is equal to that at $B$. This is the same as the stress intensity caused by a stress $p_o$ acting far away from the crack on an infinite plate.

**Example 6.2**

A crack embedded in an infinite plate is acted upon by a pressure $p_o$ for the length $c \leq x \leq a$ on both ends of the crack, as shown in Figure 6.13. Assume that the plate is of unit thickness. Calculate the stress intensities at $A$ and $B$.

*Solution* The solution in this case is quite similar to the previous example, except that the limits of integration are from $c$ to $a$. Thus,

$$K = \frac{2p_o}{\sqrt{\pi a}} \int_c^a \frac{d\xi}{\sqrt{a^2 - \xi^2}}$$

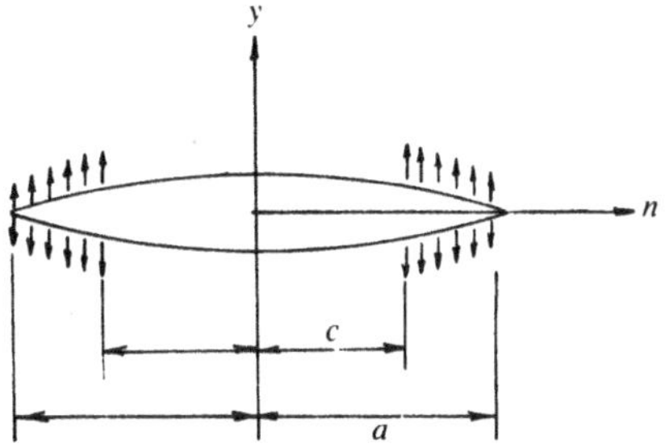


**FIGURE 6.13.** Crack face being acted upon partially by pressure.

$$= \frac{2p_o}{\sqrt{\pi a}} \arcsin\left(\frac{\xi}{a}\right)\Bigg|_c^a$$

$$= 2p_o\sqrt{\frac{a}{\pi}}\left[\frac{\pi}{2} - \arcsin\left(\frac{c}{a}\right)\right]$$

$$= p_o\sqrt{\pi a} - 2p_o\sqrt{\frac{a}{\pi}} \arcsin\left(\frac{c}{a}\right)$$

Obviously, when $c \to 0$, the results tend to that of a fully pressurized crack.

**Example 6.3**

Calculate the stress intensity factor for a centrally loaded fixed beam with a central notch on the tensile side of the beam

*Solution* A fixed-ended beam is a combination of a simply supported beam and a beam with end moment. The value of the end moment for full fixity is known to be $(PL/8)$. Thus, the stress intensity factor can be calculated according to Figure 6.14.

**Example 6.4**

Calculate the stress intensity factor for a notched bar loaded with end stresses that vary linearly over the depth as shown in Figure 6.15.

*Solution* The linear variation of applied stress from $\sigma_1$ to $\sigma_2$ can be split as a constant stress $\sigma_m = 0.5(\sigma_1 + \sigma_2)$, and a linear variation of a bending stress $\sigma_b = 0.5(\sigma_1 - \sigma_2)$. For width $B$ and depth $D$ of the beam,

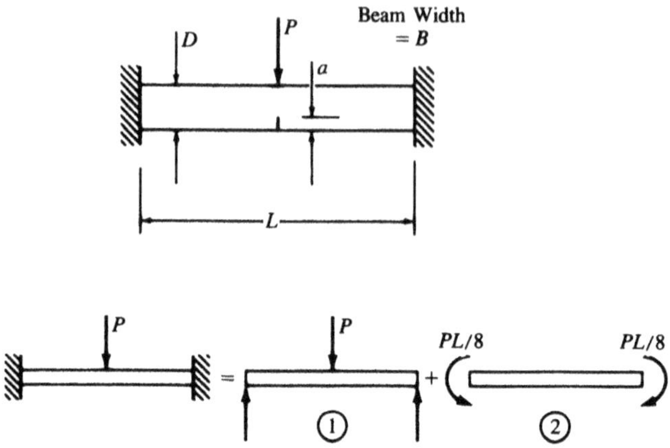


$K_I \quad = \quad K_I$ for (1) $+ K_I$ for (2)

Where $r = a/D \quad =$ Case 7 of Appendix + Case 9 of Appendix

$$K_{I,1} = \frac{PL}{BD^{3/2}}\left[2.9\sqrt{r} - 4.6(r)^{3/2} + 21.8(r)^{5/2} - 37.6(r)^{7/2} + 38.7(r)^{9/2}\right]$$

$$K_{I,2} = \frac{3PL}{4B(D-a)^{3/2}}\, g(r),$$ and $g(r)$ is read from one appendix

**FIGURE 6.14.** Calculating the $K_{Ic}$ for a fixed beam using the superposition principle.

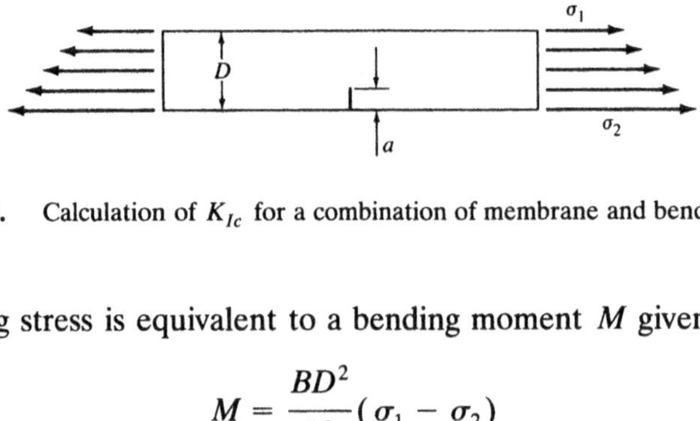


**FIGURE 6.15.** Calculation of $K_{Ic}$ for a combination of membrane and bending loads on a beam.

the bending stress is equivalent to a bending moment $M$ given by

$$M = \frac{BD^2}{12}(\sigma_1 - \sigma_2)$$

It is only necessary to treat the applied load as a combination of an axial pull and a bending moment. Thus,

$$K_I = K_I \text{ due to } \sigma_m + K_I \text{ due to } M$$

$$= C\sigma_m\sqrt{\pi a} + \frac{6M}{B(W-a)^{3/2}} \times g\left(\frac{a}{W}\right)$$

The calculation of function $g(\cdots)$ is possible when the quantities $a$ and $W$ are known. See Appendix 4.

**Example 6.5**

Find the stress intensity factor for a crack emanating from a pinhole (a loading case similar to the hanging of a heavy picture from a single peg at the top).

*Solution* The superposition principle can be used to derive the $K_I$ for this problem by inverting the same configuration and adding. Figure 6.16 provides the complete solution.

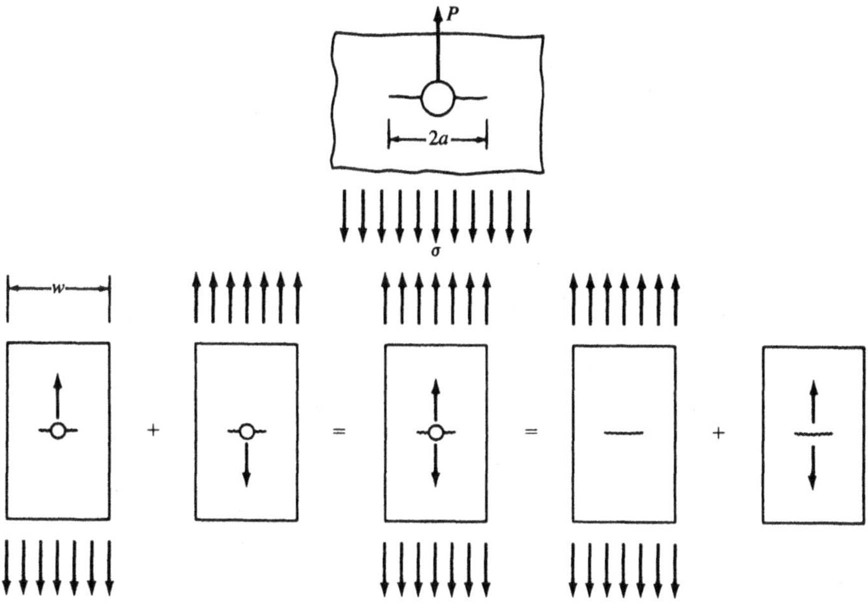


$$K_I + K_I = 2K_I = C\sigma\sqrt{\pi a} + \frac{P}{\sqrt{\pi a}}$$

$$\text{Where } C = \sqrt{\sec\left[\frac{\pi a}{W}\right]}$$

and $P = \sigma BW$, $B$ = thickness of plate

$$K_I = \frac{1}{2}\, C\sigma\sqrt{\pi a} + \frac{P}{2\sqrt{\pi a}}$$

**FIGURE 6.16.** Use of identical solutions twice to obtain the $K_I$.

**Example 6.6**

A long strip of folded plate is made of plastic and is used to support a uniformly distributed load. At the bottom of the plate are ribs running along the short side. See Figure 6.17. The plate is supported all along the long edge, so that it may be supposed to be a short beam of width $L$. If one of the ribs is cracked, discuss its failure potential.

*Solution* In the first place, there is no matching solution available for this case. We have to make a series of simplifying assumptions. As a first approximation, we can consider the cracked rib in isolation. Because it is integral with the folded border of the plate, we may suppose the plate to be clamped at its ends Next, we note that the solution for the case of a uniformly distributed load on a beam is not readily available. We may therefore suppose that the total load is concentrated at the center. The total load is the load acting over the area between the two ribs. Lastly the location of the crack is not specified. We may take it to be located at the center, because it is the location of maximum stress. Probably, if a crack develops, it would be at this location. We have effectively reduced the problem to one for which the solution is known.

The total load acting at the center of the rib, $P$ is the load acting on a typical inter-rib space.

$$P = wSL$$

The problem, as modeled here, is identical to Example 6.3, and so is the solution.

***Commentary***

*Note that every assumption made above tends to put the estimate of the rib life on the more conservative side*

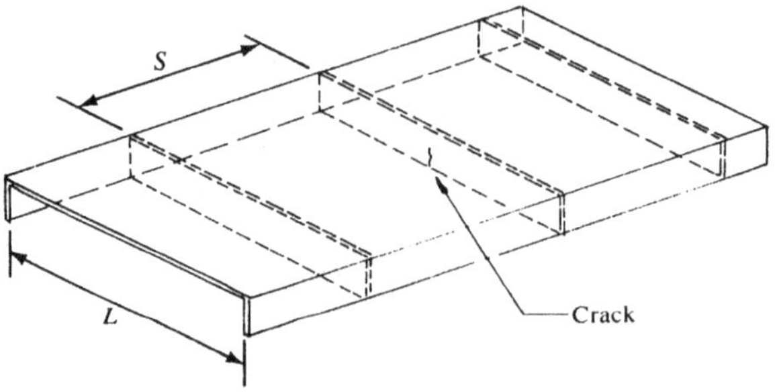


**FIGURE 6.17.** Reinforcement ribs may be treated as equivalent to fixed beams.

**Example 6.7**

A crack of 10 mm depth is present on the edge of a Nylon 6/6 strip. The strip is 25 mm thick, 75 mm wide, and 300 mm long. See Figure 6.18. It is subjected to an axial tensile force of 3000 kg. Determine $K$ for the configuration.

*Solution*

$$a = 10 \text{ mm} = 0.01 \text{ m}; \quad W = 75 \text{ mm} = 0.075 \text{ m}$$

$$\left(\frac{a}{W}\right) = 0.1333$$

$$\text{Stress } \sigma = \frac{3000 \text{ kg}}{(75 \times 25) \text{ mm}^2} = 1.6 \text{ kg/mm}^2 = 15.7 \text{ MPa}$$

The yield strength of Nylon 6/6 is about 7.00 Kg/mm$^2$. Hence the factor of safety available on the yield is $(7.0/1.6) = 4.375$.

Stress Intensity $K$ is given by $K = C\sigma\sqrt{\pi a}$ where

$$C = C_0 + C_1 r + C_2 r^2 + C_3 r^3 + C_4 r^4,$$

where

$$r = (a/W); \quad C_0 = 1.122; \quad C_1 = -0.231; \quad C_2 = 10.55;$$

$$C_3 = -21.72; \quad C_4 = 30.95$$

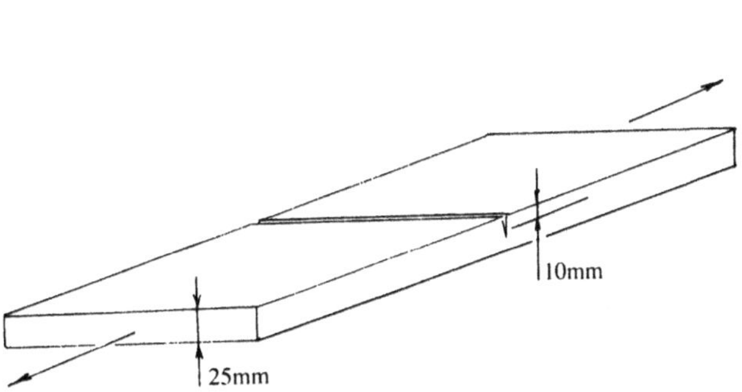


**FIGURE 6.18.** Part-through thickness crack on a finite thickness plate.

Substituting,

$$C = 1.315$$

$$K = 1.315 \times 15.7 \text{ MPa } \sqrt{\pi 0.01 \, m}$$

$$= 3.659 \text{ MPa } \sqrt{m}$$

***Commentary***
*It may be verified that the K exceeds the toughness (see Table 6.1) of the material. In actual designs, it is not enough to just carry a high FOS on yield, nor even on the stress based on the ligament of 15 mm available under the crack. The quality control measures must ensure that the defects will be detected and screened out.*

**Example 6.8**

A large polycarbonate panel has a through-the-thickness crack of length 50 mm in its middle. The fracture toughness $K_{Ic}$ of the panel is 2.75 MPa $\sqrt{m}$. (1) What is the highest allowable stress to just avoid breaking? (2) If the panel is not large, but only a strip of 250 mm width, find the highest

**Table 6.1. $K_{Ic}$ Values for Some Engineering Materials.**

| Material | $K_{Ic}$ (ksi$\sqrt{in}$) | $\sigma_Y$ (ksi) | $\left(\frac{K_{Ic}}{\sigma_Y}\right)^2$ (in) |
|---|---|---|---|
| Aluminum 2024-T851 | 24 | 66 | 0.132 |
| Aluminum 7075-T651 | 22 | 72 | 0.093 |
| Titanium Ti-6Al-4V | 105 | 132 | 0.632 |
| Titanium Ti-6Al-4V* | 50 | 150 | 0.111 |
| Steel 4340 | 90 | 125 | 0.518 |
| Steel 4350* | 55 | 220 | 0.063 |
| Steel 5210 | 13 | 300 | 0.002 |
| Polyethylene(HDPE) | 0.5 - 1 | 2 - 13.0 | 0.063 - 0.005 |
| CPVC | 2 - 6 | 1.5 - 3.0 | 1.78 - 4.0 |
| Polycarbonate | 2-3 | 8 - 10.5 | 0.063 - 0.08 |
| Nylon 6/6 | 3 | 12.5 | 0.058 |
| Polyacetal | 2 | 8 - 8.5 | 0.055 - 0.063 |
| SAN | 0.95 | 8 - 9 | 0.011 - 0.014 |
| ABS | 3 - 4 | 4 - 8 | 0.025 - 0.0563 |

* - Heat Treated to higher strength.

stress allowable. (3) If the crack is not in the middle but on the edge of the strip, find the maximum stress allowable.

*Solution* Case of a Large Panel: Since $K_{Ic} = \sigma_c\sqrt{\pi a_c}$, highest stress allowable $\sigma_c = K_{Ic}/\sqrt{\pi a_c}$. Substituting

$$\sigma_c = \frac{2.75 \text{ MPa} \sqrt{m}}{\sqrt{\pi(0.05\, m)}} = 6.939 \text{ MPa}$$

Case of a Finite Width Strip:

$$a = 50 \text{ mm}; \quad W = 250 \text{ mm}; \quad \left(\frac{a}{W}\right) = r = 0.2$$

For a central crack in a finite width plate,

$$K = C\sigma\sqrt{\pi a}$$

where,

$$C = 1 + 0.256r - 1.152r^2 + 12.20r^3$$
$$= 1.1027$$

$$\sigma_c = \frac{K_{Ic}}{1.1027\sqrt{\pi 0.05\, m}} = 6.293 \text{ MPa}$$

If we used the two alternative formulas in Appendix 4, the results are only slightly different.

For

$$C = \frac{1}{\sqrt{1 - \left(\dfrac{2a}{W}\right)^2}} = 1.0911$$

we get

$$\sigma_c = 6.360 \text{ MPa}$$

For

$$C = \sqrt{\sec\left(\frac{\pi a}{W}\right)} = 1.112$$

we get

$$\sigma_c = 6.240 \text{ MPa}$$

Case of a Narrow Strip with an Edge Crack:
The expression for $C$ is the only difference.

$$C = 1.12 - 0.231r + 10.55r^2 - 21.72r^3 + 30.95r^5$$
$$= 1.372$$
$$\sigma = 5.056 \text{ MPa}$$

### 6.12.1 More Examples of Stress Intensity Factors

When the defect tolerance of a component enters the design process as a variable, several new considerations arise. How large a defect can be tolerated by the material? Will a lower strength material be more suitable, lighter, and more economical than a high strength material? What if a crack is located somewhere else than at the location assumed in the calculation? Which location of a crack will lead to the most conservative design? Answers to some of these questions are conceptually simple but involve tedious calculations. The principal source of difficulty in the calculation of $K$ is the constant $C$ in the expression $K = C\sigma\sqrt{\pi a}$. In general $C$ is a function of the crack length, the geometry. In three dimensional cracks, more is involved, such as the elliptic integral. It is intuitive in such cases to draw the curve for the $K$ in a nondimensional form, from which information can be obtained easily.

**Example 6.9**

For a single edge crack on a narrow strip, show a calculation procedure to determine the allowable crack size for a given stress.

*Solution* Since the constant $C$ depends on crack size $a$ as a polynomial in powers of $\sqrt{\cdots a}$, it is normally difficult to determine the allowable crack size by calculation. It is convenient to draw the curve of the stress intensity factor $K$ in a nondimensional form.

$$K = C\sigma\sqrt{\pi a}$$
$$= f\left(\frac{a}{W}\right)\sigma\sqrt{W}\sqrt{\frac{\pi a}{W}}$$
$$\frac{K}{\sigma\sqrt{W}} = C\sqrt{\pi\left(\frac{a}{W}\right)}$$

Figure 6.19 gives the plot of this expression. One simply needs to calculate

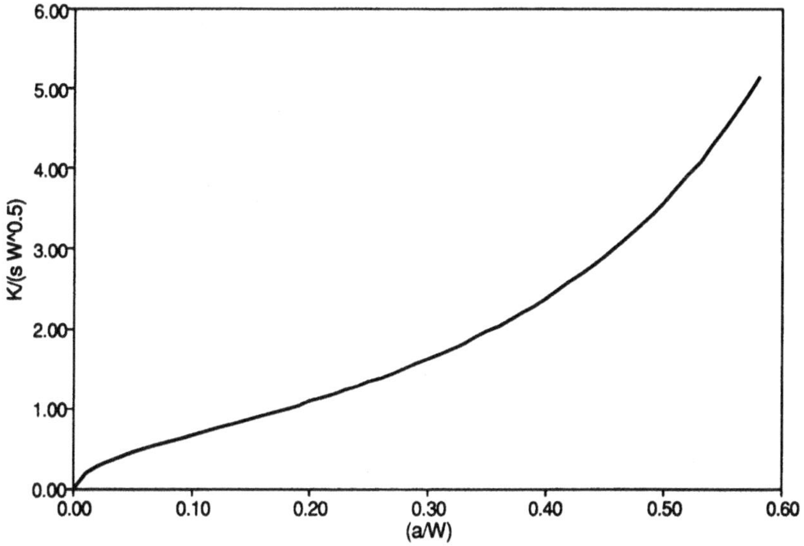


**FIGURE 6.19.** Nondimensional form of single edge crack in a finite width plate, subjected to tensile stress.

the ordinate, all quantities in it being known, and read off the corresponding allowable $(a/W)$. Any desired *FOS* may be considered conveniently by using this curve. One just needs to substitute $K_{Ic}/FOS$ in the place of $K$.

**Example 6.10**

Describe a procedure to determine the size of a crack that is in the middle of a narrow strip that would produce the same $K$ under the same stress as another crack located at the edge.

*Solution* We note from the formulas for central and edge cracks, that the forms of the equations for $K$ are the same, except that the constants are different. Hence, we can plot the curves

$$K = C_1 \sigma \sqrt{\pi a}\text{, for the edge crack, and}$$

$$K = C_2 \sigma \sqrt{\pi a}\text{, for the central crack.}$$

For the limited purposes of this problem, plotting $C_1$ and $C_2$ is enough. It is however, better to plot the non-dimensional form of these equations. Figure 6.20 shows the use of the curves to determine the crack sizes that

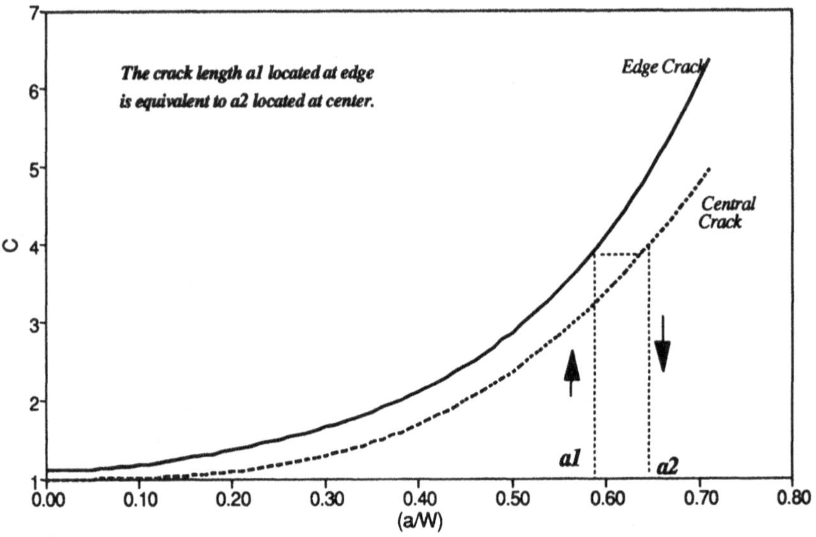


**FIGURE 6.20.** Length of edge crack equivalent to a given central crack.

cause equal $K$ at the crack tips. The plate width and stress need not be known.

## Example 6.11

A fixed beam has a single edge crack on its tensile side, and is acted upon by a central concentrated load. Give a procedure for calculating its allowable crack size for a given stress.

*Solution* The fixed beam is the superposition of two cases, (1) a simply supported beam with a central load $P$, and (2) a beam with a (counter)moment of $PL/8$ acting at the ends of the beam, in a manner that tends to reduce the $K$.

In the nondimensional form, the $K$ caused by the central load $P$, the simply supported case, is (Case 7 of Appendix 4)

$$\frac{K_1 BW\sqrt{W}}{PL} = 2.9\sqrt{r} - 4.6r\sqrt{r} + 21.8r^2\sqrt{r} - 37.6r^3\sqrt{r} + 387.7r^4\sqrt{r}$$

For the case of fixing moment $M = PL/8$, the formula for $K$ needs to be expressed in terms of $P$. (Case 9 of Appendix 4)

$$K_2 = \frac{6M}{B(W-a)^{3/2}} \times g\left(\frac{a}{W}\right)$$

$$= \frac{3PL}{4BW^{3/2}} g\left(\frac{a}{W}\right) \frac{1}{\left(1 - \frac{a}{W}\right)^{3/2}}$$

$$\frac{K_2 BW\sqrt{W}}{PL} = \frac{3}{4} g\left(\frac{a}{W}\right) \frac{1}{\left(1 - \frac{a}{W}\right)^{3/2}}$$

After superposition, we need$(K_1 - K_2)$. Figure 6.21 shows the nondimensionalized $K_1$, $K_2$, and $(K_1 - K_2)$. The problem can be solved simply by starting with a known value of $(a/W)$ and reading off a $K$ corresponding to the curve labeled "Total Effect".

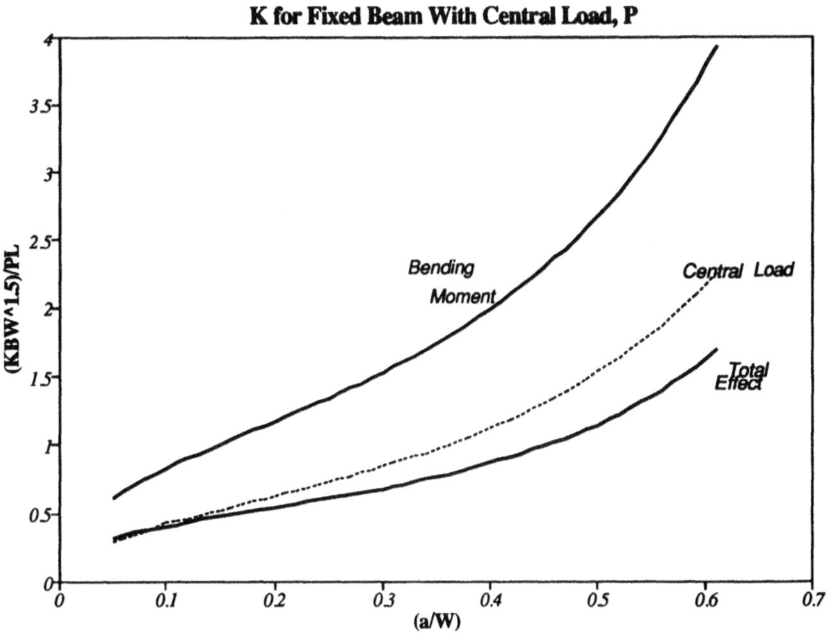


FIGURE 6.21. Stress intensity for a fixed beam with single edge crack.

**Example 6.12**

A single edge crack is loaded in bending by means of pure bending moment $M$. How would you determine whether the specimen would break or yield?

*Solution* The $K$ due to $M$ needs to be expressed as a function of the bending stress caused by $M$. This is easily done by noting that the bending stress $\sigma_b$ due to $M$ is given by

$$\sigma_b = \frac{6M}{BW^2}$$

The nondimensional curve of $K$ due to the bending moment is still the same as in the last example, but for a change in the label on the $y$-axis. Because,

$$K = \frac{6M}{B(W-a)^{3/2}} g\left(\frac{a}{W}\right) + \frac{BW^2\sigma_b}{B(W-a)^{3/2}} g\left(\frac{a}{W}\right)$$

The nondimensional equation for $K$ would have the form

$$\frac{K}{\sigma_b\sqrt{W}} = \frac{3}{4} \frac{g\left(\frac{a}{W}\right)}{\left(1 - \frac{a}{W}\right)^{3/2}}$$

Since the right side of the equation is the same as in the previous example, the curve is still the same. Thus, the curve in Figure 6.22 is the same as the curve labelled "Bending Moment" in Figure 6.21. To determine yield-or-break behavior, we consider several alternatives. Let $Y_1$, $Y_2$, and $Y_3$ be the ordinates as defined below.

$$Y_1 = \frac{K}{\sigma_b\sqrt{W}}, \text{ where } \sigma_b \text{ is the applied bending stress,}$$

$$Y_2 = \frac{K_{Ic}}{\sigma_b\sqrt{W}}, \text{ where } K_{Ic} \text{ is the fracture toughness,}$$

$$Y_3 = \frac{K_{Ic}}{\sigma_y\sqrt{W}}, \text{ where } \sigma_y \text{ is the yield stress.}$$

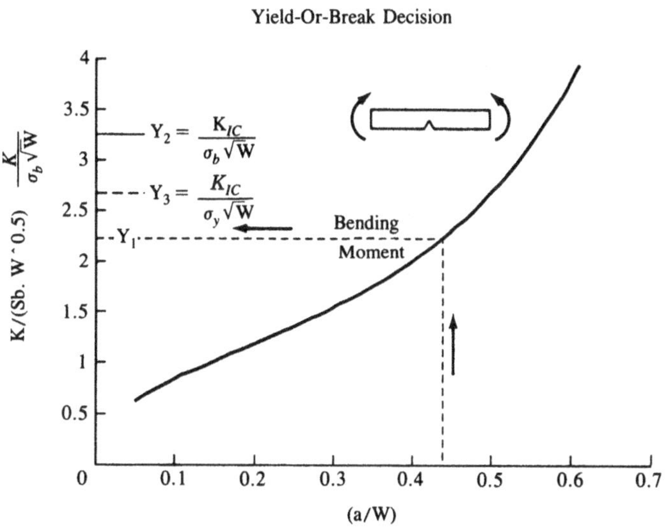


**FIGURE 6.22.** Nondimensional stress intensity due to pure bending moment on a notched beam.

- If $Y_1 < Y_2$ and $Y_1 < Y_3$, neither yielding, nor fracture occurs.
- If $Y_2 < Y_3$, fracture would occur first as a result of increasing the bending moment.
- If $Y_3 < Y_2$, then yielding would occur first as a result of increasing the bending moment.
- If $Y_2 < Y_1 < Y_2$, then fracture occurs the first time the bending moment is applied.
- If $Y_3 < Y_1 < Y_3$, then yielding occurs the first time the bending moment is applied.

## 6.13 CONCEPT OF LEAK-BEFORE-BREAK

The concept of "Leak-Before-Break"evolved in the pressure vessel industry. It is a protection against the sudden failures of components without warning. We may consider this concept closer to material selection rather than to design itself. The following example illustrates the concept with real-life numbers.

### Example 6.13

Calculate the thickness required for a CPVC pipe for an internal pressure of 515 KPa. The internal diameter of the pipe is 250 mm. The yield strength of the material is 17.25 MPa, and the fracture toughness is $K_{Ic} = 2.75\ \text{MPa}\sqrt{\text{m}}$. Consider a factor of safety of 2.5 on yield strength. Also estimate the deepest semi-elliptic shaped crack that the pipe can sustain. Neglect time effects.

*Solution*

$$\text{Thickness } t = \frac{pd(FOS)}{2\sigma_y}$$

$$= \frac{0.515\ \text{KPa} \times 250\ \text{mm} \times 2.5}{2 \times 17.25\ \text{MPa}} = 9.33\ \text{mm}$$

The stress intensity factor $K$ for this problem is straightforward. Referring to Figure 6.23, we use $a/2c = 0.25$. For an FOS = 2.5, the ratio $\sigma/\sigma_y = 0.4$ we read off $Q = 1.43$. We need a constant $M_k$, to take into account the curved nature of the pipe wall.

$$M_k = 1 + 1.2\left(\frac{a}{t} - 0.5\right)$$

$$= (0.4 + 128.62a)$$

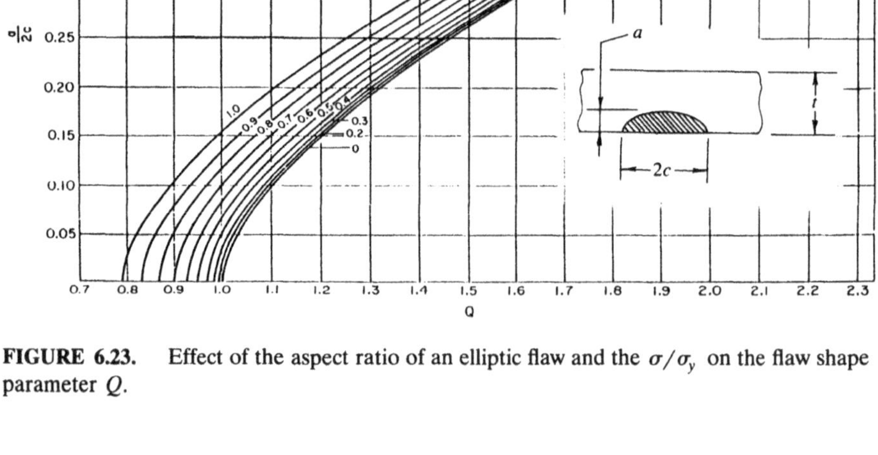


**FIGURE 6.23.** Effect of the aspect ratio of an elliptic flaw and the $\sigma/\sigma_y$ on the flaw shape parameter $Q$.

Use these results in the expression,

$$K_{Ic} = 1.12\sigma_h M_k \sqrt{\frac{\pi a}{Q}}$$

$$2.75 = 1.12 \times 6.9 \text{ MPa}(0.4 + 128.62a)\sqrt{\frac{\pi a}{1.43}}$$

$$= 11.454\,(0.4 + 128.62a)\sqrt{a}$$

Or, $128.62a\sqrt{a} + 0.4\sqrt{a} = 0.24$

Solving, $a = 0.01316 \text{ m} = 13.16 \text{ mm}$

***Commentary***

*Since the depth of the crack that can be sustained exceeds the pipe wall thickness, the crack will open up a hole and let out the contents before catastrophic failure will occur. Thus, the warning of failure comes before the failure. This is a condition of leak-before-break. This example, however, needs to be analyzed further for a more complete evaluation of its leak-before-break capability.*

***Continuation of Example 6.13***

Since the crack starts from the inside, the crack faces are also subjected to pressure. This adds to the stress intensity factor. The contribution of this effect is the same as applying the pressure at an infinite distance. See also Example 6.1. Therefore,

$$K' = 1.12pM_k\sqrt{\frac{\pi a}{Q}}$$

Adding the two, the total

$$K_{Ic} = 1.12 \times (6.9 \times 0.515) \text{ MPa}(0.4 + 128.62a)\sqrt{\frac{\pi a}{1.43}}$$

That is, $2.75 = 12.309(0.4 + 128.62a)\sqrt{a}$

Or,

$$128.62a\sqrt{a} + \sqrt{a} = 0.2234$$

Solving, $a = 12.45 \text{ mm}$

This effect slightly reduces the allowable crack depth, but the pipe would still leak before it breaks. However, we do not know as yet whether the crack that had penetrated through the full thickness would remain stable in that configuration. We have a situation similar to a straight through hole in an infinite plate, the crack faces also being subjected to pressure. Only, the crack length is $2c = 4a$. For this case,

$$
\begin{aligned}
K &= \sigma_h\sqrt{\pi(4a)} + p\sqrt{\pi(4a)} \\
&= \sqrt{\pi(4a)}\,(\sigma_h + p) \\
&= \sqrt{4\pi(12.45 \times 10^{-3})}\,(6.9 + 0.515) \\
&= 2.933 \text{ MPa}\sqrt{\text{m}} \\
&> 2.75, \quad \text{Fracture toughness of the material.}
\end{aligned}
$$

Thus, although catastrophic failure does not occur until the crack grows through the entire thickness, there is very little time available after that for taking preventive action on the pipe. The remedy is to increase the thickness thereby reducing the stress. Since $K$ is proportional to stress, we can choose a thickness of $(2.933/2.75) \times 12.4$ mm $= 13.28$ mm, to just reach critical conditions. If an additional factor of safety of about 1.5 is desired on $K_{Ic}$, then the thickness needed is 20 mm.

## 6.14 FRACTURE TOUGHNESS FOR LIGHT WEIGHT DESIGNS

Since all designs aim at minimizing weight subject to a number of performance constraints, it is worthwhile considering the role of fracture toughness $K_{Ic}$, for this objective. In plastics, and metals as well, any material processing that improves the strength of the material tends to reduce the value of strain at failure. This means that the fracture toughness is reduced. This leads to a reduction in the tolerance of the structure to the presence of defects. In general, therefore, high yield strength materials tend to have low defect tolerance.

The effect of $K_{Ic}$ in sizing a stressed part can be seen by requiring equal defect tolerance on several candidate materials, and comparing the resulting thicknesses. Since $K = C\sigma\sqrt{\pi a}$, and since $a$ is required to be constant,

$$K_{Ic} = C\sigma_c\sqrt{\pi a}$$

**Table 6.2. Thickness and Weights of Components Designed for Equal Defect Tolerance.**

| Material | $K_{Ic}$ MPa$\sqrt{m}$ | $\sigma_y$ MPa | Thickness $t \propto \frac{1}{K_{Ic}}$ | Sp. Gravity, $\rho$ Kg/m$^3$ | Weight, W $W \propto \frac{\rho}{K_{Ic}}$ |
|---|---|---|---|---|---|
| CPVC | 4.40 | 17.25 | 100% | 1350 | 100% |
| HDPE | 3.00 | 48.25 | 147% | 950 | 103% |
| PC | 2.75 | 62.00 | 160% | 1200 | 142% |
| Nylon | 3.30 | 86.10 | 133% | 1150 | 114% |
| Polyacetal | 2.20 | 56.80 | 200% | 1450 | 215% |
| SAN | 1.00 | 58.60 | 440% | 1080 | 352% |
| ABS | 3.85 | 41.30 | 114% | 1050 | 88% |

Note that all materials with higher yield strength than CPVC result in higher component weight, if designed for the same defect tolerance. ABS material has a lower weight, but certainly not as low as a calculation based on yield strength only would predict.

where $\sigma_c$ is the critical stress allowable in the material, whose fracture toughness is $K_{Ic}$. Here we see the role of $K_{Ic}$ as a means of determining the allowable design stress for a material. The allowable stress is proportional to $\sigma_c$ and hence to $K_{Ic}$.

$$\sigma_{\text{des}} \propto K_{Ic}$$

$$\text{Weight} \propto pt \propto \frac{\rho}{\sigma_{\text{des}}} \propto \frac{\rho}{K_{Ic}}$$

Thus, the resulting thickness of the component is inversely proportional to the fracture toughness.

In particular, we may consider Table 6.2 comparing the relative thicknesses and weights for various materials given with PVC as the reference material. The weights obviously contradict what an elasticity-based approach will show.

## 6.15 EFFECTS OF CRACK TIP PLASTICITY

The actual stress field at the crack tip does not exactly follow the trend of Equations 6.2, 6.3, and 6.5, because infinite stresses cannot exist, and because the plastic flow of the material places a limit on the stress value. Since most thermoplastics are normally ductile, the Von Mises criterion of plastic flow can be applied to the small region around the crack tip. Recall that the Von Mises criterion states that plastic flow just occurs when and where the principal stresses $\sigma_1$, $\sigma_2$, and $\sigma_3$ satisfy the relation

$$2\sigma_y^2 = (\sigma_1 - \sigma_2)^2 + (\sigma_2 - \sigma_3)^2 + (\sigma_3 - \sigma_1)^2$$

Since most crack problems are two dimensional, let us consider the two possibilities, plane stress and plane strain conditions. In plane stress,

$$\sigma_3 = \sigma_z = 0, \qquad \text{and} \qquad \epsilon_z \neq 0$$

In plane strain,

$$\epsilon_3 = \epsilon_z = 0, \qquad \text{and} \qquad \sigma_2 = \nu(\sigma_1 + \sigma_2)$$

***Plane Stress Conditions***

In plane stress conditions, plastic flow occurs when

$$\begin{aligned} 2\sigma_y^2 &= (\sigma_1 - \sigma_2)^2 + \sigma_2^2 + \sigma_1^2 \\ &= 2(\sigma_1^2 + \sigma_2^2 - \sigma_1\sigma_2) \\ \sigma_y^2 &= (\sigma_1^2 + \sigma_2^2 - \sigma_1\sigma_2) \\ &= \frac{K^2}{2\pi r}\cos^2\frac{\theta}{2}\left(1 + 3\sin^2\frac{\theta}{2}\right) \end{aligned}$$

from Equations 6.18.

The above equation is actually a relationship between $r$ and $\theta$ that defines the boundary of incipient plasticity. Denoting by $r_p'$ the radial coordinate of this boundary from the crack tip, its shape is given by

$$r_p' = \frac{K^2}{2\pi\sigma_y^2}\cos^2\frac{\theta}{2}\left(1 + 3\sin^2\frac{\theta}{2}\right) \qquad (6.26)$$

In particular, on the plane of crack, $\theta = 0$ and hence,

$$r_p' = \frac{K^2}{2\pi\sigma_y^2} \qquad (6.27)$$

***Plane Strain Conditions***

In contrast, the plane strain conditions impose a constraint on the incipience of plastic flow.

$$2\sigma_y^2 = (\sigma_1 - \sigma_2)^2 + [\sigma_2 - \nu(\sigma_1 + \sigma_2)]^2 + [\nu(\sigma_1 + \sigma_2) - \sigma_1]^2$$

$$= 2(\sigma_1 + \sigma_2)^2(1 - \nu + \nu^2) - 6\sigma_1\sigma_2$$

$$\sigma_y^2 = \sigma_1^2\left[(1 + \beta)^2(1 - \nu + \nu^2) - 3\beta\right] \tag{6.28}$$

$$= \sigma_1^2/M^2 \tag{6.29}$$

where $\beta = (\sigma_2/\sigma_1)$

The factor$(1/M)$ may be viewed as a *constraint factor.* [11] It is called this because it tends to reduce the size of the plastic zone $r_p'$. Note that $M$, being dependent on $\beta$, is a function of $\theta$. Just as in the case of plane stress, we can also calculate the plastic zone shape for plane strain. For distinguishing it from plane stress, let us denote it by $r_p$.

$$\sigma_y = \sigma_1/M$$

$$= \frac{K}{M\sqrt{2\pi r_p}} \cos\frac{\theta}{2}\left(1 + \sin\frac{\theta}{2}\right)$$

$$r_p = \frac{K^2}{2\pi(M\sigma_y)^2} \cos^2\frac{\theta}{2}\left(1 + \sin\frac{\theta}{2}\right)^2 \tag{6.30}$$

In particular, on the plane of crack, $\theta = 0$, and also $\beta = 1$. This makes $M > 1$ for all practical values of $\nu$.

$$r_p = \frac{K^2}{2\pi(M\sigma_y)^2}$$

$$< r_p', \text{ for plane stress}$$

This observation justifies the name constraint factor. Furthermore, by comparing Equations 6.27 and 6.32, we see that

> for all plane strain calculations, we can simply replace $\sigma_y$ by $(M\sigma_y)$, and do the calculations as if it were plane stress.

In light of the role played by $M$, it is easy to see why the $K_{Ic}$ is at a minimum for plane strain conditions. In plane strain, the plastic region size is small, and so is the amount of possible plastic work. Hence the work of external forces cannot be absorbed in plastic work and is expended in opening the crack. This manifests as a reduced $K_{Ic}$ for plane strain conditions.

## 6.16 SHAPE OF PLASTIC ZONE

It is useful to visualize the shape of the plastic boundary ahead of a straight crack front. For a finite, but sufficient thickness $B$, plane stress conditions prevail at the free surfaces, and plane strain conditions prevail in the middle of the thickness. We can express $r_p$ for plane strain directly in terms of $\sigma_1$, $\sigma_2$, $\sigma_y$, and $\theta$, and remove $M$ from the expressions.

$$\sigma_y^2 = (\sigma_1 + \sigma_2)^2(1 - \nu + \nu^2) - 3\sigma_1\sigma_2$$

$$= \frac{K^2}{2\pi r_p} \cos^2 \frac{\theta}{2}\left[4(1 - \nu + \nu^2) - 3\cos^2 \frac{\theta}{2}\right]$$

Hence,

$$r_p = \frac{K^2}{2\pi\sigma_y^2} \cos^2 \frac{\theta}{2}\left[4(1 - \nu + \nu^2) - 3\cos^2 \frac{\theta}{2}\right] \tag{6.31}$$

At the plane of the crack, $\theta = 0$, and

$$r_p = \frac{K^2(1 - 2\nu)^2}{2\pi\sigma_y^2}$$

$$= (1 - 2\nu)^2 \times (r_p' \text{ for the plane stress case.}) \tag{6.32}$$

Note that $(1 - 2\nu)^{-1}$ is exactly equal to $M$ on the crack plane. Plotted in a polar coordinate system located at the crack tip, the plastic zone boundaries have the shape shown in Figure 6.24.

### 6.16.1 ASTM Requirements on Specimen Width

ASTM test standards for fracture toughness require that the specimen width must exceed $2.5 \times (K_{Ic}/\sigma_y)^2$, where $K_{Ic}$ is the plane strain fracture toughness *yet to be obtained* from the tests. This requirement often leads to the need for repetitive testing. We tentatively assume that the fracture toughness for plane strain $K_{Ic}$ is less than that for plane stress $K_{c\sigma}$ by the same ratio as the corresponding stress intensities. We then have a condi-

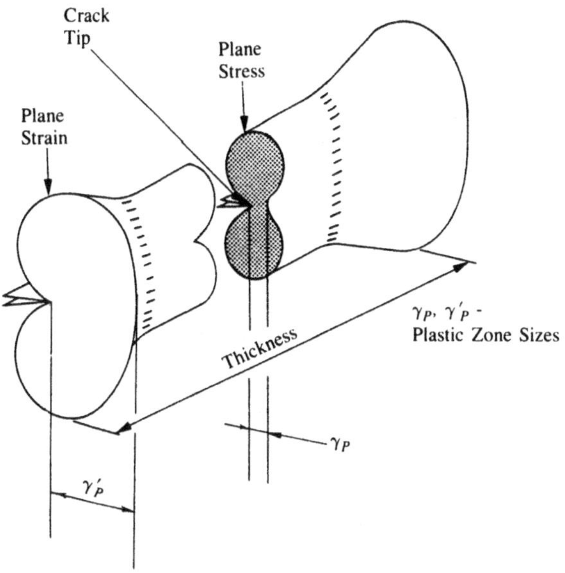


**FIGURE 6.24.** Plastic zone shape variation across the plate thickness.

tion for valid thickness of a test specimen for measuring $K_{Ic}$. It is assumed that for a depth of the plate equal to $r_p$ on either side, the plane stress slowly changes to plane strain. Since the specimen width $B$ must exceed $2 \times$ ($r_p$ for plane stress),

$$B > \frac{K_{c\sigma}^2}{\pi \sigma_y^2}$$

$$> \frac{1}{\pi}\left(\frac{K_{c\sigma}}{K_{Ic}}\right)^2 \frac{K_{Ic}^2}{\sigma_y^2}$$

$$> \frac{1}{(1-2\nu)^2 \pi} \frac{K_{Ic}^2}{\sigma_y^2} \tag{6.33}$$

| $\nu$ | 0.28 | 0.3 | 0.33 | 0.35 |
|---|---|---|---|---|
| $\left(\frac{B\sigma_y^2}{K_{Ic}^2}\right)_{\min}$ | 1.65 | 1.99 | 2.75 | 3.54 |

This equation is the basis for the requirement on the thickness of fracture toughness testing for metals in ASTM E-399, as well as in the draft standards for plastics, Reference 13. Because of this, a test for $K_{Ic}$ is valid if and only if the specimen thickness $B$ satisfies the above inequality, where $K_{Ic}$ denotes the result obtained from the test. If the inequality is not satisfied, the test is invalid and a different specimen of the appropriate thickness is to be tested again.

## 6.17 ACCOUNTING FOR PLASTIC EFFECTS

The quantity $r'_p$ for plane stress conditions can be taken to be a measure of the amount of material in plasticity. However, it is not the *real* distance of the plastic boundary. Truncating the stress value at $\sigma_y$ upsets equilibrium to the extent of the shaded area in Figure 6.25. The equilibrium is achieved by the material by spreading the plasticity over a larger area. Stress distribution follows a different curve, shown dotted on the figure. Obviously, the extended size $(r'_p + x)$ of the plastic zone is a function of the applied stress. Also, $r'_p$ provides a means of classifying crack tip plasticity into the following three cases. Referring to Figure 6.26, we can consider three cases, small, contained, and large. Note that $r_p$ provides an equally good basis for the classification, but using $r'_p$ is more conservative.

1. For $r'_p \ll$ the crack length $a$, the plasticity is considered to be small. For such cases, the linear elastic fracture mechanics rules are applicable.
2. For $r'_p < (W - a)$, but not very small compared to $a$, plasticity is considered to be "contained," that is, surrounded by material in its

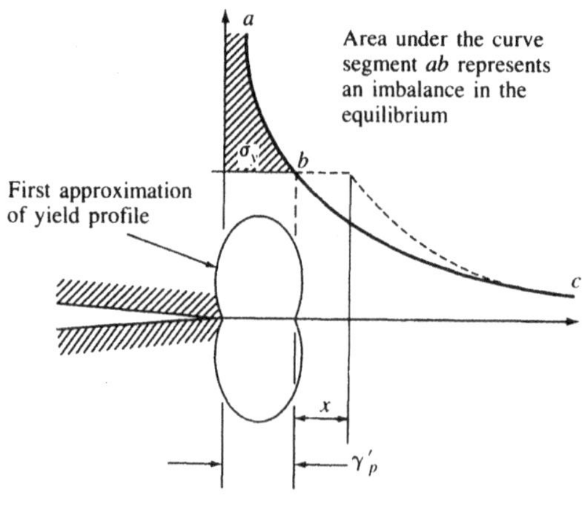


**FIGURE 6.25.** Calculation of $r_p$ by simple formulas causes imbalance of equilibrium.

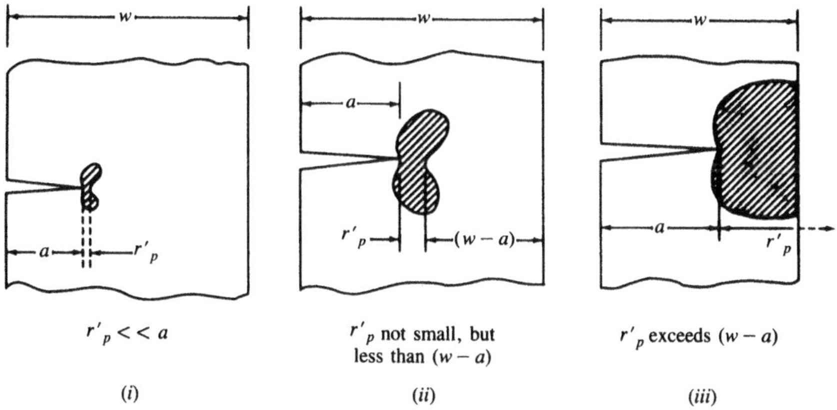


**FIGURE 6.26.** $r_p'$ is the basis for the classification of the three cases of crack tip plasticity.

elastic range. For such cases, one needs to use the modified procedures discussed below.

3. For $r_p' \geq (W - a)$, plasticity is considered to be large. Analytical tools for this category of plasticity involve advanced techniques and are outside the scope of this book. Also, a component that has reached large amounts of plasticity at the crack tip, would exhibit gross deformations and a total lack of stiffness, so it would be removed from service.

## 6.18 CONTAINED PLASTICITY

### 6.18.1 Irwin's Correction

As stated before, yielding redistributes stress, and increases the plastic zone size. Irwin [5] suggested a simple way to handle contained plasticity. He showed that on the crack plane, the crack behaves as though it were longer than its physical length. Or effectively, the crack tip appears to have been relocated inside the uncracked material. To prove this result, consider the real and fictitious crack under plane strain conditions shown in Figure 6.27. The physical crack has a length $a$, and it is moved by an amount $\delta$ along $x$. We require that the curve for $\sigma_{yy}$ drawn at the new origin is such that if we truncate the portion of the curve for which the ordinate is greater than the yield strength $\sigma_y$, we obtain the plastic zone length $x$. This shift of crack front restores the equilibrium of forces by adding more area under the curve. The shaded area under the curve

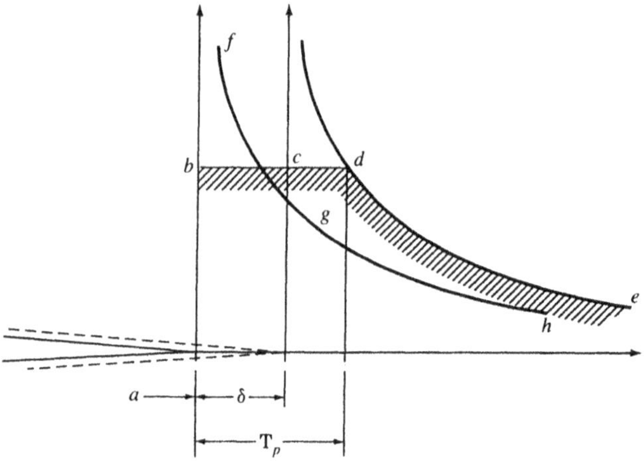


**FIGURE 6.27.** Irwin's proof of the fictitious crack length is based on advancing the crack front.

"*bcde*" should be exactly equal to that under "*fgh*." For clarity we denote the stress intensity as a function of crack size, such as $K(a)$ or $K(a + \delta)$.

$$\text{Area under "}bcde\text{"} = (M\sigma_y)r_p + \int_{r_p}^{\infty} \frac{K(a + \delta)}{\sqrt{2\pi r}}$$

$$\text{Area under "}fgh\text{"} = \int_0^{\infty} \frac{K(a)}{\sqrt{2\pi r}}$$

If the plastic zone is small in comparison with the crack size, then $K(a + \delta) = K(a)$. From equality of areas, we obtain

$$(M\sigma_y)r_p = \int_0^{\infty} \frac{K(a)\,dr}{\sqrt{2\pi r}} - \int_{r_p}^{\infty} \frac{K(a + \delta)\,dr}{\sqrt{2\pi r}}$$

$$= \int_0^{r_p} \frac{K(a)\,dr}{\sqrt{2\pi r}}$$

$$= K(a)\sqrt{\frac{2r_p}{\pi}}$$

Hence,
$$r_p = \frac{2K^2}{\pi(M\sigma_y)^2}$$

But it is seen that the first approximation of plastic zone size is exactly equal to $\delta$, we conclude that the corrected plastic zone size $r_p$ is twice the first approximation. That is,

$$r_p = 2\delta$$

### 6.18.2 Another Approach—The Dugdale Correction

Dugdale [14] approached plasticity effects in a slightly different way. His approach also involves the shifting of crack tip by an amount $r_p^*$, which denotes *the size of the plastic zone*. Hence, the amount by which the coordinate axis is shifted is $r_p^*$. (Note that in the Irwin correction, the amount of shift was one half of the plastic zone size.) Since this zone is subjected to yield stress $\sigma_y$, we can treat it as a continuously distributed line load, as in Example 6.2.

> The requirement is that the stress intensity at the physical crack tip is exactly zero because of the combination of the applied load and the yield stress acting over $r_p^*$.

Using the notations of Figure 6.13 for a crack in an infinite plate, we see that there are two components of $K$, one caused by the applied stress $\sigma$, and the other caused by the line load $\sigma_y$.
From $\sigma$:

$$K1 = \sigma\sqrt{\pi(a + r_p^*)}$$

From $\sigma_y$:
Also, we suppose that the crack length is $(a + r_p^*)$,

$$K2 = 2\sigma_y\sqrt{\frac{a + r_p^*}{\pi}} \int_a^{(a+r_p^*)} \frac{dx}{\sqrt{(a + r_p^*)^2 - x^2}}$$

$$= 2\sigma_y\sqrt{\frac{a + r_p^*}{\pi}} \arcsin\left(\frac{x}{a}\right)\Bigg|_a^{(a+r_p^*)}$$

$$= 2\sigma_y\sqrt{\frac{a + r_p^*}{\pi}} \arcsin\left(\frac{x}{a + r_p^*}\right)\Bigg|_a^{a+r_p^*}$$

$$= 2\sigma_y\sqrt{\frac{a + r_p^*}{\pi}} \left[\frac{\pi}{2} - \arcsin\left(\frac{a}{a + r_p^*}\right)\right]$$

Dugdale's approach lies in setting $K1 = -K2$. This leads to

$$\frac{a}{a + r_p^*} = \cos\frac{\pi\sigma}{2\sigma_y}$$

Or,

$$r_p^* = a\left(\sec\frac{\pi\sigma}{2\sigma_y} - 1\right)$$

$$\approx \frac{\pi^2\sigma^2 a}{8\sigma_y^2} = \frac{\pi K^2}{8\sigma_y^2} \tag{6.34}$$

We may use $(M\sigma_y)$ in the place of $\sigma_y$ to include the plane strain constraint conditions. Compare with Irwin's plastic zone size.

$$\text{Irwin's plastic zone size} = \frac{K^2}{\pi(M\sigma_y)^2}$$

$$\text{Dugdale's plastic zone size} = \frac{\pi K^2}{8(M\sigma_y)^2}$$

The apparent difference of over 26% between these two expressions is mostly removable if the sec( $\cdots$ ) is kept as such, instead of being replaced by its approximation. The relation in Equation 6.34 is good only for small values of $\sigma/\sigma_y$.

Dugdale's model of calculation will be used later when discussing "craze" stresses.

## 6.19 CRACK OPENING DISPLACEMENT (COD)

Recall that we do not as yet have a relationship between *COD* and the other parameters. With Dugdale's model for the yielding, we can derive a relationship between *COD* and *G*. It may be verified that an attempt to use the crack tip stress field to calculate displacement will lead to singularities. Further, the use of conformal mapping techniques also leads to zero displacements (not singular). However, the *COD*—the displacement normal to the crack plane calculated at the physical crack tip—is finite and nonzero because of the plasticity effects. Thus, we need to use the Dugdale model of yield stress distribution, in combination with Irwin's correction to obtain an expression for *COD*.

### 6.19.1 Displacement for Line Load Distributed Near the Tip

For the distributed line forces $\sigma_y$ shown in Figure 6.28, the expression for displacement in the range $0 \le x \le r_p$ is given by Equation 6.35. This expression is valid for a crack embedded in an infinite plate.

$$v = \frac{8(1-\nu^2)\sigma_y r_p}{\pi E}\left[\sqrt{\frac{x}{r_p}} - \left(1 - \frac{x}{r_p}\right)\ln\left(\frac{1+\sqrt{\frac{x}{r_p}}}{1-\sqrt{\frac{x}{r_p}}}\right)\right] \quad (0 \le x \le r_p) \qquad (6.35)$$

### 6.19.2 Relationship Between $G$ and COD

Since the displacement at $x = r_p$ represents the crack tip opening displacement, we can calculate it as the limit of $x \to r_p$. Such a limit is finite and can be found to be

$$\Delta = \frac{8(1-\nu^2)\sigma_y r_p}{\pi E}$$

$$= \frac{K^2(1-\nu^2)}{E\sigma_y}, \quad \text{using Equation 6.34}$$

$$= \frac{G}{\sigma_y} \qquad (6.36)$$

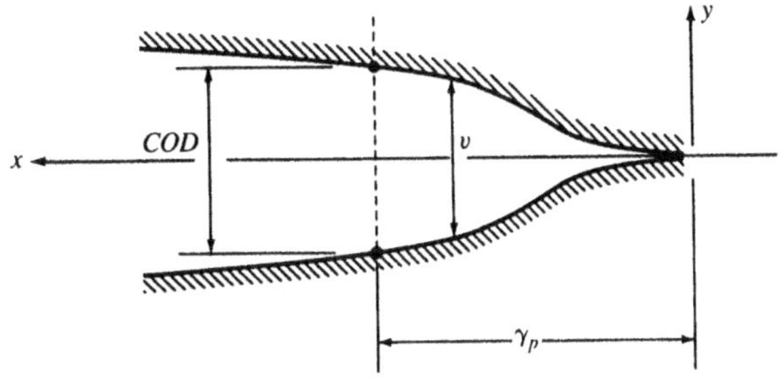


**FIGURE 6.28.** Equation 6.35 plots into a cusp shaped profile for the crack.

Note that this equation establishes the relationship among $G$, $K$, and *COD*. Such a relation at the point of instability denotes a critical value of *COD*, which may be taken as a material parameter.

In fact the term *COD* is used to refer to the critical *COD* itself and hence is understood as a material property. It inherently includes the plasticity effects, and is considered to be more realistic than the others. Equation 6.36 may also be taken to relate $G$ and $K$ with the plasticity effects included.

## 6.20 FRACTURE INITIATION PROCESS—CRAZING

From a materials science point of view, it is useful to know about how a crack initiates and propagates in a plastic material. In a microscopic scale the mechanism of crack initiation is different from that of metals although there are similarities in the overall picture. As in the metals, a crack initiates at a point of high stress concentration. At such points in glassy thermoplastics and in some semicrystalline polymers, such as polypropylene, microvoids are developed. Impurities, additives, and other particulate matter are responsible for these microvoids. Initially, yielding takes place and is usually indicated by whitening. Further, because of the high degree of orientation the small volume of material in the peak stress area is work hardened very rapidly. An increase in the applied stress leads to a concentration of the microvoids into bands of fibrils which are known to exhibit extension ratios of up to 500% before they break. Voids and fibrils make up about 50% each by volume, and constitute a highly porous material, which is termed craze. This is illustrated schematically in Figure 6.29.

Crazes do not necessarily require externally applied forces to produce them. Internal stresses such as residual stresses, differential relaxation, absorption of liquids and vapors can cause crazing too. When liquid or vapor absorption is the cause of crazing, it is called *environmental stress cracking*. The microvoids and fibrils attract liquid or vapor by capillary action and the liquid quickly moves to the crack tip where it can cause further crazing.

Precise measurements of the craze profile have confirmed the cusp shape assumed in the earlier calculations. See also Figure 6.28. This is an indirect proof for the assumption of constant stress over the face of the craze. This stress is called the "craze stress"; it may be more appropriate to call it the "average craze strength." The absolute value of the strength of the fibril material is quite different from the yield strength of the bulk

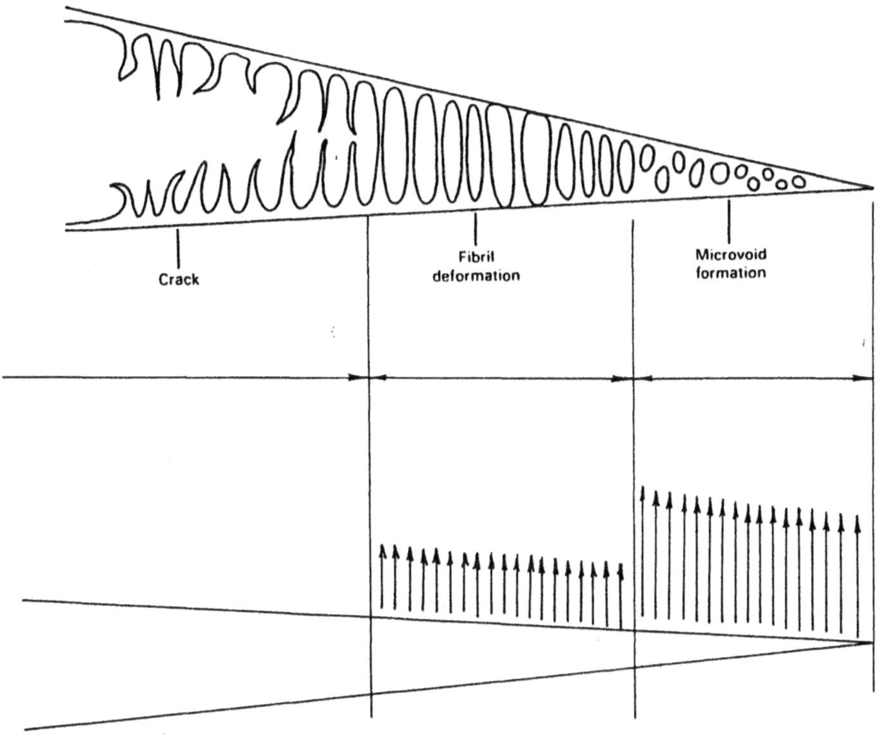


**FIGURE 6.29.** Schematic illustration of the crazing mechanism. Some authorities, (see Reference 3.) consider a craze as a two-zone Dugdale model, since at the microvoid region orientation and volume percent are both high. In the fibril region, volume percent is low.

material, (because of the severe orientation and work hardening and volume percentage) and can be of an order of magnitude of 200 MPa (29000 psi), which is close to the yield point of metals! However for the purpose of use in the Dugdale model, it is sufficiently accurate to use the yield strength of (un-oriented) bulk in the place of craze stress. Thus, all the preceding results of the craze shape, craze length, etc. derived using the Dugdale model with $\sigma_y$ are verified to be applicable for all plastics.

When the fibril ligaments rupture, a new crack surface is formed. However, at the tip of the crack the unruptured crazed material is present, and it resists the propagation of the crack with a *negative stress intensity factor K*, arising mainly from the craze stress, acting as a line load. Since the void fractions vary from its minimum value at the crack tip, to the maximum value away from the tip, the craze stress also correspondingly varies from its maximum value at the tip, to the minimum away from the tip. It is sufficiently accurate to consider the continuous variation to be composed of just two magnitudes (see Figure 6.29).

The propagation of the crack is a sequence of craze development, rupture of fibrils, or extension of the crack and new craze development. The extension of the crack can be caused by an increase in load, or by the passage of time. It is easy to see the beneficial influence the craze stress can have on the toughness of the material. The craze stress opposes the crack propagation. Hence, the greater the craze stress, the greater the fracture toughness. Similarly the size of the craze zone ahead of the crack also influences the toughness. The larger the zone size, the greater the fracture toughness. For example, polystyrene and SAN form only a few crazes, and consequently have only low fracture toughness. The rubber modified grades of these materials, on the other hand, provide more sites for craze formation, thus creating more resistance to crack propagation. The rubber particles act as concentrators which promote craze formation at low applied stresses.

The initiation of crazes has been studied with respect to the stress field causing them. It is possible to express a criterion for crazing, in a manner similar to the Von Mises criterion for yielding. For a stress state with $\sigma_1$, $\sigma_2$, and $\sigma_3$ as the principal stresses, crazing occurs if

$$\mathrm{Max}\{|\sigma_1 - \sigma_2|, |\sigma_2 - \sigma_3|, |\sigma_3 - \sigma_1|\} = A + \frac{B}{I_1} \tag{6.37}$$

where,

$$I_1 = \frac{1}{3}[\sigma_1 + \sigma_2 + \sigma_3]$$
$$> 0$$

$A$ and $B$ are material properties that are functions of time, temperature, and environment. When $I_1$ becomes zero, the stress needed to produce crazing is infinite. This means that for crazing to occur, a dilatational stress field is required. That is, under elastic conditions the stress field must be capable of producing an *increase* in material volume. Because of this, Equation 6.37 maps only on to the upper half of the stress plane. Figure 6.30 is a two dimensional version of Equation 6.37.

Apparently the parameters $A$ and $B$ are not widely available for even the popular materials. This situation can be rationalized if we consider the fact that the material property $G_c$ or the *COD* measures a toughness that includes in some way the role played by crazing in abating crack propagation. Thus the utility of $A$ and $B$ does not extend to structural computations, but simply to establishing the differences between yielding and crazing mechanisms.

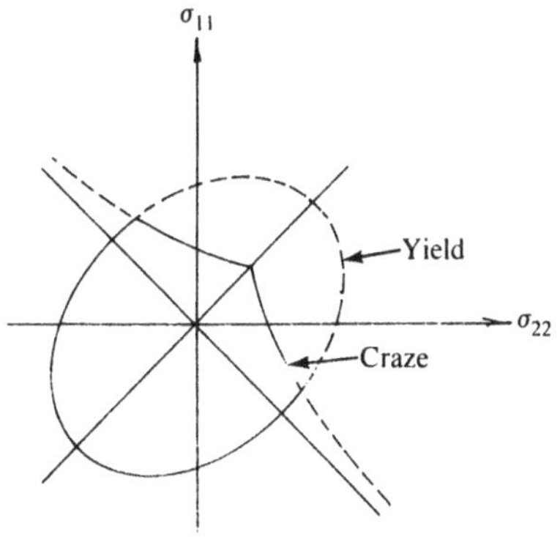


**FIGURE 6.30.** Crazing and yielding envelopes in two dimensions.

## 6.21 FATIGUE

The major success of fracture mechanics as a branch of stress analysis is its ability to explain fatigue failure analytically. Prior to the advent of fracture mechanics, fatigue failure was quantified by means of environment-specific data—called the S-N curves—each one of which was obtained by time-consuming cyclic loading tests. However, fracture mechanics enables a quantitative analysis of the useful life of structures with flaws. Given the knowledge of crazing behavior at peak stress areas, it is possible to assume that every structure has inherent flaws. This assumption is far from hypothetical, since every structure has areas of stress concentration, and sooner or later, crazes develop into cracks at these points.

However, these cracks are not of critical size. We may suppose that the cyclic loading mode is just another "environment." Because of cyclic loading, the crack extends slowly in size. This is called the *Fatigue Crack Propagation* (*FCP*) or stable crack propagation. Eventually the crack size becomes large enough that it is the critical size for the given loading. At this point instability occurs and catastrophic failure occurs.

Thus, the structurally functional life of a product is equal to the time needed for propagating an initial small crack to the critical size. We need a law that connects the number of cycles of fatigue loading, and its magnitude with the growth of the crack length.

### 6.21.1 Paris Law

The Paris law of fatigue crack propagation originated in metals, and has also been found to be valid for plastics. [3, 15] This law fits the *rate of growth of crack length per cycle with the range of stress intensity over a cycle.*

The typical form of the law is

$$\frac{da}{dN} = A(\Delta K)^m \tag{6.38}$$

where $A$ and $m$ are material constants, and are functions of temperature, environment, loading rate, frequency of loading, mean stress, loading history, and also the wave form of loading. Not all materials are sensitive to all the parameters. Hertzberg's text [15] is an excellent treatise for the materials scientist for understanding the effect of the various parameters. A few typical values of $A$ and $m$ are listed in the following table. They may be taken to be applicable for room temperature (20°C), in air, and for sinusoidal loading. The purpose of this section is to discuss the calculation procedure for determining the fatigue life using the Paris law. Since the expression for rate of growth is available, we can integrate the Paris equation to obtain the number of cycles to failure.

### 6.21.2 Life Prediction Using the Paris Law

In Equation 6.38, $\Delta K = (K_{max} - K_{min})$, where $K_{max}$ and $K_{min}$ denote the stress intensity factors corresponding to the highest and lowest stress magnitudes in the cyclic loading. Since the extension during one cycle is negligibly small, $\Delta K = C\Delta\sigma\sqrt{\pi a}$, where $\Delta\sigma$ represents the range of stress variation in a cycle. Because of its form, $K = C\sigma\sqrt{\pi a}$, it is not always easy to integrate Equation 6.38. We have to resort to numerical integration. The following examples illustrate the use of the Paris law for the prediction of component life.

**Table 6.3. Values of $A$ and $m$ for Selected Crystalline and Amorphous Polymers. (Derived from Reference 15.) Note that $\Delta K$ must be expressed in $MPa\sqrt{m}$.**

| Material | A | m |
|---|---|---|
| PC | 8.432 E-07 | 3.691 |
| PVC | 5.757 E-06 | 2.172 |
| PPO | 1.295 E-07 | 4.479 |
| NYLON 6,6 | 5.170 E-09 | 4.310 |
| $PVF_2$ | 5.955 E-09 | 4.174 |
| PMMA | 6.021 E-07 | 6.757 |
| PS | 2.804 E-05 | 2.975 |
| PSF | 4.953 E-08 | 6.757 |

**Example 6.14**

A very large PVC plate has a through-thickness crack of 1″ (0.0254 m) length. A sinusoidal load is applied. The minimum value of stress is 0 psi and the maximum is 1000 psi (6.89 MPa). Find the number of cycles for doubling the crack length.

*Solution* First of all, note that the material constants in the Paris law are in SI units. Since for a crack in an infinite plate $K = \sigma\sqrt{\pi a}$,

$$K_{\text{min}} = 0$$

$$K_{\text{max}} = 6.89\sqrt{\pi a}$$

$$\Delta K = 6.89\sqrt{\pi a}$$

For PVC,
$$\frac{da}{DN} = 5.757 \times 10^{-6}(6.89\sqrt{\pi a})^{2.172}$$

Or,
$$\int_{a_0}^{2a_0} \frac{2a}{a^{1.086}} = 0.001320\int_0^N dN$$

Since the initial crack length $a_0 = 1$ in $= 0.0254$ m, $N$ is easily calculated to be 700 cycles.

**Example 6.15**

A polycarbonate strip 40 mm thick, 100 mm wide and 300 mm long has a central crack of length 20 mm. For a sinusoidally varying applied tensile stress of 3.5 MPa, find the life of the strip until failure, expressed in terms of number of cycles. For polycarbonate, take $K_{Ic} = 3.3\ \text{MPa}\sqrt{m}$, and

$$\frac{da}{dN} = 8.432 \times 10^{-6}(\Delta K)^{3.691}, \text{ in SI units.}$$

*Solution* Since the part will break when the crack length reaches its critical length $a_c$, we must calculate the critical crack length first. We use the curve in Example 6.10 for this purpose.

$$\frac{K_{Ic}}{\sigma\sqrt{W}} = \frac{3.3}{3.5\sqrt{0.1}} = 2.9815$$

We can read off the corresponding $(a/W)$ to be 0.46. The critical crack length $a_c = 0.046$ m. The Paris equation must therefore be integrated from original crack length to the critical crack length.

$$\int_{a_0}^{a_c} \frac{da}{[C(a)\sqrt{a}]^m} = A\pi^{m/2}\,\Delta\sigma \int_0^N dN$$

$$\int_{0.01}^{0.046} \frac{da}{[C(a)\sqrt{a}]^{3.691}} = 8.432 \times 10^{-7}\pi^{3.691/2}(3.5\ \text{MPa}) \int_0^N dN$$

$$= 2.4406 \times 10^{-5} N$$

The left side of the equation is a high order polynomial and has to be integrated numerically. In this example, the trapezoidal rule was used, and results are tabulated below. An increment of $da = 1$ mm was used in the calculation, but results for every 5 mm increment are shown.

| Crack Length | $\dfrac{da}{(C(a)\sqrt{a})^m}$ | Cumulative Cycles |
|---|---|---|
| 0.010 | 4.4859 | 183804 |
| 0.015 | 2.1861 | 778901 |
| 0.020 | 1.3049 | 1107033 |
| 0.025 | 0.8722 | 1316792 |
| 0.030 | 0.6268 | 1463266 |
| 0.035 | 0.4736 | 1571759 |
| 0.040 | 0.3713 | 1655587 |
| 0.045 | 0.2995 | 1722448 |
| 0.046 | 0.2878 | 1734238 |

It is also instructive to plot the cumulative number of cycles versus the crack extension. See Figure 6.31. Since an increase in $a$ contributes to an increase in $C(a)$, which in turn, increases $a$, we should expect the crack to show an accelerated growth, which is seen in Figure 6.31.

### 6.21.3 A Remark on Life Prediction Calculations

In the above examples, the life of a product was taken to be the time to propagate a pre-existing crack. Normally, one should expect that a certain time—or number of cycles—is involved in the initiation of a crack. The two examples considered above have uniform stress fields. In these cases, if

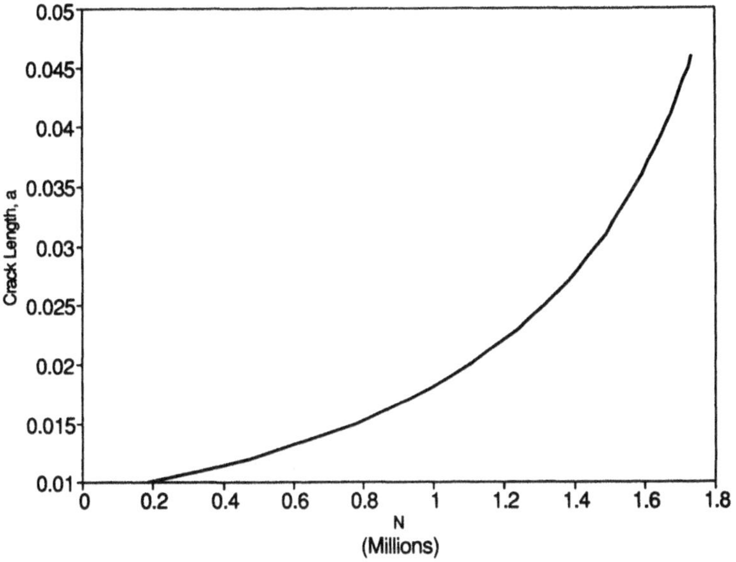


**FIGURE 6.31.** Crack growth as a function of the number of cycles, Example 6.15.

there were no pre-existing cracks, then new cracks would take some time to be created around microscopic defects. In an actual structure, stress fields are not constants and there are a number of stress raisers, where a crack can initiate quite easily.

> Therefore, the time for crack propagation can be assumed to represent the life of the product, where there are stress concentrations. In a practical situation it is conservative enough to assume an initial crack size equal to the plastic zone size of the stress concentration.

The size of the plastic zone is the region over which the Von Mises criterion shows plasticity to occur. The conservatism arises from the fact that no credit is taken for the crack initiation time, and for the beneficial orientation of the material in stress concentration areas which inhibits crack initiation. Difficulties arise when using a $\Delta K$ from experiments involving ideal conditions. Crack propagation is known to depend on the frequency of loading, waveform of load application, and whether or not a cyclic load program is interrupted by another program, etc. For such purposes, one should resort to "the worst case" approach in modeling the problem. Resorting to computational methods, such as finite elements must be done with care.

## 6.22 CONCLUSION

This chapter covers the aspects of fracture that do not involve time effects. Time effects would allow the calculation of the rate of crack growth at constant applied loads. J. G. Williams [3] uses an approach very similar to the pseudoelasticity approach to calculate the growth of strains and displacements at the crack tip with time, and at a constant applied load. The use of Boltzmann's superposition principle for calculating displacements at a crack tip under variable stress is also discussed. A real life problem involves the time taken to propagate a given crack to a critical length, based on the data of the so called "crack velocity $\dot{a}$." A simple approach to the calculation is to use the relaxation modulus to determine the displacement at the crack tip and find the time taken for this displacement to reach the critical *COD*.

Fracture toughness data for many commercial grades of plastics are not easily available. A major problem in building a data base of fracture toughness properties is the myriad of possible environmental exposures of plastics. Simplified experiments are being used to obtain data connecting environment, time of exposure, and stress level. Such data are helpful in material selection, but not in stress analysis. However, the FCP data for the specific environment is generally not available and this situation appears to be the cause of many failures of plastics.

Yet another problem in the use of pre-engineered solutions is that they are only valid for the constant loads (or stresses or moments) applied. A real-life problem has a variable stress field, and no constant stress or load is readily identifiable. This situation is not any different from elastic structural analysis. Either a conservative model of the loads or a finite element analysis has to be performed.

**References**

1. Barsom, J. M. and Rolfe, S. T., *Fatigue and Fracture Control in Structures—Applications of Fracture Mechanics*, 2nd Edn. 1987, Prentice Hall Inc., Englewood Cliffs, NJ.
2. Griffith, A. A., "The phenomenon of rupture and flow in solids," *Phil. Trans. Roy. Soc.*, London, A-221, pp 163–198, 1920.
3. Williams, J. G., *Fracture Mechanics of Polymers*, 1984, Ellis Horwood Ltd., Chichester, U.K.
4. Timoshenko, S. P. and Goodier, J. N., *Theory of Elasticity*, 3rd Edn. 1970, McGraw Hill, New York, NY.
5. Irwin, G. R, "Plastic zone near a crack and fracture toughness," *Proc. 7th Sagamore Conf.* p IV–63, 1960.

6. Sih, G. C., *Handbook of Stress—Intensity Factors—Stress Intensity Factor Solutions and Formulas for Reference*, 1973, Institute of Fracture and Solid Mechanics, Lehigh University, Bethlehem, PA.
7. Rooke, D. O. and Cartwright, D. J., *Compendium of Stress Intensity Factors*, 1974, Her Majesty's Stationery Office, London.
8. Hiroshi Tada, Paris, P. C., and Irwin, G. R., *The Stress Analysis of Cracks Handbook*, 1973, Del Research Corporation.
9. Rice, J. R., "A path independent integral and the approximate analysis of strain concentration by notches and cracks," *J. of Applied Mechanics*, Vol. 35, pp 379–386, 1968.
10. Wells, A. A., "Unstable crack propagation in metals: Damage and fast fracture," *Proceedings of the Crack Propagation Symposium*, The College of Aeronautics, Vol. 1, pp 210–230, 1962, Cranefield, U.K.
11. Williams, J. G., "Linear fracture mechanics," Chapter 2 in *Advances in Polymer Science, Number 27, Failure in Polymers* edited by Hans-Joachim Cantow, et al., 1978, Springer-Verlag, Berlin.
12. Williams, J. G., *Stress Analysis of Polymers*, Chapter 7, 2nd Edn. Ellis Horwood Ltd., Chichester, U.K.
13. ASTM Subcommittee D20.10.21 on Fracture Mechanics, *Standard Test Methods for Plane-Strain Fracture Toughness and Strain Energy Release Rate of Plastic Materials*, Project No. X-10-128 April 1990.
14. Dugdale, D. S., "Yielding of steel sheets containing slits," J. of the Mechanics and Physics of Solids, Vol. 8, p 100, 1960.
15. Hertzberg, R. W. and Manson, J. A., *Fatigue of Engineering Plastics*, 1980, Academic Press, New York, NY.
16. Broek, D., *Elementary Engineering Fracture Mechanics*, 3rd Edn. 1982, Martinus Nijhoff Publishers, Boston, MA.
17. Ewalds, H. L. and Wanhill, R. J. H., *Fracture Mechanics*, 1984, Co-published by Edward Arnold and Delftse Uitgevers Maatschappij, Delft, Netherlands.
18. Paris, P. C. and Sih, G. C., *ASTM STP 381*, American Society for Testing and Materials, Philadelphia, 1964.

**Exercises**

**6.1.** In a double cantilever test of polycarbonate, the crack is seen to advance in the midplane of the specimen. At a point in the test, it is observed that for a load of 200 N, the crack length is 25 mm. Given that the half-depth of the specimen is 35 mm, the thickness is 50 mm and the tensile modulus is 2410 MPa, find the strain energy release rate. Also given $\nu = 0.3$, and $K_{Ic}$ for the material is 2.75 $\text{MPa}\sqrt{\text{m}}$, find the load at which the test specimen would possibly come apart. Assume a linear relationship between load and deflection.

**6.2.** Glass-filled polycarbonate is used in a large water filter, which is used in swimming pool applications. The vessel is 800 mm inside diameter and is 18 mm thick. Water is found to flow out of a through-the-thickness crack 25 mm wide, under the internal pressure of 0.1 MPa. Find the stress intensity developed at the crack tip.

**6.3.** A PVC tab is adhesively bonded to the flat face of a heavy sheet, and is used for lifting it. The tab is 2 mm thick, and the weight acting on the tab is 15 kg. The hole diameter is 5 mm. What size crack on just one side of the hole would make this load critical? What would be the critical size if the crack is assumed to spread on both sides of the hole? Take $K_{Ic}$ values from Table 6.2.

**6.4.** A collet is inserted into a circular hole in a large plate. The insertion process introduces a force $F$ on the walls of the hole. The diameter of the hole is 12 mm, and the $K_{Ic}$ of the plate material is 3 MPa$\sqrt{\text{m}}$. If a crack develops on two sides of the hole, in a diametral plane, calculate and plot the relationship between $F$ and the crack length.

**6.5.** An ABS high impact element is a fixed beam of 40 mm depth. The modulus of the material is 1725 MPa, the coefficient of thermal expansion is $10^{-4}$ mm/mm/°C. A 5 mm crack is present at the center of the lower edge of the beam. This element is cooled slowly to 30°C below the environment temperature while the ends are held fixed. Calculate the stress intensity $K_I$ developed at the crack tip.

**6.6.** In the same geometry as in the last problem, the temperature is increased by 20°C above the ambient temperature. What concentrated downward force at the middle of the beam would bring the $K_I$ exactly to zero?

**6.7.** In Example 6.13 what $K_{Ic}$ would be needed to ensure a stable crack after through thickness propagation? If the material in the example was CPVC, would the crack remain stable after penetrating the full thickness of the plate?

**6.8.** A large HDPE pipe was failure-analyzed through microphotographic examination. The fracture surface shows that there was an area of poor fusion in the circumferential butt weld. This area may be treated as an elliptic flaw of size $2c = 30$ mm, and $a = 6$ mm. Given radius = 200 mm, thickness = 18 mm, and $K_{Ic} = 3$ MPa$\sqrt{\text{m}}$, what pressure was probably acting in the pipe to cause it to fail? Without the crack, was the pipe adequately sized for that pressure?

**6.9.** *A Term Paper:* For the material and geometry in the previous problem, produce a graphical representation of the relationship between the initial crack size ($2c$ and $a$), pressure $p$, and the number of cycles to failure ($N$).

**6.10.** A large plate is made of PVC and has a part-through thickness crack that penetrates 50% of the thickness. It is subjected to a stress $\sigma$. Calculate the ranges of $\sigma$ for which (1) LEFM methods are applicable, (2) LEFM can be applied with Irwin's correction for plasticity, and (3) gross yielding occurs in the specimen.

**6.11.** Two different plastics, CPVC and HDPE, are evaluated as candidates in the manufacture of a pressure vessel. The vessel inner diameter is 300 mm and the thickness is 20 mm. For a longitudinal surface flaw on the inner wall, calculate $K_I = 1.12\sigma\sqrt{\pi a}\, M_k$. Draw a family of curves of allowable critical stress versus the crack size, using the fracture toughnesses of the two plastics as parameters.

**6.12.** In the above problem, find the design of least weight, subject to the requirement that the designs will have equal factors of safety on (2) $K_{Ic}$ and (2) $\sigma_y$?

**6.13.** In Example 6.14, if the number of cycles to double the length of the crack must exceed 5000, which of the following would you do first? (1) require a smaller initial crack size, or (2) reduce stress amplitude (which itself might amount to oversizing the component thickness).

**6.14.** In general, rank the following ideas in decreasing order of effectiveness for increasing the fatigue life of a component?

(a) Start with a smaller crack size,
(b) Reduce stress amplitude (or oversize thickness),
(c) Choose a material of higher $K_{Ic}$,
(d) Choose a material with a low value of $A$ in the Paris equation,
(e) Choose a material with a low value of $n$ in the Paris equation.

# 7

# Reinforced Plastics

## LIST OF SYMBOLS USED

***Common for all sections:***

| | |
|---|---|
| $E_{11}, E_{22}, G_{12}, \nu_{12}$ — | Engineering moduli of a ply in material coordinate system. |
| $E_{xx}, E_{yy}, G_{xy}, \nu_{xy}, \nu_{yx}, \eta_{xs}, \eta_{ys}, \eta_{sx}, \eta_{sy}$ — | Engineering moduli of a ply in any general coordinate system. |
| $[\mathbf{S_{12}}], S_{11}, S_{12}, S_{22}, S_{66}$ — | Ply compliances in material coordinate system. |
| $[\mathbf{Q_{12}}], Q_{11}, Q_{12}, Q_{22}, Q_{66}$ — | Ply stiffnesses in material coordinate system. |
| $[\mathbf{S_{xy}}], S_{xx}, S_{yy}, S_{xy}, S_{xs}, S_{ys}, S_{ss}$ — | Ply compliances in any general $x$-$y$ coordinate system. |
| $[\mathbf{Q_{xy}}], Q_{xx}, Q_{yy}, Q_{xy}, Q_{xs}, Q_{ys}, Q_{ss}$ — | Ply stiffnesses in any general $x$-$y$ coordinate system. |
| $\{\boldsymbol{\sigma}\}, \sigma_1, \sigma_2, \tau_{12}$ — | Ply stresses in material coordinate system. |
| $\{\boldsymbol{\sigma}'\}, \sigma_x, \sigma_y, \tau_{xy}$ — | Ply stresses in any general $x$-$y$ coordinate system. |
| $\{\boldsymbol{\epsilon}\}, \epsilon_1, \epsilon_2, \tau_{12}$ — | Ply strains in material coordinate system. |

$\{\epsilon'\}, \epsilon_x, \epsilon_y, \tau_{xy}$—Ply strains in any general $x$-$y$ coordinate system.

*Micromechanics*

$E_{\dots}$—Young's modulus.
$G_{\dots}$—Shear modulus.
$K_{\dots}$—Bulk modulus.
$\nu_{\dots}$—Poisson's ratio.
Subscript, $c$—Used with the above, for composites.
Subscript, $m$—Used with the above, for matrix.
Subscript, $f$—Used with the above, for filler. (Filler properties may have additional subscripts to denote their orthotropy.)
$E_{33}, G_{23}, G_{13}, \nu_{23}, \nu_{13}$—Engineering moduli applicable for three-dimensional solid composites in the third plane.
$(\cdot)_m, (\cdot)_f, (\cdot)_c$—Any property associated with the Halpin-Tsai equation, $m$, for matrix; $f$, for fiber; and $c$, for composite.
$\psi, \xi, \eta$—Parameters in the Halpin-Tsai equation.
$V_f$—Volume fraction of filler in the composite.
$V_{f,\max}$—Maximum possible volume fraction.
$r, l, d, W, t$—Various geometric parameters of fibers/ribbons.
$l_c$—Critical length of a chopped fiber.

*Macromechanics*

$F_1^T, F_1^C, \epsilon_1^C, \epsilon_1^T$—Strengths of a ply along fiber, tensile and compressive, expressed as stress and strain.
$F_2^T, F_2^C, \epsilon_2^C, \epsilon_2^T$—Strengths of a ply across fiber, tensile and compressive, expressed as stress and strain.
$F_6$—Shear strength of a ply.
$S_f^-, S_f^+$—Safety factors with and without reversing the sign of stress.
$f_1, f_2, f_{11}, f_{22}, f_{12}, f_{66}$—Parameters in Tsai-Wu criterion.
$\theta$—Orientation of fibers to the $x$-$y$ coordinates.

$[\mathbf{T1}], [\mathbf{T2}], [\mathbf{R}]$—Transformation matrices for stress, strain, and moduli.

$\{\alpha_{**}\}, \{\beta_{**}\}$—Hygro-Thermal expansion coefficients, in material or *x-y* system.

***Laminates***

$n$—Total number of plies in a laminate.
Subscript, $k$—Any property of the $k$-th ply.
$t_k$—Thickness of the $k$-th ply.
$z_k$—Distance of the top of $k$-th ply from the midplane.
$T$—Total thickness of the laminate.
$\{\epsilon\}_o, \{\kappa\}_o$—Midplane strains, and curvature changes.
$\{\mathbf{N}\}, \{\mathbf{M}\}$—System of forces and moments acting on the laminate.
$[\mathbf{A}], [\mathbf{B}], [\mathbf{D}]$—Overall stiffness-related matrices of the laminate.

## 7.1 MOTIVATION

Reinforced plastics—or *composites*—are a synergetic combination of two materials, one called the *matrix*[1], and the other called the *reinforcement*. The reinforcement materials are usually much stronger and stiffer than the matrix materials. Most reinforcements come in the form of fibers, (continuous or chopped) although other forms have been used, to an extent. The other forms are particles, flakes, and ribbons. Figure 7.1 shows the various forms of reinforcements and matrices used in the industry. The reader is referred to References 2, 9, and 23. Figure 7.2 shows the possible variations in the construction of composites. In composites, the matrix material is the continuous medium, and the reinforcement is dispersed.

Fibers come in a variety of cross-sectional shapes. The characteristic size of the cross section, commonly designated as simply the *diameter*, is in the range of about 3 to 20 microns. Due to their small diameter, they cannot form a structure by themselves, unless they are bound together by another medium. The matrix binds the fibers together and helps in transferring the applied load to the fibers.

The simplest example of a composite is a thin polyester sheet, which is filled to about 50% of its volume by glass fibers, which are laid out parallel to each other. Such a sheet is called *unidirectional lamina* (*UDL*). A UDL

---

[1] The word *matrix* is used in this chapter in two different senses, one the mathematical array of numbers, and the other the material binding the fibers together. The context should make it clear which way the word is being used.

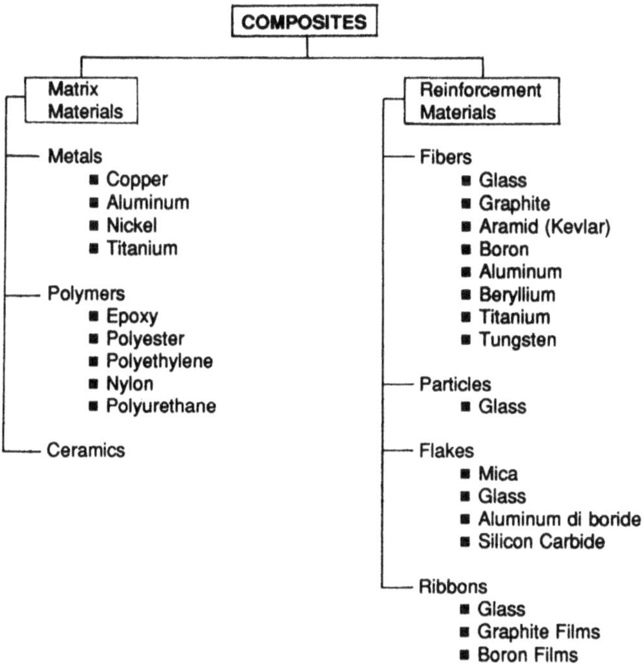


**FIGURE 7.1.** Forms of various fillers.

is a good starting point to explain several points about a composite material.

First, we note that the fibers cannot take loads by themselves. When held together by a matrix medium in the form of a composite ply, fibers take loads, not only in tension but also in compression. Furthermore, the presence of fibers in the matrix enhances the stiffness of the matrix to an extent. The composite ply resists deformation in the direction perpendicular to the fibers, better than the unreinforced matrix material. It is because of this behavior that the word "synergetic" was used.

Second, the UDL is the perfect example of *orthotropy* of moduli and strengths. By orthotropy we mean that the principal properties of the material are oriented along perpendicular directions. Obviously, the stiffness of a UDL is highest along the fibers and least across the fibers. So is its strength. None of the composite properties is the same as that of the constituents, be it stiffness or strength. If we cut tensile test specimens from the UDL at different angles as shown in Figure 7.3, each specimen would give different moduli and strengths.

However, for orthotropy-in-a-plane, the properties parallel and perpendicular to the fiber direction are sufficient to completely determine the

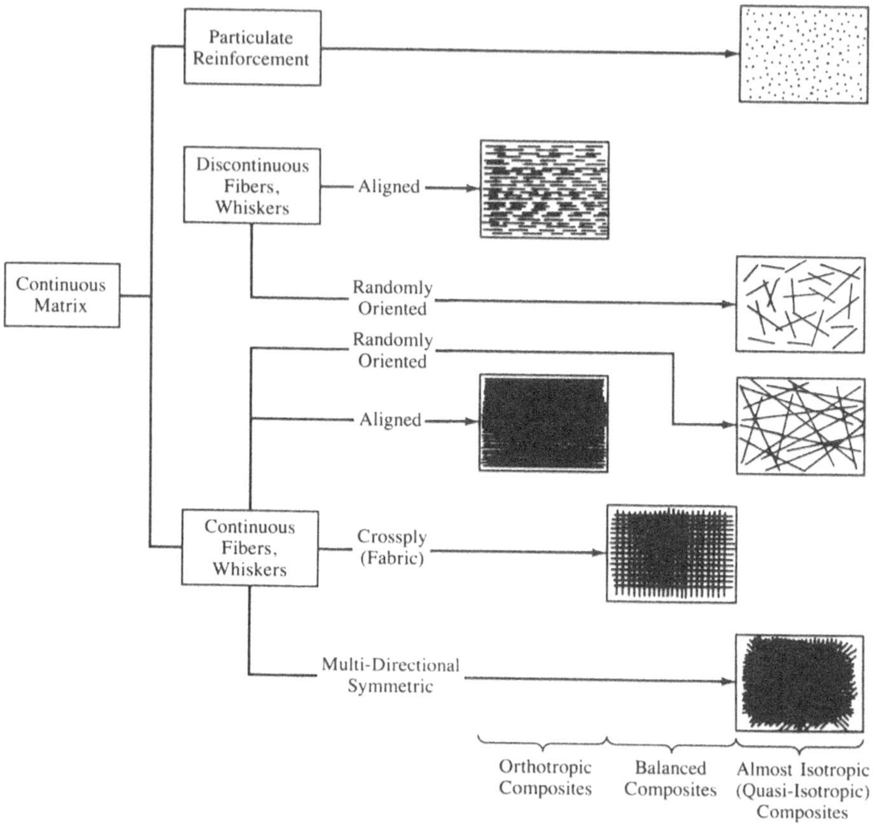


**FIGURE 7.2.** The options available in constructing composite materials.

properties along any other direction. The fiber direction is therefore significant for the composite as a reference coordinate system in the plane of the lamina. For this reason, the coordinate system 1-2 shown in Figure 7.3 is also called the *material coordinate system*.

Composites have high specific strength and specific stiffness compared to metals (see Table 7.1). In other words, the weight of material necessary to get a desired strength or stiffness is lower compared to the metals. This property results in lighter structures. Consequently, in designs where light weight is crucial, such as in aircraft and automobiles, composites are finding increasing applications.

Thus, composites bring together two (or more) materials that cannot otherwise perform structurally. The directionality of their modulus and strength can be used in designs to advantage. Consider the design of

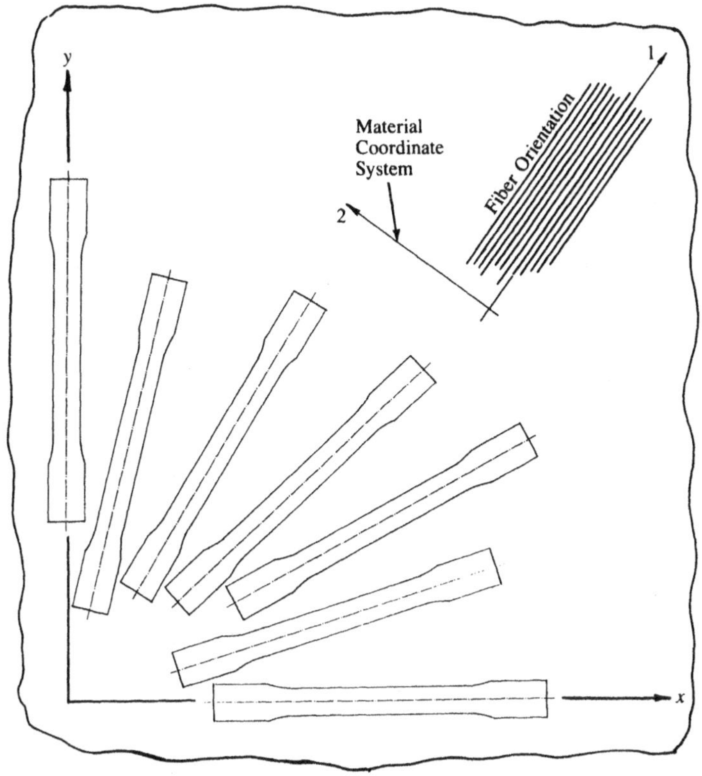


**FIGURE 7.3.** Specimens at various angles exhibit different properties.

metallic parts, in which the sizing is based on the highest stress, which applies only in a particular direction. In other directions, the stresses are lower, but the material has the same strength. Therefore, the material is under-utilized in those directions. Ideally, composites can be designed to have the right amount of stiffness and strength in the proper direction, thus putting the materials to their best use. Thus, composites permit the synthesis of material design with product design.

## 7.2 HOOKE'S LAW FOR ORTHOTROPY

It is useful at this stage to introduce Hooke's law for orthotropic materials. This is particularly helpful because we are going to discuss many different moduli of composites, which are distinguished only by subscripts. The sequence of subscripts is also important. For example, a symbol such as $\eta_{xs}$ is not the same as $\eta_{sx}$, but may be related to it.

**Table 7.1. Specific Strengths and Specific Moduli for Common Composites**

| No. | Material Name | Strength, $F$ (kg/mm$^2$) | Modulus, $E$ (kg/mm$^2$) | Density, $\rho$ (g/cc) | $\frac{F}{\rho}$ $10^6$ mm | $\frac{E}{\rho}$ $10^6$ mm |
|---|---|---|---|---|---|---|
| 1 | Carbon-Epoxy (T300/5208) | 150 | 15000 | 1.54 | 97.40 | 9740 |
| 2 | Carbon-Epoxy (t300/5208 HT) | 135 | 14000 | 1.50 | 90.00 | 9333 |
| 3 | Carbon-Epoxy (AS/3501) | 120 | 13800 | 145 | 82.76 | 9517 |
| 4 | Carbon-Epoxy (HMs/3002M) | 68 | 18500 | 1.60 | 42.50 | 11562 |
| 5 | Boron-Epoxy | 140 | 20400 | 2.00 | 70.00 | 10200 |
| 6 | Aramid-Kevlar 149 | 130 | 8800 | 1.40 | 92.86 | 6286 |
| 7 | S Glass-Epoxy | 130 | 4400 | 1.97 | 65.99 | 2234 |
| 8 | E Glass-Epoxy | 110 | 4000 | 1.97 | 55.84 | 2030 |
| 9 | Boron-Aluminum | 140 | 24000 | 2.50 | 56.00 | 9600 |
| 10 | Aluminum | 43 | 7300 | 2.70 | 15.93 | 2704 |
| 11 | Steel | 43 | 21100 | 7.80 | 5.51 | 2705 |

(Based on LAM-101, an in-house laminate analysis program of L.J. Broutman & Associates Ltd, Chicago.)

To help the reader imagine orthotropy, consider a ribbon-reinforced composite, which is known to have *full orthotropy*. The word means that principal properties are along three orthogonal directions. See Figure 7.4a. It is obvious that along each one of the axes 1, 2, and 3 in the figure the moduli are different due to the way the ribbons are arranged. Full orthotropy arises from the fact that each axis 1, 2, and 3 defines a characteristic direction of the material. They are distinguished by the material properties themselves, and hence cannot be used interchangeably. On the other hand, a bundle of glass fibers in polyester forming a rod has *plane orthotropy* and *transverse isotropy*. See Figure 7.4b. Transverse isotropy in this case means that the axes 2-3 in the transverse plane are interchangeable. Hence in the plane 2-3, directions are not distinguishable by material properties. Hooke's law for plane orthotropy is given below in the *material coordinate system*.

$$\begin{Bmatrix} \epsilon_1 \\ \epsilon_2 \\ \gamma_{12} \end{Bmatrix} = \begin{bmatrix} \dfrac{1}{E_{11}} & -\dfrac{\nu_{12}}{E_{11}} & 0 \\ -\dfrac{\nu_{21}}{E_{22}} & \dfrac{1}{E_{22}} & 0 \\ 0 & 0 & \dfrac{1}{G_{12}} \end{bmatrix} \begin{Bmatrix} \sigma_1 \\ \sigma_2 \\ \tau_{12} \end{Bmatrix}$$

$$= \begin{bmatrix} S_{11} & S_{12} & 0 \\ S_{21} & S_{22} & 0 \\ 0 & 0 & S_{66} \end{bmatrix} \begin{Bmatrix} \sigma_1 \\ \sigma_2 \\ \tau_{12} \end{Bmatrix} \tag{7.1}$$

Or simply,

$$\{\boldsymbol{\epsilon}\} = [\mathbf{S}_{12}]\{\boldsymbol{\sigma}\} \tag{7.2}$$

The matrix $[\mathbf{S}_{12}]$ is referred to as the *compliance matrix* of the ply, so called because it contains the reciprocals of Young's moduli, and hence represents the inverse of stiffness. Note the following about Equation 7.1.

- The zeros in the $[\mathbf{S}_{12}]$ matrix appear only when expressed in the material coordinate system. For other coordinate systems, they are filled by non-zero numbers.
- It can be shown that

$$S_{12} = S_{21}, \quad \text{or,} \quad \frac{\nu_{12}}{E_{11}} = \frac{\nu_{21}}{E_{22}} \tag{7.3}$$

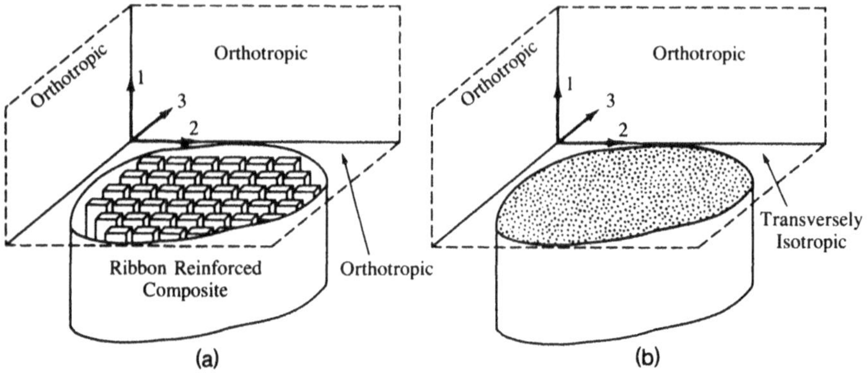


**FIGURE 7.4.** (a) Ribbon reinforced composite exhibits full orthotropy, and (b) fiber reinforced composite has plane orthotropy and transverse isotropy.

Thus, $\nu_{12}$ and $\nu_{21}$ are related to each other, so only one of them can be considered independent.

- When the ply is fully isotropic, $E_{11} = E_{22} = E$, and $\nu_{12} = \nu_{21} = \nu$; also $G_{12} = G = E/2(1 + \nu)$. This last relationship connecting $E$ and $G$ is not easy to visualize from Equation 7.1, but can be obtained from the kinematics of displacements.

The inverse of Equation 7.1 gives stress in terms of strain. Its form is

$$\begin{Bmatrix} \sigma_1 \\ \sigma_2 \\ \tau_{12} \end{Bmatrix} = \begin{bmatrix} Q_{11} & Q_{12} & 0 \\ Q_{21} & Q_{22} & 0 \\ 0 & 0 & Q_{66} \end{bmatrix} \begin{Bmatrix} \epsilon_1 \\ \epsilon_2 \\ \gamma_{12} \end{Bmatrix} \tag{7.4}$$

where,

$$Q_{11} = \frac{E_{11}}{1 - \nu_{12}\nu_{21}}$$

$$Q_{22} = \frac{E_{22}}{1 - \nu_{12}\nu_{21}}$$

$$Q_{12} = \frac{\nu_{12}E_{22}}{1 - \nu_{12}\nu_{21}} = \frac{\nu_{21}E_{11}}{1 - \nu_{12}\nu_{21}}$$

$$Q_{66} = G_{12} \tag{7.5}$$

Equation 7.4 can be written in short form as

$$\{\boldsymbol{\sigma}\} = [\mathbf{Q}_{12}]\{\boldsymbol{\epsilon}\} \tag{7.6}$$

where $[\mathbf{Q}_{12}]$ is called the *stiffness matrix* or *matrix of mathematical moduli.* The word stiffness is to be understood in this chapter in a limited sense. It is just another word for mathematical moduli, and has no connotations of structural stiffness as used in the context of finite element technique or strength of materials. The above equations are important for calculating the elastic properties of laminates.

The following examples illustrate the use of orthotropic Hooke's law.

**Example 7.1**

A plane-orthotropic composite lamina (see Figure 7.5) is known to have the following elastic properties.

$$E_{11} = 8800 \text{ kg/mm}^2$$

$$E_{22} = 560 \text{ kg/mm}^2$$

$$G_{12} = 220 \text{ kgm/mm}^2$$

$$\nu_{12} = 0.34$$

It is subjected to a biaxial stress of $\sigma_1 = 5 \text{ kg/mm}^2$, $\sigma_2 = 1 \text{ kg/mm}^2$, and $\tau_{12} = 1 \text{ kg/mm}^2$. Find $\nu_{21}$ and the in-plane strain components.

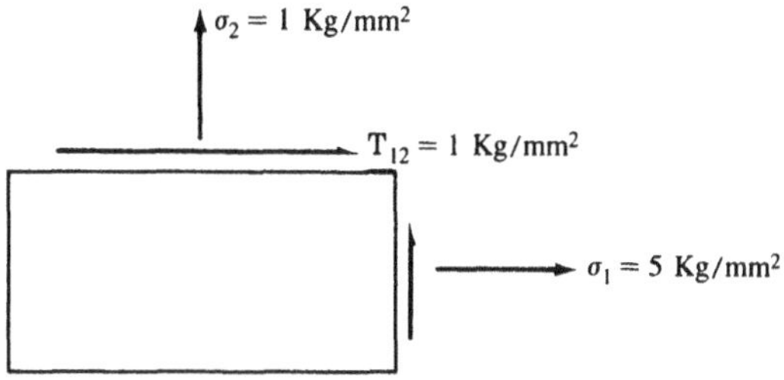


**FIGURE 7.5.** Example 7.1.

*Solution*

$$\nu_{21} = \nu_{12}\frac{E_{22}}{E_{11}}, \text{ from Equation 7.3}$$

$$= 0.34\frac{560}{8800} = 0.0216$$

$$\epsilon_1 = \frac{\sigma_1}{E_{11}} - \nu_{12}\frac{\sigma_2}{E_{11}}$$

$$= \frac{5}{8800} - 0.34\frac{1}{8800} = 0.5295 \times 10^{-3} \text{ kg/mm}^2$$

$$\epsilon_2 = -\frac{\nu_{21}}{E_{22}} + \frac{\sigma_2}{E_{22}}$$

$$= -\frac{0.0216}{560} \times 5 + \frac{1}{560} = 1.592 \times 10^{-3} \text{ kg/mm}^2$$

$$\gamma_{12} = \frac{\tau_{12}}{G_{12}}$$

$$= \frac{1}{220} = 4.545 \times 10^{-3} \text{ kg/mm}^2$$

**Example 7.2**

A unidirectional composite laminate (see Figure 7.6) is made of graphite fibers in epoxy matrix. The thickness of the laminate is 0.1 mm. The elastic

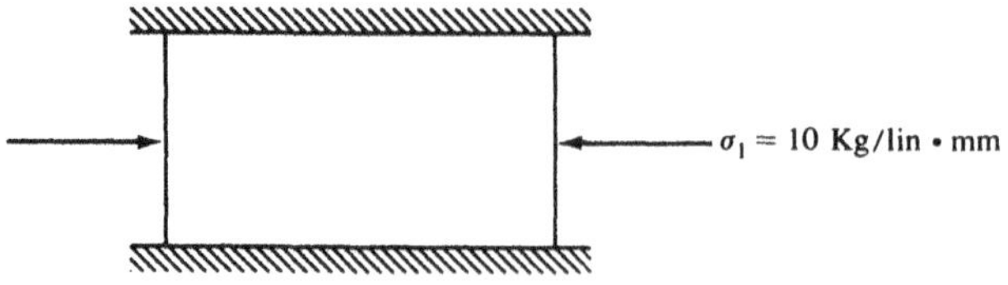


**FIGURE 7.6.** Example 7.2.

moduli of the composite are

$$E_{11} = 15000 \text{ kg/mm}^2$$

$$E_{22} = 1000 \text{ kg/mm}^2$$

$$G_{12} = 600 \text{ kg/mm}^2$$

$$\nu_{12} = 0.27$$

It is kept under a compressive load of 10 kg per lineal mm. Its sides are held from expanding by means of a pair of rigid guides. Find $\sigma_2$, $\epsilon_2$ and $\nu_{21}$.

*Solution*

$$\nu_{21} = \nu_{12}\frac{E_{22}}{E_{11}} = 0.27 \times \frac{1000}{15000} = 0.018$$

$$\sigma_1 = (\text{Load/Area}) = \frac{10}{0.1} = 100 \text{ kg/mm}^2$$

$$\epsilon_2 = -\nu_{21}\frac{\sigma_1}{E_{22}} + \frac{\sigma_2}{E_{22}} \text{ from Eq. 7.1}$$

But $\epsilon_2 = 0$, because of the rigid restraint.

Hence,

$$\sigma_2 = \nu_{21}\sigma_1$$

$$= 0.018 \times 100 = 1.8 \text{ kg/mm}^2$$

$$\epsilon_1 = \frac{\sigma_1}{E_{11}} - \nu_{12}\frac{\sigma_2}{E_{11}} \text{ from Eq. 7.1}$$

$$= \frac{100}{15000} - 0.27\frac{1.8}{15000}$$

$$= 0.00665 = 6.65 \times 10^{-3} \text{ mm/mm}$$

**Example 7.3**

For the graphite-epoxy composite in previous example, which of the following conditions requires a higher restraining force?

1. Apply a stress of 1 kg/mm$^2$ along the fiber in compression and hold the lamina from expanding across the fiber,
2. Apply a stress of 1 kg/mm$^2$ across the fiber in compression and hold the lamina from expanding along the fiber.

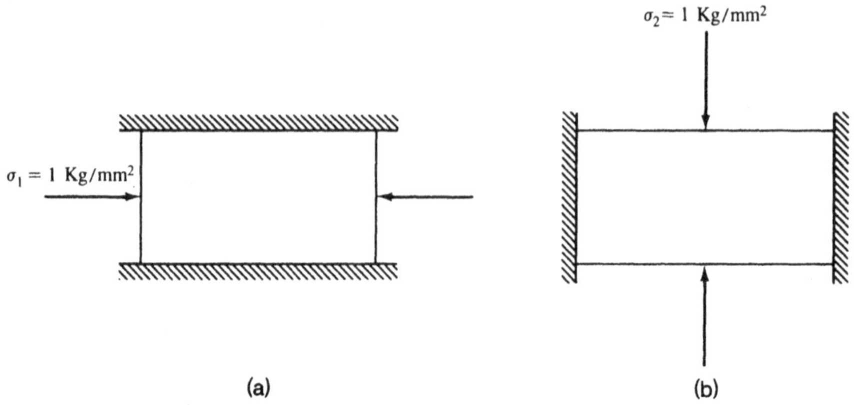


**FIGURE 7.7.** Example 7.3.

*Solution*

*Case #1:* $\epsilon_2 = 0$

Hence,

$$\sigma_2 = \nu_{21}\sigma_1$$
$$= \nu_{21} \times 1$$
$$= 0.018 \text{ kg/mm}^2$$

*Case #2:* $\epsilon_1 = 0$

Hence,

$$\sigma_1 = \nu_{12}\sigma_2$$
$$= \nu_{12} \times 1$$
$$= 0.27 \text{ kg/mm}^2$$

Obviously, restraining along fibers requires more force.

**Example 7.4**

A cylindrical pressure vessel is made using a filament winding process. Therefore, the vessel material has the following moduli.

$$E_{\text{hoop}} = 1400 \text{ kg/mm}^2; \quad E_{\text{axial}} = 700 \text{ kg/mm}^2$$

The major Poisson's ratio $\nu_{ha} = 0.28$. In order to pressure test the vessel, a test-rig is constructed which consists of a coaxial steel pipe of a smaller size as shown in Figure 7.8. The rationale is to minimize the material needed for withstanding the pressure on the ends. The modulus of steel $E_s$

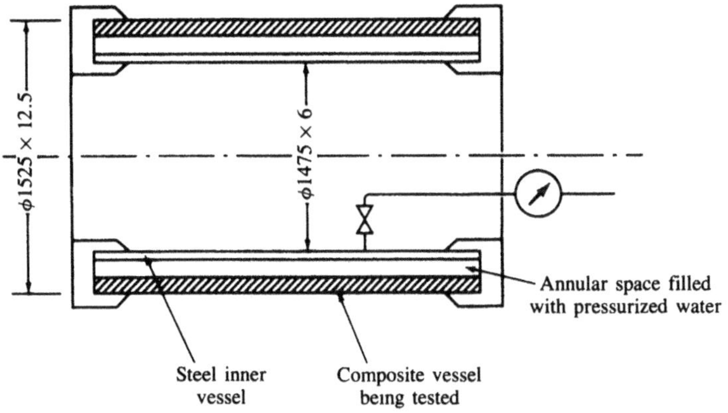


**FIGURE 7.8.** Example 7.4.

is 21100 kg/mm$^2$, and Poisson's ratio, $\nu_s$ is 0.3. Calculate the axial stresses in the composite and steel pipe, for an applied test pressure of 10 kg/cm$^2$.

*Solution* Hoop tensile stress in the composite pipe,

$$\sigma_{h,1} = \frac{pR_{i1}}{2t_1} = \frac{0.1 \times 1500}{2 \times 12.5} = 6.0 \text{ kg/mm}^2$$

Hoop compressive stress in the steel pipe,

$$\sigma_{h,2} = \frac{pR_{o2}}{2t_2} = -\frac{0.1 \times 1475}{2 \times 6} = 12.30 \text{ kg/mm}^2$$

Axial stresses in the pipes are to be obtained by using the condition that they are rigidly held together at the ends. Hence they have equal axial elongation. That is,

$$\epsilon_{a,1} = \epsilon_{a,2}$$

Also,

$$\nu_{21} = \nu_{12}\frac{E_{22}}{E_{11}} = 0.14$$

By Hooke's laws,

$$\epsilon_{a,1} = \frac{\sigma_{a,1}}{E_{22}} - \nu_{21}\frac{\sigma_{h,1}}{E_{22}} \tag{7.7}$$

$$\epsilon_{a,2} = \frac{\sigma_{a,2}}{E_s} - \nu_s\frac{\sigma_{h,2}}{E_s} \tag{7.8}$$

Putting together the condition of rigid-jointed ends,

$$(\sigma_{a,1} - 0.14 \times 6.0) = \frac{E_{22}}{E_s}(\sigma_{a,2} + 0.3 \times 12.3)$$

$$\begin{aligned} \sigma_{a,1} - 0.84 &= 0.0332(\sigma_{a,2} + 3.69) \\ &= 0.0332\sigma_{a,2} + 0.1225 \end{aligned}$$

$$\sigma_{a,1} - 0.0332\sigma_{a,2} = 0.9625$$

For equilibrium, we have the condition that the axial stresses acting over the respective pipe cross sectional areas must add up to the end force due to the pressure in the annular area.

$$\sigma_{a,1}(\pi \times 1512.5 \times 12.5) + \sigma_{a,2}(\pi \times 1469 \times 6) = p.\frac{\pi}{4}(1500^2 - 1475^2)$$

That is, $$\sigma_{a,1} + 0.4662\sigma_{a,2} = 0.09835$$

Solving for $\sigma_{a,1}$ and $\sigma_{a,2}$,

$$\sigma_{a,1} = 0.9050 \text{ kg/mm}^2$$

$$\sigma_{a,2} = -1.730 \text{ kg/mm}^2$$

Although the axial load on the ends tends to push the pipe ends apart, the steel pipes actually goes into compression. This is another peculiarity of the orthotropy of the composite pipe material.

## 7.3 MICROMECHANICS—MODULI OF COMPOSITES

*Micromechanics* is the set of techniques for *predicting* the composite properties as functions of the constituent properties. This section lists certain formulas for predicting properties of different types of composites, rather than describing the actual techniques of micromechanics. Methods of micromechanics involve extensive experimentation, and curve-fitting on the one hand and analytical models, bounding techniques, and also computational models on the other. Reference 3 outlines briefly the analytical modeling methods adopted in the micromechanics of composites of various filler forms.

In composites, the reinforcing material (most commonly fibers) takes up almost all the applied load. However, in reality, it is not possible to apply the loads directly on the reinforcements. Loads have to be applied primarily through the matrix material. The matrix transfers the load to the reinforcements. The overall response of composite to applied loads is influenced by the manner in which such microscopic load transfers take place. Thus, the strengths and stiffnesses of the constituent materials influence those of the composite. In the microscopic scale in composites, the load transfer and the failure mechanisms involve more factors than just the apparent properties of the constituents. In general, the factors that influence the composite properties are,

- properties of the reinforcement,
- properties of the matrix,
- the volume fraction of the reinforcement,
- the geometry of packing, whether aligned, or random, etc.,
- the cross sectional shape of the reinforcement,
- the aspect ratio of the reinforcement, and
- properties of the interface layer bonding the reinforcement with the matrix.

A few of the above factors are extremely difficult to study experimentally. For example, a true random arrangement of fibers in two or three dimensions is nearly impossible to achieve by any fabrication technique, in the laboratory or in a production shop. So is the study of the interface bonding layer. Therefore, the properties which are influenced strongly by such factors cannot be predicted accurately by micromechanics.

A composite ply can have properties that are not easy to expect on the basis of the properties of the constituents. For example, they have orthotropy of strength and modulus, although the constituents are isotropic. Also composites have apparent thermal expansion coefficients $\alpha_{**}$, which are functions of the constituent $\alpha$'s *as well as the constituent moduli*. (See Section 7.17)

All composite forms of practical use fall into two categories with regard to mechanical properties. They can be (i) fully isotropic, or (ii) orthotropic in a plane with transverse isotropy. (In other words, *composites are never fully anisotropic*.) Accordingly, the elastic constants that apply for the orthotropic composites are,

- $E_{11}$, $E_{22}$, $G_{12}$, and $\nu_{12}$ in the plane of reinforcement, (plane 1-2 or 1-3 in Figure 7.4a) and
- $E_{33}$, $G_{13}$, $G_{23}$, $\nu_{13}$, and $\nu_{23}$ in the plane transverse to the reinforcements, (plane 2-3 in Figure 7.4a).

However, the number of independent elastic constants are not nine as listed above, but only five. Referring to Figure 7.4b, we see that in the plane 2-3, there is complete isotropy. This means that the subscripts 2 and 3 can be interchanged without any effect. That leads to the following relations.

$$E_{33} = E_{22} \tag{7.9}$$

$$G_{12} = G_{13} \tag{7.10}$$

$$\nu_{12} = \nu_{13} \tag{7.11}$$

$$G_{23} = \frac{E_{22}(\text{or } E_{33})}{2(1 + \nu_{23})} \tag{7.12}$$

Therefore, only five of the constants are independent, namely $E_{11}$, $E_{22}$, $G_{12}$, $\nu_{12}$, and $\nu_{23}$.

### 7.3.1 Fiber Reinforcement, Long and Short, Aligned

Several in-plane properties of continuous and long discontinuous aligned fiber composites can be obtained from a single formula, popularly known as the Halpin-Tsai formula. This formula is presented below as modified by Nielsen. It can be written in a general form as [22]

$$\frac{(\cdot)_c}{(\cdot)_m} = \frac{1 + \xi\eta V_f}{1 - \psi\eta V_f}. \tag{7.13}$$

where,

$\xi$ = A measure of reinforcement as given in Table 7.2

$\psi$ = A factor that accounts for the maximum packing possible.

$$= 1 + \left(\frac{1 - V_{f,\max}}{V_{f,\max}^2}\right) Vf, \text{ (see Table 7.3)}$$

$$\eta = \frac{r - 1}{r + \xi}$$

$$r = \frac{(\cdot)_f}{(\cdot)_m}$$

**Table 7.2. Parameter $\xi$ for the Halpin-Tsai Equations. [2, 5, 20, 22]**

| | $E_{11}$ | $E_{22}$ | $\nu_{12}$ | $G_{12}$ |
|---|---|---|---|---|
| Continuous Fibers | $\infty$ | Hexagonal Array, 1<br>Square Array, 2 | $\infty$ | 1 |
| Chopped Fibers | $\frac{2l}{d}$ | Hexagonal Array, 1<br>Square Array, 2 | $\infty$ | 1 |

$V_{f,\max}$ is simply the physical maximum packing possible, and is obtained from the geometry of packing for the condition when the fillers touch each other.

The following points are worth noting about the Halpin-Tsai formulas.

1. In its original form Equation 7.13 did not have the term $\psi$ in it. In other words, Halpin and Tsai used $\psi = 1$. From its definition above, we see that $\psi > 1$, and hence Nielsen's formulas predict composite properties somewhat higher than do the Halpin-Tsai formulas. The difference is negligible at low $V_f$, and considerable at high $V_f$.
2. The properties of the composite and the constituents $(\cdot)_c$, $(\cdot)_f$ and $(\cdot)_m$ are intentionally left unspecified. In fact it is possible to have the symbol mean any of the quantities, $E_{11}$, $E_{22}$, $G_{12}$, or $\nu_{12}$.
3. When $\xi \to \infty$, the Halpin-Tsai formulas lead to a form popularly called the rule of mixtures. That is,

$$(\cdot)_c = V_f(\cdot)_f + (1 - V_f)(\cdot)_m$$

**Table 7.3. Maximum Packing Fraction, $V_{f,max}$**

| Particles | Type of Packing | $V_{f,max}$ |
|---|---|---|
| Spheres | Hexagonal Close Packing | 0.7405 |
| Spheres | Face Centered Cubic | 0.7405 |
| Spheres | Body Centered Cubic | 0.60 |
| Spheres | Simple Cubic | 0.5236 |
| Spheres | Random Close Packing | 0.637 |
| Spheres | Random Loose Packing | 0.601 |
| Fibers | Parallel Hexagonal Packing | 0.907 |
| Fibers | Parallel Cubic Packing | 0.82 |
| Fibers | Random Orientation | 0.52 |

4. When $\xi \to 0$, the Halpin-Tsai formulas lead to the inverse rule of mixtures. The equation takes the form

$$\frac{1}{(\cdot)_c} = \frac{V_f}{(\cdot)_f} + \frac{(1 - V_f)}{(\cdot)_m}$$

5. The parameter $\xi$ is obtained experimentally and by curve-fitting techniques. So the values of $\xi$ in Table 7.2 are valid strictly for the cases indicated.
6. Orthotropic constituent properties are also covered by the Halpin-Tsai formulas. For example, graphite fibers are known to be orthotropic. The formulas will have the corresponding orthotropic properties for the constituents.

Examples below illustrate the use of Halpin-Tsai formulas.

**Example 7.5**

A unidirectional composite is made of long continuous glass fibers embedded in a matrix that has a modulus 0.008 times that of the fiber. Assume that both the fiber and matrix have the same Poisson's ratio, and that they are both isotropic. Take $\psi = 1$.

1. What is the expected ratio of $(G_{12}/G_m)$ as a function of $V_f$?
2. What is the expected ratio of $(E_{22}/E_m)$ as a function of $V_f$?
3. What is the expected ratio of $(E_{11}/E_m)$ as a function of $V_f$?

*Solution* Since the fiber and matrix are both isotropic, and also have the same Poisson's ratio, we have

$$r = \frac{G_f}{G_m} = \frac{E_f}{E_m} = \frac{1}{0.008} = 125$$

**Case of $G_{12}/G_m$:**
From Table 7.2, we find that $\xi = 1$. Also $\psi = 1$.

$$\eta = \frac{r - 1}{r + \xi} = \frac{125 - 1}{125 + 1} = 0.984$$

Substituting in the Halpin-Tsai equation,

$$\frac{G_{12}}{G_m} = \left(\frac{1 + 0.984V_f}{1 - 0.984V_f}\right)$$

**Case of $E_{22}/E_m$:**
Two cases arise with regard to the packing arrangement, hexagonal and square. Let us do the calculations for each of them. First, let us consider the square arrangement, with $\xi = 2$.

$$\eta = \frac{r - 1}{r + \xi} = \frac{125 - 1}{125 + 2} = 0.976.$$

$$\frac{E_{22}}{E_m} = \frac{1 + 2 \times 0.976V_f}{1 - 0.976V_f} = \frac{1 + 1.952V_f}{1 - 0.976V_f}$$

In the case of hexagonal arrangement, $\xi = 1$, so $\eta = 0.984$. Substituting,

$$\frac{E_{22}}{E_m} = \frac{1 + 0.984V_f}{1 - 0.984V_f}$$

**Case of $E_{11}/E_m$:**
The factor $\xi \to \infty$, and hence the Halpin-Tsai equation reduces to simply the rule of mixtures.

$$\frac{E_{11}}{E_m} = V_f\frac{E_f}{E_m} + (1 - V_f) = 124V_f + 1$$

All the ratios are plotted in Figure 7.9. It is seen that $E_{11}$ and $G_{12}$ are not affected by the packing arrangement, but the $E_{22}$ is considerably affected by the packing arrangement.

**Example 7.6**

Repeat the previous problem with the modified Halpin-Tsai equations, taking $V_{f,\max} = 0.82$.

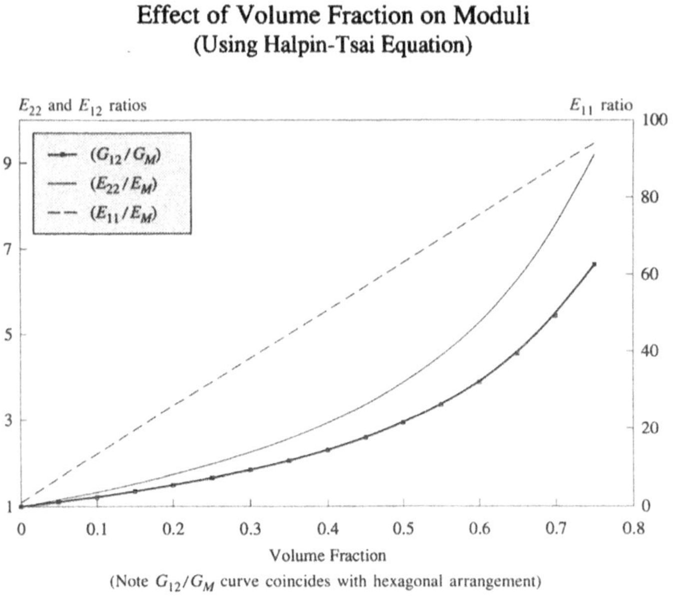


**FIGURE 7.9.** Example 7.5: Moduli of composites expressed as a multiple of the matrix modulus.

*Solution*

Since

$$V_{f,\max} = 0.82, \text{ we get } \psi = 1 + \left(\frac{1-0.82}{0.82^2}\right)V_f = (1 + 0.268V_f).$$

**Case of $G_{12}/G_m$:**

As in the previous example, $\eta = 0.984$. Substituting in the Halpin-Tsai equation,

$$\frac{G_{12}}{G_m} = \left(\frac{1 + 0.984V_f}{1 - (1 + 0.268V_f)0.984V_f}\right) = \left(\frac{1 + 0.984V_f}{1 - 0.984V_f - 0.264V_f^2}\right)$$

**Case of $E_{22}/E_m$:**

Consider $\xi = 2$ (Square packing). As before,

$$\eta = 0.976.$$

$$\frac{E_{22}}{E_m} = \frac{1 + 2 \times 0.976V_f}{1 - (1 + 0.268V_f)0.976V_f} = \frac{1 + 1.952V_f}{1 - 0.976V_f - 0.262V_f^2}$$

For the hexagonal arrangement, $\xi = 1$, so $\eta = 0.984$. Substituting,

$$\frac{E_{22}}{E_m} = \frac{1 + 0.984V_f}{1 - 0.984V_f - 0.264V_f^2}$$

A plot of the above two ratios is shown in Figure 7.10, with the results of the previous example superimposed.

**Case of $E_{11}/E_m$:**

In this case, the modified Halpin-Tsai equation reduces to the rule of mixtures, and we obtain

$$\frac{E_{11}}{E_m} = 124V_f + 1$$

In the transverse plane 2-3, we need only one property $\nu_{23}$. It is found using a formula proposed by Hashin [20]. The formula applies for aligned fiber composites, both continuous and long chopped. Many problems in

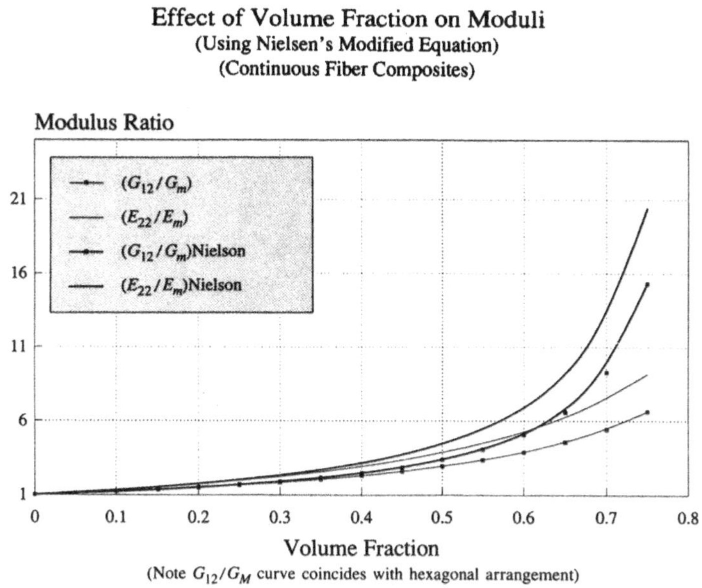


**FIGURE 7.10.** Comparison of Halpin-Tsai equations with Nielsen's modification thereto. Note that the $G_{12}$/Gm curve coincides with $E_{22}$/EM for the hexagonal arrangement.

composites make use of conventional thin shell theory, in which the properties of the composite in the transverse plane are not used.

Let $K_{2f} = \begin{cases} \text{Plane strain bulk modulus of (orthotropic) fiber} \\ \text{in the plane normal to the fiber axis.} \end{cases}$

$K_{2m} = \begin{cases} \text{Plane strain bulk modulus of (orthotropic) matrix} \\ \text{in the plane normal to the fiber axis. (Note that the} \\ \text{matrix is assumed aligned parallel to the fiber.)} \end{cases}$

$G_{23m} = \begin{cases} \text{Shear modulus of matrix in the plane} \\ \text{normal to the fiber.} \end{cases}$

$$K_2 = \frac{(K_{2f} + G_{23m})K_{2m} + (K_{2f} - K_{2m})G_{23m}V_f}{(K_{2f} + G_{23m}) - (K_{2f} - K_{2m})V_f} \tag{7.14}$$

$$\nu_{23} = \frac{2E_{11}K_2 - E_{11}E_{22} - 4\nu_{12}^2 K_2 E_{22}}{2E_{11}K_2} \tag{7.15}$$

The other transverse properties are calculated using equations of isotropy.

### 7.3.2 Random Orientation of Fibers

Fibers may be embedded in the matrix in random orientations. For *continuous fibers* oriented in random directions, the in-plane properties are isotropic. It is only necessary to determine the $E_c$ and $G_c$ for such a composite. There are two different approaches to this problem, giving results that are close to each other (but not quite equal).

The first approach for random orientation in two dimensions is due to Tsai [2, 16, 22]. In two dimensions,

$$E_c = \frac{3}{8}E_{11} + \frac{5}{8}E_{22} \tag{7.16}$$

$$G_c = \frac{1}{8}E_{11} + \frac{1}{4}E_{22} \tag{7.17}$$

where $E_{11}$ and $E_{22}$ are the moduli along and across the fibers in a unidirectional composite of the same volume fraction of fibers.

The second approach is due to Christensen, [16] and is based on the transformation of moduli of an unidirectional composite for a rotation of $\theta$. The transformed moduli are obtained for this direction. The angle is then varied continuously and the values of the moduli are averaged on $\theta$ in the range of 0 to 360 degrees. This approach is equivalent to assuming an equal probability of $\theta$ in a random composite. This formula is more complicated only in its appearance.

$$E_c = \left[\frac{E_{11} + E_{22} + 2\nu_{12}E_{22}}{1 - \nu_{12}\nu_{21}}\right] \times \left[\frac{E_{11} + E_{22} - 2\nu_{12}E_{22} + 4(1 - \nu_{12}\nu_{21})G_{12}}{3(E_{11} + E_{22}) + 2\nu_{12}E_{22} + 4(1 - \nu_{12}\nu_{21})G_{12}}\right] \quad (7.18)$$

$$G_c = \left[\frac{E_{11} + E_{22} - 2\nu_{12}E_{22}}{8(1 - \nu_{12}\nu_{21})}\right] + \frac{1}{2}G_{12} \quad (7.19)$$

### 7.3.3 Randomly Oriented Chopped Fibers

Let, $E_f$ = Fiber tensile modulus,
$d_f$ = Diameter of the fiber,
$\epsilon_{y,m}$ = Yield strain of the matrix,
$\tau_{y,m}$ = Shear strength of the matrix,
$l$ = Actual length of the fibers.

It was mentioned that the applied load is transferred from the matrix to the fibers. This takes place over a certain length of the fiber. See Figure 7.11. The length of the fiber which is just enough for the complete transfer of load is called the *critical length*, $l_c$. The critical length is such that the load-transfer would take place without causing matrix failure or fiber break.

$$l_c = d_f \frac{E_f \epsilon_m}{2\tau_{y,m}}$$

When $l < l_c$ we have a *short fiber* and a *long fiber* when $l > l_c$. For random orientation of long fibers in three dimensions,

$$E_c = \frac{1}{6}V_f E_f \left(1 - \frac{l_c}{2l}\right) + (1 - V_f)E_m \quad (7.20)$$

$$G_c = \frac{1}{16}V_f E_f \left(1 - \frac{l_c}{2l}\right) + (1 - V_f)E_m \quad (7.21)$$

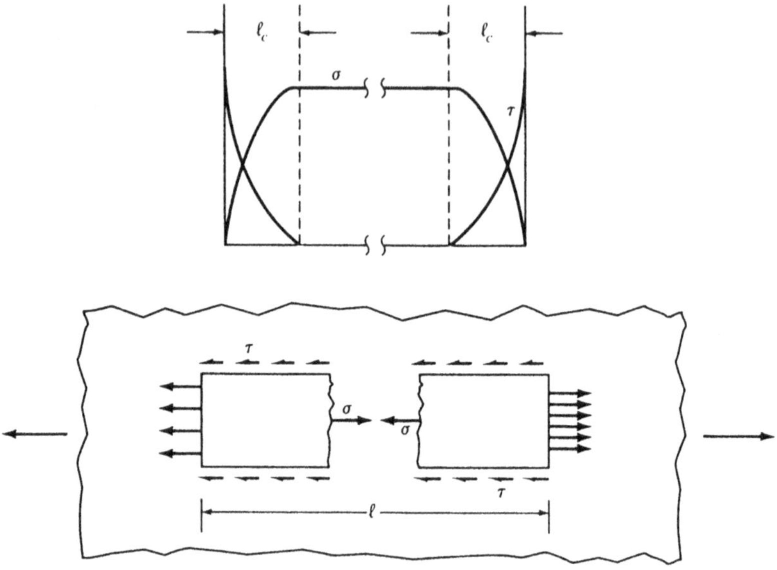


**FIGURE 7.11.** Simple model for load transfer mechanism from matrix to a short fiber and vice versa. Near the ends of the fiber, the load transfer is in shear mode. A small, negligible amount is transferred through the end faces of the fiber. $l_c$ is the critical length needed to transform all shear stress into all fiber tensile stress.

For random orientation of short fibers in three dimensions,

$$E_c = \frac{1}{6} V_f E_f \frac{l}{2l_c} + (1 - V_f) E_m \tag{7.22}$$

$$G_c = \frac{1}{16} V_f E_f \frac{l}{2l_c} + (1 - V_f) E_m \tag{7.23}$$

For random orientation of long fibers in two dimensions,

$$E_c = \frac{1}{3} V_f E_f \left(1 - \frac{l_c}{2l}\right) + (1 - V_f) E_m \tag{7.24}$$

$$G_c = \frac{1}{8} V_f E_f \left(1 - \frac{l_c}{2l}\right) + (1 - V_f) E_m \tag{7.25}$$

For random orientation of short fibers in two dimensions,

$$E_c = \frac{1}{3} V_f E_f \frac{l}{2l_c} + (1 - V_f) E_m \tag{7.26}$$

$$G_c = \frac{1}{8} V_f E_f \frac{l}{2l_c} + (1 - V_f) E_m \tag{7.27}$$

Other elastic constants of the composite can be obtained by using the property of isotropy.

### 7.3.4 Particle Filled Composites

Particle fillers in polymers are generally spherical in shape. Such fillers result in fully isotropic composites. Therefore. only two properties are needed. The modified Halpin-Tsai equations are also applicable in this case.

$$\frac{(\cdot)_c}{(\cdot)_m} = \frac{1 + \xi \eta V_f}{1 - \psi \eta V_f}. \tag{7.28}$$

where
$$\xi = \frac{7 - \nu_m}{8 - 10\nu_m}$$

$$\psi = 1$$

$$\eta = \frac{r - 1}{r + \xi}$$

$$r = \frac{(\cdot)_f}{(\cdot)_m}$$

As before, the symbol $(\cdot)$ represents the appropriate property of the matrix, filler, and the composite. We can use this for bulk modulus and Young's modulus, as well as shear modulus. Therefore, one may find $E_c$ and $G_c$ using the Halpin-Tsai equation and compute the value of $\nu_c$ from that. Of course, one can calculate $\nu_c$ starting with different pairs of properties. The value of $\nu_c$ obtained by each procedure is bound to be different from that obtained by another, by a small amount.

The example below illustrates the prediction of isotropic particle filled composites.

**Example 7.7**

A component uses polyester reinforced by small spherical glass beads. The polyester has a modulus of 0.31 GPa, and the filler material has a modulus of 72 GPa. Their Poisson's ratios are 0.35 and 0.28, respectively. Estimate the modulus and Poisson's ratio that can be achieved in the reinforced material for a practical range of volume fractions.

*Solution*

$$\text{Polyester: } E_m = 0.31 \text{ GPa};\ \nu_m = 0.35$$

$$\text{Glass: } E_f = 72 \text{ GPa};\ \nu_f = 0.28$$

Using the Halpin-Tsai formulas for modulus (Equations 7.28) we get,

$$r = \frac{E_f}{E_m} = \frac{72}{0.31} = 231.9$$

$$\xi = \frac{(7 - \nu_m)}{(8 - 10\nu_m)} = \frac{6.65}{4.5} = 1.478$$

$$\eta = \frac{(r - 1)}{(r + \xi)} = \frac{230.9}{232.4} = 0.9935$$

$$\psi = 1$$

$$\frac{E_c}{E_m} = \frac{1 + \xi\eta V_f}{1 - \psi\eta V_f}$$

$$= \frac{[1 + (1.478)(0.9935)V_f]}{(1 - 1 \times 0.9935V_f)} = \frac{(1 + 1.4684V_f)}{(1 - 0.9935V_f)}$$

The relationship is calculated for various values of $V_f$, from 0 to 0.8.

Using the Halpin-Tsai formulas for Poisson's ratio (Equations 7.28) we have,

$$r = \frac{\nu_f}{\nu_m} = \frac{0.28}{0.35} = 0.8$$

$\xi = 1.478$, as before.

$$\eta = \frac{(r-1)}{(r+\xi)} = \frac{(0.8-1)}{(0.8+1.478)} = -0.088$$

$$\psi = 1$$

$$\frac{\nu_c}{\nu_m} = \frac{1-\xi\eta V_f}{1-\psi\eta V_f}$$

$$= \frac{[1+(1.478)(-0.088)V_f]}{[1-(0.088)V_f]} = \frac{1-0.13V_f}{1+0.088V_f}$$

We may tabulate the results as a function of $V_f$.

| $V_f$ | $\frac{\nu_c}{\nu_m}$ | $\nu_c$ | $\frac{E_c}{E_m}$ | $E_c$ (GPa) |
|---|---|---|---|---|
| 0 | 1 | 0.35 | 1 | 0.31 |
| 0.1 | 0.978 | 0.274 | 1.273 | 0.395 |
| 0.2 | 0.957 | 0.268 | 1.615 | 0.500 |
| 0.3 | 0.936 | 0.262 | 2.052 | 0.636 |
| 0.4 | 0.916 | 0.256 | 2.634 | 0.817 |
| 0.5 | 0.896 | 0.251 | 3.446 | 1.068 |
| 0.6 | 0.876 | 0.245 | 4.657 | 1.444 |
| 0.7 | 0.856 | 0.240 | 6.659 | 2.064 |
| 0.8 | 0.837 | 0.232 | 10.598 | 3.285 |

The above values are also plotted in Figure 7.12.

***Commentary***

*Note that the actual size of the particle does not figure in the calculation of the modulus and Poisson's ratio. Therefore, it may seem that any size of the particle would result in the same* $E_c$ *and* $\nu_c$ *provided the volume fraction* $V_f$ *is*

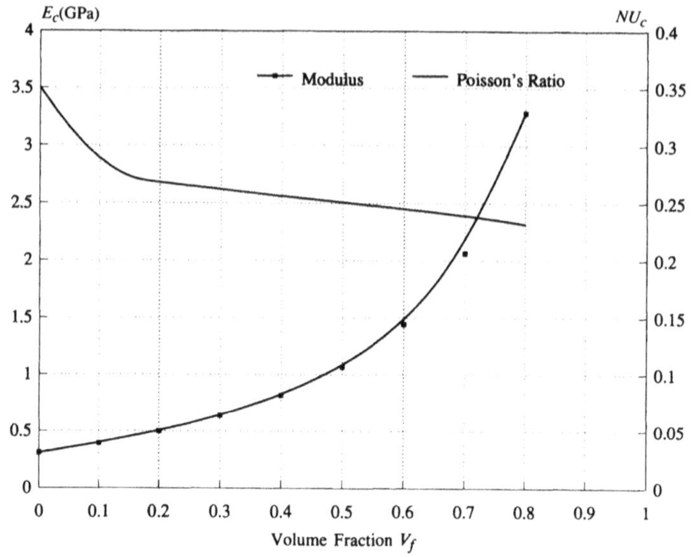


**FIGURE 7.12.** Variation of $E_c$ and $G_c$ of a particle filled composite, with volume fraction, $V_f$.

*the same. By and large this is true subject to the condition that the modulus is observed on a scale that is very large compared to the particle size.*

### 7.3.5 Ribbon Reinforced Composites (Aligned)

Ribbons are reinforcements whose cross sectional shape has an aspect ratio. Although the most common shape is a rectangle, a long elliptical shape is also categorized as a ribbon. Referring to Figure 7.4a we see that a ribbon-reinforced composite is *fully orthotropic*. With no plane of isotropy, full orthotropy requires nine elastic constants, namely

$$\text{Young's moduli,} \quad E_{11}, E_{22}, E_{33},$$

$$\text{Shear moduli,} \quad G_{12}, G_{23}, G_{31}$$

$$\text{Poisson's ratios,} \quad \nu_{12}, \nu_{23}, \nu_{31}$$

Unfortunately, there is very little information available for the three

**Table 7.4. Values of $\xi$ for Ribbon Composites.**

| To Compute | Use this $\xi$ |
|---|---|
| $E_{11}$ | $\infty$ |
| $E_{22}$ | $\frac{2W}{t}$ |
| $E_{33}$ | 0 |
| $G_{12}$ | $\infty$ |
| $G_{23}$ | 0 |
| $G_{31}$ | $\left(\frac{W}{t}\right)^{\sqrt{3}}$ |

Poisson's ratios. The Young's and shear moduli can be determined from the Halpin-Tsai equations, with $\xi$ determined from Table 7.4.

$$\frac{(\cdot)_c}{(\cdot)_m} = \frac{1 + \xi\eta V_f}{1 - \psi\eta V_f}.$$

where, $\xi = A$ constant given in Table 7.4.

$$\psi = 1$$

$$\eta = \frac{r - 1}{r + \xi}$$

$$r = \frac{(\cdot)_f}{(\cdot)_m}$$

Recall that $\xi \to \infty$ means rule of mixtures and that $\xi \to 0$ means inverse rule of mixtures.

## 7.4 MICROMECHANICS—A SUMMARY

Techniques of micromechanics, whether experimental or theoretical, are advanced. Since composites are a class of engineered materials, it is impossible to understate the importance of micromechanics. However, the variables involved are too numerous to permit derivation of closed form formulas for all possible situations. In order to make a point about the number of variables, consider the following

***Variables Involved in Design of Composites Materials:***

| | |
|---|---|
| Filler Materials | Filler Forms (shapes) |
| Filler Aspect Ratios | Filler Sizing |
| Orientation of Fillers | Type of composite (whether or not the two phases penetrate each other) |
| Volume Fraction | Matrix Materials |
| Interface | Two- or Multi-Material Composites |
| Two Dimensional Composites | Three Dimensional Composites |

***Properties to Predict:***

| | |
|---|---|
| Engineering moduli | —Five of them for two dimensional composites, and nine for three dimensional composites. |
| Strengths | —Five of them for two dimensional composites. |
| Thermal conductivity | —Two of them in two dimensions. |
| Thermal expansion coefficient | —Two of them in two dimensions. |
| Hygral expansion coefficient | —Two of them in two dimensions. |
| Impact properties | |
| Fracture toughness | |

The task of deriving or curve-fitting predictive formulas for all of the properties, for all situations is therefore formidable. One would find that micromechanical formulas appear in the literature for highly specialized conditions. It is understandable that a formula of broad applicability such as the Halpin-Tsai formula, can result only from extensive experimentation. The engineer needs to be aware of the limitations and applicability of the formulas being used. For instance, some formulas represent the lower bounds of the property, while some others represent upper bounds. Moreover, certain assumptions in the derivations of the formulas may be conservative. If a finite element approach was used, then there could be a difference between what the word failure means in calculations and a physical failure of a specimen in a laboratory test. Overall, it is essential to examine and understand the basis of the formulas before using it in actual applications.

## 7.5 MACROMECHANICS OF A UDL

The term *macromechanics* denotes the study of the mechanics of the ply with regard to its overall elastic responses to stresses and strains but without regard to the details of its constituents. It treats the composite ply as one continuous medium with orthotropic properties. It ignores the presence of the matrix, the fiber, the orientation, etc. In macromechanics, we are concerned about

1. the elastic properties of the ply referred to any arbitrary direction,
2. the stresses along and across the fiber in response to a set of in-plane loads applied in an arbitrary direction, and
3. a failure criterion for biaxial stresses in the orthotropic composite lamina, to determine whether or not a stress state will cause failure of the ply.

## 7.6 TRANSFORMATION OF ELASTIC MODULI

It was mentioned in Chapter 1 that elastic moduli are fourth order tensors. They do not transform from one set of axes to another like vectors, or even like stresses. The rule for their transformation is obtained as follows.

> Strains and stresses in 1-2 axes satisfy Hooke's laws, generalized for orthotropy (Equation 7.1). If strains and stresses are transformed in accordance with Section 1.12, to another axes system *x-y*, then the elastic moduli must transform such that the transformed strains and transformed stresses will formally fit once again in Hooke's laws, in an *x-y* system.

The above statement is equivalent to the requirement, that material behavior must follow Hooke's laws regardless of the coordinate system chosen for measurement of stress and strain.

It is convenient to use the matrix representation here. Things are easier to understand if we limit the discussions to two dimensions. For a rotation of axes by an angle $\theta$, (Figure 7.13) we have the transformation equations in Section 1.12. For rotation of axes *from the x-y system to the* 1-2 *system*,

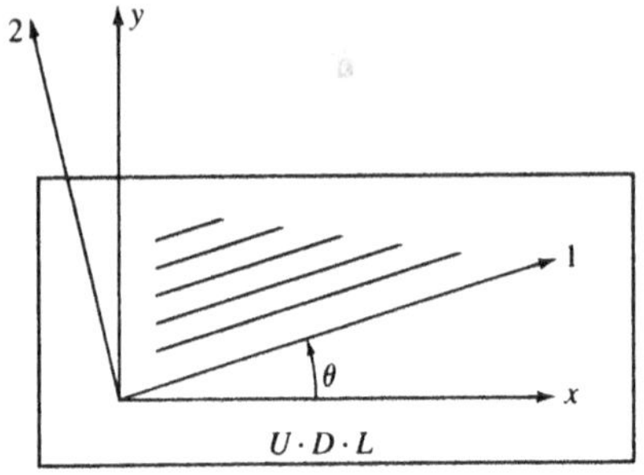


**FIGURE 7.13.** Rotation of coordinate axes changes the observed values of moduli.

we have the following. Recall that 1-2 is the material axes system and $x$-$y$ is the arbitrary coordinate system.

$$m = \cos\theta$$

$$n = \sin\theta$$

$$[\mathbf{T1}] = \begin{bmatrix} m^2 & n^2 & 2nm \\ n^2 & m^2 & -2nm \\ -nm & nm & (m^2 - n^2) \end{bmatrix} \tag{7.29}$$

$$\begin{Bmatrix} \sigma_1 \\ \sigma_2 \\ \tau_{12} \end{Bmatrix} = [\mathbf{T1}] \begin{Bmatrix} \sigma_x \\ \sigma_y \\ \tau_{xy} \end{Bmatrix} \tag{7.30}$$

Or simply,

$$\{\boldsymbol{\sigma}\} = [\mathbf{T1}]\{\boldsymbol{\sigma}'\} \tag{7.31}$$

The transformation rules for strains need some more care. From Section 1.12, we know that shear strains carry a factor $1/2$ which makes them a tensor. Accordingly,

$$\begin{Bmatrix} \epsilon_1 \\ \epsilon_2 \\ \dfrac{\gamma_{12}}{2} \end{Bmatrix} = \begin{bmatrix} m^2 & n^2 & 2mn \\ n^2 & m^2 & -2mn \\ -mn & nm & (m^2 - n^2) \end{bmatrix} \begin{Bmatrix} \epsilon_x \\ \epsilon_y \\ \dfrac{\gamma_{xy}}{2} \end{Bmatrix}$$

We can rewrite this equation in terms of engineering shear strains by removing the factor 1/2.

$$\begin{Bmatrix} \epsilon_1 \\ \epsilon_2 \\ \gamma_{12} \end{Bmatrix} = \begin{bmatrix} m^2 & n^2 & mn \\ n^2 & m^2 & -mn \\ -2mn & 2nm & (m^2 - n^2) \end{bmatrix} \begin{Bmatrix} \epsilon_x \\ \epsilon_y \\ \gamma_{xy} \end{Bmatrix} \tag{7.32}$$

Or simply, $$\{\epsilon\} = [\mathbf{T2}]\{\epsilon'\} \tag{7.33}$$

In the material coordinate system, Hooke's law is given by

$$\{\boldsymbol{\epsilon}\} = [\mathbf{S}_{12}]\{\boldsymbol{\sigma}\}$$

Expressing this in terms of *x*-*y* components,

$$[\mathbf{T2}]\{\boldsymbol{\epsilon}'\} = [\mathbf{S}_{12}][\mathbf{T1}]\{\boldsymbol{\sigma}'\}$$

$$\{\boldsymbol{\epsilon}'\} = [\mathbf{T2}]^{-1}[\mathbf{S}_{12}][\mathbf{T1}]\{\boldsymbol{\sigma}'\}$$

$$= [\mathbf{S}_{xy}]\{\boldsymbol{\sigma}'\} \tag{7.34}$$

Note that in the last step above, we need $[\mathbf{T2}]^{-1}$. This is easily obtained by changing $\theta$ to $-\theta$ in [**T2**]. This is the same as changing $n$ to $-n$ in [**T2**]. *This is so because, inversion is physically equivalent to simply rotating back to the* 1-2 *system by an angle* $-\theta$. See Figure 7.13. The matrix $[\mathbf{S}_{xy}]$ on the right side of Equation 7.35 is the matrix of transformed moduli, because it fits *formally* into Hooke's law. Therefore, the transformation rule for moduli is

$$[\mathbf{S}_{xy}] = [\mathbf{T2}]^{-1}[\mathbf{S}_{12}][\mathbf{T1}] \tag{7.35}$$

The general form of $[\mathbf{S}_{xy}]$ is a $3 \times 3$ matrix as follows. In general, no element in it is zero.

$$[\mathbf{S}_{xy}] = \begin{bmatrix} \dfrac{1}{E_{xx}} & -\dfrac{\nu_{xy}}{E_{xx}} & \dfrac{\eta_{xs}}{E_{xx}} \\ -\dfrac{\nu_{yx}}{E_{yy}} & \dfrac{1}{E_{yy}} & \dfrac{\eta_{ys}}{E_{yy}} \\ \dfrac{\eta_{sx}}{G_{xy}} & \dfrac{\eta_{sy}}{G_{xy}} & \dfrac{1}{G_{xy}} \end{bmatrix} \tag{7.36}$$

It is interesting to note that the stiffness matrix [**Q**] also transforms very similar to the compliance matrix. The rule of transforming the stiffness matrix is obtained as follows. Starting with Equation 7.6, we have,

$$\{\sigma\} = [\mathbf{Q}_{12}]\{\epsilon\}$$

That is,

$$[\mathbf{T1}]\{\sigma'\} = [\mathbf{Q}_{12}][\mathbf{T2}]\{\epsilon'\}$$

$$\{\sigma'\} = [\mathbf{T1}]^{-1}[\mathbf{Q}_{12}][\mathbf{T2}]\{\epsilon'\}$$

$$= [\mathbf{Q}_{\mathbf{xy}}]\{\epsilon'\}$$

Hence,

$$[\mathbf{Q}_{\mathbf{xy}}] = [\mathbf{T1}]^{-1}[\mathbf{Q}_{12}][\mathbf{T2}] \tag{7.37}$$

Note that Equation 7.37 is simply the inverse of Equation 7.35. In general, neither $[\mathbf{S}_{\mathbf{xy}}]$ nor $[\mathbf{Q}_{\mathbf{xy}}]$ will contain zeros. Because of their symmetry, only six out of the nine terms in these two matrices are independent.

However, the triple matrix product as above is not a convenient way to work problems. A more convenient way to work out transformations is to precalculate the triple multiplication of matrices. Such a procedure is summarized, again in matrix notation, in Equation 7.38. (Note that for this purpose, the independent elements of the $[\mathbf{Q}_{\mathbf{xy}}]$, and $[\mathbf{S}_{\mathbf{xy}}]$ matrices are stretched out and written as two column vectors, six rows each). We define a matrix [**R**], which is effectively the same as the triple multiplication.

$$[\mathbf{R}]_{6\times 4} = \begin{bmatrix} m^4 & n^4 & 2m^2n^2 & n^2m^2 \\ n^4 & m^4 & 2n^2m^2 & m^2n^2 \\ m^2n^2 & m^2n^2 & (m^4+n^4) & -(m^2n^2) \\ 4m^2n^2 & 4m^2n^2 & -8m^2n^2 & (m^2-n^2)^2 \\ 2m^3n & -2mn^3 & 2(mn^3-m^3n) & (mn^3-m^3n) \\ 2mn^3 & -2m^3n & 2(m^3n-mn^3) & (m^3n-mn^3) \end{bmatrix}$$

$$\begin{Bmatrix} S_{xx} & Q_{xx} \\ S_{yy} & Q_{yy} \\ S_{xy} & Q_{xy} \\ S_{ss} & 4Q_{ss} \\ S_{xs} & 2Q_{xs} \\ S_{ys} & 2Q_{ys} \end{Bmatrix} = [\mathbf{R}]_{6\times 4} \times \begin{Bmatrix} S_{11} & Q_{11} \\ S_{22} & Q_{22} \\ S_{12} & Q_{12} \\ S_{66} & 4Q_{66} \end{Bmatrix} \tag{7.38}$$

**Example 7.8**

Show that for isotropic materials the [**R**] matrix leaves the moduli unaltered, for all directions.

*Solution* We note that for isotropic materials, $E_{11} = E_{22} = E$, $\nu_{12} = \nu_{21} = \nu$, and $G = E/2(1 + \nu)$. Also,

$$S_{11} = S_{22} = \frac{1}{E}; \quad S_{12} = -\frac{\nu}{E}; \quad S_{66} = \frac{1}{G}$$

$$m = \cos\theta$$

$$n = \sin\theta$$

$$S_{xx} = S_{yy} = \frac{1}{E}(m^4 + n^4 - 2m^2n^2\nu) + \frac{2(1+\nu)m^2n^2}{E}$$

$$= \frac{m^4 + n^4 + 2m^2n^2}{E}$$

$$= \frac{(m^2 + n^2)^2}{E} = \frac{1}{E}$$

$$S_{xy} = \frac{2m^2n^2}{E} - \frac{(m^4 + n^4)\nu}{E} - \frac{2m^2n^2(1+\nu)}{E}$$

$$= -\frac{1}{E}[(m^4 + n^4)\nu + 2m^2n^2(1+\nu) - 2m^2n^2]$$

$$= -\frac{\nu(m^2 + n^2)^2}{E}$$

$$= -\frac{\nu}{E}$$

$$S_{ss} = \frac{1}{E}\left[4m^2n^2 + 4m^2n^2 - 8m^2n^2\nu + 2(1+\nu)(m^2 - n^2)^2\right]$$

$$= \frac{2(1+\nu)}{E}\left[(m^2 - n^2)^2 + 4m^2n^2\right] = \frac{2(1+\nu)}{G}(m^2 + n^2)^2$$

$$= \frac{2(1+\nu)}{E} = \frac{1}{G}$$

The above equations show that the transformation leaves the moduli unchanged, whatever the angle.

### 7.6.1 Calculation of Engineering Moduli From $\{\mathbf{S}_{xy}\}$

Equation 7.38 is a conversion of *mathematical moduli* and compliances from one system to another. We need the formulas for obtaining the *engineering moduli* (Young's moduli, Shear moduli, and Poisson's ratios). The formulas are as follows. Note that there are six independent elastic constants for in-plane orthotropy.

$$E_{xx} = \frac{1}{S_{xx}}; \quad \nu_{xy} = -\frac{S_{xy}}{S_{xx}}$$

$$G_{xy} = \frac{1}{S_{ss}}; \quad \eta_{xs} = \frac{S_{xs}}{S_{xx}}$$

$$E_{yy} = \frac{1}{S_{yy}}; \quad \eta_{ys} = \frac{S_{ys}}{S_{yy}} \tag{7.39}$$

Additional, but not independent constants may be obtained as below.

$$\nu_{yx} = -\frac{S_{xy}}{S_{yy}}$$

$$\eta_{sx} = \frac{S_{sx}}{S_{ss}}$$

$$\eta_{sy} = \frac{S_{ys}}{S_{ss}} \tag{7.40}$$

One can obtain the same elastic constants by considering the $[\mathbf{Q}_{xy}]$ matrix also. But the expressions are not simple, and they take much longer to calculate.

**Example 7.9**

Show that for isotropic materials, $\eta$'s are identically zero in all directions.

*Solution* To show that the $\eta$'s are zero, we look at the equations for $S_{xs}$ and $S_{ys}$.

$$S_{xs} = \frac{1}{E}(2mn^3 - 2m^3n) - 2(mn^3 - m^3n)\frac{\nu}{E} + (mn^3 - m^3n)\frac{2(1+\nu)}{E}$$

$$= 0, \quad \text{after simplification, for all } \theta.$$

So also $S_{ys}$ becomes identically equal to zero, for all $\theta$. Since $\eta$'s are only a multiple of $S_{xs}$ and $S_{ys}$, they are also zero, for all $\theta$.

**Example 7.10**

A unidirectional ply is made of graphite continuous fibers in epoxy. Its moduli are given by $E_{11} = 13800\ \text{kg/mm}^2$, $E_{22} = 1000\ \text{kg/mm}^2$, $G_{12} = 700$ kg/mm$^2$, and $\nu_{12} = 0.26$. Find the transformed moduli of the material along (1) 30° clockwise to the material axis, (2) 45° clockwise to the material axis.

*Solution* We may recall the solution procedure in Figure 7.14. As the first step, we calculate the compliance matrix in the material coordinate system, $\{\mathbf{S}_{12}\}$.

$$S_{11} = \frac{1}{E_{11}} = \frac{1}{13800} = 0.00072$$

$$S_{22} = \frac{1}{E_{22}} = \frac{1}{1000} = 0.001$$

$$S_{12} = -\frac{\nu_{12}}{E_{11}} = -\frac{0.26}{13800} = -1.90 \times 10^{-5}$$

$$S_{66} = \frac{1}{G_{12}} = \frac{1}{700} = 0.001429$$

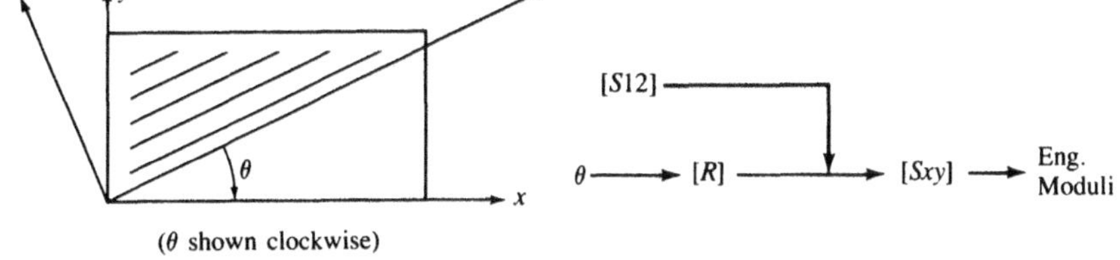


**FIGURE 7.14.** Flow chart for calculating moduli components at an angle to the material axes.

The second step is to calculate the [**R**] matrix. Let us first consider the case of 30°.

$$m = \cos 30° = \frac{\sqrt{3}}{2}$$

$$n = \sin 30° = \frac{1}{2}$$

$$[\mathbf{R}] = \begin{bmatrix} 0.5625 & 0.0625 & 0.375 & 0.1875 \\ 0.0625 & 0.5625 & 0.375 & 0.1875 \\ 0.1875 & 0.1875 & 0.625 & -0.1875 \\ 0.75 & 0.75 & -1.5 & 0.25 \\ 0.6495 & -0.2165 & -0.433 & -0.2165 \\ 0.2165 & -0.6495 & 0.433 & 0.2165 \end{bmatrix}$$

The third step is to obtain the compliance matrix $\{\mathbf{S_{xy}}\}$ transformed to the new axes system.

$$\{\mathbf{S_{xy}}\} = \begin{Bmatrix} S_{xx} \\ S_{yy} \\ S_{xy} \\ S_{ss} \\ S_{xs} \\ S_{ys} \end{Bmatrix} = [\mathbf{R}]\{\mathbf{S_{12}}\} = \begin{Bmatrix} 0.000354 \\ 0.000828 \\ -7.90 \times 10^{-5} \\ 0.00119 \\ -0.00047 \\ -0.00033 \end{Bmatrix}$$

As the last step, we use Equations 7.39 to retrieve the moduli in the $x$-$y$ coordinate system.

$$E_{xx} = \frac{1}{S_{xx}} = 2747 \text{ kg/mm}^2$$

$$E_{yy} = \frac{1}{S_{yy}} = 1208 \text{ kg/mm}^2$$

$$G_{xy} = \frac{1}{S_{ss}} = 841 \text{ kg/mm}^2$$

$$\nu_{xy} = -\frac{S_{xy}}{S_{xx}} = 0.216$$

$$\eta_{xs} = \frac{S_{xs}}{S_{xx}} = -1.293$$

$$\eta_{ys} = \frac{S_{ys}}{S_{yy}} = -0.402$$

We can repeat the same process for the case of 45° also. We note that $\{\mathbf{S}_{12}\}$ remains the same.

$$[\mathbf{R}] = \begin{bmatrix} 0.25 & 0.25 & 0.50 & 0.25 \\ 0.25 & 0.25 & 0.50 & 0.25 \\ 0.25 & 0.25 & 0.50 & -0.25 \\ 1.00 & 1.00 & -2.00 & 0 \\ 0.50 & -0.50 & 0 & 0 \\ -0.50 & 0.50 & 0 & 0 \end{bmatrix}$$

$$\{\mathbf{S}_{xy}\} = [\mathbf{R}]\{\mathbf{S}_{12}\} = \begin{Bmatrix} 0.000616 \\ 0.000616 \\ -9.80 \times 10^{-5} \\ 0.00111 \\ -0.00046 \\ -0.00046 \end{Bmatrix}$$

Hence,

$$E_{xx} = 1624 \text{ kg/mm}^2$$

$$E_{yy} = 1624 \text{ kg/mm}^2$$

$$G_{xy} = 901 \text{ kg/mm}^2$$

$$\nu_{xy} = 0.16$$

$$\eta_{xs} = -0.753$$

$$\eta_{ys} = -0.753$$

**Example 7.11**

A composite ply with in-plane orthotropy has the following properties along the material coordinates in its own plane.

$$E_{11} = 4400 \text{ kg/mm}^2$$

$$E_{22} = 900 \text{ kg/mm}^2$$

$$\nu_{12} = 0.27$$

$$G_{12} = 460 \text{ kg/mm}^2$$

Find (1) the mathematical moduli in material coordinates, (2) the mathematical moduli along axes inclined at 30°, 45°, 60°, and 90°, (3) the engineering moduli along 30°, 45°, 60°, and 90°, and (4) the engineering constants $E_{xx}$, $E_{yy}$, $G_{xy}$, $\nu_{xy}$, $\eta_{xs}$, and $\eta_{ys}$ as a function of direction and plot them.

*Solution*

Part (1) Mathematical moduli:

Mathematical moduli are simply $Q_{**}$'s given in Equations 7.4 and 7.5.

$$\nu_{21} = \nu_{12}\frac{E_{22}}{E_{11}} = 0.27\frac{900}{4400} = 0.0552$$

$$(1 - \nu_{12}\nu_{21}) = [1 - (0.27)(0.0552)] = 0.9851$$

$$Q_{11} = \frac{E_{11}}{(1 - \nu_{12}\nu_{21})} = \frac{4400}{0.9851} = 4467 \text{ kg/mm}^2$$

$$Q_{22} = \frac{E_{22}}{(1 - \nu_{12}\nu_{21})} = \frac{900}{0.9851} = 914 \text{ kg/mm}^2$$

$$Q_{12} = \frac{\nu_{12}E_{22}}{(1 - \nu_{12}\nu_{21})} = \frac{0.27 \times 900}{0.9851} = 247 \text{ kg/mm}^2$$

$$Q_{66} = G_{12} = 460 \text{ kg/mm}^2$$

Part (2) Mathematical moduli along given directions:

This is done by constructing the [**R**] matrix for the given angles, and multiplying it by the $Q_{**}$ arranged as a vector. A programmable calculator can be used. The procedure is explicitly shown only for $\theta = 30°$. The other cases are left to the reader to verify.

$$m = \cos 30° = \frac{\sqrt{3}}{2}$$

$$n = \sin 30° = \frac{1}{2}$$

$$[\mathbf{R}] = \begin{bmatrix} 0.5625 & 0.0625 & 0.375 & 0.1875 \\ 0.0625 & 0.5625 & 0.375 & 0.1875 \\ 0.1875 & 0.1875 & 0.625 & -0.1875 \\ 0.75 & 0.75 & -1.5 & 0.25 \\ 0.6495 & -0.2165 & -0.433 & -0.2165 \\ 0.2165 & -0.6495 & 0.433 & 0.2165 \end{bmatrix}$$

$$\begin{Bmatrix} Q_{xx} \\ Q_{yy} \\ Q_{xy} \\ Q_{ss} \\ Q_{xs} \\ Q_{ys} \end{Bmatrix} = [\mathbf{R}] \begin{Bmatrix} Q_{11} \\ Q_{22} \\ Q_{12} \\ Q_{66} \end{Bmatrix} = \begin{Bmatrix} 3007 \\ 1230 \\ 818 \\ 1031 \\ 1099 \\ 439 \end{Bmatrix}$$

By a similar procedure we can obtain the values for other angles.

| $\theta$ | $Q_{xx}$ | $Q_{yy}$ | $Q_{xy}$ | $Q_{ss}$ | $Q_{xs}$ | $Q_{ys}$ |
|---|---|---|---|---|---|---|
| 0.000 | 4467 | 914 | 247 | 460 | 0 | 0 |
| 30.000 | 3007 | 1230 | 818 | 1031 | 1099 | 439 |
| 45.000 | 1928 | 1928 | 1008 | 1222 | 888 | 888 |
| 60.000 | 1230 | 3007 | 818 | 1031 | 439 | 1099 |
| 90.000 | 914 | 4467 | 247 | 460 | 0 | 0 |

Part (3) Engineering moduli along given directions:

The engineering moduli are obtained in accordance with the procedures in Section 7.2.1. The calculations are shown explicitly for $\theta = 30°$ only as before. The other results are left to the reader to verify. Results are tabulated for every 5°, for use in part (4).

$$S_{11} = \frac{1}{E_{11}} = \frac{1}{4400} = 0.2272 \times 10^{-3}\ (\text{kg/mm}^2)^{-1}$$

$$S_{22} = \frac{1}{E_{22}} = \frac{1}{900} = 1.111 \times 10^{-3}\ (\text{kg/mm}^2)^{-1}$$

$$S_{12} = -\frac{\nu_{12}}{E_{11}} = -\frac{0.27}{4400} = -0.0614 \times 10^{-3}\ (\text{kg/mm}^2)^{-1}$$

$$S_{66} = \frac{1}{G_{12}} = \frac{1}{460} = 2.174 \times 10^{-3}\ (\text{kg/mm}^2)^{-1}$$

[**R**] for 30° is available above. Hence, we get

$$\begin{Bmatrix} S_{xx} \\ S_{yy} \\ S_{xy} \\ S_{ss} \\ S_{xs} \\ S_{ys} \end{Bmatrix} = [\mathbf{R}] = \begin{Bmatrix} S_{11} \\ S_{22} \\ S_{12} \\ S_{66} \end{Bmatrix} = \begin{Bmatrix} 0.5818 \\ 1.0237 \\ -0.1951 \\ 1.6393 \\ -0.5370 \\ -0.2283 \end{Bmatrix} 10^{-3}$$

The engineering moduli are obtained as below.

$$E_{xx} = \frac{1}{S_{xx}} = \frac{10^3}{0.5818} = 1718 \text{ kg/mm}^2$$

$$E_{yy} = \frac{1}{S_{yy}} = \frac{10^3}{1.0237} = 976 \text{ kg/mm}^2$$

$$G_{xy} = \frac{1}{S_{ss}} = \frac{10^3}{1.6393} = 610 \text{ kg/mm}^2$$

$$\nu_{xy} = \frac{-S_{xy}}{S_{xx}} = \frac{0.1951}{0.5818} = 0.335$$

$$\eta_{xs} = \frac{S_{xs}}{S_{xx}} = -\frac{0.5370}{0.5818} = -0.923$$

$$\eta_{yx} = \frac{S_{ys}}{S_{yy}} = -\frac{0.2283}{1.0237} = -0.223$$

| $\theta$ | $E_{xx}$ | $E_{yy}$ | $Q_{xy}$ | $\nu_{xy}$ | $-\eta_{xs}$ | $-\eta_{ys}$ |
|---|---|---|---|---|---|---|
| 0 | 4400 | 900 | 460 | 0.270 | 0 | 0 |
| 5 | 4178 | 901 | 465 | 0.279 | 0.575 | 0.014 |
| 10 | 3639 | 905 | 478 | 0.299 | 0.967 | 0.033 |
| 15 | 3021 | 912 | 501 | 0.320 | 1.134 | 0.061 |
| 20 | 2473 | 925 | 532 | 0.334 | 1.137 | 0.100 |
| 25 | 2042 | 945 | 570 | 0.339 | 1.050 | 0.154 |
| 30 | 1719 | 977 | 610 | 0.335 | 0.923 | 0.223 |
| 35 | 1481 | 1023 | 647 | 0.324 | 0.784 | 0.308 |
| 40 | 1307 | 1088 | 674 | 0.306 | 0.648 | 0.407 |
| 45 | 1180 | 1180 | 684 | 0.283 | 0.522 | 0.522 |
| 50 | 1088 | 1307 | 674 | 0.255 | 0.407 | 0.648 |
| 55 | 1023 | 1481 | 647 | 0.224 | 0.308 | 0.784 |
| 60 | 977 | 1719 | 610 | 0.190 | 0.223 | 0.923 |
| 65 | 945 | 2042 | 570 | 0.157 | 0.154 | 1.050 |
| 70 | 925 | 2473 | 532 | 0.125 | 0.100 | 1.137 |
| 75 | 912 | 3021 | 501 | 0.097 | 0.061 | 1.134 |
| 80 | 905 | 3639 | 478 | 0.074 | 0.033 | 0.967 |
| 85 | 901 | 4178 | 465 | 0.060 | 0.014 | 0.575 |
| 90 | 900 | 4400 | 460 | O.055 | 0 | 0 |

Note that the last two columns are for negative $\eta$'s.

Part (4) Plotting:

The moduli and coupling constants are plotted in two separate Figures 7.15a and 7.15b, respectively.

## 7.7 CALCULATION OF STRESSES IN A 1-2 SYSTEM

Determination of stresses in the material coordinate system is necessary primarily for applying the failure criteria. It is also needed for finding the fiber stress and strain. The calculation is nothing but an application of Equations 7.31 and 7.33 (or, simply the Mohr's circle). However, for the sake of completeness, we repeat them here, in explicit form.

***Stress Along Fiber Direction***

$$\sigma_1 = \sigma_x m^2 + \sigma_y n^2 + 2\tau_{xy} mn$$

***Stress Across Fiber Direction***

$$\sigma_2 = \sigma_x n^2 + \sigma_y m^2 - 2\tau_{xy} nm$$

***Shear Stress in the Plane of the Ply***

$$\tau_{12} = (\sigma_y - \sigma_x) nm + \tau_{xy}(m^2 - n^2)$$

## 7.8 THE MEANING OF $\nu$'s AND $\eta$'s

The elastic constants denoted by $\nu$'s and $\eta$'s are new. To the student, who works primarily with isotropic materials, they may be unfamiliar. Their meaning will become clear if we look at the response of the orthotropic material to applied stress, one stress at a time. The meaning of $\nu_{xy}$ is similar to the $\nu_{12}$ and $\nu_{21}$ illustrated in Examples 7.1 and 7.2. They are simply the lateral strains that occur when loading in a longitudinal direction. Consider Equations 7.35 and 7.36 together, and put $\sigma_x = 1$ and

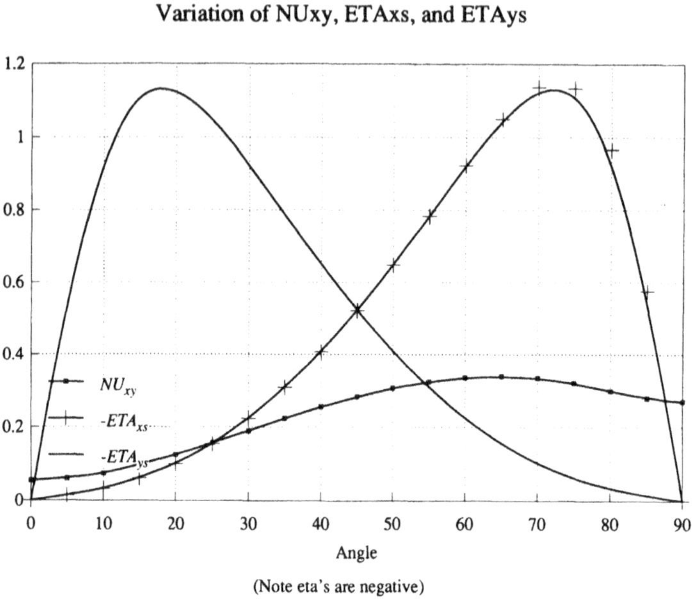


(a)

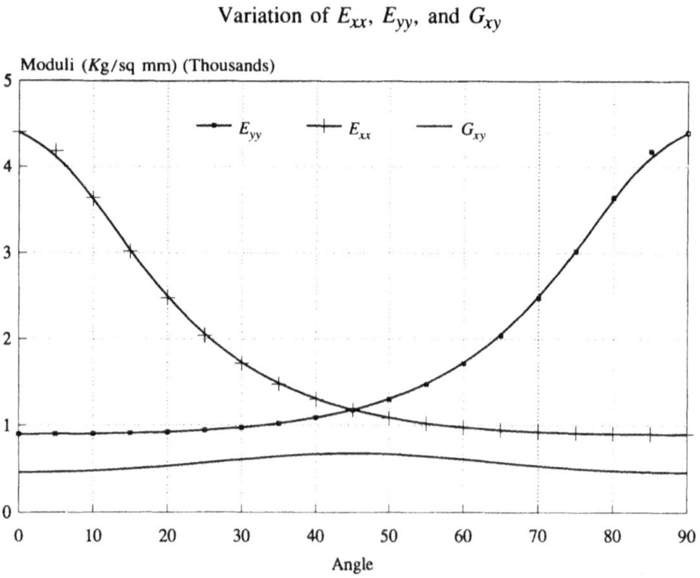


(b)

**FIGURE 7.15.** Variation of moduli and coupling coefficients of an UDL with angle.

$\sigma_y = \tau_{xy} = 0$. We obtain,

$$\epsilon_x = \frac{\sigma_x}{E_{xx}} = \frac{1}{E_{xx}}$$

$$\epsilon_y = -\frac{\nu_{yx}}{E_{yy}} = -\frac{\nu_{yx}}{E_{yy}}$$

$$\gamma_{xy} = \frac{\eta_{sx}\sigma_x}{G_{xy}} = \frac{\eta_{sx}}{G_{xy}}$$

This shows that $\eta_{sx}$ is a measure of the *shear strain in response to a normal stress*. Further, by considering a stress state of $\sigma_x = \sigma_y = 0$, and $\tau_{xy} = 1$, it is possible to take a view point that the $\eta$'s are also a measure of normal strain in response to applied shear stress. Therefore, we may call the $\eta$'s the *normal-to-shear couplings*. Figure 7.16 gives a graphical explanation of these factors, and also explains the meaning of the two subscripts of $\eta$ and $\nu$.

**Example 7.12**

Is it meaningful to refer to the top and bottom of a unidirectional ply?

*Solution* In general, there is a significance to those words, although they are not the commonly accepted terminology in composites. Consider a UDL specimen whose loading axis is inclined to the fiber direction at 30°, as shown in Figure 7.17.

A review of the **[R]** matrix shows that the terms $S_{xx}$, $S_{yy}$, $S_{xy}$, and $S_{ss}$ do not change sign if the sign of $\theta$ is reversed. They are even functions of $\theta$. The other two terms $S_{xs}$ and $S_{ys}$ do change signs. They are odd functions of $\theta$. As such, they can be taken to define the top and bottom sides for the UDL The distinction between top and bottom surfaces of a lamina disappears if the loading axis and the material axis coincide, since $S_{xs} = S_{ys} = 0$.

> The top side is such that the fiber axis is obtained by counter-clockwise rotation from the loading axis.

This rule is in agreement with the sign conventions adopted for stress and strain, the Mohr's circle and other transformations including **[R]**. In

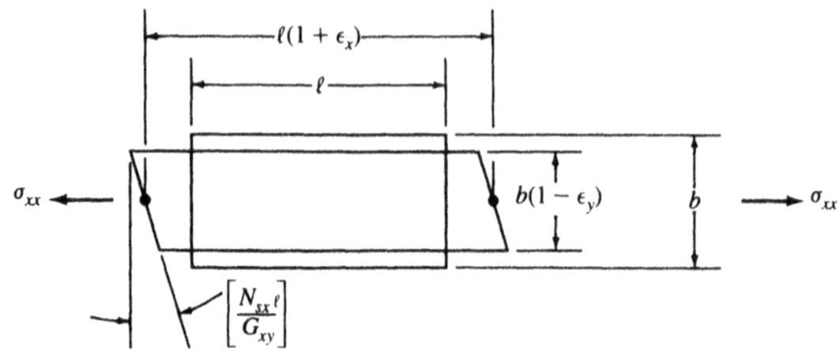


Shear strains are produced in an off-axis specimen in response to a normal stress.

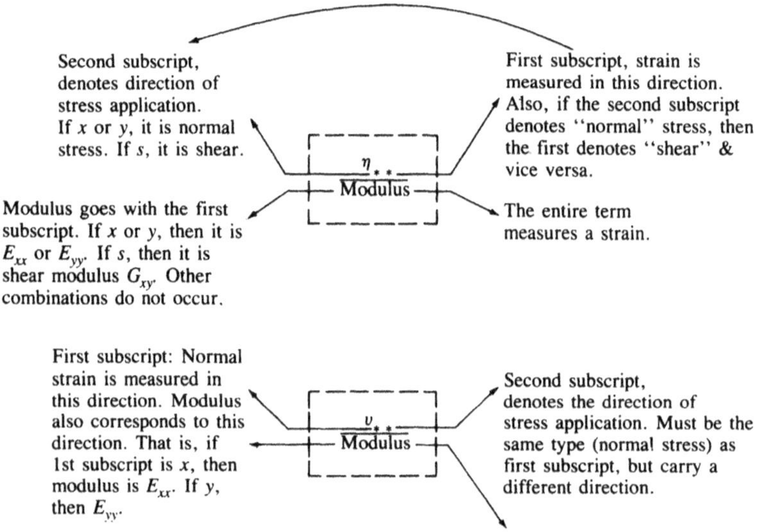


| PERMITTED | NOT PERMITTED | |
|---|---|---|
| $\frac{\eta_{sx}}{G_{sx}}$ | $\frac{\eta_{yx}}{G_{xy}}$ | , one of the subscripts of $\eta$ must be shear. |
| | $\frac{\eta_{sx}}{E_{xx}}$ | , first subscript of $\eta$ does not match $E$. |
| | $\frac{\eta_{xs}}{E_{xx}}$ | , $\upsilon$ must not have a shear subscript. |

**FIGURE 7.16.** Physical meaning of $\eta$ and $\nu$.

practice, the terms top and bottom of plies are equivalently replaced by positive and negative angles of orientation.

## Example 7.13

A tension test sample is prepared from a unidirectional ply, with the material 1 axes making an angle of 30° with the loading direction. The

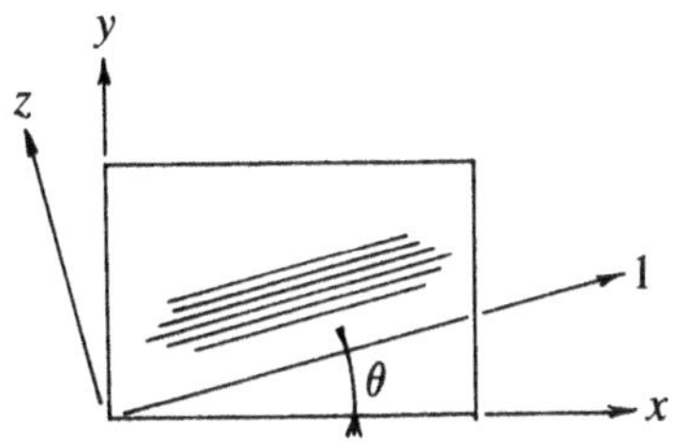


**FIGURE 7.17.** Example 7.12

material of the ply is E-Glass-Epoxy which has the following properties in the 1-2 axes system.

$$E_{11} = 4000 \text{ kg/mm}^2; \quad E_{22} = 870 \text{ kg/mm}^2;$$

$$G_{12} = 380 \text{ kg/mm}^2; \quad \nu_{12} = 0.28$$

It is loaded by a pure tensile load of 20 kg/mm$^2$. Draw a not-to-scale sketch to illustrate the deformations in the sample.

*Solution* The aim of this example is to show that direct tensile stress along the $x$-axis produces shear strains also. Hence we need to calculate the $\eta$'s. As in Example 7.6, we proceed to find the $\{\mathbf{S}_{xy}\}$.

$$S_{11} = 0.00025; \quad S_{22} = 0.001149; \quad S_{12} = -7.0 \times 10^{-5}; \quad S_{66} = 0.002632$$

The [**R**] matrix for 30° is available from Example 7.6.

$$[\mathbf{R}] = \begin{bmatrix} 0.5625 & 0.0625 & 0.375 & 0.1875 \\ 0.0625 & 0.5625 & 0.375 & 0.1875 \\ 0.1875 & 0.1875 & 0.625 & -0.1875 \\ 0.75 & 0.75 & -1.5 & 0.25 \\ 0.6495 & -0.2165 & -0.433 & -0.2165 \\ 0.2165 & -0.6495 & 0.433 & 0.2165 \end{bmatrix}$$

Multiplying [**R**] by $\{\mathbf{S}_{12}\}$, we get the compliances, as below.

$$S_{xx} = 0.00068$$

$$S_{yy} = 0.001129$$

$$S_{xy} = -0.00027$$

$$S_{ss} = 0.001812$$

$$S_{xs} = -0.00063$$

$$S_{ys} = -0.00015$$

Use of Equations 7.39 leads to the following moduli along the *x-y* system.

$$E_{xx} = 1471 \text{ kg/mm}^2; \quad E_{yy} = 885 \text{ kg/mm}^2; \quad G_{xy} = 552 \text{ kg/mm}^2$$

$$\nu_{xy} = 0.404; \quad \eta_{xs} = -0.921; \quad \eta_{ys} = -0.135$$

We can calculate the strains caused in the material by a stress of $\sigma_x = 20 \text{ kg/mm}^2$, $\sigma_y = \tau_{xy} = 0$.

$$\epsilon_x = \frac{\sigma_x}{E_{xx}} - \nu_{xy}\frac{\sigma_y}{E_{yy}} + \eta_{xs}\frac{\tau_{xy}}{E_{xx}}$$

$$= 0.01359$$

$$\epsilon_y = -\nu_{21}\frac{\sigma_x}{E_{yy}} + \frac{\sigma_y}{E_{yy}} + \eta_{ys}\frac{\tau_{xy}}{E_{yy}}$$

$$= -0.0055$$

$$\gamma_{xy} = \eta_{xs}\frac{\sigma_x}{E_{xx}} + \eta_{ys}\frac{\sigma_y}{E_{yy}} + \frac{\tau_{xy}}{G_{xy}}$$

$$= -0.03338$$

A considerable fraction of the *x*-strain appears also in the shear mode. This is because of the rather large value of $\eta_{xs}$ as compared to the Poisson's ratios.

The three strains are marked (not to scale) in Figure 7.18.

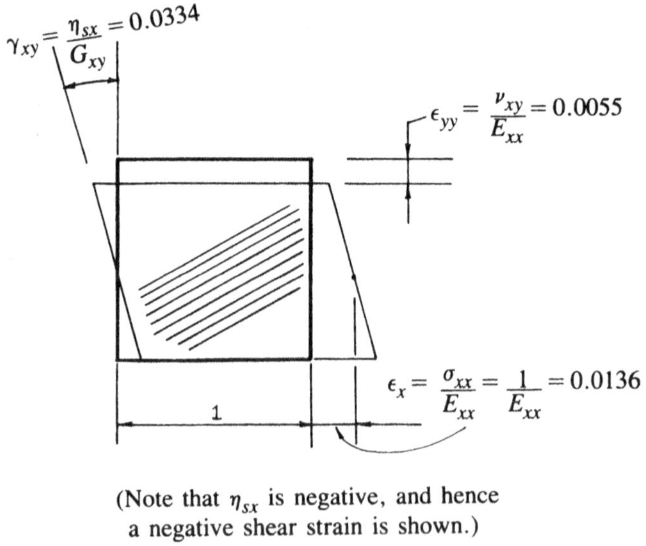


**FIGURE 7.18.** A not-to-scale plot of the deformed shape of the off-axis specimen.

## 7.9 FAILURE CRITERIA

The reader is already familiar with the Tresca and Von Mises failure criteria for metals. The use of such criteria in metals is in the context of multiaxial stress states. For example, the strength of structural steel is measured in the laboratory under a uniaxial stress state—that is, when all but one stress are zero. In real life situations, it is almost impossible to get uniaxial stress conditions. Hence the question arises, Under what combinations of stresses can a yielding mode of failure occur? The failure criteria set forth the rules governing the determination of failure or otherwise.

In the case of composites failure criteria serve a similar purpose. However, the situation is further complicated by the orthotropy of their strengths. Furthermore, the modes of failure of a composite ply are numerous in contrast to the metals where yielding is the only mode of failure. In the laboratory the composites are tested in the uniaxial stress condition. The similarity between metals and composites ends here. A ply can fail under one of the following mechanisms (Figure 7.19),

- fiber breaking,
- fiber pull-out,
- matrix cracking,
- micro-buckling.

In general, the mode of failure of a ply under a given load cannot be predicted. But it can be ascertained by microscopic examination, after the failure. For the structural designer, however, the details of the mode of failure are unimportant. One simply looks for a single number to denote the so-called strength. Usually such a number is associated with a total failure of the test specimen into two parts.

The strengths of composites are usually referred to the material coordinate system. They vary with the direction of loading and with the sign of the stress. As such, there are five basic strengths associated with a UDL:

$F_1^T$—Strength along fiber, in tension.
$F_1^C$—Strength along fiber, in compression.
$F_2^T$—Strength across fiber, in tension.
$F_2^C$—Strength across fiber, in compression.
$F_6$—Strength in shear.

Because the loading directions in real life are in general not along

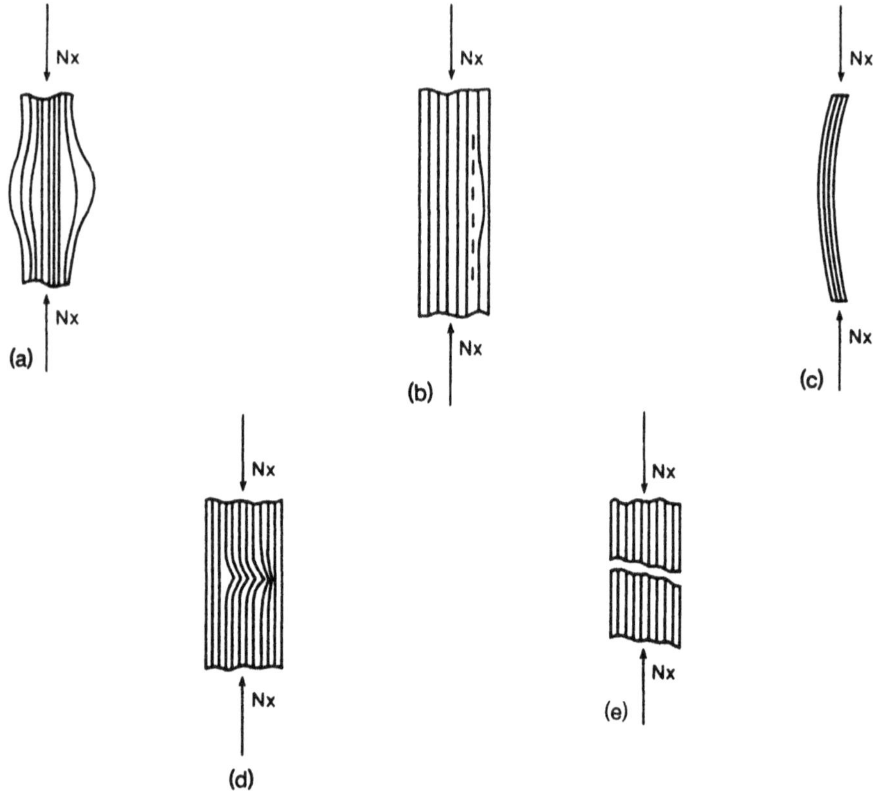


**FIGURE 7.19.** The various ways in which composite ply can fail. Failure modes of unidirectional composites subjected to longitudinal tensile load: (a) brittle failure, (b) brittle failure with fiber pullout, and (c) brittle failure with debonding and/or matrix failure. In-plane compression failure modes: (d) induced transverse tensile failure, (e) compressive delamination failure, (f) Evier failure, (g) microbuckling, and (h) strength failure.

material coordinates, and because stress states are biaxial, we need failure criteria to reconcile the real life situations with uniaxial test results. Over the years, a number of criteria have been proposed [14]. Each criterion has its own advantages and level of complexity. The following four criteria have gained popularity due to their ease of use and applicability. All of them refer to the stresses in the material coordinate system.

*Maximum Stress Criterion:* This criterion postulates that failure will occur if any one of the stresses in the 1-2 system just exceeds the corresponding strength. In other words, failure occurs if any one of the following conditions is satisfied.

$$\sigma_1 \text{ just reaches } F_1^T \text{ or } F_1^C,$$

$$\sigma_2 \text{ just reaches } F_2^T \text{ or } F_2^C,$$

$$\tau_{12} \text{ just reaches } F_6.$$

*Maximum Strain Criterion:* This criterion postulates that failure will occur if any one of the strains in the 1-2 system just exceeds the corresponding value at failure in a uniaxial test. In other words, this criterion states that failure will occur if one of the following conditions is satisfied.

$$\epsilon_1 \text{ just reaches } \frac{F_1^T}{E_{11}} \text{ or } \frac{F_1^C}{E_{11}}$$

$$\epsilon_2 \text{ just reaches } \frac{F_2^T}{E_{22}} \text{ or } \frac{F_2^C}{E_{22}}$$

$$\gamma_{12} \text{ just reaches } \frac{F_6}{G_{12}}$$

*Maximum Work Criterion (Tsai-Hill):* This criterion is stated in terms of a single equation. It states that failure occurs when the equality below is just satisfied.

$$g(\boldsymbol{\sigma}) \equiv \left(\frac{\sigma_1}{F_1^*}\right)^2 - \left(\frac{\sigma_1}{F_1^*}\right)\left(\frac{\sigma_2}{F_2^*}\right) + \left(\frac{\sigma_2}{F_2^*}\right)^2 + \left(\frac{\tau_{12}}{F_6}\right)^2 = 1$$

where,

$$F_1^* = F_1^T \text{ if } \sigma_1 \text{ is tensile,}$$

$$F_1^* = F_1^C \text{ if } \sigma_1 \text{ is compressive,}$$

$$F_2^* = F_2^T \text{ if } \sigma_2 \text{ is tensile,}$$

$$F_2^* = F_2^C \text{ if } \sigma_2 \text{ is compressive.}$$

Failure does not occur if $g(\boldsymbol{\sigma}) < 1$.

*Tsai-Wu Criterion:* This criterion states that failure occurs if the following equation is just satisfied.

$$g(\boldsymbol{\sigma}) \equiv f_1\sigma_1 + f_2\sigma_2 + f_{11}\sigma_1^2 + f_{22}\sigma_2^2 + f_{66}\tau_{12}^2 + 2f_{12}\sigma_1\sigma_2 = 1,$$

where the $f$'s are functions of the uniaxial strengths of the composite, defined in the following.

$$f_1 = \frac{1}{F_1^T} - \frac{1}{F_1^C}$$

$$f_{11} = \frac{1}{F_1^T F_1^C}$$

$$f_2 = \frac{1}{F_2^T} - \frac{1}{F_2^C}$$

$$f_{22} = \frac{1}{F_2^T F_2^C}$$

$$f_{66} = \frac{1}{F_6^2}$$

$$f_{12} = -\frac{1}{2}\sqrt{f_{11}f_{22}} \tag{7.41}$$

Failure does not occur if $g(\boldsymbol{\sigma}) < 1$.

A few examples are given below to elucidate the application of the different criteria of failure. They are in lieu of a more detailed explanation of the theories. For a formal discussion, References 2, 3, and 21 must be consulted.

**Example 7.14**

Use the maximum stress criterion to investigate the strength of an S-Glass Epoxy ply for loading angles varying from 0° to 90°. Consider tensile as well as compressive strengths. Take

$$E_{11} = 4400; \quad E_{22} = 900; \quad G_{12} = 460; \quad \nu_{12} = 0.27$$

$$F_1^T = 130; \quad F_1^C = 70; \quad F_2^T = 5; \quad F_2^C = 16; \quad F_6 = 7$$

$$\text{Units: kg/mm}^2$$

*Solution*

$$m = \cos\theta; \quad n = \sin\theta$$

$$\sigma_1 = \sigma_x m^2; \quad \sigma_2 = \sigma_x n^2; \quad \tau_{12} = \sigma_x mm$$

For failure, $\sigma_1 = F_1^T$ or $F_1^C$ etc. That is,

$$\sigma_x = \min\left\{\frac{F_1^T}{m^2}, \frac{F_2^T}{n^2}, \frac{F_6}{mn}\right\}$$

$$= \min\left\{\frac{130}{m^2}, \frac{5}{n^2}, \frac{7}{mn}\right\}$$

When $\sigma_x$ is compressive, the same procedure can be used. In this case,

$$\sigma_x = \min\left\{\frac{F_1^C}{m^2}, \frac{F_2^C}{n^2}, \frac{F_6}{mn}\right\}$$

$$= \min\left\{\frac{70}{m^2}, \frac{16}{n^2}, \frac{7}{mn}\right\}$$

Figure 7.20 shows the plots of the stress at failure for off-axis loading, or in other words, the figure shows the off-axis strengths according to Maximum stress criterion.

***Commentary***
*There is a sharp decrease in the strength in the range* $\approx 10 \le \theta \le 25$. *This indicates the penalty for changing the orientation. This is the fundamental problem in building up laminates of desired strength.*

**Example 7.15**

Use the maximum strain criterion to calculate the strength of an S-Glass Epoxy ply (1) for a loading angle $\theta = 70$, (2) as a function of the loading

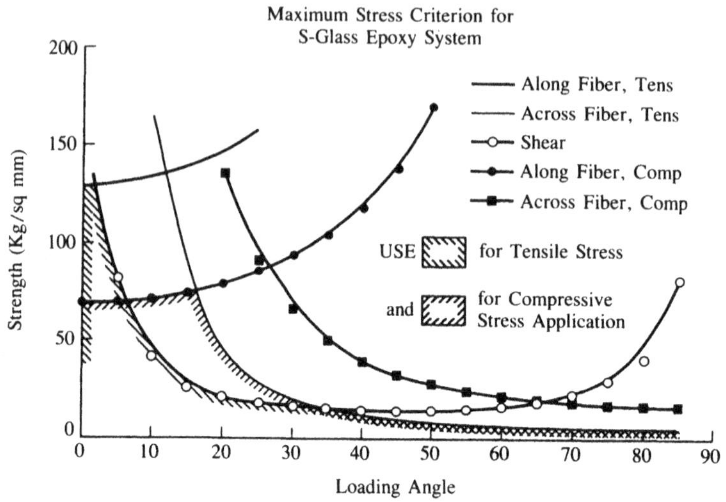


**FIGURE 7.20.** Off-axis strength as a function of angle, using the maximum stress criterion.

angle, $\theta$. Take

$$E_{11} = 4400; \quad E_{22} = 900; \quad G_{12} = 460; \quad \nu_{12} = 0.27$$

$$F_1^T = 130; \quad F_1^C = 70; \quad F_2^T = 5; \quad F_2^C = 16; \quad F_6 = 7$$

$$(\text{Units: kg/mm}^2)$$

*Solution* Case (1) $\theta = 70°$.
Given $\sigma_x = \sigma$; $\sigma_y = \tau_{xy} = 0$.

$$\sigma_1 = \sigma_x \cos^2 70 = 0.117\sigma_x$$

$$\sigma_2 = \sigma_x \sin^2 70 = 0.883\sigma_x$$

We can calculate the strength by supposing that $\epsilon_1$ is exceeded. Hence,

$$\epsilon_1 = \frac{\sigma_1}{E_{11}} - \nu_{12}\frac{\sigma_2}{E_{11}}$$

$$= -0.1214\frac{\sigma_x}{E_{11}} \quad \text{(Note negative strain for positive stress!)}$$

For failure to occur, we must have

$$0.1214\frac{\sigma_x}{E_{11}} = \frac{F_1^C}{E_{11}} \quad \text{(Note compressive strength was used!)}$$

Or,
$$\sigma_x = \frac{F_1^C}{0.1214} = 576.53 \text{ kg/mm}^2$$

Next, we assume that $\epsilon_2$ is exceeded

$$\begin{aligned} \epsilon_2 &= -\nu_{21}\frac{\sigma_1}{E_{11}} + \frac{\sigma_2}{E_{22}} \\ &= \frac{\sigma_x}{E_{22}}(-0.055 \times 0.883 + 0.117) \\ &= 0.068\frac{\sigma_x}{E_{22}} \quad \text{(Note positive } \epsilon_2\text{)} \end{aligned}$$

For failure,

$$0.068\frac{\sigma_x}{E_{22}} = \frac{F_2^T}{E_{22}}$$

Or,
$$\sigma_x = \frac{5}{0.068} = 73.28 \text{ kg/mm}^2$$

Lastly, we calculate the stress needed to produce shear failure.

$$\gamma_{12} + \frac{\tau_{12}}{G_{12}} = \frac{\sigma_x}{G_{12}} \sin 70 \cos 70 = 0.3214\frac{\sigma_x}{G_{12}}$$

For failure,

$$0.3214\sigma_x = F_6 \text{ that is, } \sigma_x = 21.78 \text{ kg/mm}^2$$

Comparing the three modes of failure, we see that the shear mode governs the failure, the strength being the lowest in that mode. Hence, the ply strength at 70° = 21.78 kg/mm$^2$.

**Case (2): Strength as a function of θ.**
Given that

$$\sigma_x = \sigma_{xf}$$

$$\sigma_y = \tau_{xy} = 0$$

$$m = \cos\theta \quad \text{and} \quad n = \sin\theta$$

We can evaluate the strength of the composite as a function of $\theta$, by supposing, in sequence, that the failure strains $\epsilon_{1f}, \epsilon_{2f}, \gamma_{12f}$ are exceeded.

Suppose that $\epsilon_{1f}$ is exceeded

$$\epsilon_{1f} = \left(\frac{\sigma_1}{E_{11}} - \nu_{12}\frac{\sigma_2}{E_{22}}\right) = \frac{\sigma_{xf}}{E_{11}}(m^2 - \nu_{12}n^2)$$

Since

$$\epsilon_{1f} = \frac{F_1^T}{E_{11}}, \quad \text{or,} \quad \frac{F_1^C}{E_{11}}$$

We require that for failure to occur,

$$\sigma_{xf} \geq \left(\frac{F_1^T}{m^2 - \nu_{12}n^2}\right) \quad \text{or} \quad \left(\frac{F_1^C}{m^2 - \nu_{12}n^2}\right)$$

In a similar manner, by supposing that $\epsilon_{2f}$ is exceeded, we can obtain another condition as follows.

$$\sigma_{xf} \geq \left(\frac{F_2^T}{-\nu_{21}m^2 + n^2}\right) \quad \text{or} \quad \left(\frac{F_2^C}{-\nu_{21}m^2 + n^2}\right)$$

Lastly, by supposing that $\gamma_{12}$ is exceeded, we can obtain a third condition on $\sigma_{xf}$, as

$$\sigma_{xf} \geq \frac{F_6}{mn}$$

Evidently, the strength of the ply in any direction $\theta$ is its lowest value that satisfies all three of the conditions. This is easily obtained by plotting the conditions as curves, and taking the lowest of the curves. Figure 7.21 gives the plot of the equations. As in the previous example, the shaded area is different for tensile and compressive loading.

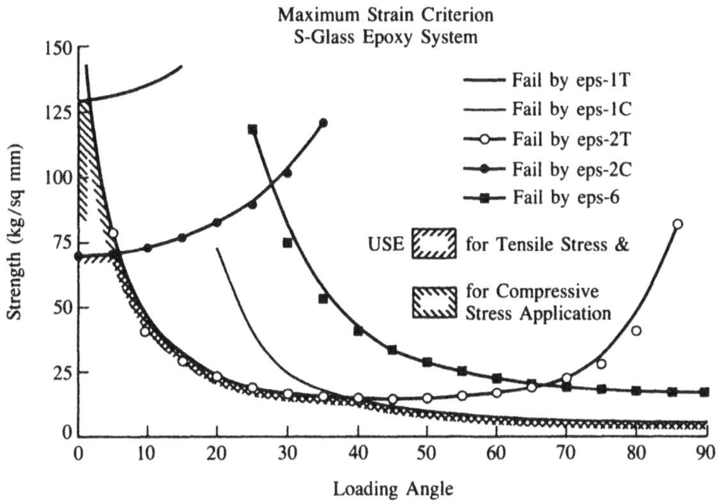


**FIGURE 7.21.** Off-axis strengths according to the maximum strain criterion.

## Example 7.16

Use the Tsai-Hill criterion to determine whether the given biaxial stress state causes failure in the Kevlar-Epoxy composite ply. Find the available safety factor $S_f$. The fiber is at $\theta = 35°$ to the loading $x$ axis.

$$\sigma_x = 3; \quad \sigma_y = 3; \quad \tau_{xy} = 1$$

$$F_1^T = 130; \quad F_1^C = 34; \quad F_2^T = 3; \quad F_2^C = 16; \quad F_6 = 5$$

All stress and strength are in kg/mm$^2$ units.

*Solution*

$$m = \cos 35°; \quad n = \sin 35°$$

$$\sigma_1 = \sigma_x m^2 + \sigma_y n^2 + 2mn\tau_{xy}$$

$$= 3.94 \text{ kg/mm}^2$$

$$\sigma_2 = \sigma_x n^2 + \sigma_y m^2 - 2mn\tau_{xy}$$

$$= 2.06 \text{ kg/mm}^2$$

$$\tau_{12} = (\sigma_y - \sigma_x)mn + (m^2 - n^2)\tau_{xy}$$

$$= 0.342 \text{ kg/mm}^2$$

The Tsai-Hill criterion requires that for failure,

$$\frac{\sigma_1^2}{F_1^{T^2}} + \frac{\sigma_2^2}{F_2^{T^2}} + \frac{\tau_{12}^2}{F_6^2} - \frac{\sigma_1\sigma_2}{F_1^T F_2^T} = 1$$

$$\frac{3.94^2}{130^2} + \frac{2.06^2}{3^2} + \frac{0.342^2}{5^2} - \frac{3.94 \times 2.06}{130 \times 5} = 0.465 < 1$$

Hence no failure occurs. Since $S_f$ is the factor by which all stresses must be scaled to just avoid failure, we consider $S_f\sigma_1$, $S_f\sigma_2$, and $S_f\tau_{12}$ such that the criterion is just $= 1$. That leads to

$$S_f^2\left(\frac{3.94^2}{130^2} + \frac{2.06^2}{3^2} + \frac{0.342^2}{5^2} - \frac{3.94 \times 2.06}{130 \times 5}\right) = 1$$

Available factor of safety $S_f = \sqrt{1/0.465} = 1.467$.

## 7.10 FACTOR OF SAFETY (FOS)

The term *Factor of Safety* in composites is used in a rather special sense and needs explanation. Let us suppose that a stress state represented by $\{\boldsymbol{\sigma}\} = [\sigma_x, \sigma_y, \tau_{xy}]$ is below the failure point. Then it is possible to scale the stresses up by a factor *Sf*, such that $(Sf) \times \{\boldsymbol{\sigma}\}$ will just produce failure of the ply. The factor *Sf* is called the *factor of safety*. In other words, each of the stress components is increased proportionately until failure.

It is important to note that the factor *Sf* has nothing to do with the *design safety factor* required by the product designer in order to carry a conservative margin on stress, strain, etc. The design safety factor (*DSF*) is based on experience, and is applied to the strength of a ply in order to get a design stress. On the contrary, $S_f$ is the margin by which an actual stress in a ply differs from failure.

The factor *Sf* can be positive or negative. When positive, it means that the stresses are magnified while retaining their sign. When negative, it means that the stresses are magnified while reversing their sign.

For example, when applied to the Tsai-Wu criterion, it leads to the quadratic equation,

$$(Sf)^2(\sigma_1^2 f_{11} + \sigma_2^2 f_{22} + \tau_{12}^2 f_{66} + 2\sigma_1\sigma_2 f_{12})$$
$$+ (Sf)(\sigma_1 f_1 + \sigma_2 f_2) - 1 = 0$$

If we set

$$A = (\sigma_1^2 f_{11} + \sigma_2^2 f_{22} + \tau_{12}^2 f_{66} + 2\sigma_1\sigma_2 f_{12})$$

$$B = (\sigma_1 f_1 + \sigma_2 f_2),$$

then the solutions for *Sf* are obtained as

$$(Sf)^+ = \frac{-B + \sqrt{B^2 + 4A}}{2A} \tag{7.42}$$

$$(Sf)^- = \frac{-B + \sqrt{B^2 + 4A}}{2A} \tag{7.43}$$

Obviously, $(Sf)^-$ is numerically larger than $(Sf)^+$.

The calculation of the factor of safety according to other criteria is tricky because of the way the strengths $F_1^*$ and $F_2^*$ are used.

**Example 7.17**

Using Tsai-Wu criterion, determine whether or not failure occurs in the Boron-Epoxy ply, when the following stress state is applied at a point at $\theta = 25°$. Also find the available safety factor *Sf*.

$$\sigma_x = 10; \quad \sigma_y = 5; \quad \tau_{xy} = 2$$

$$F_1^T = 140; \quad F_1^C = 240; \quad F_2^T = 8; \quad F_2^C = 22; \quad F_6 = 10$$

(All stresses and strengths are in kg/mm$^2$ units.)

*Solution* First, the material constants in the Tsai-Wu criterion are calculated.

$$f_1 = \frac{1}{F_1^T} - \frac{1}{F_1^C} = \frac{1}{140} - \frac{1}{240} = 0.002976$$

$$f_2 = \frac{1}{F_2^T} - \frac{1}{F_2^C} = \frac{1}{8} - \frac{1}{22} = 0.07954$$

$$f_{11} = \frac{1}{F_1^T F_1^C} = \frac{1}{140 \times 240} = 0.2976 \times 10^{-4}$$

$$f_{22} = \frac{1}{F_2^T F_2^C} = \frac{1}{8 \times 22} = 0.5682 \times 10^{-2}$$

$$f_{66} = \frac{1}{F_6^2} = \frac{1}{10^2} = 0.01$$

$$f_{12} = -\frac{1}{2}\sqrt{f_{11} f_{22}} = -\frac{1}{2}\sqrt{0.002976 \times 0.5682 \times 10^{-4}} = -0.2056 \times 10^{-3}$$

The stress components along the material axes are calculated.

$$\sigma_1 = \sigma_x \cos^2 25 + \sigma_y \sin^2 25 + \tau_{xy} 2 \cos 25 \sin 25$$

$$= 10.639 \text{ kg/mm}^2$$

$$\sigma_2 = \sigma_x \sin^2 25 + \sigma_y \cos^2 25 - \tau_{xy} 2 \cos 25 \sin 25$$

$$= 4.631 \text{ kg/mm}^2$$

$$\tau_{12} = (\sigma_y - \sigma_x) \cos 25 \sin 25 + \tau_{xy} (\cos^2 25 - \sin^2 25)$$

$$= -0.630$$

According to the Tsai-Wu criterion, failure occurs if

$$g(\boldsymbol{\sigma}) \equiv f_1 \sigma_1 + f_2 \sigma_2 + f_{11} \sigma_1^2 + f_{22} \sigma_2^2 + f_{66} \tau_{12}^2 + 2 f_{12} \sigma_1 \sigma_2 = 1$$

Substituting, we obtain,

$$g(\boldsymbol{\sigma}) = 0.4748 < 1$$

Hence, no failure occurs.

To calculate the available factor of safety, we scale up all stresses by a factor $S_f$. That leads to the quadratic

$$A S_f^2 + B S_f - 1 = 0$$

where, $A = f_{11} \sigma_1^2 + f_{22} \sigma_2^2 + f_{66} \tau_{12}^2 + 2 f_{12} \sigma_1 \sigma_2 = 0.0648$

$B = f_1 \sigma_1 + f_2 \sigma_2 = 0.410$

Solving the quadratic, $S_f = 1.8803$ or $-8.201$. The negative sign indicates the $S_f$ available if all the stresses were reversed in sign.

### 7.10.1 Another Interim Summary

We have already introduced a number of concepts and formulas for the composite ply. At this stage, it is a good idea to summarize the results for the ply.

- Given data:

$$\{\mathbf{E}_{12}\}, \{\mathbf{E}_{12}\}, E_{11}, E_{22}, \nu_{12}, G_{12}\ F_1^T, F_1^C, F_2^T, F_2^C, F_6$$

- To find mathematical moduli in 1-2 system, use Equations 7.4 through 7.5. They give $\{\mathbf{Q}_{12}\}$ in terms of $\{\mathbf{E}_{12}\}$.
- To find compliances in 1-2 system, use Equations 7.1, and 7.2, which give $\{\mathbf{S}_{12}\}$ in terms of $\{\mathbf{E}_{12}\}$
- To find mathematical moduli and compliances in any $x$-$y$ system, proceed according to the flow-chart in Figure 7.22.
- To find the engineering moduli in the $x$-$y$ coordinate system, do all the calculations as in the figure, and then use Equations 7.39.
- To find stresses along the fiber directions back again, use Section 1.7 or Mohr's circle.
- To apply a failure criterion, study Examples 14, 15, and 16 above, and use a failure theory from Section 7.9.

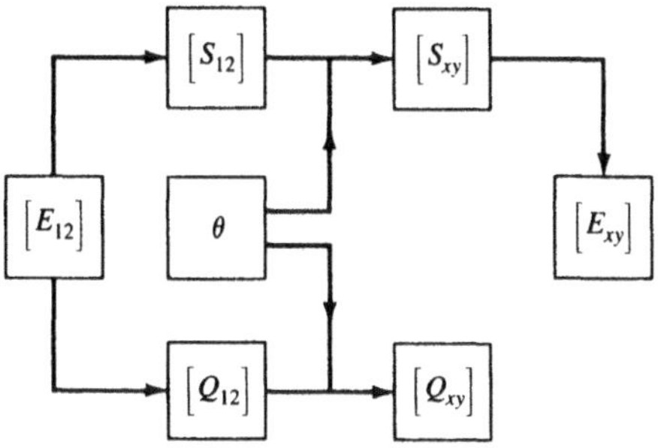


**FIGURE 7.22.** Procedure for computing moduli and compliances in any general $x$-$y$ coordinate system.

- To calculate the available factor of safety, use Equations 7.42 and 7.43 of Section 7.10.

## 7.11 FAILURE ENVELOPES

In practice, the stress state is not uniaxial, as is the case with all structures. A ply can take up two-dimensional stress, with $\sigma_1$, $\sigma_2$, and $\tau_{12}$ being independent. In such cases, the occurrence or absence of failure can be represented as a set of curves in a two-dimensional $\sigma_1$-$\sigma_2$ plane, with $\tau_{12}$ as a parameter. One can view the failure envelope as the curve for which $S_f = 1$ exactly. So, a point inside the envelope is a safe point and those on or outside it unsafe.

***Failure Envelope for Maximum Stress Criterion***

The failure envelope for the maximum stress criterion is trivially simple. It is a rectangle bounded by the lines $\sigma_1 = F_1^T$, $\sigma_2 = F_2^T$, $\sigma_1 = F_1^C$, and $\sigma_2 = F_2^C$. The presence or absence of shear stress $\tau_{12}$ does not affect these boundaries. See Figure 7.23.

***Failure Envelope for Maximum Strain Criterion***

The envelope for the maximum strain criterion is a parallelogram, the edges of which are not parallel to any of the axes. The slopes of the sides are $\nu_{12}$ and $\nu_{21}$. To draw these lines, we go from the first quadrant on the

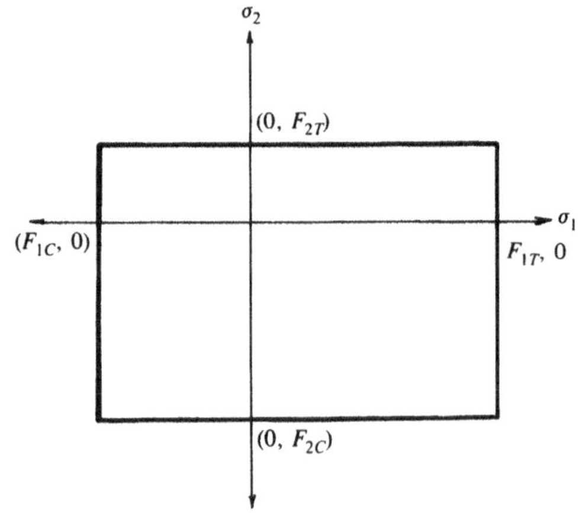


**FIGURE 7.23.** Failure envelope for two dimensional stress state in a composite, using the maximum stress criterion.

$\sigma_1$-$\sigma_2$ plane. In the first quadrant,

$$\frac{\sigma_1}{E_{11}} - \nu_{12}\frac{\sigma_2}{E_{11}} = \frac{F_1^T}{E_{11}}$$

Hence,

$$\sigma_1 - \nu_{12}\sigma_2 = F_1^T$$

This is a straight line passing through $(F_1^T, 0)$, and is marked #1 in Figure 7.24. So also, we have the relation for transverse strain, which is

$$-\nu_{21}\sigma_1 + \sigma_2 = F_2^T$$

This straight line passes through $(0, F_2^T)$ and is marked #2 in the diagram. We can consider the third quadrant next, where both the stresses are negative. We would likewise obtain two straight lines parallel to #1 and #2, and passing through the points $(F_1^C, 0)$ and $(0, F_2^C)$, respectively. Together the lines enclose a parallelogram, which is the envelope for the maximum strain criterion.

In the second and fourth quadrants, the criterion will be governed by $\epsilon_1^T$, and $\epsilon_2^T$ for some combinations of normal stresses and by $\epsilon_1^C$, and $\epsilon_2^C$ for the rest.

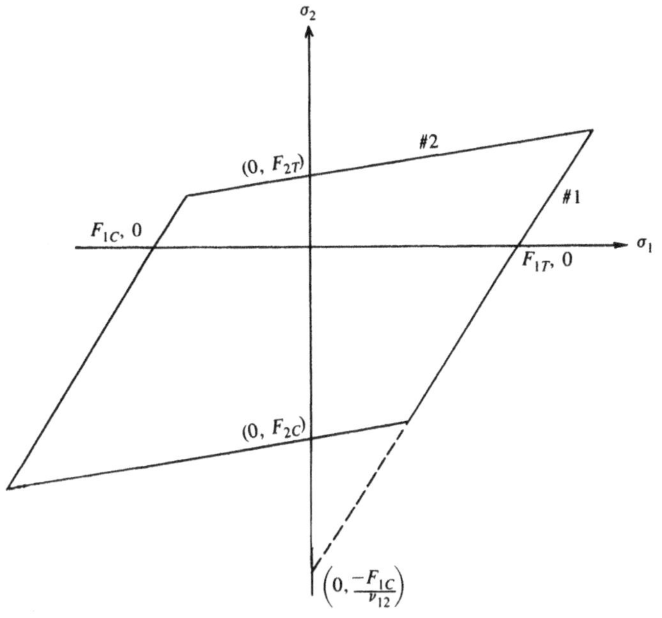


**FIGURE 7.24.** Failure envelope according to the maximum strain criterion.

***Failure Envelope for the Tsai-Hill Criterion***
The form of the equation for the Tsai-Hill criterion represents an ellipse. However, the constants for the ellipse are the uniaxial strengths $F_1^T, \ldots$, etc., which are different for different quadrants. Thus, the Tsai-Hill envelope is really one of four different elliptical quadrants patched together. The larger the magnitude of the shear stress $\tau_{12}$, the smaller the size of the envelope.

$$g(\sigma) \equiv \left(\frac{\sigma_1}{F_1^*}\right)^2 - \left(\frac{\sigma_1}{F_1^*}\right)\left(\frac{\sigma_2}{F_2^*}\right) + \left(\frac{\sigma_2}{F_2^*}\right)^2 = 1 - \left(\frac{\tau_{12}}{F_6}\right)^2$$

Obviously, for larger values of $\tau_{12}$, the size of the ellipse shrinks. This means that the permissible combinations of normal stresses is limited. Figure 7.25 shows the failure envelope. Also, the major and minor axes of the ellipse are at 45° and 13.5°, respectively.

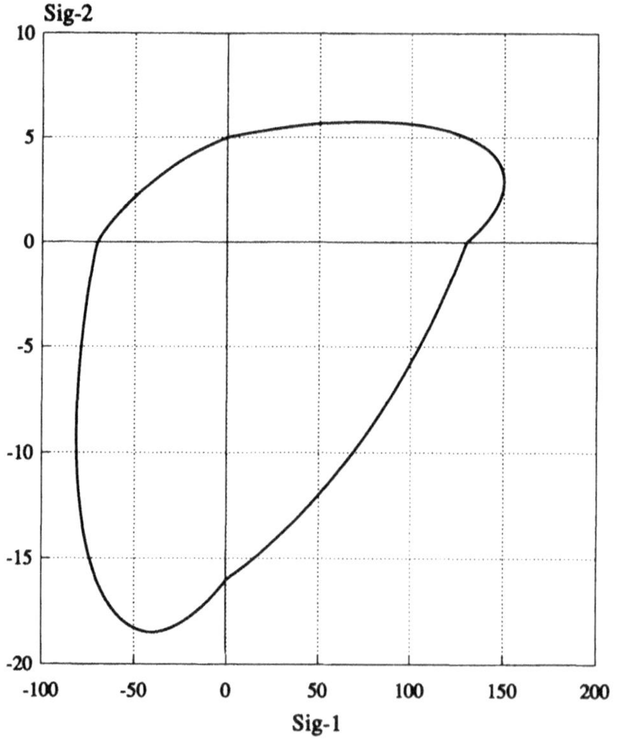


**FIGURE 7.25.** Failure envelope represented by Tsai-Hill criterion.

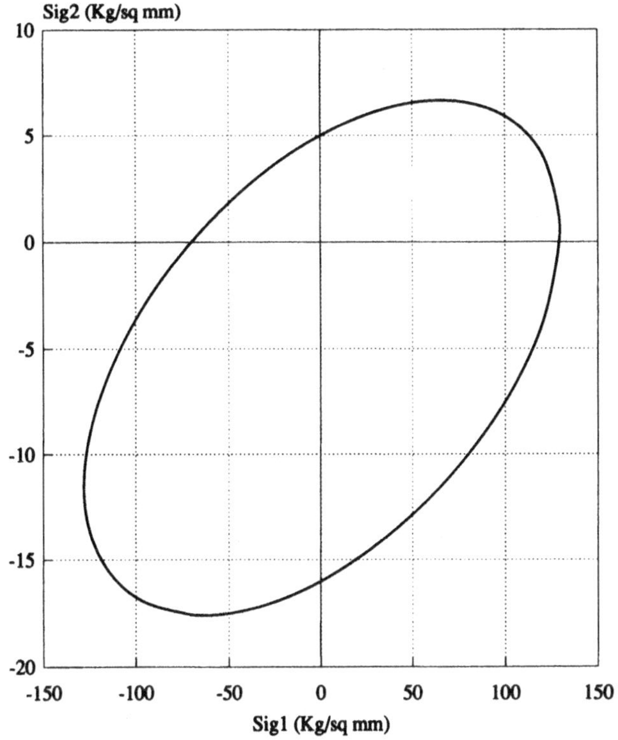


FIGURE 7.26. Failure envelope represented by Tsai-Wu criterion.

***Failure Envelope for the Tsai-Wu Criterion***

The failure envelope for Tsai-Wu criterion is an exact ellipse. It passes through the same set of points on the $x$ and $y$ axes as the other criteria. Figure 7.26 represents the failure envelope using the Tsai-Wu criterion. This criterion uses different constants in the four quadrants and hence appears as an ellipse with discontinuous slopes.

Sometimes, it is advantageous to know the mode of failure. For instance, if we know that the matrix cracked in a ply due to $\sigma_2$, the fibers are still intact and functional. Hence, they will withstand some more load along the fibers, even after matrix cracking. Knowing this can be useful in the design. The maximum stress and maximum strain criteria have the advantage of predicting what constituent would fail in the ply—whether it is the matrix or the fiber. The Tsai-Hill and Tsai-Wu criteria cannot predict which constituent will fail.

### 7.11.1 The Most Restrictive Envelope

It must be recognized that the failure criteria are rules of observation, supported by some marginal theoretical consideration. The ultimate proof

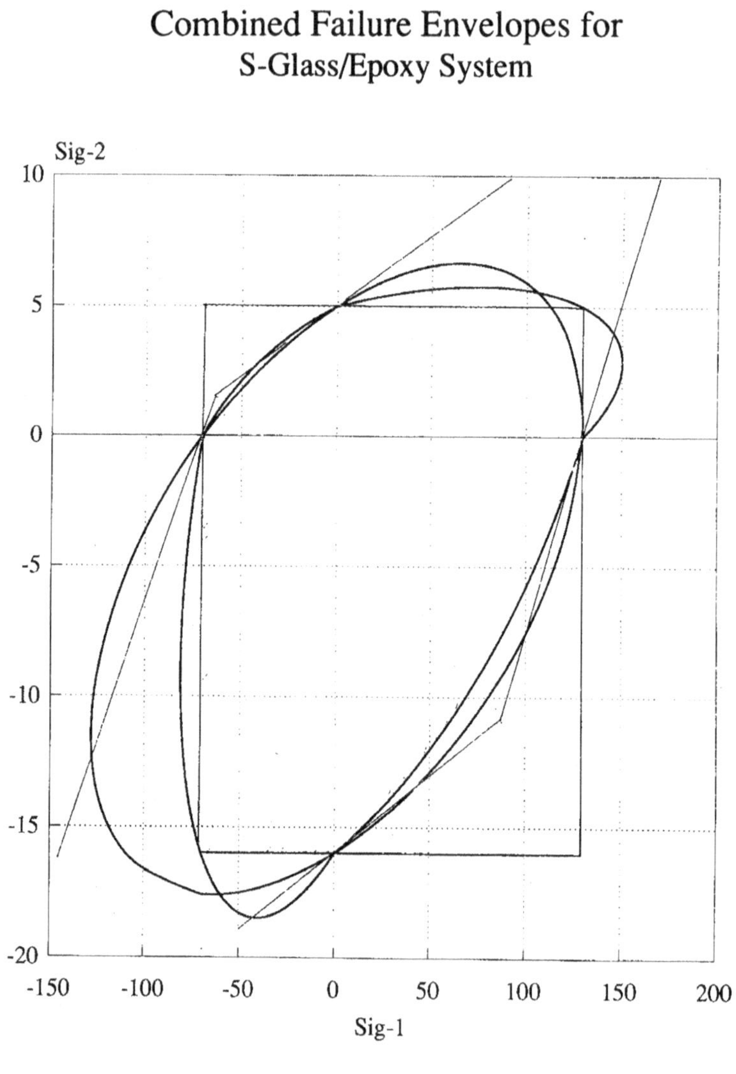


**FIGURE 7.27.** Innermost failure envelope represents maximum conservatism.

of the criteria comes from experimental verification. Since composites can fail in a variety of different modes, different composites tend to behave differently. Depending upon the combination of properties (such as ductility of matrix, strength of fiber, etc.) a composite may conform better with one theory than with another. In general, the Tsai-Wu criterion is known to fit with experimental results best, for many commonly used composites. However, experts believe that in order to be most conservative, the most restrictive envelope may be used as the design criterion. It is obtained as the innermost boundary, when all the four criteria are superimposed on the same diagram. See Figure 7.27.

## 7.12 MECHANICS OF LAMINATED PLATES

A *laminate* is a plate-like structure constructed by stacking several plies of composites, which are bonded together. As a result of the bonding, no ply can slip past the adjacent ones. In other words, the laminate bends and stretches as one thick plate. In a laminate, each ply can have its own thickness, orientation, and material. However, in practice laminates are made of the same material and thickness. The major design variable is the angle of orientation of the plies. A basic premise of the following discussion is that laminates are used as a plate or shell type of structure.

The important point about the theory of laminated plates is that the elastic behavior of a ply-in-laminate is not the same as if it were an independent ply. It is affected by the other plies in the laminate.

As a result of this, it would be incorrect to calculate the ply stress using the overall strains and overall modulus of the laminate. For example, consider a laminate made of eight plies, with a thickness of 0.0025 mm each, and each having a different orientation. The total thickness is 0.020 mm. Let us also suppose that the laminate is subjected to a tensile load of 1.5 kg/lineal mm of the laminate. The *average stress in the laminate* is $1.5/0.02 = 75\ \text{kg/mm}^2$. However, this is not necessarily the stress in any of the individual plies. Nor is the stress state in any ply one-dimensional. It is, in general, two dimensional, although the loading is only one-dimensional.

The reason for this is that plies share the applied load unequally, in proportion to their stiffnesses in the direction of the load. Since the plies have different stiffnesses because of their fiber orientation, the ply stresses are different in the direction of the load.

Also, since the laminate deflects without slippage of the plies, we may suppose the entire plate to be a homogeneous plate of a single material. However, this assumption is good only up to a certain stage in the calculation. For example, if we want to calculate the ply stresses from the laminate strains, this model is not adequate.

As a contrast to the isotropic plates, the stress in a laminate is *not necessarily* maximum at the outer surfaces. Furthermore, the ply carrying the maximum stress may be different for different loading directions. A subsurface ply can fail at a lower load than the outermost plies. In order to track such failures, one needs to compute the stresses in the individual plies, apply a failure criterion to each of them, and then assess the available factor of safety in each.

## 7.13 STRESS ANALYSIS OF A LAMINATE POINT

In this section we will develop a procedure to calculate the stresses in the individual plies in a laminate which arise as a result of the forces and moments applied at that point. For this, we need to make the following three assumptions, which are similar to the assumptions made in the beam theory, or the classical plate/shell theories.

1. There is a neutral plane, on which the bending strains are zero in all directions. However, extensional strains may be non-zero.
2. The neutral plane is the midplane of the laminate.
3. Planes which are normal to the midplane before deformation, remain plane and normal after deformation.

The stresses in the plies at a point in a laminate must satisfy the equilibrium conditions. That means, the stresses in all the plies together must be in equilibrium with the applied forces and moments in each direction.

Consider a laminate as shown in Figure 7.28, whose downward deflections along $z$ are considered positive. The original positions are in dotted lines and the displaced positions are in solid lines.

Let $U_o$ = $x$-Displacement of the point $A$ on the neutral axis,
$U$ = $x$-Displacement of any point such as $C$, located at a distance $z$ from the neutral axis.

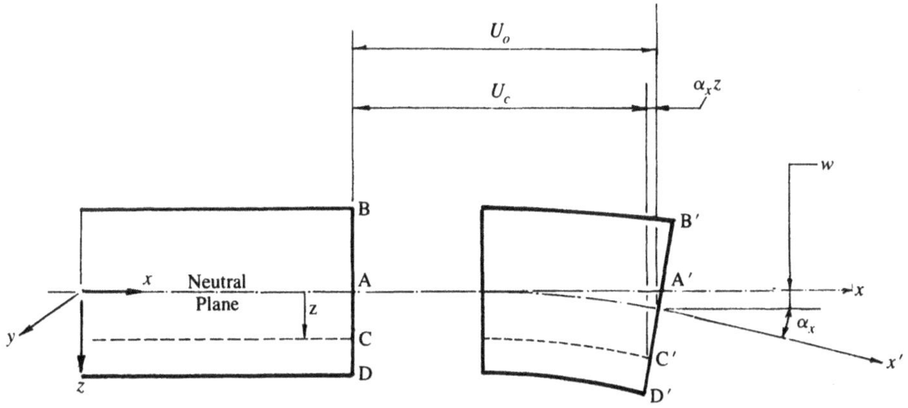


**FIGURE 7.28.** Bending of a laminated composite.

$V_o$ = $y$-Displacement of the point $A$ on the neutral axis.
$V$ = $y$-Displacement of the general point $C$.
$w$ = $z$-Displacement of the point $A$, at the midplane.
$\alpha_x, \alpha_y$ = Angles made by the tangent to the surface, with the $x$ and $y$ axes, as in Figure 7.28.

From the figure, it is obvious that

$$\alpha_x = \frac{\partial w}{\partial x}$$

$$\alpha_y = \frac{\partial w}{\partial y}$$

Also

$$U = (U_o - z\alpha_x) = \left(U_o - z\frac{\partial w}{\partial x}\right)$$

$$V = (V_o - z\alpha_y) = \left(V_o - z\frac{\partial w}{\partial y}\right)$$

The strain components at $C$ along the $x$ and $y$ axes are obtained from the definitions, Equations 1.2, 1.3, and 1.4.

$$\epsilon_x = \frac{\partial U}{\partial x} = \left(\frac{\partial U_o}{\partial x} - z\frac{\partial^2 w}{\partial x^2}\right) = \left(\epsilon_{x,o} - z\frac{\partial^2 w}{\partial x^2}\right)$$

$$\epsilon_y = \frac{\partial U}{\partial y} = \left(\frac{\partial U_o}{\partial y} - z\frac{\partial^2 w}{\partial y^2}\right) = \left(\epsilon_{y,o} - z\frac{\partial^2 w}{\partial x^2}\right)$$

$$\gamma_{xy} = \left(\frac{\partial U}{\partial y} + \frac{\partial V}{\partial x}\right) = \left(\frac{\partial U_o}{\partial y} + \frac{\partial V_o}{\partial x} - 2z\frac{\partial^2 w}{\partial x \partial y}\right) = \left(\gamma_{xy,o} - 2z\frac{\partial^2 w}{\partial x \partial y}\right)$$

where

$\epsilon_{x,o}, \epsilon_{y,o}, \gamma_{xy,o}$ = Components of in-plane extensional strains at the midsurface.

Putting the above equations in matrix form, we can write

$$\begin{Bmatrix} \epsilon_x \\ \epsilon_y \\ \gamma_{xy} \end{Bmatrix} = \begin{Bmatrix} \epsilon_{x,o} \\ \epsilon_{y,o} \\ \gamma_{xy,o} \end{Bmatrix} + z \begin{Bmatrix} -\dfrac{\partial^2 w}{\partial x^2} \\ -\dfrac{\partial^2 w}{\partial y^2} \\ -2\dfrac{\partial^2 w}{\partial x \partial y} \end{Bmatrix}$$

$$\{\boldsymbol{\epsilon}\} = \{\boldsymbol{\epsilon}\}_o + z\{\boldsymbol{\kappa}\}_o \tag{7.44}$$

Equation 7.44 states that the strains vary linearly with $z$. This conforms to the assumption that "plane and normal cross sections remain plane and normal." A consequence of this equation is that ply stresses must vary in proportion to their stiffnesses (moduli).

The in-plane stress components in the ply are simply equal to (Stiffness) × (Strains) . Also, we add a subscript $k$ to the strain vector, to denote the $k$-th ply.

$$\{\boldsymbol{\sigma}\}_k = [\mathbf{Q}_{xy}]_k\{\boldsymbol{\epsilon}\}_k$$
$$= [\mathbf{Q}_{xy}]_k\{\boldsymbol{\epsilon}\}_o + z_k[\mathbf{Q}_{xy}]_k\{\boldsymbol{\kappa}\}_o \quad (7.45)$$

where,

$[\mathbf{Q}_{xy}]_k$ = Stiffness matrix for the $k$-th ply, along the $x$-$y$ axes

$\{\boldsymbol{\sigma}\}_k$ = Stress in the $k$-th ply in the the $x$-$y$ system.

$Z_k$ = Distance of the $k$-th ply from the neutral axis.

As the last step, we consider the equilibrium of the stresses in Equation 7.45 with applied external forces and moments. It is the practice in plate theories to consider force and moment at a point *per unit length*, rather than absolute force or moment. This is so because the force and moment have a lineal distribution in a plate structure. The forces and moments that can cause in-plane stresses in the plies are marked in Figure 7.29. They are

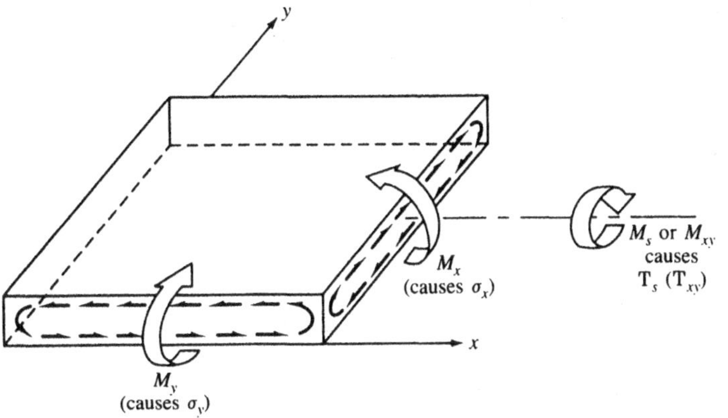


**FIGURE 7.29.** The laminate subjected to bending moments and extensional forces.

respectively, $N_x$, $N_y$, $N_s$, $M_x$, $M_y$ and $M_s$. From Figure 7.29 it is seen that

$$T = \sum t_k$$

$$z_o = -\frac{T}{2}$$

$$z_1 = z_o + t_1$$

$$z_2 = z_1 + t_2$$

$$z_3 = z_2 + z_3$$

$$\vdots$$

$$z_n = z_{n-1} + t_n = \frac{T}{2} \tag{7.46}$$

where, $n$ = total number of plies in the laminate.

***Force Equilibrium***
For force equilibrium,

$$\left\{\begin{matrix}\text{The total of ply} \\ \text{forces summed up} \\ \text{over the plies}\end{matrix}\right\} = \left\{\begin{matrix}\text{Applied force} \\ \text{on laminate} \\ \text{per unit length}\end{matrix}\right\}$$

In quantitative terms this means that stress, integrated through the laminate thickness must be equal to the corresponding force. The force contribution for any single ply is obtained by integrating the stress in that ply, over the thickness of that ply.

$$\text{Force contribution of the } k\text{-th ply}\} = \int_{z_{k-1}}^{z_k} \{\boldsymbol{\sigma}\}_k \, dz$$

$$= \int_{z_{k-1}}^{z_k} \left\{\begin{matrix}\sigma_{x,k} \\ \sigma_{y,k} \\ \tau_{xy,k}\end{matrix}\right\} dz$$

The total force is calculated by adding similar contributions from all the plies. Thus,

$$\begin{Bmatrix} N_x \\ N_y \\ N_s \end{Bmatrix} = \sum_{k=1}^{n} \int_{z_{k-1}}^{z_k} \begin{Bmatrix} \sigma_{x,k} \\ \sigma_{y,k} \\ \tau_{xy,k} \end{Bmatrix} dz$$

$$= \sum_{k=1}^{n} \int_{z_{k-1}}^{z_k} \{\boldsymbol{\sigma}\}_k \, dz$$

$$= \sum_{k=1}^{n} \int_{z_{k-1}}^{z_k} \left[ [\mathbf{Q_{xy}}]_k \{\boldsymbol{\epsilon}\}_o + z[\mathbf{Q_{xy}}]_k \{\boldsymbol{\kappa}\}_o \right] dz \tag{7.47}$$

Since $[\mathbf{Q_{xy}}]$, $\{\boldsymbol{\epsilon}\}_o$, and $\{\boldsymbol{\kappa}\}_o$ are constants for a ply, they can be treated as constants in the integration. Therefore,

$$\{\mathbf{N}\} = \left[ \sum_{k=1}^{n} [\mathbf{Q_{xy}}]_k \int_{z_{k-1}}^{z_k} dz \right] \{\boldsymbol{\epsilon}\}_o + \left[ \sum_{k=1}^{n} [\mathbf{Q_{xy}}]_k \int_{z_{k-1}}^{z_k} z \, dz \right] \{\boldsymbol{\kappa}\}_o$$

$$= \left[ \sum_{k=1}^{n} (z_k - z_{k-1}) [\mathbf{Q_{xy}}]_k \right] \{\boldsymbol{\epsilon}\}_o + \left[ \sum_{k=1}^{n} \frac{1}{2} (z_k^2 - z_{k-1}^2) [\mathbf{Q_{xy}}]_k \right] \{\boldsymbol{\kappa}\}_o$$

$$= [\mathbf{A}]\{\boldsymbol{\epsilon}\}_o + [\mathbf{B}]\{\boldsymbol{\kappa}\}_o \tag{7.48}$$

where,

$$[\mathbf{A}] = \sum_{k=1}^{n} (z_k - z_{k-1}) [\mathbf{Q_{xy}}]_k \tag{7.49}$$

and,

$$[\mathbf{B}] = \sum_{k=1}^{n} \frac{1}{2} (z_k^2 - z_{k-1}^2) [\mathbf{Q_{xy}}]_k \tag{7.50}$$

Looking at Equation 7.48 we find that in general, the tensile or compressive forces $\{\mathbf{N}\}$ can produce changes in curvatures $\{\boldsymbol{\kappa}\}$, (that is, bending strains) in addition to stretching and squeezing. This is a peculiarity of composites behavior.

***Moment Equilibrium***

For equilibrium of moments we have a similar relation.

$$\begin{Bmatrix} \text{The total of the moments} \\ \text{due to the ply stresses,} \\ \text{taken about the mid-plane} \end{Bmatrix} = \begin{Bmatrix} \text{Applied moment} \\ \text{per unit length} \\ \text{of the laminate} \end{Bmatrix}$$

As before, we consider the moment contribution of the $k$-th ply, and sum similar contributions over all the plies.

$$\begin{Bmatrix} \text{Moment Contribution} \\ \text{of the } k\text{-th ply} \end{Bmatrix} = \int_{z_{k-1}}^{z_k} \begin{Bmatrix} \sigma_{x,k} \\ \sigma_{y,k} \\ \tau_{xy} \end{Bmatrix} z\,dz$$

The total moments are got by adding similar contributions from all the plies, by summing up over $k$. Thus,

$$\begin{Bmatrix} M_x \\ M_y \\ M_s \end{Bmatrix} = \sum_{k=1}^{n} \int_{z_{k-1}}^{z_k} \begin{Bmatrix} \sigma_{x,k} \\ \sigma_{y,k} \\ \tau_{xy} \end{Bmatrix} z\,dz$$

$$= \sum_{k=1}^{n} \int_{z_{k-1}}^{z_k} \{\boldsymbol{\sigma}\}_k z\,dz$$

$$= \sum_{k=1}^{n} \int_{z_{k-1}}^{z_k} \left[ z[\mathbf{Q_{xy}}]\{\boldsymbol{\epsilon}\}_o + z^2[\mathbf{Q_{xy}}]\{\kappa\}_o \right] dz$$

As before, $[\mathbf{Q_{xy}}]$, $\{\kappa\}_o$, and $\{\kappa\}_o$ are constants for the ply, and hence,

$$\{\mathbf{M}\} = \left[ \sum_{k=1}^{n} [\mathbf{Q_{xy}}]_k \int_{z_{k-1}}^{z_k} z\,dz \right] \{\boldsymbol{\epsilon}\}_o + \left[ \sum_{k=1}^{n} [\mathbf{Q_{xy}}]_k \int_{z_{k-1}}^{z_k} z^2 dz \right] \{\boldsymbol{\kappa}\}_o$$

$$= \left[ \sum_{k=1}^{n} \frac{1}{2}(z_k^2 - z_{k-1}^2)[\mathbf{Q_{xy}}]_k \right] \{\boldsymbol{\epsilon}\}_o + \left[ \sum_{k=1}^{n} \frac{1}{3}(z_k^3 - z_{k-1}^3)[\mathbf{Q_{xy}}]_k \right] \{\boldsymbol{\kappa}\}_o$$

$$= [\mathbf{B}]\{\boldsymbol{\epsilon}\}_o + [\mathbf{D}]\{\boldsymbol{\kappa}\}_o \qquad (7.51)$$

where,

$$[\mathbf{D}] = \sum_{k=1}^{n} \frac{1}{3}(z_k^3 - z_{k-1}^3)[\mathbf{Q_{xy}}]_k \qquad (7.52)$$

Looking at the Equation 7.51 we note that in general, moments can produce stretching and squeezing strains, $\{\boldsymbol{\epsilon}\}_o$ in addition to changes in curvatures.

Combining Equations 7.48 and 7.51, we get the following equation for a general laminate.

$$\begin{Bmatrix} \mathbf{N} \\ \mathbf{M} \end{Bmatrix} = \begin{bmatrix} \mathbf{A} & \mathbf{B} \\ \mathbf{B} & \mathbf{D} \end{bmatrix} \begin{Bmatrix} \boldsymbol{\epsilon}_o \\ \boldsymbol{\kappa}_o \end{Bmatrix} \qquad (7.53)$$

**Example 7.18**

A laminate is made of four graphite-epoxy plies, each 1 mm thick. They are stacked and bonded to form a laminate. From bottom to top, they are kept at angles 0°, 90°, 90°, and 0°. The moduli in kg/mm$^2$ units of the material of the plies are

$$E_{11} = 15000; \quad E_{22} = 1000; \quad G_{12} = 600; \quad \nu_{12} = 0.27$$

Construct the $[ABD]$ matrix along the 0°–90° axes system.

*Solution* It is a good practice to start with a tabular statement to organize the given data

| Ply No. | Angle | Thickness (mm) | z (mm) |
|---|---|---|---|
| 1 | 0 | 1 | −1 |
| 2 | 90 | 1 | 0 |
| 3 | 90 | 1 | 1 |
| 4 | 0 | 1 | 2 |

$$\nu_{21} = \frac{E_{22}}{E_{11}} \nu_{12} = 0.018$$

$$Q_{11} = \frac{E_{11}}{1 - \nu_{12}\nu_{21}} = 15073 \text{ kg/mm}^2$$

$$Q_{22} = \frac{E_{22}}{1 - \nu_{12}\nu_{21}} = 1005 \text{ kg/mm}^2$$

$$Q_{12} = \frac{\nu_{12}E_{22}}{1 - \nu_{12}\nu_{21}} = 271 \text{ kg/mm}^2$$

$$Q_{66} = G_{12} = 600 \text{ kg/mm}^2$$

The **[R]** matrix is trivially simple in this example. For the first and fourth plies, no transformation is needed, since $\theta = 0$.

$$[\mathbf{Q}_{xy}]_1 = [\mathbf{Q}_{xy}]_4 = \begin{bmatrix} 15073 & 271 & 0 \\ 271 & 1005 & 0 \\ 0 & 0 & 600 \end{bmatrix}$$

For the second and third plies, due to the 90° rotation, $Q_{11}$ and $Q_{22}$ interchange their values, and others remain as they are.

$$[\mathbf{Q}_{xy}]_2 = [\mathbf{Q}_{xy}]_3 = \begin{bmatrix} 1005 & 271 & 0 \\ 271 & 15073 & 0 \\ 0 & 0 & 600 \end{bmatrix}$$

$$[\mathbf{A}] = \sum_{k=1}^{4} [\mathbf{Q}_{xy}]_k (z_k - z_{k-1})$$

$$= t\{[\mathbf{Q}_{xy}]_1 + [\mathbf{Q}_{xy}]_2 + [\mathbf{Q}_{xy}]_3 + [\mathbf{Q}_{xy}]_4\}$$

$$= (1)\begin{bmatrix} 32156 & 1084 & 0 \\ 1084 & 32156 & 0 \\ 0 & 0 & 2400 \end{bmatrix} \text{kg/mm} \leftarrow \text{(Note units)}$$

$$[\mathbf{B}] = \frac{1}{2}\sum_{k=1}^{4} [\mathbf{Q}_{xy}]_k (z_k^2 - z_{k-1}^2)$$

$$= [\mathbf{Q}_{xy}]_1 \left\{\frac{(-1)^2 - (-2)^2}{2} + \frac{(2)^2 - (1)^2}{2}\right\}$$

$$+[\mathbf{Q}_{xy}]_2 \left\{\frac{(0)^2 - (-1)^2}{2} + \frac{(1)^2 - (0)^2}{2}\right\}$$

$$= [0] + [0] = [0]$$

$$[\mathbf{D}] = \frac{1}{3}\sum_{k=1}^{4} [\mathbf{Q}_{xy}]_k (z_k^3 - z_{k-1}^3)$$

$$= [\mathbf{Q}_{xy}]_1 \left\{\frac{(-1)^3 - (-2)^3}{3} + \frac{(2)^3 - (1)^3}{2}\right\}$$

$$+[\mathbf{Q}_{xy}]_2 \left\{\frac{(0)^3 - (-1)^3}{3} + \frac{(1)^3 - (0)^3}{3}\right\}$$

$$= \begin{bmatrix} 71011 & 1445 & 0 \\ 1445 & 14739 & 0 \\ 0 & 0 & 3200 \end{bmatrix} \text{kg/mm} \leftarrow \text{(Note units)}$$

The $[ABD]$ matrix is simply an arrangement of the above matrices in the form

$$\begin{bmatrix} \mathbf{A} & \mathbf{B} \\ \mathbf{B} & \mathbf{D} \end{bmatrix}.$$

In the construction of the $[ABD]$ matrix, the $x$ and $y$ axes may not be interchanged since, the $[\mathbf{B}]$ and $[\mathbf{D}]$ matrices are proportional to $(z\ldots)^2$ and $(z\ldots)^3$, respectively. Given the fact that $[\mathbf{Q}_{xy}]$ depends on the angle, the interchange affects the values of the $[\mathbf{B}]$ and $[\mathbf{D}]$ matrices.

**Example 7.19**

For the laminate in Example 18, find the $[ABD]$ matrix along the 45°-135° axes system.

*Solution* In the first place we have to transform the $[\mathbf{Q}_{12}]$.

| Ply No. | Angle | Thickness (mm) | z (mm) |
|---|---|---|---|
| 1 | −45 | 1 | −1 |
| 2 | 45 | 1 | 0 |
| 3 | 45 | 1 | 1 |
| 4 | −45 | 1 | 2 |

The $[\mathbf{R}]$ matrix is easy to obtain.

$$[\mathbf{R}]_{-45^\circ} = \begin{bmatrix} 0.25 & 0.25 & 0.50 & 0.25 \\ 0.25 & 0.25 & 0.50 & 0.25 \\ 0.25 & 0.25 & 0.50 & -0.25 \\ 1.00 & 1.00 & -2.00 & 0 \\ -0.50 & 0.50 & 0 & 0 \\ -0.50 & 0.50 & 0 & 0 \end{bmatrix}$$

$$[\mathbf{R}]_{45^\circ} = \begin{bmatrix} 0.25 & 0.25 & 0.50 & 0.25 \\ 0.25 & 0.25 & 0.50 & 0.25 \\ 0.25 & 0.25 & 0.50 & -0.25 \\ 1.00 & 1.00 & -2.00 & 0 \\ 0.50 & -0.50 & 0 & 0 \\ 0.50 & -0.50 & 0 & 0 \end{bmatrix}$$

Note carefully that the terms $Q_{66}$, $Q_{ss}$, $Q_{xs}$, $Q_{ys}$ carry factors (4 or 2) with them. When written in the vector form, we have

$$\begin{Bmatrix} Q_{xx} \\ Q_{yy} \\ Q_{xy} \\ 4Q_{ss} \\ 2Q_{xs} \\ 2Q_{ys} \end{Bmatrix} = [\mathbf{R}] \ldots \begin{Bmatrix} Q_{11} \\ Q_{22} \\ Q_{12} \\ 4Q_{66} \end{Bmatrix}$$

We take the values of $Q_{11}, \cdots$ from Example 17, premultiply by [**R**] and rewrite the result in the square matrix notation.

$$[\mathbf{Q_{xy}}]_1 = [\mathbf{Q_{xy}}]_r = \begin{bmatrix} 4755 & 3555 & -3517 \\ 3555 & 4755 & -3517 \\ -3517 & -3517 & 3884 \end{bmatrix}$$

$$[\mathbf{Q_{xy}}]_2 = [\mathbf{Q_{xy}}]_3 = \begin{bmatrix} 4755 & 3555 & 3517 \\ 3555 & 4755 & 3517 \\ 3517 & 3517 & 3884 \end{bmatrix}$$

As before,

$$[\mathbf{A}] = t\{[\mathbf{Q_{xy}}]_1 + [\mathbf{Q_{xy}}]_2 + [\mathbf{Q_{xy}}]_3 + [\mathbf{Q_{xy}}]_4\}$$

$$= \begin{bmatrix} 19020 & 14220 & 0 \\ 14220 & 19020 & 0 \\ 0 & 0 & 15536 \end{bmatrix}$$

$$[\mathbf{B}] = \frac{1}{2} \sum_{k=1}^{4} [\mathbf{Q_{xy}}]_k (z_k^2 - z_{k-1}^2)$$

$$= [\mathbf{Q_{xy}}]_1 \left\{ \frac{(-1)^2 - (-2)^2}{2} + \frac{(2)^2 - (1)^2}{2} \right\}$$

$$+ [\mathbf{Q_{xy}}]_2 \left\{ \frac{(0)^2 - (-1)^2}{2} + \frac{(1)^2 - (0)^2}{2} \right\}$$

$$= [0] + [0] = [0]$$

$$[\mathbf{D}] = \frac{1}{3}\sum_{k=1}^{4} [\mathbf{Q}_{xy}]_k (z_k^3 - z_{k-1}^3)$$

$$= [\mathbf{Q}_{xy}]_1 \left\{ \frac{(-1)^3 - (-2)^3}{3} + \frac{(2)^3 - (1)^3}{2} \right\}$$

$$+ [\mathbf{Q}_{xy}]_2 \left\{ \frac{(0)^3 - (-1)^3}{3} + \frac{(1)^3 - (0)^3}{3} \right\}$$

$$= \begin{bmatrix} 25360 & 18961 & 14068 \\ 18961 & 25360 & 14068 \\ 14068 & 14068 & 20715 \end{bmatrix} \text{kg/mm}$$

Equation 7.53 is very important in the analysis of laminates, and is fundamental to several types of laminate calculations. We want to call the square matrix on the right side simply the $[ABD]$ matrix. It can be used for

- calculating ply-stress and ply-strains for a given laminate load,
- calculating ply-stress and laminate loads for a given mid-plane strain,
- calculating the overall laminate elastic properties,
- finding out whether or not an extensional load would produce a bending strain, and vice versa,
- determining the variation of laminate properties along different directions,
- comparative evaluation of different lay-ups,
- the optimization of lay-ups,
- calculating the hygral and thermal stresses in plies,
- estimating the stresses in the laminate due to curing,
- calculating the equivalent thermal expansion coefficient of the laminate,
- calculating the equivalent swelling coefficient of the laminate.

### 7.13.1 Apparent Moduli of Laminates

The apparent moduli of a laminate are those that are observable in a tension test of the laminate. They are also useful in describing the overall stress-strain (or Hooke's law) of the laminate. That, in turn can be used in a finite element analysis as the property of the "homogeneous" laminate, thus ignoring the presence of plies.

For symmetric laminates, (discussed in Section 7.14) the apparent moduli can be derived exactly from the [**A**] matrix. We may consider a unidirectional load, say $N_x = \sigma_x T$. Inverting Equation 7.48,

$$\{\boldsymbol{\epsilon}_0\} = [\mathbf{A}]^{-1}\{\boldsymbol{\sigma}\}T = [\mathbf{a}]\{\boldsymbol{\sigma}\}T$$

In particular

$$\epsilon_x = a_{xx}\sigma_x T$$

Hence,

$$\sigma_x = \frac{1}{a_{xx}T}\epsilon_x = \bar{E}_{xx}\epsilon_x$$

Hence,

$$\bar{E}_{xx} = \frac{1}{a_{xx}T} \tag{7.54}$$

Similar relationships can be obtained for the other directions as well.

In the following, we will review a few examples of how to construct the [$ABD$] matrix, its properties, and also use it to calculate some of the items listed above.

**Example 7.20**

Find out the modulus of the laminate in Examples 18 and 19 along the $x$ and $y$ direction.

*Solution* In 0°–90° directions:
From Example 7.17, we have the [A] matrix.

$$[\mathbf{A}] = \begin{bmatrix} 32156 & 1084 & 0 \\ 1084 & 32156 & 0 \\ 0 & 0 & 2400 \end{bmatrix} \text{kg/mm}$$

$$[\mathbf{A}]^{-1} = [\mathbf{a}] = 10^{-6}\begin{bmatrix} 31.1338 & -1.0495 & 0 \\ -1.0495 & 31.1338 & 0 \\ 0 & 0 & 416.67 \end{bmatrix}$$

Laminate moduli are obtained by Equations 7.54

$$\bar{E}_{xx} = \frac{1}{Ta_{xx}} = \frac{1}{4 \times 31.1338 \times 10^{-6}} = 8029 \text{ kg/mm}^2$$

$$\bar{E}_{yy} = \frac{1}{Ta_{xx}} = \frac{1}{4 \times 31.1338 \times 10^{-6}} = 8029 \text{ kg/mm}^2$$

$$\bar{\nu}_{xy} = -\frac{a_{xy}}{a_{xx}} = \frac{1.0495 \times 10^{-6}}{31.1338 \times 10^{-6}} = 0.0337$$

$$\bar{G}_{xy} = \frac{1}{Ta_{ss}} = \frac{1}{4 \times 416.67 \times 10^{-6}} = 600 \text{ kg/mm}^2$$

In the 45°–135° directions:

From Example 7.18, we have the [**A**] matrix along these axes.

$$[\mathbf{A}] = \begin{bmatrix} 19020 & 14220 & 0 \\ 14220 & 19020 & 0 \\ 0 & 0 & 15336 \end{bmatrix} \text{kg/mm}$$

$$[\mathbf{A}]^{-1} = [\mathbf{a}] = 10^{-6} \begin{bmatrix} 119.21 & -89.12 & 0 \\ -89.12 & 119.21 & 0 \\ 0 & 0 & 64.367 \end{bmatrix}$$

As before, moduli are obtained by Equations 7.54

$$\bar{E}_{xx} = \frac{1}{Ta_{xx}} = \frac{1}{4 \times 119.21 \times 10^{-6}} = 2097 \text{ kg/mm}^2$$

$$\bar{E}_{yy} = \frac{1}{Ta_{xx}} = \frac{1}{4 \times 119.21 \times 10^{-6}} = 2097 \text{ kg/mm}^2$$

$$\bar{\nu}_{xy} = -\frac{a_{xy}}{a_{xx}} = -\frac{89.12 \times 10^{-6}}{119.21 \times 10^{-6}} = 0.7476$$

$$\bar{G}_{xy} = \frac{1}{Ta_{ss}} = \frac{1}{4 \times 64.367 \times 10^{-6}} = 3884 \text{ kg/mm}^2$$

*NOTE*
In this example, the inverse of matrix [**A**] was calculated without regard to the whole [*ABD*] matrix. This was possible only because the [**B**] matrix was identically zero. When it is not the case, the definition of apparent moduli becomes obscured by the coupling effects, which are removable in the calculations, but not in practice.

**Example 7.21**

A laminate is constructed of eight plies of E-Glass-Epoxy composite, each of 1.5 mm thickness. The stacking sequence from bottom up is 0°, 45°, −45°, 90°, 90°, −45°, and 0°. The laminate is loaded with a force distribution of 100 kg/mm along the $x$-axis (0°). The material of the ply has the following moduli in the 1-2 coordinate system. $E_{11} = 4000\ \text{kg/mm}^2$; $E_{22} = 870\ \text{kg/mm}^2$; $G_{12} = 380\ \text{kg/mm}^2$; and $\nu_{12} = 0.28$. Find

a. Laminate equivalent tensile moduli.
b. Laminate equivalent flexural moduli.
c. Laminate mid-plane strains.

*Solution*

*Step 1.* The table:

| Ply No. | Angle | Thickness (mm) | z (mm) |
|---|---|---|---|
| 1 | 0 | 1.5 | −4.5 |
| 2 | 45 | 1.5 | −3.0 |
| 3 | −45 | 1.5 | −1.5 |
| 4 | 90 | 1.5 | 0 |
| 5 | 90 | 1.5 | 1.5 |
| 6 | −45 | 1.5 | 3.0 |
| 7 | 45 | 1.5 | 4.5 |
| 8 | 0 | 1.5 | 6.0 |

*Step 2.* The $Q_{**}$'s in material coordinates:
(Details of calculations not shown)

$$Q_{11} = 4069\ \text{kg/mm}^2; \quad Q_{12} = 248\ \text{kg/mm}^2$$

$$Q_{22} = 885\ \text{kg/mm}^2; \quad Q_{66} = 380\ \text{kg/mm}^2$$

*Step 3.* The transformed $Q_{**}$'s
The details of the calculations are not shown. It is similar to the previous examples.

$$[\mathbf{Q_{xy}}]_{0^\circ} = \begin{bmatrix} 4069 & 248 & 0 \\ 248 & 885 & 0 \\ 0 & 0 & 380 \end{bmatrix}$$

$$[\mathbf{Q_{xy}}]_{-45^\circ} = \begin{bmatrix} 1743 & 983 & -796 \\ 983 & 1743 & -796 \\ -796 & -796 & 1115 \end{bmatrix}$$

$$[\mathbf{Q_{xy}}]_{45^\circ} = \begin{bmatrix} 1743 & 983 & 796 \\ 983 & 1743 & 796 \\ 796 & 796 & 1115 \end{bmatrix}$$

$$[\mathbf{Q_{xy}}]_{0^\circ} = \begin{bmatrix} 885 & 248 & 0 \\ 248 & 4065 & 0 \\ 0 & 0 & 380 \end{bmatrix}$$

*Step 4.* Assemble the **[A]**, **[B]**, and **[D]** matrices:

$$[\mathbf{A}] = t \sum_{k=1}^{8} [\mathbf{Q_{xy}}]_k$$

$$= \begin{bmatrix} 25320 & 7385 & 0 \\ 7386 & 25320 & 0 \\ 0 & 0 & 8970 \end{bmatrix}$$

**[B]** = 0°, due to symmetry.

$$[\mathbf{D}] = \frac{1}{3} \sum_{k=1}^{8} [\mathbf{Q_{xy}}]_k (z_k^3 - z_{k-1}^3)$$

$$= \frac{2}{3}\{[\mathbf{Q_{xy}}]_{0^\circ}(6^3 - 4.5^3)$$

$$+ [\mathbf{Q_{xy}}]_{45^\circ}(4.5^3 - 3.0^3) + [\mathbf{Q_{xy}}]_{-45^\circ}(3.0^3 - 1.5^3) + [\mathbf{Q_{xy}}]_{90^\circ}(1.5^3)\}$$

$$= \frac{2}{3}\{[\mathbf{Q_{xy}}]_{0^\circ}83.25 + [\mathbf{Q_{xy}}]_{45^\circ}42.75 + [\mathbf{Q_{xy}}]_{-45^\circ}15.75 + [\mathbf{Q_{xy}}]_{90^\circ}3.375\}$$

$$= \begin{bmatrix} 442706 & 78667 & 21492 \\ 78667 & 184778 & 21492 \\ 21492 & 21492 & 9700 \end{bmatrix}$$

*Step 5.* Invert the [**A**] and [**D**] matrices:

$$[\mathbf{a}] = [\mathbf{A}]^{-1} = 10^{-6}\begin{bmatrix} 43.167 & -12.586 & 0 \\ -12.586 & 43.167 & 0 \\ 0 & 0 & 111.50 \end{bmatrix}$$

$$\bar{E}_{xx} = \bar{E}_{yy} = \frac{1}{Ta_{xx}} = \frac{1}{12 \times 41.76 \times 10^{-6}} = 1931 \text{ kg/mm}^2$$

$$\bar{\nu}_{xy} = -\frac{a_{xy}}{a_{xx}} = \frac{12.586 \times 10^{-6}}{43.167 \times 10^{-6}} = 0.292$$

$$\bar{G}_{xy} = \frac{1}{Ta_{ss}} = \frac{1}{12 \times 111.48 \times 10^{-6}} = 7474.5 \text{ kg/mm}^2$$

$$[\mathbf{d}] = [\mathbf{D}]^{-1} = 10^{-6}\begin{bmatrix} 2.453 & -1.007 & 0.318 \\ -1.007 & 5.968 & -1.091 \\ -0.318 & -1.091 & 10.545 \end{bmatrix}$$

$$\bar{E}_{f,xx}\left(\neq \bar{E}_{f,yy}\right) = \frac{12}{T^3 d_{xx}} = \frac{12}{12^3 \times 2.453 \times 10^{-6}} = 2831 \text{ kg/mm}^2$$

$$\bar{E}_{yy} = \frac{12}{T^3 d_{yy}} = \frac{12}{12^3 \times 5.968 \times 10^{-6}} = 1164 \text{ kg/mm}^2$$

*Step 6.* Mid-plane strains:
The mid-plane strains are found by Equation 7.53. We note that $N_x = 100$ kg/mm, and all other loads are zero.

$$\{\boldsymbol{\epsilon}\}_0 = \begin{bmatrix} \mathbf{A} & \mathbf{B} \\ \mathbf{B} & \mathbf{D} \end{bmatrix}^{-1} \begin{Bmatrix} \mathbf{N} \\ \mathbf{M} \end{Bmatrix}$$

$$= \begin{bmatrix} \mathbf{A} & \mathbf{0} \\ \mathbf{0} & \mathbf{D} \end{bmatrix}^{-1} \begin{Bmatrix} \mathbf{N} \\ \mathbf{M} \end{Bmatrix}$$

$$= \begin{bmatrix} \mathbf{a} & \mathbf{0} \\ \mathbf{0} & \mathbf{d} \end{bmatrix} \begin{Bmatrix} \mathbf{N} \\ \mathbf{M} \end{Bmatrix}$$

$$= 10^{-6}\begin{bmatrix} 43.17 & -12.59 & 0 & 0 & 0 & 0 \\ -12.59 & 43.17 & 0 & 0 & 0 & 0 \\ 0 & 0 & 111.50 & 0 & 0 & 0 \\ 0 & 0 & 0 & 2.45 & -1.01 & 0.32 \\ 0 & 0 & 0 & -1.02 & 5.97 & -1.10 \\ 0 & 0 & 0 & -0.32 & -1.10 & 10.55 \end{bmatrix}$$

$$\times \begin{Bmatrix} 100 \\ 0 \\ 0 \\ 0 \\ 0 \\ 0 \end{Bmatrix}$$

$$= \begin{Bmatrix} 43.17 \\ -12.59 \\ 0 \\ 0 \\ 0 \\ 0 \end{Bmatrix} 10^{-4} \text{ mm/mm}$$

**Example 7.22**

Continue from Example 7.21, and find the following.

a. Ply stresses along the $x$-$y$ axes,
b. Ply stresses along the material coordinate system of each ply.
c. Fiber stresses in the plies, assuming a volume fraction $V_f = 0.55$.

*Solution* The ply stresses are obtained by multiplying the mid-plane strains (which are the strains for each ply) by the ply [**Q**] along the $x$-$y$ axes. Thus,

$$\{\sigma\}_1 = \{\sigma\}_8 = [\mathbf{Q}_{xy}]_{0°}\{\epsilon\}_o$$

$$= \begin{bmatrix} 4069 & 248 & 0 \\ 248 & 885 & 0 \\ 0 & 0 & 380 \end{bmatrix} \begin{Bmatrix} 43.17 \\ -12.59 \\ 0 \end{Bmatrix} 10^{-4}$$

$$= \begin{Bmatrix} 17.25 \\ 0 \\ 0 \end{Bmatrix} \text{ kg/mm}^2$$

$$\{\sigma\}_2 = \{\sigma\}_7 = [\mathbf{Q}_{xy}]_{45°}\{\epsilon\}_o$$

$$= \begin{bmatrix} 1743 & 983 & 796 \\ 983 & 1743 & 796 \\ 796 & 796 & 1115 \end{bmatrix} \begin{Bmatrix} 43.17 \\ -12.59 \\ 0 \end{Bmatrix} 10^{-4}$$

$$= \begin{Bmatrix} 6.29 \\ 2.05 \\ 2.43 \end{Bmatrix} \text{ kg/mm}^2$$

$$\{\boldsymbol{\sigma}\}_3 = \{\boldsymbol{\sigma}\}_6 = [\mathbf{Q}_{\mathbf{xy}}]_{-45^\circ}\{\boldsymbol{\epsilon}\}_o$$

$$= \begin{bmatrix} 1743 & 983 & -796 \\ 983 & 1743 & -796 \\ -796 & -796 & 1115 \end{bmatrix} \begin{Bmatrix} 43.17 \\ -12.59 \\ 0 \end{Bmatrix} 10^{-4}$$

$$= \begin{Bmatrix} 6.6 \\ 2.05 \\ -2.43 \end{Bmatrix} \text{kg/mm}^2$$

$$\{\boldsymbol{\sigma}\}_4 = \{\boldsymbol{\sigma}\}_5 = [\mathbf{Q}_{\mathbf{xy}}]_{90^\circ}\{\boldsymbol{\epsilon}\}_o$$

$$= \begin{bmatrix} 885 & 248 & 0 \\ 248 & 4065 & 0 \\ 0 & 0 & 380 \end{bmatrix} \begin{Bmatrix} 43.17 \\ -12.59 \\ 0 \end{Bmatrix} 10^{-4}$$

$$= \begin{Bmatrix} 3.51 \\ -4.05 \\ 0 \end{Bmatrix} \text{kg/mm}^2$$

Ply stresses in the material coordinate system is obtained by transforming the stresses. For plies 1, 4, 5 and 8, the $x$-$y$ system and the material system are coincident. For the other four plies, a transformation is needed. Equations 1.25 and 1.26 are used.

Details of calculations are shown only for ply 2. The other plies differ from this only in respect of the angle $\theta$.

$$\text{Stress along fibers} = \sigma_x \cos^2 45 + \sigma_y \sin^2 45 + 2\tau_{xy} \cos 45 \sin 45$$

$$= \frac{6.29}{2} + \frac{2.05}{2} + \frac{2 \times 2.43}{2}$$

$$= 6.6 \text{ kg/mm}^2$$

$$\text{Stress across fibers} = \sigma_x \cos^2 45 + \sigma_y \sin^2 45 - 2\tau_{xy} \cos 45 \sin 45$$

$$= \frac{6.29}{2} + \frac{2.05}{2} - \frac{2 \times 2.43}{2}$$

$$= 1.74 \text{ kg/mm}^2$$

$$\text{Shear stress} = (\sigma_y - \sigma_x)\cos 45 \sin 45 + \tau_{xy}(\cos^2 45 - \sin^2 45)$$

$$= (2.05 - 6.29)\frac{1}{2}$$

$$= -2.12 \text{ kg/mm}^2$$

The final results for all the plies are tabulated below.

| Ply No. | $\sigma_1$ | $\sigma_2$ | $\tau_{12}$ |
|---|---|---|---|
| 1 | 17.25 | 0 | 0 |
| 2 | 6.6 | 1.732 | −2.118 |
| 3 | 6.6 | 1.732 | 2.118 |
| 4 | −4.05 | 3.508 | 0 |
| 5 | −4.05 | 3.508 | 0 |
| 6 | 6.6 | 1.732 | 2.118 |
| 7 | 6.6 | 1.732 | −2.118 |
| 8 | 17.25 | 0 | 0 |

For calculating the fiber stresses in the plies we may simply assume that all the stress is taken by the fibers. Hence, it is sufficiently accurate to simply divide the ply stress along the fiber direction, namely $\sigma_1$ by $V_f$. Thus, fiber stresses in the plies are

$$31.36, 12, 12, -7.36, -7.36, 12, 12, 31.36 \text{ kg/mm}^2, \text{respectively}.$$

## 7.14 SYMMETRIC LAMINATES

From a practical point of view, there is an important class of laminates called the *symmetric laminates*. Laminates are said to be symmetric, if all the following conditions are met. Plies at equal distance from the middle plane,

- have the same thickness.
- have the same material.
- have the same orientation.

For example, laminate with the same ply material and thickness stacked as in the following symmetric laminate.

$$0°, -30°30° - 60°, 60°,|60°, -60°, 30°, -30°, 0°$$

The most important consequence of symmetric lay-up is that all the elements in the **[B]** matrix become zero. This eliminates the undesirable coupling of in-plane and bending effects. In other words, symmetry ensures that unknown residual stresses and warping deformations are not introduced in the laminate.

Symmetric laminates are however, not free of the $D_{xs}$ and $D_{ys}$ terms. That is, bending moments produce torsional deflections and vice versa. The general form of the $[ABD]$ matrix for a symmetric laminate is as follows:

$$\begin{bmatrix} * & * & 0 & 0 & 0 & 0 \\ * & * & 0 & 0 & 0 & 0 \\ 0 & 0 & * & 0 & 0 & 0 \\ 0 & 0 & 0 & * & * & * \\ 0 & 0 & 0 & * & * & * \\ 0 & 0 & 0 & * & * & * \end{bmatrix}$$

(Symmetric lay-up)

$$(A_{xs} = A_{ys} = 0;\ D_{xs} \neq D_{ys} \neq 0;\ B_{ij} = 0)$$

As opposed to this, we can also have *anti-symmetric laminates* which satisfy the following conditions. In anti-symmetric lay-ups, plies at equal distance from the middle plane,

- have the same thickness,
- have the same material,
- have the opposite orientation.

For example, if the fourth ply below the middle plane is at −30° to the $x$ axis, then the fourth ply above the middle plane must be at 30°. Thus, the following is an example of an anti-symmetric laminate.

$$0°, -30°30° - 60°, 60°,| - 60°, 60°, -30°, 30°, 0°$$

The consequence of anti-symmetry is that $D_{xs} = D_{ys} = 0$. But the [**B**] matrix is not zero. The general form of the $[ABD]$ matrix for an anti-symmetric laminate is as follows:

$$\begin{bmatrix} * & * & 0 & * & * & * \\ * & * & 0 & * & * & * \\ 0 & 0 & * & * & * & * \\ * & * & * & * & * & 0 \\ * & * & * & * & * & 0 \\ * & * & * & 0 & 0 & * \end{bmatrix}$$

(Anti-symmetric lay-up)

$$(A_{xs} = A_{ys} = 0;\ D_{xs} = D_{ys} = 0;\ B_{ij} \neq 0)$$

The table below summarizes the salient properties of the $[ABD]$ matrix for symmetric and anti-symmetric laminates.

| | Symmetry | Anti-Symmetry |
|---|---|---|
| $\left.\begin{matrix} A_{xs} = 0 \\ A_{ys} = 0 \end{matrix}\right\}$ | True | True |
| All $B_{ij} = 0$ | True | False |
| $\left.\begin{matrix} D_{xs} = 0 \\ D_{ys} = 0 \end{matrix}\right\}$ | False | True |

Symmetric laminates are structurally well-behaved. They do not warp excessively because of in-plane loads, temperature changes, or moisture absorption. Hence structures made of symmetric laminates maintain dimensional stability over long periods of time. They are also easy to analyze.

We can explain this table by considering Equation 7.53, we note that

$$[\mathbf{A}] = \sum [\mathbf{Q_{xy}}]_k t_k.$$

Also $Q_{xx}$, $Q_{yy}$, $Q_{xy}$ and $Q_{ss}$ are even functions of $\theta$. Their contribution to the [**A**] matrix is always positive. On the other hand, $Q_{xs}$ and $Q_{ys}$ are odd functions of $\theta$. Their summation can be made equal to zero, if for each positive $Q_{*s}$, there is a negative $Q_{*s}$ in the laminate. In other words, for each ply at $\theta$ orientation, there must be a ply at $-\theta$ orientation. Symmetric laminates meet this requirement, and hence $A_{xs} = A_{ys} = 0$ for symmetric laminates.

For this reason, symmetric laminates are called "orthotropic" for in-plane (membrane) loads. This nomenclature is based on the similarity to the orthotropic plies having $Q_{xs} = Q_{ys} = 0$.

Next, we turn to the [**B**] matrix.

$$[\mathbf{B}] = \frac{1}{2} \sum [\mathbf{Q_{xy}}](z_k^2 - z_{k-1}^2)$$

The geometry term in the above may be positive or negative. For plies below the mid-plane it is negative, since $|z_k| < |z_{k-1}|$. For a similar reason, it is positive for plies above the mid-plane. So, all the $B$'s will become zero, if the $Q_{**}$'s are identical for corresponding plies. This requirement is met by symmetry. Hence, the [**B**] matrix is identically zero for symmetric laminates.

Lastly, we consider the terms in the [**D**] matrix.

$$[\mathbf{D}] = \frac{1}{3} \sum [\mathbf{Q}_{xy}](z_k^3 - z_{k-1}^3)$$

The geometry term is always positive in this formula. The only way in which the $D_{*s}$ terms can vanish is by making the corresponding $Q_{*s}$'s of opposite signs. This is possible if the plies have opposite orientations (anti-symmetric). Thus, it is impossible to have simultaneously, $B_{ij} = 0$ *and* $D_{xs} = D_{ys} = 0$.

## 7.15 QUASI-ISOTROPIC LAMINATES

Another class of laminates of practical importance is the *quasi-isotropic laminates*. Such laminates have (approximately) the same [**A**] matrix in all directions. This property is useful when the direction of applied loading is unknown or is known to vary. It is preferable to make quasi-isotropic laminates also symmetric. A laminate can be made quasi-isotropic by

- having equal thickness for all plies,
- setting the increment in the direction of adjacent plies to $2\pi/n$, where $n$ is the total number of plies in the laminate, and
- having a large number of plies, at least four.

The prefix "quasi" is used because there is an inevitable small variation of the $A_{**}$'s (or equivalently the laminate moduli) between $2\pi i/n$ and $2\pi(i+1)/n$. Figure 7.30 shows qualitatively a typical variation of $E_{xx}$, $E_{yy}$, $G_{xy}$, and $\nu_{xy}$ of a $(\pm 2\pi i/n)_s$ laminate made of E-Glass-Epoxy.

### 7.15.1 First Ply and Ultimate Laminate Failures

From the examples in the preceding sections, we see that in a laminate, the in-plane stresses do not vary smoothly through the thickness; rather, they jump from one value on a ply to another value on the next. This is true regardless of the loading direction and the stress direction. Therefore, there is no similarity between the stress distribution in an isotropic beam/plate and that in a composite laminate. There is no guarantee that the stresses in the outermost ply is the maximum. It could well be that a subsurface ply is closer to its failure point than the outermost ply. For a given direction of loading, each ply can carry a factor of safety $S_f$ that is a function of the entire laminate construction and its position in the laminate. However, the ply with the least $S_f$ is closest to its failure, and is likely to fail first, notwithstanding its position in the lay-up.

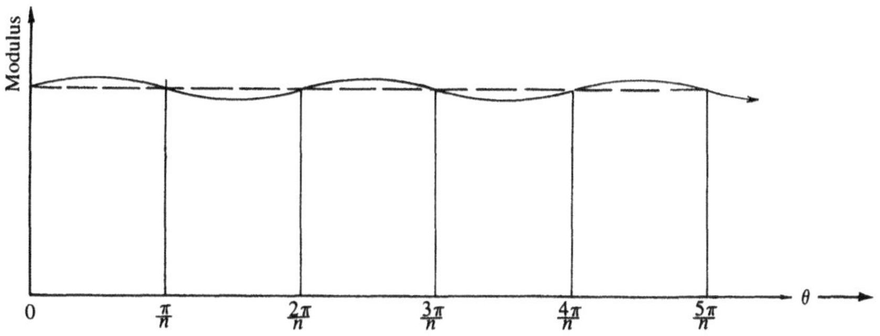


**FIGURE 7.30.** Quasi-isotropic laminates exhibit only a minor variation in apparent moduli with direction.

A philosophy of design prevalent in the composites is to consider the laminate to have failed, when *any single ply fails by calculation.* In other words, when the first failure of a ply occurs, the laminate is deemed to have failed. Of course, this philosophy is too restrictive since it does not take credit for the load-bearing capability available in the remaining plies. This philosophy is called the "FPF-based design." FPF stands for *First Ply Failure*. The load acting on the laminate at the FPF event is called the *FPF load.* In order to compensate for the too restrictive philosophy, the design safety factor (*DSF*) applied to the FPF load is small. This is in keeping with our discussions in Chapter 4.

As opposed to this we can take a different view of the design philosophy. Since FPF is only the *beginning of failure*, we can take credit for the load bearing capability of the remaining plies. In fact, the laminate theory can be applied on the remaining configuration, and we may investigate the FPF failure of the remainder laminate, at a higher level of load. This procedure can be repeated until enough number of plies fail so it is no longer possible for the unfailed plies to sustain the load. At this point all the remaining plies fail too. The load at this point represents the absolute maximum load that the ply can take. This state of failure is the "*ultimate laminate failure*" (*ULF*), and corresponds to the "ultimate strength" in the metals. The average stress (based on original thickness of the laminate) at ULF may be called the "ultimate strength" of the laminate. The design philosophy requires a higher *DSF* on the ULF strength of the laminate.

It is important to note that both FPF and ULF strengths are functions not only of the laminate construction, but also of the direction of loading.

It is neither possible nor correct to globally adopt one philosophy in preference to the other. In practical situations, there are many examples where the FPF design is the right one to do,[2] and equally numerous examples for the ULF. For the same laminate, for a particular operating environment, an FPF-based design may be appropriate, whereas for a different environment ULF may be right.

> Since composite strengths and stiffnesses diminish with time just like the un-reinforced plastics, it is imperative that long term properties be considered in the design, right at the stage of deciding the design stress.

***FPF Load Calculation***

The procedure for FPF load calculation is outlined below. However, the calculation is long and one resorts to a computer or a programmable calculator. Example 7.23 in FPF load calculation was performed on a personal computer.

1. Define the layup.
2. Calculate the $[ABD]$ matrix, and invert it.
3. From the applied load vector, find the mid-plane strains and curvature changes.
4. Calculate the ply strains, and ply stresses along the material axes.
5. Using any chosen failure criterion, find the safety factor $S_f$, for every ply.
6. Find the least of $S_f$, and on which ply this occurs.
7. Multiply the applied load vector by $S_f$, and that gives the FPF load.

***ULF Load Calculation***

The procedure for ULF is much longer, and is impossible without some computing facility. It is technically an iterative application of the FPF load calculation. It is outlined below.

1. Find the FPF load, and the ply to fail first. Let the ply number be $n_1$.
2. Replace the failed ply $n_1$ by a dummy material of very low moduli, and very high strength. Low modulus, because that way the failed ply will not pick up any load. High strength, because the failed ply will not fail

[2] Underground composite pipes used for water transport are often given a gel coat as a barrier for water ingression. Gel coat, being brittle, can take up only a limited amount of strain, and is liable to fail first. When it has failed, the pipe has, in fact, failed functionally, although not structurally.

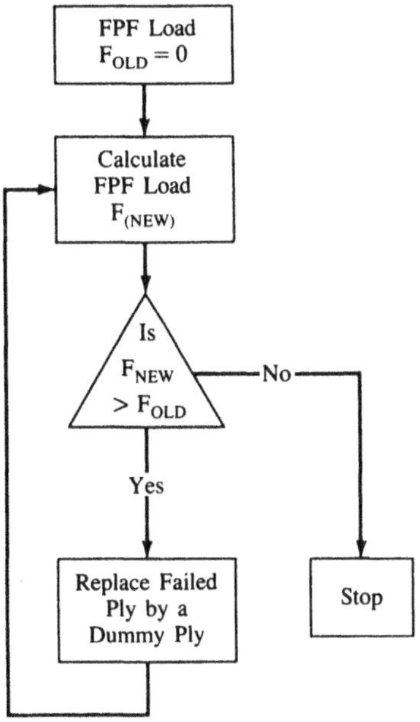


**FIGURE 7.31.** Procedure for calculating the ULF strength of a composite laminate.

again and confuse the subsequent calculations. We wish to find out which of the remaining good plies will fail next.

3. Hopefully, this new laminate (with a failed ply embedded in it) will support the FPF load, and probably more. Hence, repeat the entire calculations to find out the new FPF load.
4. If a higher FPF load is obtained then the new laminate in Step 3 supports a higher load, or else the most recent FPF load is the ULF load for the laminate.
5. If the most recent FPF load was higher than its previous value, then continue the calculation cycle one more time. Or else, calculation terminates.

The flowchart in Figure 7.31 summarizes the ULF load calculation procedures.

**Example 7.23**

A symmetric laminate is constructed from E-Glass-Epoxy plies each 0.1 mm thick. It consists of 10 layers and is laid up in the sequence ($\pm 30$,

$\pm 60, 0)_s$. Find the maximum moment $M_x$ that the laminate can withstand. The following material properties may be used.

| Units—kg/mm² | | |
|---|---|---|
| $E_{11} = 4000$ | $E_{22} = 870$ | $G_{12} = 380$ |
| $\nu_{12} = 0.28$ | $F_{1T} = 110$ | $F_{1C} = 63$ |
| $F_{2T} = 6$ | $F_{2C} = 13$ | $F_6 = 9$ |

*Solution* The following is a computer solution, with annotations. All the steps involved have been explained in detail in the earlier examples. The procedure for finding the strength in a particular direction consists of assuming a unit load (or moment) acting in that direction, and calculating the safety factor in each ply. The lowest such safety factor is the load (or moment) that the laminate can withstand. It may be noted that (1) the safety factor can vary depending on the failure criterion used, and (2) that the strength calculated is the FPF strength, and not a ULF This particular example uses the Tsai-Wu failure criterion.

**Step 1: Find [Q₁₂] of the plies**

$$Q_{11} = 4069.39; \quad Q_{22} = 885.09; \quad Q_{12} = 247.83; \quad Q_{66} = 380$$

**Step 2: Find [$\mathbf{Q_{xy}}$] and construct [ABD]**

$$[\mathbf{ABD}] = \begin{bmatrix} 2354 & 689 & 0 & 0 & 0 & 0 \\ 689 & 1718 & 0 & 0 & 0 & 0 \\ 0 & 0 & 821 & 0 & 0 & 0 \\ 0 & 0 & 0 & 200 & 66 & 19 \\ 0 & 0 & 0 & 66 & 122 & 14 \\ 0 & 0 & 0 & 19 & 14 & 77 \end{bmatrix}$$

Note the symmetry of the [**ABD**] matrix and that the [**B**] is entirely zero.
**Step 3: Invert the [ABD] matrix, multiply by the load vector**

$$[\mathbf{ABD}]^{-1}$$

$$= 10^{-4} \begin{bmatrix} 4.810 & -1.928 & 0 & 0 & 0 & 0 \\ -1.928 & 6.594 & 0 & 0 & 0 & 0 \\ 0 & 0 & 12.183 & 0 & 0 & 0 \\ 0 & 0 & 0 & 61.585 & -32.450 & -9.342 \\ 0 & 0 & 0 & -32.450 & 101.080 & -10.304 \\ 0 & 0 & 0 & -9.342 & -10.304 & 133.681 \end{bmatrix}$$

The load vector in this case is trivially simple.

$$\begin{Bmatrix} \mathbf{N} \\ \mathbf{M} \end{Bmatrix} = \begin{Bmatrix} 0 \\ 0 \\ 0 \\ 1 \\ 0 \\ 0 \end{Bmatrix}$$

Note that the forces and moments here are per mm. Multiplying the $[\mathbf{ABD}]^{-1}$ by the load vector gives the strain matrix. Accordingly,

$$\begin{Bmatrix} \boldsymbol{\epsilon}_0 \\ \boldsymbol{\kappa}_0 \end{Bmatrix} = 10^{-3} \begin{Bmatrix} 0 \\ 0 \\ 0 \\ 6.158 \\ -3.245 \\ -0.934 \end{Bmatrix}$$

**Step 4: Find ply stresses and FOS in the material coordinate system**
The stresses are to be calculated first in the *x-y* system and then converted to the material coordinate system. We quote the final result.

| Ply no. | $\sigma_1$ | $\sigma_2$ | $\tau_{12}$ | $SF^+$ | $SF^-$ |
|---|---|---|---|---|---|
| 1 | −6.863 | −0.205 | 1.636 | 4.966 | −5.255 |
| 2 | −6.728 | 0.004 | −1.167 | 5.460 | −7.680 |
| 3 | 1.272 | −1.022 | 0.875 | 9.209 | −3.400 |
| 4 | 0.230 | −0.578 | −0.654 | 14.345 | −5.837 |
| 5 | −2.426 | 0.135 | 0.004 | 15.092 | −37.715 |
| 6 | | | | | |
| 7 | | Due to the symmetry of the lay-up, | | | |
| 8 | | the stresses change sign, and | | | |
| 9 | | $SF^+$ becomes $-SF^-$ and vice versa. | | | |
| 10 | | | | | |

**Step 5: Look for the minimum *SF***
The minimum safety factor × the applied moment represents the failure

moment corresponding to FPF. In this case, because of symmetry again,

$$|SF^{+}| = |SF^{-}| = 3.4$$

Hence, the failure moment of the laminate is $\pm 3.4$ kgmm/mm.

### 7.15.2 Design of Laminate Thickness

In practical situations, the thickness of a part is unknown, and is to be determined at first. However, from the considerations of direction and the nature of loading the lay-up sequence is known. The procedure for determining the required thickness is as follows.

*Given:* The number of plies, the ply material(s), the stacking sequence, and the orientation of each ply. Laminate is required to have a "design factor of safety" of *DSF* against first ply failure.

*Required:* The thickness of the laminate.

*Procedure:*

1. Assume a *fictitious* thickness $T_0$ of the plies, so that the total laminate thickness is unity.
2. Carry out the calculations for laminate elastic properties.
3. Apply the *actual, real life* load on this laminate.
4. Find the available factor of safety $S_f$ for first ply failure.
5. If the load is a purely in-plane membrane type, then the required thickness is simply

$$\text{Required thickness } T_{req} = T_0 \frac{DSF}{S_f}$$

6. If the load is purely bending type, then the required thickness is given by

$$\text{Required thickness } T_{req} = T_0 \left( \frac{DSF}{S_f} \right)^{1/3}$$

7. If the load is a combination, an iterative process is needed to determine the exact thickness needed. A suggested approach is to treat it like pure bending loads, find the required thickness, and scale up (or down) the

ply thicknesses, and repeat from Step 2 above, until two iterations give close enough thickness.

Care must be taken to use the same failure criterion throughout the calculation.

## 7.16 HYGROTHERMAL (HT) EFFECTS

### 7.16.1 HT Effects in a UDL

Temperature changes in a composite ply produce changes in its linear dimensions. Unlike the case of metals, composites may expand or *even contract* due to an increase in temperature. The change in length is observed to be linear with temperature change. Figure 7.32 shows typical unit changes in the length of a composite. It is seen that the relationship is linear. The slope of the line is the rate of change of length per unit temperature change, which is termed as the coefficient of linear expansion, $\alpha$. It is a physical property of the ply. All composites exhibit orthotropy of thermal expansion. That is, the unit changes of length are different along and across fibers for the same $\Delta T$.

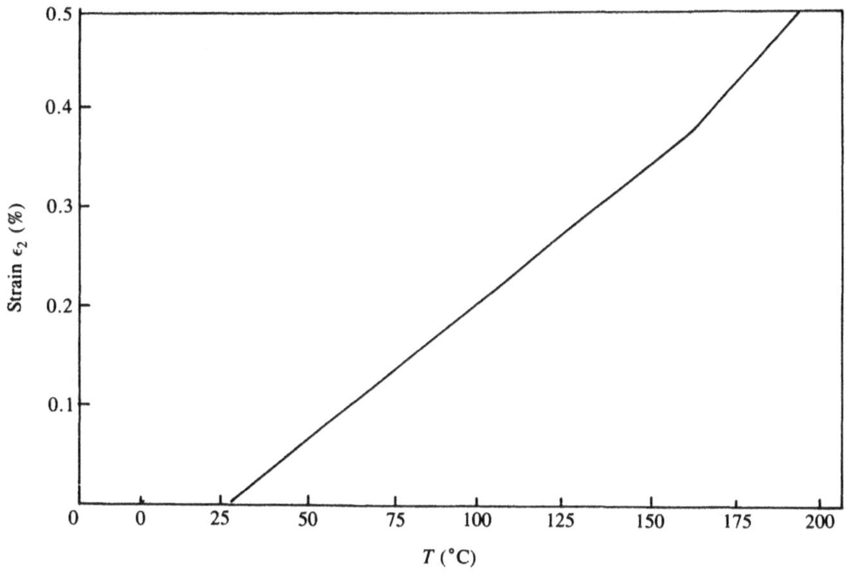


**FIGURE 7.32.** Transverse thermal expansional strain for Graphite-Epoxy AS/3501-5 UDL as a function of temperature. The slope of this line is $\alpha$.

Besides this, composites can absorb moisture from the environment, which causes them to expand. Figure 7.33 shows the typical relationship between the amount of moisture absorbed, $\Delta c$ and the consequent change in length $\Delta l$. Although the trend is bi-linear, for practical purposes it can be treated as linear passing through origin. The slope of the straight line represents the $\Delta l$ produced per unit $\Delta c$, is denoted by $\beta$, and is a property of the ply. This property $\beta$ is called the "hygroscopic expansion coefficient." Composites exhibit orthotropy of $\beta$'s too. That is, for the same $\Delta c$ the unit expansions along fiber and across fiber give two different characteristic values, $\beta_1$ and $\beta_2$. Once again, as in the temperature effects, $\Delta T$ does not produce a shear strain in the material coordinate systems. For many material systems, $\beta_1 = 0$. From $\Delta T$ only,

$$\Delta l_1 = l_o \alpha_1 \Delta T, \quad \text{along fibers,}$$

$$\Delta l_2 = l_o \alpha_2 \Delta T, \quad \text{across fibers.}$$

From $\Delta c$ only,

$$\Delta l_1 = l_o \beta_1 \Delta c, \quad \text{along fibers,}$$

$$\Delta l_2 = l_o \beta_2 \Delta c, \quad \text{across fibers.}$$

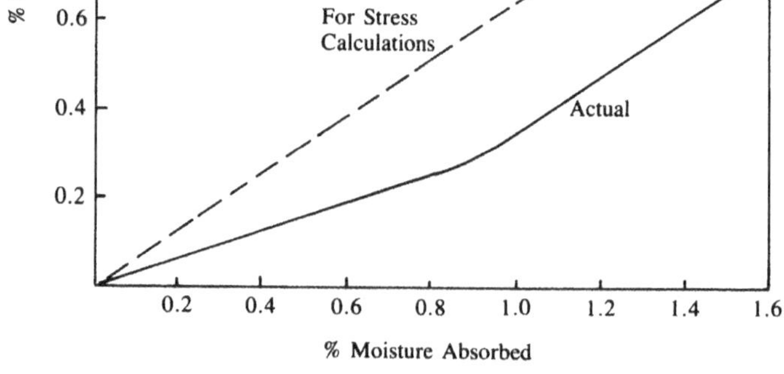


**FIGURE 7.33.** Transverse swelling strain due to moisture absorption for AS/3501-5 Graphite-Epoxy UDL. The dotted line is a conservative estimate for use in hygral stress calculations. The slope of this line gives $\beta$.

Their combined effect, commonly called the *hygro-thermal effect*, is simply obtained by adding the two.

$$\Delta l_1 = l_o \alpha_1 \Delta T + l_o \beta_1 \Delta c,$$

$$\Delta l_2 = l_o \alpha_2 \Delta T + l_o \beta_2 \Delta c$$

This can be written in matrix form as,

$$\begin{Bmatrix} \Delta l_1 \\ \Delta l_2 \\ 0 \end{Bmatrix} = l_o \left\{ \begin{Bmatrix} \alpha_1 \\ \alpha_2 \\ 0 \end{Bmatrix} \Delta T + \begin{Bmatrix} \beta_1 \\ \beta_2 \\ 0 \end{Bmatrix} \Delta c \right\} \tag{7.55}$$

$$= l_o\{\boldsymbol{\alpha}_{12}\}\Delta T + l_o\{\boldsymbol{\beta}_{12}\}\Delta c. \tag{7.56}$$

where

$$\{\boldsymbol{\alpha}_{12}\}, \{\boldsymbol{\beta}_{12}\} = \begin{cases} \text{Hygro-thermal coefficients in the} \\ \text{material coordinate system.} \end{cases} \tag{7.57}$$

The form of Equation 7.55 suggests that the quantities $\alpha\Delta T$, and $\beta\Delta c$ represent strains. Therefore they are called the hygro-thermal strains. They can be transformed from one coordinate system to another by

$$\{\boldsymbol{\alpha}_{12}\} = [\mathbf{T2}]\{\boldsymbol{\alpha}_{xy}\}, \tag{7.58}$$

and

$$\{\boldsymbol{\beta}_{12}\} = [\mathbf{T2}]\{\boldsymbol{\beta}_{xy}\}, \tag{7.59}$$

where

$$\{\boldsymbol{\alpha}_{xy}\}, \{\boldsymbol{\beta}_{xy}\} = \begin{cases} \text{Hygro-thermal coefficients in} \\ \text{any general coordinate system.} \end{cases} \tag{7.60}$$

The meaning of **[T2]** is the same as in Equation 7.33. Note that the shear HT coefficients in a general *x-y* system have a nonzero $\alpha_s$ and $\beta_s$.

## 7.17 HYGRO-THERMAL STRESSES

Hygro-thermal stresses are produced in a ply if its natural or free expansions are prevented. No stress is produced if the natural expansions are allowed to take place freely. In Figure 7.34, the actual expansion of a ply falls short of the natural expansion by a certain amount. This is possible only if a stress acts in the plane of the ply and brings it to the common line

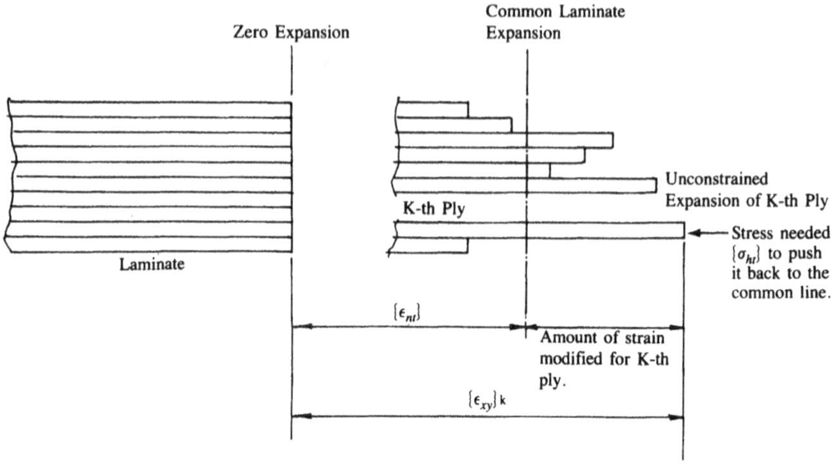


**FIGURE 7.34.** Hygro-Thermal stress arises because the natural expansion of the plies is not allowed to take place.

of expansion. Thus, the hygro-thermal stress is proportional to the amount of strain arrested or modified. In simple terms,

$$\text{HT Stress} = (\text{Modulus}) \times (\text{Actual strain} - \text{Free unit expansion})$$

In a laminate natural expansions of the plies are not allowed to take place freely, because the plies have to deform without slipping. This results in a system of ply stresses, which together are in self-equilibrium. Thus the basic cause of hygro-thermal stresses in laminates is that the plies have different orientations.

*Step 1.* Calculate the free HT strains in the plies, (that is, by treating them as completely free).

*Step 2.* Calculate the force that would produce the same strain mechanically. Sum up such forces for all plies.

*Step 3.* Apply the total force on the laminate, and calculate the ply strains using the laminate theory. This way we calculate the hygro-thermal strains in the plies taking into account the mutual restraints of the plies.

*Step 4.* The strain in Step 3 − free strain is the amount of strain modified. Hygro-thermal stress is proportional to this strain. If free strain is arrested, then this quantity is negative, and the stress is compressive; and vice versa.

### 7.17.1 HT Stresses in a Symmetric Laminate

Symmetric laminates are the most widely used in advanced structural applications. Besides their good behavior under mechanical loads, symmetric laminates are also free from warping under HT loads. There is only a mid-plane extensional strain, which is common to all the plies.

Let us suppose that the $[ABD]$ matrix for the laminate is calculated and is available.

$$\left.\begin{array}{r}\text{Free HT strain in the } k\text{-th}\\ \text{ply, along 1-2 axes.}\end{array}\right\} = \{\mathbf{e}_{12}\}_k$$

$$= \{\boldsymbol{\alpha}_{12}\}_k \Delta T + \{\boldsymbol{\beta}_{12}\}_k \Delta c$$

$$= \{\mathbf{e}_{12}\}_k^{\alpha} + \{\mathbf{e}_{12}\}_k^{\beta}$$

$$\left.\begin{array}{r}\text{Free HT strain in the } k\text{-th}\\ \text{ply, along } x\text{-}y \text{ axes.}\end{array}\right\} = \{\mathbf{e}_{xy}\}_k$$

$$= [\mathbf{T2}]_k\left\{\{\mathbf{e}_{12}\}_k^{\alpha} + \{\mathbf{e}_{12}\}_k^{\beta}\right\}$$

$$= \{\mathbf{e}_{xy}\}_k^{\alpha} + \{\mathbf{e}_{xy}\}_k^{\beta}$$

$$\left.\begin{array}{r}\text{Ply force which would produce}\\ \text{the same HT strains mechanically.}\end{array}\right\} = [\mathbf{Q}_{xy}]_k\{\mathbf{e}_{xy}\}_k t_k$$

$$\left.\begin{array}{r}\text{Total HT force for the}\\ \text{laminate.}\end{array}\right\} = \{\mathbf{N}_{ht}\}$$

$$= \sum_{k=1}^{n} [\mathbf{Q}_{xy}]_k t_k \left\{\{\mathbf{e}_{xy}\}_k^{\alpha} + \{\mathbf{e}_{xy}\}_k^{\beta}\right\}$$

$$= \{\mathbf{N}_{ht}\}^{\alpha} + \{\mathbf{N}_{ht}\}^{\beta}$$

Since we are considering a symmetric laminate, there are no warping and bending forces present in the laminate. Only in-plane strains need to be considered. Hence only [**A**] is involved in the subsequent calculations, and not [**D**]. Continuing from the above,

$$\left.\begin{array}{r}\text{The actual strains produced in}\\ \text{the plies.}\end{array}\right\} = \{\boldsymbol{\epsilon}_{ht}\}$$

$$= [\mathbf{A}]^{-1}\{\mathbf{N}_{ht}\} = [\mathbf{a}]\{\mathbf{N}_{ht}\}$$

$$\left.\begin{array}{r}\text{Amount of strain prevented}\\ \text{or modified in the }k\text{-th ply}\end{array}\right\} = \{\boldsymbol{\epsilon}_{\mathbf{ht}}\} - \{\mathbf{e}_{\mathbf{xy}}\}_k^{\alpha} - \{\mathbf{e}_{\mathbf{xy}}\}_k^{\beta}$$

The hygro-thermal stress in the plies is given by

$$\{\boldsymbol{\sigma}_{\mathbf{ht}}\} = [\mathbf{Q}_{\mathbf{xy}}]_k \left\{\{\boldsymbol{\epsilon}_{\mathbf{ht}}\} - \{\mathbf{e}_{\mathbf{xy}}\}_k^{\alpha} - \{\mathbf{e}_{\mathbf{xy}}\}_k^{\beta}\right\} \tag{7.61}$$

### 7.17.2 Apparent HT Expansion Coefficients

The apparent HT expansion coefficients of a symmetric laminate are functions of the construction of the laminate. *They involve not only the HT coefficients of the plies, but also their stiffnesses*. From the last section, we see that the actual strains are given by,

$$\{\boldsymbol{\epsilon}_{\mathbf{ht}}\} = [\mathbf{a}]\left[\{\mathbf{N}_{\mathbf{ht}}\}^{\alpha} + \{\mathbf{N}_{\mathbf{ht}}\}^{\beta}\right]$$

The apparent expansion coefficients is readily available from this.

$$\{\bar{\boldsymbol{\alpha}}\} = \left\{\begin{array}{l}\text{Apparent thermal expansion}\\ \text{coefficient of the laminate}\end{array}\right.$$

$$= [\mathbf{a}]\{\mathbf{N}_{\mathbf{ht}}\}^{\alpha} \tag{7.62}$$

$$\{\bar{\boldsymbol{\beta}}\} = \left\{\begin{array}{l}\text{Apparent hygroscopic expansion}\\ \text{coefficient of the laminate}\end{array}\right.$$

$$= [\mathbf{a}]\{\mathbf{N}_{\mathbf{ht}}\}^{\beta} \tag{7.63}$$

The reason for considering only symmetric laminates here is now evident. The warping deformations associated with unsymmetric laminates make it difficult to separate the expansion coefficients.

## 7.18 CONCLUSION

It is very important to note that the material in this chapter covers the properties of the ply and the laminate, *at a point*. All the formulae were obtained as aggregates of the ply properties through the thickness of the laminate. The behavior of a *structure* composed of composite material is considerably more involved. In practice we use a thin shell theory as modified for orthotropic materials.

Practical problems in composite structural analysis begin with the laminate properties discussed in this chapter. Such analyses yield the distribution of forces {**N**} and moments {**M**} at every point of the structure. Then we return to the laminate analysis to translate this information into stresses, and factors of safety.

It may be recalled that in laminates—symmetric or anti-symmetric—it is impossible to have $B_{ij} = 0$ and $D_{ij} = 0$ simultaneously. This means that a laminate cannot be made to behave truly orthotropically, when bending and membrane loads coexist. Most structural analysis software on the market do not account for the fact that composite laminates can be strictly orthotropic only under membrane loads, and that the equivalent extensional moduli are not the same as the equivalent flexural moduli. In technical language, the effect of non-zero coupling coefficients would remain un-accounted for in the analysis. The analyst must be aware of this fact.

Furthermore, everything in this chapter is valid at a point in the laminate, located far from the free edges. Near the free edges of a composite laminate, shear stresses exist between the layers of ply, in order to keep them from sliding past one another. Such stresses, called the *interlaminar shear stresses*, are of considerable interest in the design and analysis of composite structures and their joints. Interlaminar shear stresses are prevalent only over a very short distance from the free edges. It is also regarded as another mechanism of laminate failure. Hence, alongside with interlaminar shear stresses, we also need to consider *interlaminar shear strengths* of the ply, which is a fundamental laminate property. ASTM D-2344 provides the test method for measuring this property. Except for the special, custom-designed software, the classical shell theory, and most of the commercial finite element techniques are inadequate to address the problem of calculating the interlaminar stress as a field confined to edges. Rather they take a simpler approach that the interlaminar shear is what is needed for balancing the bending stresses which jump from one value to another over adjacent plies.

In many practical situations of the repair of composite materials, a bulk of composite patch is applied over the repair area. This portion is not easy to treat as a laminate, because there is no portion of constant thickness, and therefore the mathematics of bending theory that was initially used in the laminate theory is not valid. One needs to consider this patch as a three-dimensional composite, and resort to finite elements. Yet another example is the filament-wound composite, in which the continuous fibers may smoothly change directions around openings, nozzles, etc. The volume fraction of the composite changes continuously in the neighborhood of openings. Consequently the composite properties become intractable for computation by FEA or any other means.

Composites is a fast expanding field in terms of materials, forms of fillers, structural applications, fabrication techniques, design and stress analysis. As of the 1990s, experience plays a more vital role in the design of parts than does theory. In advanced composite applications, the prototype test continues to be the final word for the integrity of composites.

**References**


1. Rosen, B. W., Analysis of material properties in *Engineered Materials Handbook, Vol. 1, COMPOSITES*, 1987, published by ASM (Am. Soc. of Metals) International, pp 185–205.
2. Agarwal, B. D. and Broutman, L. J., *Analysis and Performance of Fiber Composites*, 1980, John Wiley & Sons.
3. Ashton, J. E., Halpin, J. C., and Petit, P. H., *Primer on Composite Materials—Analysis*, 1969, Technomic Publishing Co., Stamford, CT.
4. Chawla, K. K., *Composite Materials—Science and Engineering*, 1987, Springer-Verlag, New York.
5. Berlin, A. A., Volfson, S. A., Enikopian, N. S., and Negamatov, S. S., *Principles of Polymer Composites*, 1986, Springer-Verlag, Berlin.
6. Tsai, S. W., *Composites Design*, 4th Ed. 1988, Think Composites, Dayton, OH.
7. Johnson, W. S., (editor) *Delamination and Debonding of Materials*, 1985, ASTM Special Technical Publication STP 876, Published by Am. Soc. for Testing Materials, Philadelphia, PA.
8. Hussein, R. M., *Composites Panels/Plates—Design and Analysis*, 1986, Technomic Publishing Co., Inc., Lancaster, PA.
9. Vinson, J. R. and Sierakowski, R. L., *The Behavior of Structures Composed of Composite Materials*, 1986, Martinus Nijhoff Publishers, Dordrecht, Netherlands.
10. Dorgham, M. A. and Rosato, D. V., *Designing With Plastics and Advanced Plastic Composites*, Technological Advances in Vehicle Design, Special Publication SP6, 1986, Published by Interscience Enterprises Ltd., Geneva, Switzerland.
11. Whitney, J. M., (editor) *Analysis of the Test Methods for High Modulus Fibers and Composites*, 1973, ASTM STP 521, Published by Am. Soc. for Testing and Materials, Philadelphia, PA.
12. Sendeckyj, G. P., *Fracture Mechanics of Composites*, 1975, ASTM STP 593, Published by Am. Soc. for Testing and Materials, Philadelphia, PA.
13. Richardson, T., *Composites—A Design Guide*, 1987, Industrial Press Inc., New York.
14. Nahas, M. N., *Survey of Failure and Post-Failure Theories of Laminated Fiber-Reinforced Composites*, Journal of Composites Technology and Research, Vol. 8, No. 4, Winter, 1986, pp 138–153.
15. *Engineered Materials Handbook—Vol. 1, COMPOSITES*, 1987, Published by Am. Soc. for Metals International, OH.
16. Hull, D., *An Introduction to Composite Materials*, 1981, Cambridge University Press, Cambridge, UK.
17. Baker, E. H., Kovalevsky, L., and Rish, F. L., *Structural Analysis of Shells*, 1972, McGraw Hill Book Co., New York.

18. Humphreys, E. A., and Rosen, B. W., Properties analysis of laminates, in *Engineered Materials Handbook, Vol. 1, COMPOSITES*, 1987, Published by ASM International, pp 218–235.
19. Wang, A. S. D., Strength, failure and fatigue analysis of laminates, in *Engineered Materials Handbook, Vol. 1, COMPOSITES*, 1987, Published by ASM International, pp 218–235.
20. Whitney, J. M., Daniel, I. M., and Byron Pipes, R., *Experimental Mechanics of Fiber Reinforced Composite Materials*, 1982, Published by The Society For Experimental Stress Analysis, Brookfield, MA.
21. Jones, R. M., *Mechanics of Composite Materials*, 1975, McGraw Hill Book Company, New York.
22. Nielsen, L. W., *Mechanical Properties of Polymers and Composites*, Vol. 2, 1974, Marcel Dekker, Inc., New York.
23. *Handbook of Fillers and Reinforcements for Plastics*, Ed. by Katz, H. S. and Milewski, J. V., 1978, Van Nostrand Reinhold Company, New York.

**Exercises**

**7.1.** There exist a total of six mathematical moduli for an orthotropic material with transverse isotropy. True or false?

**7.2.** There exist . . . independent moduli for a fully orthotropic material. a. 6, b. 36, c. 9, d. 15, e. 21.

**7.3.** For a typical thin composite panel with randomly oriented short fiber reinforcement, which two properties are identical?

a. $E_{11}$ and $G_{12}$, b. $E_{22}$ and $G_{12}$, c. $E_{11}$ and $E_{22}$, d. $\nu_{12}$ and $\nu_{21}$, e. $G_{12}$ and $G_{31}$, f. $G_{12}$ and $G_{21}$

**7.4.** For a typical, thin orthotropic composite ply which of the following is a realistic ascending sequence of magnitude of moduli?

a. $E_{11}, E_{22}, G_{12}$, b. $E_{11}, G_{12}, E_{22}$, c. $E_{22}, E_{11}, G_{12}$, d. $G_{12}, E_{22}, E_{11}$, e. $G_{12}, E_{11}, E_{22}$.

**7.5.** For a three-dimensional chunk of a composite material, it is found that $\nu_{23} = \nu_{13}$. Is it possible that the material is a ribbon reinforced composite? Yes or no?

**7.6.** For a three-dimensionally orthotropic material it is found that $\nu_{23} = \nu_{13}$. Which of the following is true?

a. $E_{22} = E_{11}$, b. $E_{11} = E_{33}$, c. $\nu_{32} = \nu_{31}$, d. $G_{23} = G_{13}$.

**7.7.** For a transversely isotropic composite rod, $\nu_{13} = \nu_{23}$. True or false?

**7.8.** Prove that in the Halpin-Tsai formula,

$$\text{when } \xi \to \infty, (\cdot)_c = (\cdot)_f V_f + (\cdot)_m V_m$$

$$\text{when } \xi \to 0, \frac{1}{(\cdot)_c} = \frac{V_f}{(\cdot)_f} + \frac{V_m}{(\cdot)_m}$$

**7.9.** A stainless steel cafeteria tray is made of 1 mm thick plate. It is to be replaced by E-Glass-Epoxy composite material with randomly oriented glass fibers, 35% by weight. If the stiffness is kept the same, what is the weight saved, if any? (Hint: Tray stiffness $\alpha$ $Et^3$.)

**7.10.** Use the principle of similarity to set up a finite element model to predict $E_{22}, E_{33}, \nu_{12}$ of a ribbon reinforced composite, with a square pattern of filling. How do you consider the volume fraction $V_f$ in the model?

**7.11.** Repeat Problem 10 for a staggered arrangement of fibers.

**7.12.** Repeat Problem 10 for spherical particles of equal size.

**7.13.** For a typical orthotropic material, which of the following is/are realistic descending sequences of strengths?

a. $F_{1T}, F_{1C}, F_{2T}, F_{2C}, F_6$
b. $F_{1T}, F_{1C}, F_{2C}, F_{2T}, F_6$
c. $F_{1T}, F_6, F_{2C}, F_{2T}, F_{1C}$
d. $F_{1T}, F_{2T}, F_{2C}, F_{1C}, F_6$
e. $F_{1T}, F_{1C}, F_{2C}, F_6, F_{2T}$.

**7.14.** On the edges of an off-axis square specimen of unit size, a shear stress of unit magnitude is applied ($\tau_{xy} = 1$). What is the change in length of the edges along the $x$ and $y$ directions?

**7.15.** $E$, $G$, $\nu$ are the only three independent elastic constants of a randomly oriented fiber reinforced composite panel. True or false?

**7.16.** Prove that it is impossible to stress a plane orthotropic lamina in such a manner as to cause no change in the shape of the lamina. Show this is the case for isotropic materials also.

**7.17.** Write Hooke's law in matrix form for fully orthotropic material in three dimensions. Define the meaning of the symbols used.

**7.18.** A square unidirectional laminate is squeezed in its plane, across fibers, between the flat faces of a vise. The size of the laminate is 100 mm × 100 mm × 10 mm. If the in-plane elastic constants are $E_{11}$, $E_{22}$, $G_{12}$, and $\nu_{12}$ (kg/mm$^2$ units), find the maximum stress that can be applied without causing failure of the laminate. Assume the strengths to be $F_{1T}$, $F_{1C}$, $F_{2T}$, $F_{2C}$, and $F_6$ respectively. Use maximum strain criterion.

**7.19.** Repeat Problem 18 using the Tsai-Hill and Tsai-Wu criteria.

**7.20.** A Kevlar/Epoxy ply is known to be in a state of biaxial stress $\sigma_x = 3$, $\sigma_y = 3$, and $\tau_{xy} = 1$. Use the following strengths and calculate the available $SF$ if all the stresses were reversed. Assume kg/mm$^2$ units.

$$F_{1T} = 130; \quad F_{1C} = 34; \quad F_{2T} = 3; \quad F_{2C} = 16; \quad F_6 = 5$$

$$E_{11} = 8800; \quad E_{22} = 560; \quad G_{12} = 220; \quad \nu_{12} = 0.34$$

Use the Tsai-Wu criterion.

**7.21.** Plot the apparent strength of Kevlar-Epoxy ply in Problem 20 as a function of the loading angle. Use the maximum strain failure criterion.

**7.22.** Lamina with $E_{11} = E_{22}$ and $Q_{xs} = Q_{ys} = 0$ is called ...... lamina.

**7.23.** For an orthotropic composite ply, $\nu_{21} \geq \nu_{12}$. True or false?

**7.24.** Generally orthotropic materials are those for which there are no coordinate axes with any of the $Q$'s equal to zero. True or false?

**7.25.** The utilization of the fiber strength in a UDL is likely to be better with a thermoset matrix than with a thermoplastic matrix. True or false?

**7.26.** The reinforcement of a matrix with fibers tends to reduce its Poisson's ratio. True or false?

**7.27.** The mathematical modulus $Q_{11}$ is always ...... than $E_{11}$. Greater or less?

**7.28.** Define a symmetric laminate. A large off-axis prepreg is rolled in the form of a cylinder several times over, flattened to a plane. We obtain a ...... laminate. Symmetric? Unsymmetric? Neither? (Verify your answer with a sheet of ruled paper.)

**7.29.** For a symmetric laminate, fill in the values.

$$0, \cdots, -60, 90, \cdots, -60, 60, \cdots, 90, \cdots, -60, 0, 90, \cdots, 60, 0$$

**7.30.** Symmetry causes the following elements in the [**ABD**] matrix to become zero. True or false?

$$B_{ij} = D_{xs} = D_{ys} = 0 \quad \text{for all } i \text{ and } j.$$

**7.31.** Which one of the following is a quasi-isotropic laminate?
a. $(0, \pm 45)_{2s}$, b. $(0, \pm 45)_s$, c. $(0, 90, 90, 0)_{2s}$, d. $(0, \pm 30, \pm 60, 90)_{2s}$.

**7.32.** Which one of the following is the closest to true isotropy?
a. $(0, \pm 45, 90)_s$, b. $(0, \pm 30, \pm 60, 90)_s$, c. $(0, \pm 15, \pm 30, \pm 60, \pm 75, 90)_s$, d. $(0, \pm 60)_s$.

**7.33.** Just by examining the lay-ups, state which one would produce the maximum apparent tensile modulus $\bar{E}_{xx}$ along the 0° direction?
a. $(0, 90, 90, 0)$
b. $(0, 90, 0, 90, 90, 0, 90, 0)$
c. $(0, 60, -60, -60, 60, 0)$
d. $(0, 45, -45, -45, 45, 0)$

**7.34.** A unidirectional composite is made of E-Glass-Epoxy. It is square in shape and the fibers are at 30° to the $x$ axis. The temperature is raised by 45°C, and all edges are left free. Calculate and plot (not-to-scale) the natural expanded shape of the laminate. Take $\alpha_1 = 7 \times 10^{-6}\epsilon/°\text{C}$ and $\alpha_2 = 21 \times 10^{-6}\epsilon/°\text{C}$ respectively.

**7.35.** E-Glass-Epoxy unidirectional pre-preg is used to construct a unidirectional laminate. Test specimens are cut at 30° and 60° to the fiber. Given

$$E_{11} = 4000; \quad E_{22} = 870; \quad G_{12} = 380; \quad \nu_{12} = 0.28$$

$$F_{1T} = 110; \quad F_{1C} = 63; \quad F_{2T} = 4; \quad F_{2C} = 13; \quad F_6 = 9$$

in kg/mm$^2$ units, what are the expected test values for 1. $E_{xx}$, 2. $E_{yy}$, 3. $G_{xy}$, 4. tensile strength, 5. compressive strength for these specimens. Use Tsai-Wu criterion for 4. and 5.

**7.36.** An E-Glass-Epoxy laminate of size 600 mm × 500 mm, is constructed with $[(0/\pm 30/\pm 60/90)_n]_s$ layup. It is used as a baffle in an air-flow path. Occasionally, it may be simply supported on all its edges. Under this condition, it is subjected to a maximum pressure of 0.0176 $kg/mm^2$, (about 0.25 psi). Design the thickness of the laminate if the deflection is not to exceed 0.5 mm. Assume all ply properties to be as in Problem 35. (You will need a computer program to solve this problem.)

**7.37.** An off-axis unidirectional composite laminate is gripped in the vise of a tensile testing machine. Because of rigid grips, the natural shear deformations are prevented, thus creating forces and moments in the specimen other than the intended axial tensile load. Identify these additional loads, and their effects on

a. the measurement of moduli,

b.the measurement of strengths.

**7.38.** A $(\pm\theta)_n$ laminate is constructed (by filament winding process) for a pressure vessel under pressure. Show that the thickness needed to withstand a given internal pressure is at a minimum when $\theta = 55°$ to the axis of the vessel. (You will need a computer program to solve this problem.)

# 8

# Finite Element Method: An Introduction

The purpose of computation is not numbers, but insight—R. W. Hamming

## 8.1 MOTIVATION

The major change that occurred in the 1980s in the field of plastics parts design was the introduction of the stress analysis of components as a formal activity. This has led to a change in the profile of the design engineer, who in addition to having to perform the hitherto routine material selection, quantity estimation, fabrication drawing production, now needs to be a stress analyst too. It is about the only engineering field that is distinctly technical, and needs specialized knowledge. Having discussed earlier the stress categories and their significance at length, it is clear that the finite element stress analysis cannot be used as a black box, notwithstanding the way such software packages are marketed. The current generation of plastics designers do experience difficulty when working with finite elements, simply because of the different engineering discipline in which they have experience. Also, it is a matter of computational insight to determine what factors of material behavior are significant in what types of problems.

This chapter and the next are both motivated by the need for the plastics engineer to use the finite elements method more knowledgeably. With a minimum amount of mathematics and an optimum amount of verbal explanations, the following two chapters should be able to give the reader a clear understanding of the mechanics of finite elements. Understanding finite elements—not expertise—is the objective of this chapter and the next.

## 8.2 OVERVIEW OF FEM

The Finite Element Method (FEM) is a computational technique that started in the early 1950s as a tool for the structural and stress analysis of complex shapes. Later in the 1960s, its connections with the fundamental principles of mechanics were established. As a result, FEM has become a versatile computational tool in several other branches of engineering. See historical outline below for the development of the FEM in engineering and applied mathematics. Presently, it is widely applied in the areas of mechanics, stress analysis, fluid flow, heat conduction, geomechanics, magnetics, etc. The variety of problems to which it is applied has also expanded. Originally it was applied to steady state problems or static problems only. Quickly, its application has extended to time dependent problems, wave propagation problems, instability problems, etc. The formative years of its development were in the 1950s to mid 1960s. Being computation intensive, FEM was available as a large, general purpose software, running mainly on mainframe computers, which were themselves of a lower order of capability as compared to those of the present day. Consequently, its availability was limited and its development slow. The era of personal computers that started in the early 1980s sharply increased the commercial availability, applications, and use of FEM. In the meantime, that is, from the 1960s to 1980s, FEM matured into a well-known, well-established, and well-researched technique. In that period, finite element approximation techniques became a formal topic in applied mathematics. Today, every school offers FEM courses (both hands-on and theoretical courses) at the undergraduate and graduate levels. Quite a large number of text books have been written to address the needs of the beginner as well as the needs of the research student.

***Brief History of Development of FEA***

1941 Hrenikoff pointed out a limited similarity between the behavior of shell structures to that of a framework. [1]

1943 Courant suggested a solution scheme for field problems, which is essentially similar to the present day FEM. He used piece-wise continuous functions inside a triangular region as the basis of solution. [2]

1949 Newmark refined the suggestion of Hrenikoff for computation. [3]

1954 Argyris established the connection between the matrix structural analysis methods and the energy theorems of elasticity. [4]

1956 Turner, Clough, Martin, and Topp wrote the landmark paper outlining the finite element method. (The name was not used yet). Experts consider that this paper, and that of Argyris [4] were the genesis of FEM. [5]

1960 Clough used the term "Finite Elements" to describe his new method, which had come to stay. [6]

1965 Zienkiewicz and Cheung saw the fundamental connection between FEM and the formulation of all kinds of field problems. This constituted a major extension of the power of FEM as a general purpose method. [7]

1966 Irons formulated the "parametric" elements, which substantially improved the capability of finite element to model real-life shapes. [8]

1967 Zienkiewicz and Cheung published the first text book on the finite element Method. [9]

1967 The STRUDL program was released. It was the first program to use its own Problem Oriented Language (POL). POL dispensed with traditional input methods of typing in a stream of numbers, or "data cards." Instead, it employed common engineering language, which facilitates data preparation as well as data checking. [10]

1969 ASKA, a general purpose finite element analysis program was released commercially. [11]

1970 Several other general purpose finite element analysis software packages were released commercially. (ANSYS, SAP 4, MARC, PAFEC). [11]

1971 The finite element method extends itself to another new science in mechanical engineering, namely, fracture mechanics. [12]

1974 NISA was released commercially.

1975 ADINA released. [11]

1978 ABAQUS released. [11]

1981 COSMOS 7 released.

Presently, many finite element software packages for stress analysis are on the market. In a very short period of five years, 1985–1990, the personal computer market expanded in terms of quality, size, and computing speed. The effect on finite elements was that the number of packages of software increased; the quality of user-interface improved; their ability to handle large problems was enhanced; and they also became capable of handling higher levels of complexity of engineering problems.

Originally, because of the paucity of core memory space available in computers, FEM research was oriented to the development of memory-efficient algorithms. Such algorithms involve much movements of data to storage locations and reading of the data in and out of memory many times. Known as "input-output," or I/O operations, such read-write operations slowed down the calculation speed, especially in smaller computers. On the other hand, such operations have the advantage that they allow large size problems to be solved with small computers. As it turned out, in the late 1980s cost per bit of memory declined continuously, and full in-core procedures became affordable. The advantage of in-core procedures is that they do not involve any unnecessary I/O operations or bookkeeping. The computation takes place fully in-core, and hence the speed is high.

A finite element analysis (FEA) calls for enormous amount of input data to be prepared. Likewise, the computer gives an enormous volume of data as the output. In the early days of its development, one had to look through piles of voluminous paper output, and compile it the way one wanted. Thanks to computer color graphics, the chore of input preparation and output reading have been greatly simplified. Most operations can be done by visual, on-screen, or interactive techniques. Output data are presented as whole-field pictures on the screen. Such pictures are more meaningful to the engineer as an output than are tables of numbers.

All FEA software have three distinct parts:

- *The Pre-processor*, which helps the engineer to build a file of input data. In this part, the engineer interacts with the computer very closely. The engineer builds up the shape of the product being analyzed, the materials, the sizes, loads, and supports. Strictly speaking, this activity is not finite elements per se, but is solid modeling (or the equivalent). Thus, a number of CAD vendors are able to provide "pre-processors" as an adjunct to the FEA software.
- *The Solver*, which proceeds almost entirely without user intervention. This part uses the file of input data and options prepared by the pre-processor. It solves the problem, and also generates a database of output results. Everything in this phase is strictly finite elements, and is computation intensive.
- *The Post-processor*, which helps the user to view the output results graphically in a variety of ways. It also enables hard copying and the archiving of graphic outputs on to different kinds of media. Once again, this activity belongs mainly in the area of computer graphics. The engineer interacts with the computer very closely.

It is obvious that the complexity of the software increases with its capability. Large FEA programs, currently available on the market, may have as many as a thousand specialized commands, and each command may have, on the average, five qualifiers/options. With numerous options available at the pre- and post-processing stages, the FEA user needs to be aware of the mechanics of the FEM itself to an extent. Formal training is recommended when working with large general purpose FEA codes like NASTRAN[1], ANSYS[2], etc.

Complexity of shapes in the analysis has led to the marriage of Computer Aided Design (CAD) with FEM. CAD is an enigmatic term. In the present context, let us consider it to mean a bunch of graphics tools used to create the image of an assembly of complex three-dimensional objects. Also, we assume that the scope of CAD stops with the creation of the three-dimensional image. More specifically, the analysis of the structural response of an object to applied loads is not possible within the capability of a CAD. Hence, from the point of the creation of the geometry, a CAD-to-FEM interface program takes over. Essentially the interface program does the preprocessing job for the engineer, however, with his/her interaction. The interface program finally creates a file of input data that is acceptable to the FEM software. From this point, the FEM software takes over, produces a database of input, solves the problem, and produces the output database which is ready for viewing. At this stage, the user interaction takes place once again. The user can view the output data graphically in several different ways, and save the graphics image for records.

To summarize, FEM started as a numerical technique for stress analysis, and it has now been established as the most popular technique for other fields as well. Its close association with, and dependence on, computer graphics have made the latter a necessary adjunct to FEM. Furthermore, the computer-graphics techniques of the late 1980s have brought about a cultural change in the way engineers think of input preparation and output reading. The CAD-Interface-FEM chain of software has come to stay as the state of the art for stress analysis. Considerable effort is being expended to make the transition from CAD to FEM "seamless." In the early 1980s there was no single software developer who packaged the preprocessor, the solver, and the postprocessor. This gave rise to a confusing range of choice of software, hardware, and vendors for the lay user. However, this situation is changing. Presently, there are at least a handful of companies that can deliver all three in one package. However, in the context of stress analysis, the full range of CAD capabilities far exceeds

---

[1] NASTRAN is a product of MacNeal Schwendler Corporation.
[2] ANSYS is a product of Swanson Analysis Corporation, Inc.

| | Shape Design | Geometry Manipulation | Mesh Generation | Finite Elem. Solution | Post-Processing |
|---|---|---|---|---|---|
| CAD Software | Yes | Yes | Yes | No | Yes |
| FEA-CAD Interface | No | No | Yes | No | Yes |
| FEA Software | No | Yes | Yes | Yes | Yes |
| Modeler Modules | Yes | Yes | Yes | No | No |

In the mid 80 s ————————

In the late 80 s - - - - - - - - - - -

**FIGURE 8.1.** The typical steps involved in creating a model, meshing, solving, and post-processing of a finite element analysis problem.

the modeling needs. Hence, the current trend in the 1990s is to replace the CAD and interface by "modeling module," which has CAD-like capabilities, but does not have the drafting features. An interface program is offered as an option to cater to the needs of users who are already set up with CAD facilities. See Figure 8.1.

## 8.3 BASICS OF FEM STRESS ANALYSIS

### 8.3.1 What is Degree of Freedom (DOF)?

The term *Degree of Freedom* (DOF) simply refers to the unknowns for which a problem is to be solved. In practical engineering problems, the number of unknowns is not self evident and depends on the way the problem is modeled.

In the viewpoint of elasticity theory, there are an infinite number of points in any problem. In the FEM, however, there are only a finite number of points or *nodes*, at which displacements are sought. The number of nodal displacements to be solved for is the DOF for the problem. Since, there are as many linear equations in the solution process as the DOF, one may also think of DOF as simply the number of equations in the problem. For the same physical problem, one may vary the DOF by simply remodeling it with a different number of nodes.

For example, if a three-dimensional problem has 2500 nodes, then there are (2500 nodes $\times$ 3 displacements per node) = 7500 unknowns to be solved for. The number 3 is the *degrees of freedom per node* and the number 7500 is the *total number of degrees of freedom* for the entire problem. In

FEA practice, the terms DOF number and equation number are used interchangeably.

### 8.3.2 Degrees of Freedom Per Node

While the DOF for a problem can be varied by the analyst, the DOF per node is a given. For each element type DOF per node is determined at the stage of element formulation, that is, at the time of software development.

In the case of plate problems, it is customary to think of displacements and rotations as basic unknowns. For a point in the plate, the out-of-plane displacement $w$, and the rotation $\theta$ in the plane of the plate are the two unknowns. In the FEM language, we say that the DOF per node is 2.

As another example, in plane two-dimensional problems, it is sometimes preferable to think of in-plane displacements $u$, $v$ as well as the in-plane rotations $\theta$ as the basic unknowns (although this is not done at all in the elasticity theory). Thus, in this case, the DOF per node (or the *nodal DOF*) is 3, although it is only a two-dimensional problem.

Another example is the case of a general three-dimensional problem. Normally, three displacements $u$, $v$, and $w$ are the basic unknowns. However, it is possible in FEM to consider the three rotations about the $x$, $y$, and $z$ axes as additional unknowns at each point. In such a case the total nodal DOF becomes 6. Such an approach is not in conflict with the theory of elasticity. Hence,

> the term degree of freedom is simply another name for the basic unknowns that are to be solved for.

### 8.3.3 What Is A Finite Element?

Until the finite element method was developed, the popular numerical method available for solving field problems, was the Finite Difference Method (FD), which is still a popular technique for transient field problems. In the FD, differential equations are approximated by their corresponding difference equations. The procedure involves taking uniform sized increments parallel to the coordinate axes. See Figure 8.2a. Difference equations are set up for each increment. As the entire area (or volume) is completely swept out, a large system of linear equations is obtained. The solution of this system is the solution to the problem. In this method, the role of the nodes is to define a grid of equally spaced points. But there are no elements; no properties are attributed to the space between the nodes. Rather such properties are contained in the differential equation itself.

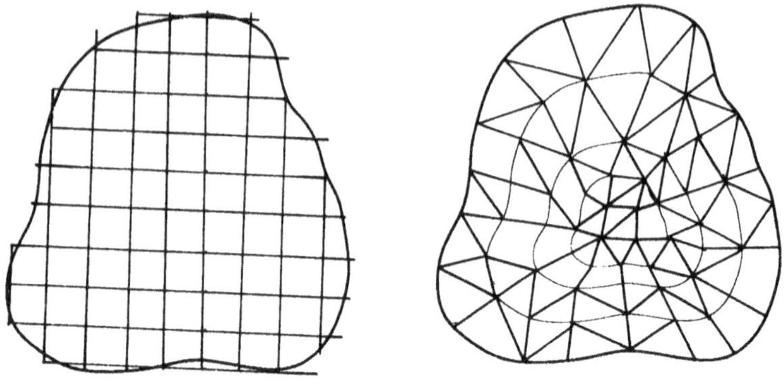


**FIGURE 8.2.** Modeling a problem in finite differences and finite elements. The finite element concept allows more flexibility in the modeling of the shape.

In the finite difference method, the points *must* be laid out parallel to the coordinate axes. Also, the size of the increments must be uniform. This presents problems when trying to model a curved boundary.

In finite elements too, the problem domain is divided into small regions. The corner points of such regions do not have to be laid out parallel to the coordinate axes. The shapes of the elements themselves may be straight lines, triangles, rectangles, bricks, etc. Furthermore, there is a greater freedom in the choice of their size, shape, orientation, mix, etc. In the finite elements, besides the nodes, the space spanned by the nodes—called the *finite element*—also plays a role. Rather than writing the approximate form of the field equation, a different approach is taken, as discussed below.

It is known that behind every differential field equation (such as the plates equation in structural analysis, the Navier-Stokes equation in fluid mechanics, the Laplace equation and the Helmholtz equation in several engineering idealizations), a more fundamental "minimum principle" is operable. For example, the principle of virtual work, or the principle of minimum potential energy may be considered to be the working principle behind the field equations. A variety of functionals have been constructed in the theory of engineering mechanics. These functionals are all candidate expressions for minimization.[3] When the minimization is performed

---

[3] Some of these functionals represent the exact solution, while some others lead to an approximate solution. Those that represent the approximate solution are of interest in engineering. They usually correspond to different levels of accuracy.

over a finite element, we obtain a set of finite element equations. The number of equations obtained is equal to the number of nodes in the element × nodal DOF.

> Thus, a "finite element" is a miniature domain to which a minimum principle of mechanics is applied, and from which a set of algebraic equations for the nodal unknowns may be obtained.

In general, an approximate displacement pattern—displacements, because they are taken as the basic unknowns—is assumed inside a prototype finite element to describe the variation of $u$, $v$, and $w$. The expressions for $u$, $v$, and $w$ are simple polynomials, which is a good assumption, *locally, inside an element*. The choice of these polynomials is the basic technique in the formulation of the finite elements theory. Some of the topics of importance in the formulation of elements are the type of polynomials, the shape of the elements, the DOF associated with the nodes, the number of nodes in an element, the simulation of a real-life material behavior, to state a few. The study of these topics constitutes a subdiscipline in the finite elements called the *element technology*.

A minimum principle will work only for elastic problems, and more generally, for problems in which energy is conserved. For problems involving dissipation of energy, such as those involving plasticity, friction, or viscous flow, other methods are necessary. Such methods are classified as a group of methods called the *methods of weighted residuals*. For a classical treatment of the methods the reader is referred to Finlayson [13].

The application of the methods of weighted residuals to a finite element (rather than to the whole problem domain) has helped establish FEM as a formal topic in applied mathematics.

### 8.3.4 The Basic Steps in FEA

The ultimate objective of an FEA is the analysis of stress. The overall strategy used for this is to

- calculate nodal displacements $u$.
- calculate strains $\{\epsilon\}$ from displacements.
- calculate stresses $\{\sigma\}$ from strains.

Thus, calculation of nodal displacements is a vital step in the analysis of stresses and strains in a structure. For this purpose, we construct a large

**Table 8.1. The Computational Steps in the Finite Element Method.**

| *Operation Performed* | *Mathematical Equivalent* |
|---|---|
| Discretization | None |
| Interpolate Displacements | $\{\mathbf{u}\} = [\mathbf{N}]\{\mathbf{u}\}^{\mathbf{e}}$ |
| Gradient Matrix | $[\mathbf{B}] = \dfrac{\partial[\mathbf{N}]}{\partial\{x\}}$ |
| Elasticity Matrix | Construct [D] |
| Element Stiffness | $[\mathbf{K}]^{\mathbf{e}} = \int_{\mathbf{V}} [\mathbf{B}]^{\mathbf{T}}[\mathbf{D}][\mathbf{B}]d\mathbf{V}$ |
| Element Load | $\{\mathbf{f}\}^{\mathbf{e}} = \int_{\mathbf{V}} [\mathbf{N}]^{\mathbf{T}}\{\mathbf{f_b}\}d\mathbf{V} + \int_{\mathbf{S}} [\mathbf{N}]^{\mathbf{T}}\{\mathbf{f_s}\}d\mathbf{S}$ |
| Global Stiffness | $[\mathbf{K}] = \sum_e [\mathbf{K}]^{\mathbf{e}}$ |
| Global Force Vector | $\{\mathbf{f}\} = \sum_e \{\mathbf{f}\}^{\mathbf{e}}$ |
| Global Displ. Vector | Obtained by Gauss elimination. |
| Element Displ. Vector | Retrieved by bookkeeping. |
| Element Strain | $\{\epsilon\}^{\mathbf{e}} = [\mathbf{B}]\{\mathbf{u}\}^{\mathbf{e}}$ |
| Element Stress | $\{\sigma\}^{\mathbf{e}} = [\mathbf{D}]\{\epsilon\}^{\mathbf{e}}$ |
| Nodal Anything | Spatial Interpolation |

system of linear equations in the form

$$[\mathbf{K}]\{\mathbf{u}\} = \{\mathbf{f}\},$$

where [**K**] is known as the *global stiffness matrix*, {**u**} is the *global displacement vector*, and {**f**} is the *global vector of nodal forces*. This equation is formally similar to that of a loaded spring. The difference is just in the number of forces and displacements in the equation. In an actual analysis the basic steps of the stress calculation are the ones below. Not all of the steps may be apparent to the user. Some involve an interaction between the user and the computer. Many do not.

1. Discretization of the problem domain into elements. (Pre-processing stage)

2. Interpolation of displacements. (Solution stage)
3. Calculation of element stiffness. (Solution stage)
4. Calculation of element load vectors. (Solution stage)
5. Assembly of the global stiffness matrix equation. (Solution stage)
6. Application of boundary conditions. (Solution stage)
7. Solution of the unknown displacements. (Solution stage)
8. Computation of strains from displacements. (Solution stage)
9. Computation of stresses from strains. (Solution stage)
10. Interpolation of stresses and strains for on-screen display and hard copy outputs. (Post-processing stage)

Table 8.1 provides the mathematical equivalents of the above steps. We consider each one of these steps in some detail.

## 8.4 DISCRETIZATION

The problem domain is divided into smaller regions—the finite elements —which can have a variety of shapes such as triangles, rectangles, bricks, tiles, etc. In order to model curvilinear boundaries of the problem domain, the elements themselves may be provided with curved boundaries. Alternatively, a curved boundary of the problem domain may also be covered with flat tiles, or other straight-sided elements, without much deviation from the local geometry. FEM permits each element to have its own shape, size, material, and geometric constants. Further, a model can mix any number of element types. This affords a far greater flexibility in modeling a problem than is true of any other method.

Discretization consists of the preparation of a computer input file, which will contain (1) coordinates of all the nodal points, (2) the element types, (3) the element connectivity, (4) applicable element constants, and (5) material properties. Usually, the nodes are numbered sequentially, as are the elements. The elements are identified by the element number. *Element connectivity*—sometimes also called the *element topology*—is defined by the node numbers that bound the element.

Basic characteristics of a prototype finite element can be tailor-made. Elements can be "designed" to simulate any desired structural behavior. All FEA software vendors provide a wide range of choice of pre-engineered elements. See Figure 8.3. Elements may have mid-side nodes in addition to the corner nodes which are required. Every edge need not have an equal number of nodes. Elements can be one-dimensional, two-dimensional, three-dimensional, axisymmetric, plane strain, plane stress, thin shell, thick shell, layered shell, etc. Each element or group of elements can

| Element Type | | | | Loads | Outputs | Description |
|---|---|---|---|---|---|---|
| **Linear Stress, Vibration and Mode Shape Elements** | | | | | | |
| **1** | *3-D* | *2 nodes* | 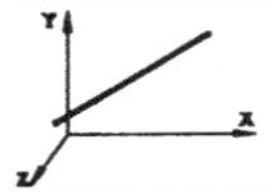 | Nodal forces, thermal loading and uniform acceleration (g's) in three directions. | Axial forces and stresses as well as displacements. | Supports translational degrees of freedom. This pin-jointed element is used to model a variety of structures, such as towers, bridges, and buildings. |
| **Truss** | | | | | | |
| **2** | *3-D* | *3 nodes* | 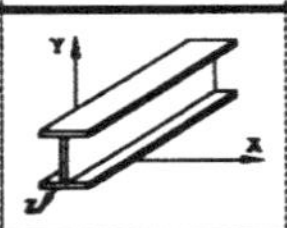 | Moments and forces at nodes and intermediate locations, thermal loading, continuous and intermediate distributed loads, fixed-end forces and uniform accelerations (g's) in three directions. | Two moments, torque, two shear forces, and axial forces at end nodes. Axial, bending, and shear stresses. Displacements, maximum absolute normal stresses produced by combined axial and bending area calculations. Beam end forces and moments. | Supports three translational and three rotational degrees of freedom. The third node is used to define the orientation. This element is used in modeling a variety of structures, such as towers, bridges, and buildings. |
| **Beam** | | | | | | |
| **3** | *3-D* | *3 or 4 nodes* | 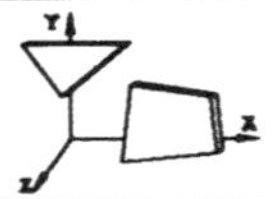 | Nodal forces, edge pressures, thermal loading and accelerations (g's) in three directions. | In-plane shear and two normal stresses, two principal stresses. | Supports three translational degrees of freedom and in-plane (membrane) loading. Orthotropic material properties may be temperature dependent. An incompatible mode is available. Membrane elements are used to model "fabric-like" structures, such as tents, cots, domed stadiums, etc. |
| **Membrane Plane Stress** | | | | | | |
| **4** | *2-D* | *3 or 4 nodes* | 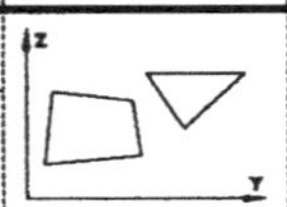 | Nodal forces, edge pressures, thermal loading and uniform accelerations (g's) in two directions. | In-plane local coordinate normal and shear stresses, three principal stresses. | Supports two translational degrees of freedom. Orthotropic material properties may be temperature dependent. An incompatible mode is available. The element supports plane strain, plane stress, and axisymmetric formulations. |
| **2-D Elasticity** | | | | | | |
| **5** | *3-D* | *6 or 8 nodes* | 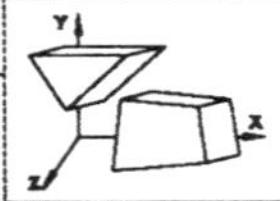 | Nodal forces, uniform distributed regular and hydrostatic pressure, thermal loading and accelerations (g's) in three directions. | Six stresses and displacements at nodes. | Supports three translational degrees of freedom as well as isotropic material properties and incompatible displacement modes. Applications include solid objects, such as wheels, turbine blades, flanges, etc. |
| **Solid Elasticity** | | | | | | |
| **6** | *3-D* | *3 or 4 nodes* | 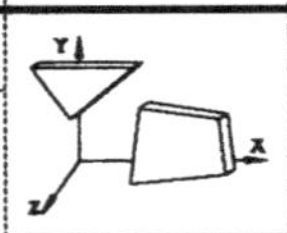 | Nodal forces, normal face pressure, accelerations in three directions, thermal loading (both nodal temperatures and a temperature gradient through the thickness). | Two normal and one shear membrane stress and bending stress at nodes. | Supports three translational and two rotational degrees of freedom as well as orthotropic material properties. An optional rotational stiffness around the perpendicular axis is automatically added to the node of each element. This element is used in the design of pressure vessels, electronic enclosures, automobile body parts, etc. |
| **Plate/Shell** | | | | | | |

**FIGURE 8.3.** Typical choice of pre-engineered element types offered in FEA software. The figure not only gives the element shapes and element DOFs, it also indicates the types of problems in which each element type is suitable. *Reproduced with kind permission from ALGOR Interactive Systems, Inc., Pittsburgh, PA.*

| Element Type (No.) | Dimension | Nodes | Name | Diagram | Loads | Outputs | Description |
|---|---|---|---|---|---|---|---|
| ***Linear Stress, Vibration and Mode Shape Elements*** | | | | | | | |
| 7 | 2/3-D | 2 or 5 nodes | **Boundary** | 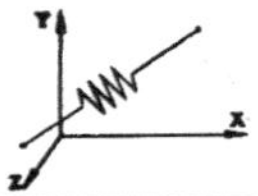 | Attached between a node and ground. | Force or moment. | Used in conjunction with other elements, a boundary element rigidly or elastically supports a model and enables the extraction of support reactions. Also used to impose a specified rotation or translation. |
| 10 | 2/3-D | -n/a- | **Stiffness Matrix** | [k] | | | A convenient method of creating your own element. With one stiffness matrix it is possible to connect up to 48 degrees of freedom directly into the system stiffness matrix. |
| 12 | 3-D | 2 or 3 nodes | **Pipe** |  | Nodal forces, internal pressure, thermal loading, uniform accelerations (g's) in three directions. | Forces, moments, and torsional and maximum transverse shear stress at nodes. Axial, bending, and hoop normal stresses. For elbow, data is also reported at the bend. | Supports three translational and rotational degrees of freedom. Temperature-dependent material properties are supported. Typical applications include the combined analysis of piping and structural systems. For piping system analysis, see PipePlus, Part 150. |
| 13 | 3-D | 2 to 8 nodes | **Rigid Element** | 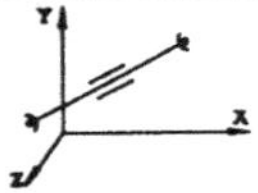 | | | Used to rigidly connect up to seven secondary nodes to one primary node. May be used in conjunction with type 1, 2, 6, 7 and 12 elements. |
| 14 | 2/3-D | 2 nodes | **Gap / Cable Element** | 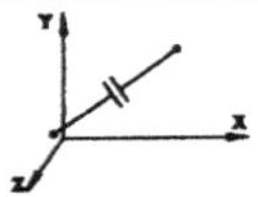 | Attach between any two nodes or between a node and ground. | Axial force. | "Gap distance" available for compression gap or tension cable. |
| ***Composite Elements*** | | | | | | | |
| 15 | 3-D | 3 or 4 nodes | **Thin Plate** | 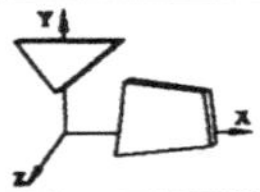 | Nodal forces, normal face pressure, thermal loading and temperature difference in curing (for thermal curing residual stress). | Stress (in-plane, 2 normal, 1 shear) for every layer (top and bottom) at nodes. Also, laminate failure factor, laminate curing stress and core crushing stresses. | Formulation is based on Kirchhoff theory and supports up to 25 layers with no limitations regarding orientation or stacking sequence. Supports the Tsai-Wu, maximum stress and maximum strain failure criteria. Used in models such as athletic equipment, bicycle frames, etc. |
| 16 | 3-D | 3 or 4 nodes | **Thick Plate** | 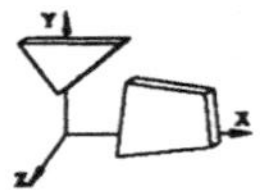 | Nodal forces, normal face pressure, thermal loading and temperature difference in curing (for thermal curing residual stress). | Stress (in-plane, 2 normal, 1 shear) for every layer (top and bottom) at nodes. Also, laminate failure factor, laminate curing stress and core crushing stresses as well as crushing stress and two transverse shear stresses at the core layer. | Formulation based on Mindlin theory and supports up to 25 layers with no limitations regarding orientation or stacking sequence. Supports the Tsai-Wu, maximum stress and maximum strain failure criteria. Used in models such as honeycomb sandwich structures, aerospace products, etc. |

**FIGURE 8.3.** (Continued)

| Element Type | | | | Loads | Outputs | Description |
|---|---|---|---|---|---|---|
| **Nonlinear Elements** | | | | | | |
| **21** | 3-D | 2 nodes | 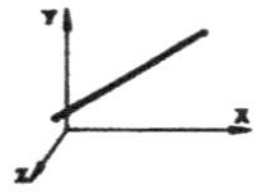 | | Displacements and axial stresses. | Supports three translational degrees of freedom and constant area. May be used in linear elastic as well as material and geometrical (large displacement) nonlinear analyses. Also possible to be used as a boundary element to specify displacement. Applications include modeling of towers, pre-tensioned cables, etc. |
| **Truss** | | | | | | |
| **22** | 3-D | 3 nodes | 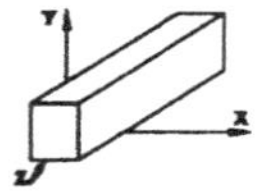 | | Displacements and two ends' forces, moments and torque. Beam's axial stress and two shear stresses are available on two end cross-sections and between. Shear and moment distribution diagrams are available in Superview. | Supports three translational and three rotational degrees of freedom. User definable cross-section types (linear analysis). Large deformations and/or elasto-plastic nonlinear effects are supported for rectangular and round cross-sections. Useful in 3-D beam, member or skeletal systems. |
| **Beam** | | | | | | |
| **24** | 2-D | 3 to 8 nodes | 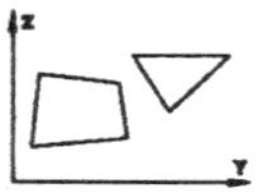 | Static nodal forces with step-loading function and time-dependent nodal forces. | 2-D stress state, hoop stress, principal stresses, yield function stress, displacements. Vibration analysis results include nodal velocities and accelerations. | Supports two translational degrees of freedom and plane stress, plane strain and axisymmetric formulations. Similar applications as type 4 element with the addition of nonlinear effects. |
| **Solid** | | | | | | |
| **25** | 3-D | 6 to 20 nodes | 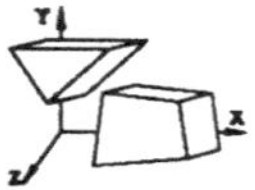 | | Complete 3-D stress tensor (3 normal and 3 shear) and displacements. Vibration analysis results include nodal velocities and accelerations. | Supports three translational degrees of freedom as well as a wide range of material and nonlinear behavior formulations. This element is used in the analysis of three-dimensional solids and thick shells where nonlinear effects are essential. |
| **Brick** | | | | | | |
| **26** | 3-D | 3 to 8 nodes | 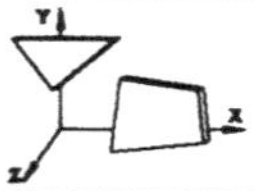 | | Complete 3-D stress tensor (3 normal and 3 shear) and displacements. Vibration analysis results include nodal velocities and accelerations. | Supports three translational and two or three rotational degrees of freedom as well as a wide range of material and nonlinear behavior formulations. This element is used in the analysis of three-dimensional shell structures where nonlinear effects are essential. |
| **Shell** | | | | | | |

**FIGURE 8.3.** (Continued)

be assigned to belong to a predefined material, which may be linear elastic, elastic-plastic, nonlinear elastic, etc. In fact, element types are simply miniature models of a macroscopic structural behavior. Thus, a beam element behaves like a beam, and a shell element behaves like a shell. *Elements may also be designed to simulate abstract structural entities*, such as in the case of boundary elements. Such elements can be used to represent a spring support or a rigid support. Other abstract elements are gap-friction elements which are single-acting members. That is, a reaction develops between two parts if the gap between them closes and they come into contact; if the gap is open, no force is developed. Such vast range of choice of element types makes FEM a truly general purpose method. In terms of element technology, each item of choice represents a certain specific computational purpose, accuracy, and advantage.

As explained earlier, discretization is accomplished by a CAD software, or by an interface. With minor variations, all such software permit the problem geometry to be broadly divided into smaller regions which can be later "meshed" automatically into finite elements.

## 8.5 INTERPOLATION OF DISPLACEMENTS

We mentioned earlier that the displacements at each node represent an economical way of solving stress problems in FEM. However, in order to "formulate" an element, it is necessary to express the displacements interior to an element as continuous functions of coordinates. Since the displacements are not yet known, it is customary to express the interior displacements as the weighted average of nodal displacements—or an interpolation of nodal displacements. Note that nodal displacements are also unknowns as yet.

For simplicity, we discuss the interpolation of displacements for a two dimensional triangular element. In Figure 8.4 we have a triangle whose vertices have been labelled arbitrarily as 1, 2, and 3.[4] The displacements, which are as yet unknown, at 1, 2, and 3 are $(u_1, v_1)$, $(u_2, v_2)$ and $(u_3, v_3)$ respectively. A point $P$ inside the element has a displacement $(u, v)$, which is also unknown. However, we can set up a relationship between the displacements at $P$ and those at the vertices. In FEM we assume that the displacements inside a finite element must follow simple patterns, and are

[4] In a real-life finite element mesh the nodes have arbitrary numbers, $i$, $j$, and $k$. The numbers 1, 2, and 3 are actually the internal numbering assigned by the computer.

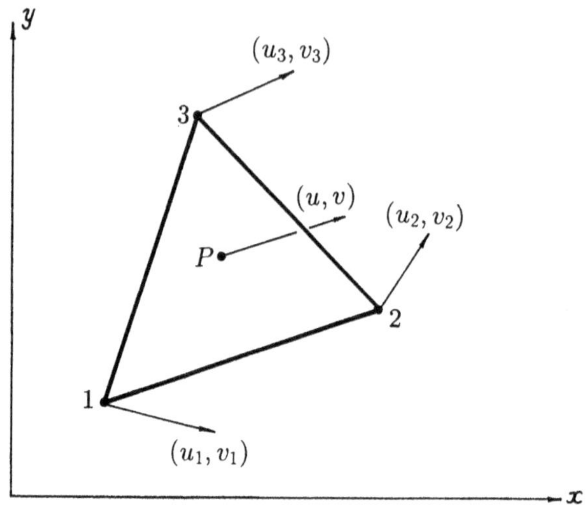


**FIGURE 8.4.** Interpolation of displacements of an interior point in terms of nodal displacements. Simplest of all types of elements is the plane triangle shown here. The vertices are internally numbered as 1, 2, and 3, but may be any arbitrary $i$, $j$, and $k$ when embedded in the global model.

*well behaved*. As a first approximation, one may assume,

$$\begin{aligned} u &= N_1 u_1 + N_2 u_2 + N_3 u_3 \\ v &= N_1 v_1 + N_2 v_2 + N_3 v_3 \end{aligned} \tag{8.1}$$

where the $N$'s are simply weights applied to the $u$'s and $v$'s. They are also called the *interpolation functions*. Or, in matrix notation,

$$\begin{Bmatrix} u \\ v \end{Bmatrix} = \begin{bmatrix} N_1 & 0 & N_2 & 0 & N_3 & 0 \\ 0 & N_1 & 0 & N_2 & 0 & N_3 \end{bmatrix} \begin{Bmatrix} u_1 \\ v_1 \\ u_2 \\ v_2 \\ u_3 \\ v_3 \end{Bmatrix} \tag{8.2}$$

This can be abbreviated as,

$$\{u\} = [\mathbf{N}]\{u\}^e, \tag{8.3}$$

In plain English, Equation 8.2 states that the displacements at $P$ are some kind of weighted average of the displacements at 1, 2, and 3. Many of the later calculations depend on the definition of the $N$'s. If the $N$'s are linear, then the interpolation is linear. If they are quadratic, then the interpolation is quadratic and so on.

### 8.5.1 The Form of the Functions $N_1$, $N_2$, and $N_3$

Consider the triangle 123 again, as shown in Figure 8.5. The point $P$ inside the triangle, divides it into three smaller triangles, namely, $23P$, $31P$, and $12P$, whose areas are $A_1$, $A_2$, and $A_3$ respectively. Let the area of the triangle 123 be $A$. We may define three dimensionless ratios,

$$L_1 = \frac{A_1}{A}; \quad L_2 = \frac{A_2}{A}; \quad L_3 = \frac{A_3}{A}.$$

Evidently, for any position of the point $P$ inside the triangle, the ratios of the areas are unique. Thus, there is a one-to-one correspondence between the ordered set of numbers $(L_1, L_2, L_3)$ and the position of $P$. By virtue of its uniqueness, the set $(L_1, L_2, L_3)$ can be taken as the *natural coordinates* of the point $P$. The following statements are easy to verify.

- When $P$ is at the vertex 1, its coordinates are $(1, 0, 0)$.
- When $P$ is at the vertex 2. its coordinates are $(0\,1, 0)$.
- When $P$ is at the vertex 3, its coordinates are $(0, 0, 1)$.
- When $P$ is at the centroid of the triangle, its coordinates are $(1/3, 1/3, 1/3)$.
- When $P$ is at the mid-point of the side 23, its coordinates are $(0, 1/2, 1/2)$.

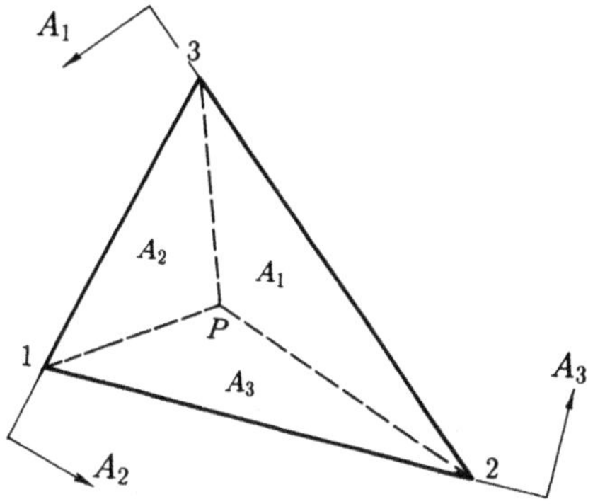


**FIGURE 8.5.** Natural coordinate system for a triangular element. Note that the areas of the smaller triangles bear a definite ratio to the area of 123. Since these ratios are unique to the location of $P$, they can be treated as the coordinates of $P$.

- When $P$ is at the mid-point of the side 31, its coordinates are $(1/2, 0, 1/2)$.
- When $P$ is at the mid-point of the side 12, its coordinates are $(1/2, 1/2, 0)$,

Since an interpolation function must be expressed as a function of coordinates, the form of the $N$'s is in general

$$N_1 = N_1(L_1, L_2, L_3)$$

$$N_2 = N_2(L_1, L_2, L_3)$$

$$N_3 = N_3(L_1, L_2, L_3)$$

However, several restrictions apply as far as the nature of the $N$'s are concerned. $N$'s may be polynomials in $L$'s. The simplest of all are the linear functions.

$$N_1 = L_1; \quad N_2 = L_2; \quad N_3 = L_3$$

A quadratic polynomial for a 6-noded triangular element is discussed in Section 8.14.3. With these *assumed* interpolation functions, and with Equation 8.1, it is possible to see that the following statements are valid inside the triangle.

- When the point $P$ is at vertex 1, the displacement at $P$ is simply $(u_1, v_1,)$; when at 2, or at 3, its displacement coincides with that of the vertex 2, or 3, respectively.
- When the point $P$ is at the mid-point of the side 23, the displacements are

$$\begin{Bmatrix} u \\ v \end{Bmatrix} = \left\{ \frac{(u_2 + u_3)}{2}, \frac{(v_2 + v_3)}{2} \right\}$$

- When the point $P$ is at the centroid of the triangle, then its displacements are

$$\begin{Bmatrix} u \\ v \end{Bmatrix} = \left\{ \frac{(u_1 + u_2 + u_3)}{3}, \frac{(v_1 + v_2 + v_3)}{3} \right\}$$

We can also define quadratic functions in the natural coordinates and thereby render the interpolation a quadratic one. The point is that the above linear equations are just one of the many legitimate ways of interpolating the displacements.

**Example 8.1**

A triangular element embedded in a mesh has three corners, whose coordinates in a two-dimensional Cartesian system are,

$$\begin{aligned} &1: (2, 2.5) \\ &2: (3, 3.5) \\ &3: (2.5, 4) \end{aligned}$$

Determine the equations for displacements in terms of interpolation functions. Show that along the edges of the triangle, and therefore over the interior of the triangle, the interpolation is linear.

*Solution* The point $P(x, y)$ is always inside the element. The coordinates are shown in Figure 8.6, but not the axes. We utilize the well known

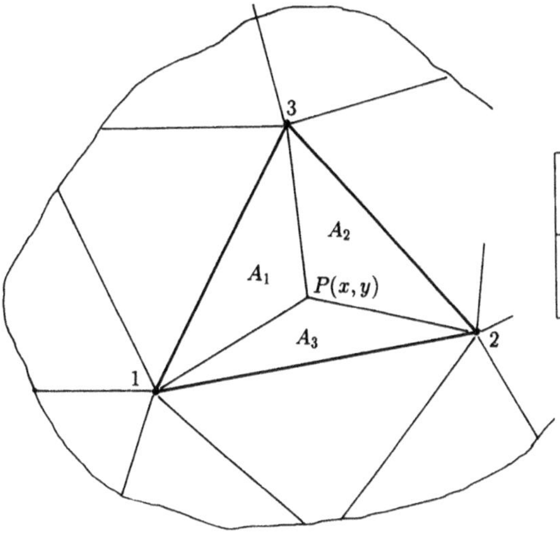


| Internal Node Number | Global Node Number (arbitrary) | Coordinates |
|---|---|---|
| 1 | 10 | (2.0, 2.5) |
| 2 | 18 | (3.0, 3.5) |
| 3 | 27 | (2.5, 4.0) |

**FIGURE 8.6.** Example 8.1.

principle in analytic geometry that an area can be expressed as a determinant involving coordinates of the vertices.

$$A = \frac{1}{2} = \begin{vmatrix} 1 & 2 & 2.5 \\ 1 & 3 & 3.5 \\ 1 & 2.5 & 4 \end{vmatrix} = \frac{1}{2}$$

$$A_1 = \frac{1}{2} = \begin{vmatrix} 1 & x & y \\ 1 & 3 & 3.5 \\ 1 & 2.5 & 4 \end{vmatrix} = \frac{1}{2}(3.25 - 0.5x - 0.5y) \tag{8.4}$$

$$A_2 = \frac{1}{2} = \begin{vmatrix} 1 & 2 & 2.5 \\ 1 & x & y \\ 1 & 2.5 & 4 \end{vmatrix} = \frac{1}{2}(-1.75 - 1.5x - 0.5y) \tag{8.5}$$

$$A_3 = \frac{1}{2} = \begin{vmatrix} 1 & 2 & 2.5 \\ 1 & 3 & 3.5 \\ 1 & x & y \end{vmatrix} = \frac{1}{2}(-0.5 - x + y) \tag{8.6}$$

$$L_1 = \frac{A_1}{A} = (3.25 - 0.5x - 0.5y) \tag{8.7}$$

$$L_2 = \frac{A_2}{A} = (-1.75 - 1.5x - 0.5y) \tag{8.8}$$

$$L_3 = \frac{A_3}{A} = (-0.5 - x + y) \tag{8.9}$$

In particular, when the point $P$ is at the centroid of the element, we have

$$x = 2.500; \quad y = 3.333 \tag{8.10}$$

$$A_1 = A_2 = A_3 = A/3 \tag{8.11}$$

$$L_1 = L_2 = L_3 = 1/3 \tag{8.12}$$

$$u = \frac{u_1 + u_2 + u_3}{3}; \quad v = \frac{v_1 + v_2 + v_3}{3} \tag{8.13}$$

Since only linear terms in $x$ and $y$ appear in Equations 8.7, 8.8, and 8.9, the interpolation in $u \ldots$ and $v \ldots$ is linear over the entire triangle.

***Displacements Along Edges***
In order to examine the nature of the interpolation over the edges, let us consider the edge 1-2. On this edge, the following relation holds good.

$$x + y = 6.5.$$

Hence, $(3.25 - 0.5x - 0.5y) = 0$ over this edge. This can also be seen from the fact that all over this edge, $A_1 = 0$, and hence $N_1 = 0$. Thus, on the edge 1-2, the expression for displacements reduce to

$$N_1 = 0$$

$$u = N_2 u_2 + N_3 u_3 \tag{8.14}$$

$$= (-1.75 + 1.5x - 0.5y)u_2 + (-0.5 - x + y)u_3$$

$$v = N_2 v_2 + N_3 v_3 \tag{8.15}$$

$$= (-1.75 + 1.5x - 0.5y)v_2 + (-0.5 - x + y)v_3$$

Note that the displacements along the edge 2-3 do not involve those at node 1, namely $(u_1, v_1)$. Once again it is easily seen that the interpolation is also linear over the edges. The reader may verify this for the other two edges.

***Commentary***

- *Note that the sum* $N_1 + N_2 + N_3 = 1$, *because the sum of the areas of the smaller triangle is the area of the whole. This condition is a fundamental property to be satisfied by the interpolation, for any element type.*
- *Note that the expression for displacement over the edges involves the displacements of only the nodes on that edge (see Equations 8.14 and 8.15). This is another essential condition to be satisfied by the functions* $N\ldots$. This way, the displacement along an edge will depend only on those of the nodes on that edge. *This ensures, that for two neighboring elements sharing the edge, the edge displacements will be identical, whether the interpolation is performed from one element or the other. This uniqueness is an important characteristic of the interpolation functions. This condition is referred to as* inter-element compatibility.

This example illustrates the linear variation of displacements. The methods used in this example are not used in actual computer programs since better algorithms are available.

**Example 8.2**

Generalize Example 8.1 for a plane linear triangular element. Obtain formulas for the following in a form suitable for programming.

- Area coordinates,
- Interpolation functions, and
- Expression for displacements.

*Solution* Let the triangle be formed by the three nodes $i(xi, y_i)$, $j(x_j, y_j)$, and $k(x_k, y_k)$. See Figure 8.7. Let $P(x, y)$ be a point interior to the element $ijk$, at which we wish to determine the area coordinates, the interpolation functions, and the displacements in terms of the nodal displacements.

*1. The Area Coordinates*

$$A_i = \frac{1}{2}\begin{vmatrix} 1 & x & y \\ 1 & x_j & y_j \\ 1 & x_k & y_k \end{vmatrix}$$

$$= \frac{1}{2}\{(x_j y_k - x_k y_j) + x(y_j - y_k) + y(x_k - x_j)\}$$

$$= \frac{1}{2}(a_i + b_i x + c_i y)$$

where
$$a_i = (x_j y_k - x_k y_j)$$
$$b_i = (y_j - y_k)$$
$$c_i = (x_k - x_j)$$

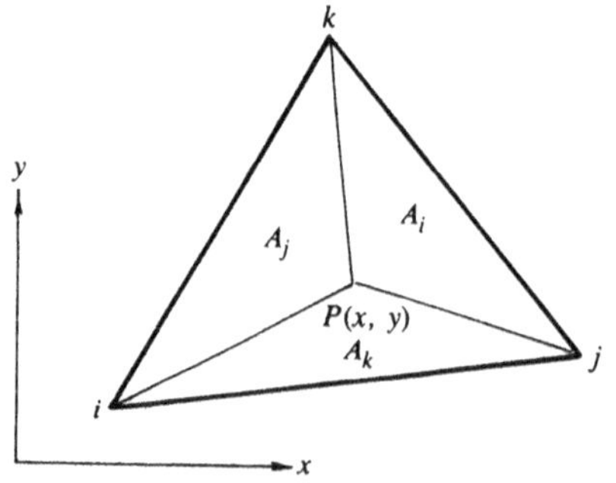


**FIGURE 8.7.** A general plane linear triangle.

By a cyclic permutation of the subscripts, we can get the expressions for the other two areas. The complete expressions are:

$$A_i = \frac{1}{2}(a_i + b_i x + c_i y) \tag{8.16}$$

$$A_j = \frac{1}{2}(a_j + b_j x + c_j y) \tag{8.17}$$

$$A_k = \frac{1}{2}(a_k + b_k x + c_k y) \tag{8.18}$$

where

$$\begin{aligned} a_i &= (x_j y_k - y_j x_k); & b_i &= (y_j - y_k); & c_i &= (x_k - x_j) \\ a_j &= (x_k y_i - y_k x_i); & b_j &= (y_k - y_i); & c_j &= (x_i - x_k). \\ a_k &= (x_i y_j - y_i x_j); & b_k &= (y_i - y_j); & c_k &= (x_j - x_i) \end{aligned}$$

Obviously, the area $A$ of the triangle is given by $A = (a_i + a_j + a_k)/2$.

*2. Natural Coordinates—$L_i$, $L_j$, and $L_k$*

The intermediary symbols $L\ldots$ for natural coordinates are very simple, and may not appear significant in the linear triangle because of its simplicity. However, in quadratic and higher order elements, these symbols are significant and are useful in the calculus of elements. For the present example,

$$L_i = (A_i/A); \quad L_j = (A_j/A); \quad L_k = (A_k/A).$$

$$L_i = \frac{(a_i + b_i x + c_i y)}{2A}; \quad L_j = \frac{(a_j + b_j x + c_j y)}{2A};$$

$$L_k = \frac{(a_k + b_k x + c_k y)}{2A};$$

For a linear triangle, the interpolation functions are simply,

$$N_i = L_i; \quad N_j = L_j; \quad N_k = L_k.$$

It is easy to verify that $N_i + N_j + N_k = 1$.

*3. Displacements at $P(x, y)$*

Let the displacements at $i$, $j$, and $k$ be $(u_i, v_i)$, $(u_j, v_j)$, and $(u_k, v_k)$ respectively. Let the displacements at $P$ be $(u, v)$. We know that $(u, v)$ are given by

$$u = N_i u_i + N_j u_j + N_k u_k \tag{8.19}$$

$$v = N_i v_i + N_j v_j + N_k v_k \tag{8.20}$$

For programming in a computer, the following flow chart may be used as a guide.

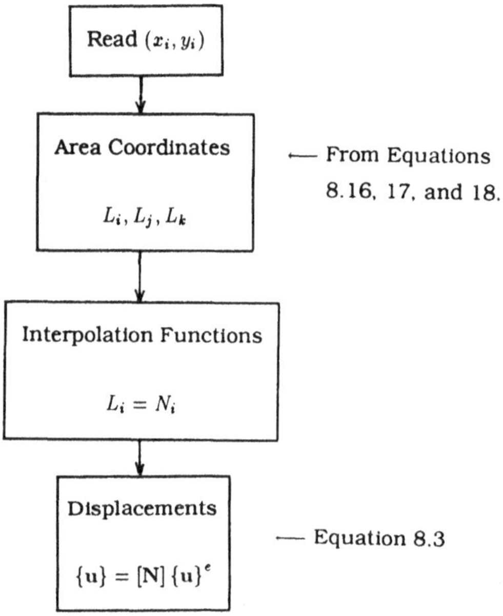


## 8.6 CALCULATION OF ELEMENT STIFFNESS

### 8.6.1 What is Stiffness?

Consider again the triangle 123, as shown in Figure 8.8. Let us denote the nodal displacements by $u_1, u_2, u_3, \cdots, u_6$. (Note that the symbol $v$ is replaced by $u$. This change in notation is for easy programming.) We adopt the point of view that *displacements are the cause of forces*. We ask for example, the following question in general terms.

> A unit displacement is applied at $i$ in the $k$-direction. All other displacements are maintained at zero. What force will the displacement produce at $j$ in the $m$-direction?

The nodes $i$ and $j$, and the directions $k$ and $m$ are completely arbitrary

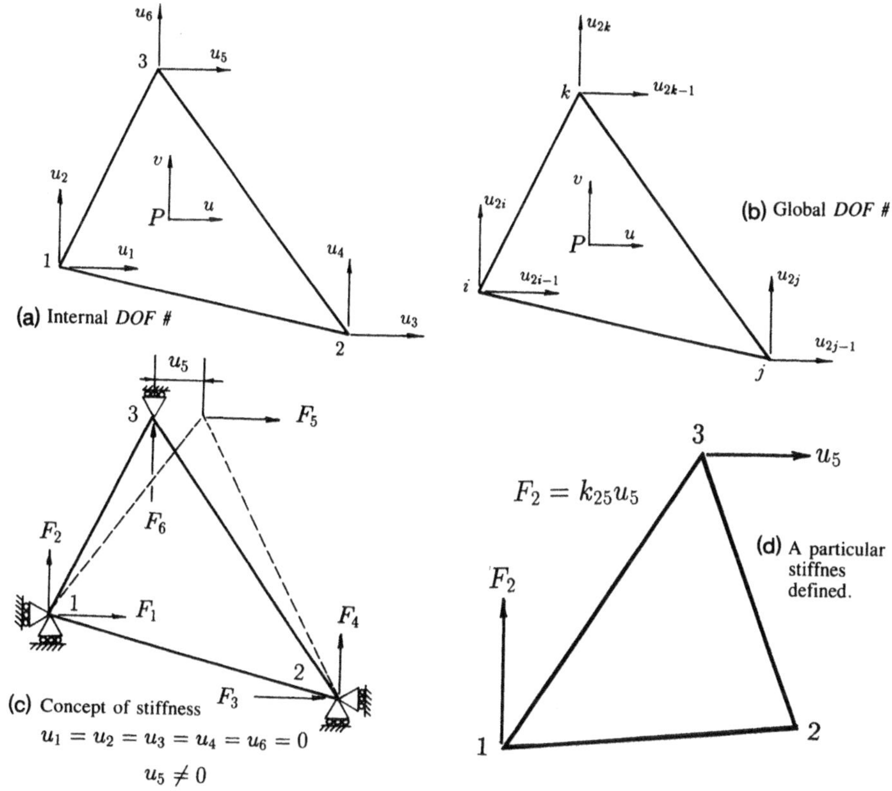


**FIGURE 8.8.** Definition of a stiffness component involves the application of a single displacement and observation of the resulting forces.

and are meant as examples. In any case the answer to this question is a linear expression for linear elastic materials. Typically, in Figure 8.8, let us suppose that $u_5$ = displacement applied along the direction of the fifth DOF, and that $F_2$ = force caused along the direction of the second DOF. Then for linear behavior of the material,

$$F_2 = k_{25}u_5,$$

where $k_{25}$ is a constant, which must obviously be linear in the elastic modulus of the material. Note that the applied displacement may cause other forces (or reactions) to be produced at the other supports, such as $F_1$, $F_3$, $F_4$, $F_5$, *and* $F_6$ shown in the figure. However, our question is about what a particular force is. Since there are six possible nodal forces, (3 nodes × 2 directions), and six possible displacements, there are a total of 6 × 6 = 36 stiffness components.

In general, for an element of $n$ nodes, with $d$ degrees of freedom each, there are $nd$ possible displacements, and $nd$ possible forces per element. Hence there are $(nd)^2$ stiffness components connecting each displacement with each force. They are normally arranged in the form of a square matrix $[K]^e$, of size $(nd) \times (nd)$. To summarize,

> the stiffness component in a particular direction is the force caused by an applied unit displacement.[5]

### 8.6.2 The [B] Matrix

Calculation of the stiffness matrix uses two other matrices, one of which is the [**B**] matrix, as explained below. It is also called the *gradient matrix*. Essentially, the [**B**] matrix is a property of the element shape. It arises in the process of expressing the strain-displacement equations in matrix form. We had earlier defined displacements at any point inside a finite element,

$$\{u\} = [\mathbf{N}]\{u\}^e.$$

If we wish to get the strains, we need to obtain the derivatives of $\{u\}$. Consider again the equation

$$\begin{Bmatrix} u \\ v \end{Bmatrix} = \begin{bmatrix} N_1 & 0 & N_2 & 0 & N_3 & 0 \\ 0 & N_1 & 0 & N_2 & 0 & N_3 \end{bmatrix} \begin{Bmatrix} u_1 \\ v_1 \\ u_2 \\ v_2 \\ u_3 \\ v_3 \end{Bmatrix}$$

Also, we know that

$$\epsilon_x = \frac{\partial u}{\partial x}$$

$$\epsilon_y = \frac{\partial v}{\partial y}$$

[5] If we adopted the point of view that forces cause displacements, then the definition of stiffness would be "the force needed along a particular direction at a particular point to produce a unit displacement at another given point in a given direction." The difficulty with this definition is that it does not reflect the fact that all other displacements are held at zero. To do so is not easy with the point of view adopted.

Putting this in matrix form, we obtain the definition of [**B**].

$$\begin{Bmatrix} \epsilon_x \\ \epsilon_y \\ \gamma_{xy} \end{Bmatrix} = \begin{bmatrix} \dfrac{\partial N_1}{\partial x} & 0 & \dfrac{\partial N_2}{\partial x} & 0 & \dfrac{\partial N_3}{\partial x} & 0 \\ 0 & \dfrac{\partial N_1}{\partial y} & 0 & \dfrac{\partial N_2}{\partial y} & 0 & \dfrac{\partial N_3}{\partial y} \\ \dfrac{\partial N_1}{\partial y} & \dfrac{\partial N_1}{\partial x} & \dfrac{\partial N_2}{\partial y} & \dfrac{\partial N_2}{\partial x} & \dfrac{\partial N_3}{\partial y} & \dfrac{\partial N_3}{\partial x} \end{bmatrix} \begin{Bmatrix} u_1 \\ v_1 \\ u_2 \\ v_2 \\ u_3 \\ v_3 \end{Bmatrix} \tag{8.21}$$

Or, more briefly,

$$\{\mathbf{e}\} = [\mathbf{B}]\{u\}^e.$$

We see therefore that the [**B**] matrix is simply the matrix of derivatives of the interpolation functions. Hence the name gradient matrix. Because the $N$'s are assumed, their derivatives are known right at the beginning. The following example illustrates the point.

**Example 8.3**

For the triangular element in the previous example, derive the [**B**] matrix.

*Solution* From the previous example, we take the interpolation functions, $N_1$, $N_2$, and $N_3$.

$$N_1 = (3.25 - 0.5x - 0.5y)$$

Hence,
$$\frac{\partial N_1}{\partial x} = -0.5; \quad \frac{\partial N_1}{\partial y} = -0.5.$$

$$N_2 = (-1.75 + 1.5x - 0.5y)$$

Hence,
$$\frac{\partial N_2}{\partial x} = 1.5; \quad \frac{\partial N_2}{\partial y} = -0.5.$$

$$N_3 = (-0.5 - x + y)$$

Hence,
$$\frac{\partial N_3}{\partial x} = -1; \quad \frac{\partial N_3}{\partial y} = 1.$$

Therefore, from the definition of the [**B**] matrix (Equation 8.21),

$$\begin{Bmatrix} \epsilon_x \\ \epsilon_y \\ \gamma_{xy} \end{Bmatrix} = \begin{bmatrix} -0.5 & 0 & 1.5 & 0 & -1 & 0 \\ 0 & -0.5 & 0 & -0.5 & 0 & 1 \\ -0.5 & -0.5 & -0.5 & 1.5 & 1 & -1 \end{bmatrix} \begin{Bmatrix} u_1 \\ v_1 \\ u_2 \\ v_3 \\ u_3 \\ v_3 \end{Bmatrix}$$

$$= [\mathbf{B}]\{u\}^e$$

***Commentary***
*The* [**B**] *matrix happens to be constant because the element is a linear triangle. Higher order elements would involve higher powers of N's and hence* [**B**] *would in general be a function of coordinates.*

### 8.6.3 The [D] Matrix

The other ingredient of the stiffness matrix is the elasticity matrix or the [**D**] matrix. In contrast to the [**B**] matrix, the [**D**] matrix is a property of the material of the element. For the two-dimensional element being discussed, there are only three strains $\epsilon_x, \epsilon_y, \gamma_{xy}$ and three stresses $\sigma_x, \sigma_y, \tau_{xy}$. Their relationship can be expressed as a matrix equation.

$$\begin{Bmatrix} \sigma_x \\ \sigma_y \\ \tau_{xy} \end{Bmatrix} = \frac{E}{(1-\nu^2)} \begin{bmatrix} 1 & \nu & 0 \\ \nu & 1 & 0 \\ 0 & 0 & \dfrac{(1-\nu)}{2} \end{bmatrix} \begin{Bmatrix} \epsilon_x \\ \epsilon_y \\ \gamma_{xy} \end{Bmatrix} \tag{8.22}$$

Or simply

$$\{\sigma\} = [\mathbf{D}]\{\epsilon\}$$

We have earlier seen in the theory of elasticity for all two-dimensional problems, there must be a stress or a strain present in the $z$-direction, depending on whether the problem represents a plane strain or a plane stress condition. Under plane stress conditions, $\epsilon_z = -(\nu/E)(\sigma_x + \sigma_y) \neq 0$. Under plane strain conditions, $\sigma_z = -\nu(\sigma_x + \sigma_y) \neq 0$. In either case, the $\epsilon_z$ or the $\sigma_z$ can be calculated externally after evaluating the stresses on the left side of Equation 8.22.

### 8.6.4 The Element Stiffness Matrix

The stiffness matrix for an element is defined as

$$[K]^e = \int_V [\mathbf{B}]^T[\mathbf{D}][\mathbf{B}]\, dV, \tag{8.23}$$

for three-dimensional elements, and

$$[K]^e = \int_V [\mathbf{B}]^T[\mathbf{D}][\mathbf{B}] t\, dA, \tag{8.24}$$

for plane two-dimensional elements. The integration in each case is carried out over the volume of the element. In the case of plane elements, the thickness of the element is a constant, $t$. Hence $dV = t\,dA$, $dA$ being an small area of the element.

Note that $[\mathbf{B}]^T$ is of size $6 \times 3$, $[\mathbf{D}]$ is of size $3 \times 3$, and $[\mathbf{B}]$ is of size $3 \times 6$. Therefore, the resulting $[K]^e$ matrix has a size $6 \times 6$, as discussed earlier in Section 8.6.1.

**Example 8.4**

Given that the modulus and Poisson's ratio of the material of the plane triangular element of the previous examples are 245 kg/mm$^2$ and 0.35, respectively, and that the thickness of the element is 1.25 mm, and is uniform throughout the element, calculate the element stiffness matrix. Assume the area of the triangle is 325 mm$^2$.

*Solution* We construct the [**D**] matrix first.

$$[\mathbf{D}] = \frac{E}{1-\nu^2}\begin{bmatrix} 1 & \nu & 0 \\ \nu & 1 & 0 \\ 0 & 0 & \dfrac{1-\nu}{2} \end{bmatrix}$$

$$= \frac{245}{1-0.35^2}\begin{bmatrix} 1 & 0.35 & 0 \\ 0.35 & 1 & 0 \\ 0 & 0 & 0.325 \end{bmatrix}$$

$$= 279.2\begin{bmatrix} 1 & 0.35 & 0 \\ 0.35 & 1 & 0 \\ 0 & 0 & 0.325 \end{bmatrix}$$

We note that the [**B**] matrix is available from the previous example.

Therefore, we can proceed to calculate the stiffness matrix. The

$$[K]^e = \int [\mathbf{B}]^{\mathrm{T}}[\mathbf{D}][\mathbf{B}]\,dV$$

$$= \int [\mathbf{B}]^{\mathrm{T}}[\mathbf{D}][\mathbf{B}]t\,dA$$

$$= [\mathbf{B}]^{\mathrm{T}}[\mathbf{D}][\mathbf{B}]tA$$

where $t$ is the thickness of the plane triangular element, and $A$ is its area. Hence the calculation of $[K]^e$ simply reduces to an exercise in matrix multiplication. We can do it in two steps as shown below. We may temporarily keep aside the factor outside $[\mathbf{D}]$.

$$[\mathbf{D}][\mathbf{B}] = \begin{bmatrix} 1 & 0.35 & 0 \\ 0.35 & 1 & 0 \\ 0 & 0 & 0.325 \end{bmatrix}$$

$$\times \begin{bmatrix} -0.5 & 0 & 1.5 & 0 & -1 & 0 \\ 0 & -0.5 & 0 & -0.5 & 0 & 1 \\ -0.5 & -0.5 & -0.5 & 1.5 & 1 & -1 \end{bmatrix}$$

$$= \begin{bmatrix} -0.50 & -0.18 & 1.50 & -0.18 & -1.00 & 0.35 \\ -0.18 & -0.50 & 0.53 & -0.50 & -0.35 & 1.00 \\ -0.16 & -0.16 & -0.16 & 0.49 & 0.33 & 0.33 \end{bmatrix}$$

$$[\mathbf{B}]^{\mathrm{T}}[\mathbf{D}][\mathbf{B}] = \begin{bmatrix} -0.5 & 0 & -0.5 \\ 0 & -0.5 & 0 \\ 1.5 & 0 & -0.5 \\ 0 & -0.5 & 1.5 \\ -1 & 0 & 1 \end{bmatrix}$$

Hence,

$$[K]^e = (325\ \mathrm{mm}^2)(1.25\ \mathrm{mm})(279.2\ \mathrm{kg/mm}^2)$$

$$\times \begin{bmatrix} 0.33 & 0.17 & -0.67 & -0.16 & 0.34 & -0.013 \\ 0.17 & 0.33 & -0.18 & 0.007 & 0.013 & -0.34 \\ -0.67 & -0.18 & 2.33 & -0.51 & -1.66 & 0.69 \\ -0.16 & 0.007 & -0.51 & 0.98 & 0.66 & -0.99 \\ 0.34 & 0.013 & -1.66 & 0.66 & 1.33 & -0.68 \\ -0.013 & -0.34 & 0.69 & 0.99 & -0.68 & 1.33 \end{bmatrix}$$

It can be seen that the $[K]^e$ matrix is symmetric. Also, its determinant is zero. This can be seen from the fact that every row (or every column) adds up to zero.

## 8.7 CALCULATION OF ELEMENT LOAD VECTORS

Loads on an element may be *body forces*, or *surface forces*. Body forces are those that are distributed volumetrically over all the points of the body, including the interior points. Examples of body forces are centrifugal forces, gravity, and magnetic attraction forces. Surface forces are those that are applied on the boundary surface of the body. Examples of surface forces are mechanical contact forces, fluid pressure. Most applied forces can only be surface forces. Surface forces can be distributed or concentrated.

In FEM, forces of all kinds—body forces, surface forces, distributed forces, and concentrated forces—are converted to equivalent concentrated nodal forces.

If a force is a concentrated force in reality, then there is no need to convert it; rather, we take care of this in the modeling, by managing to get a node at the point of force application. Other types need to be converted, and that requires a rule for converting distributed forces to concentrated nodal forces. The rule is stated below.

$$\{f\}^e = \int_S [\mathrm{N}]^{\mathrm{T}} \{f_s\}^e \, dS, \quad \text{for surface forces,} \tag{8.25}$$

$$\{f\}^e = \int_V [\mathrm{N}]^{\mathrm{T}} \{f_b\}^e \, dV, \quad \text{for body forces.} \tag{8.26}$$

In the presence of body forces, all the elements in a model will experience the force. In the case of surface tractions, only those elements near the boundary of the problem domain will, if at all, experience the forces. Thus, not all elements will call for a calculation of $\{f\}^e$ vector. Hence, computational effort may be saved by some bookkeeping about which elements need the element vector calculation. Obviously, the $\{f_s\}^e$ is of size $(2 \times 1)$, the [N] is of size $2 \times (nd)$ and hence, the $\{f\}^e$ vector is of size $(nd) \times 1$.

**Example 8.5**

For the same linear plane triangular element described in Example 8.1, assume that two of its sides are acted upon by pressures inclined to the sides at an angle, so there are both $x$ and $y$ components of pressure. The values of pressures are shown in Figure 8.9. Assume consistent units and

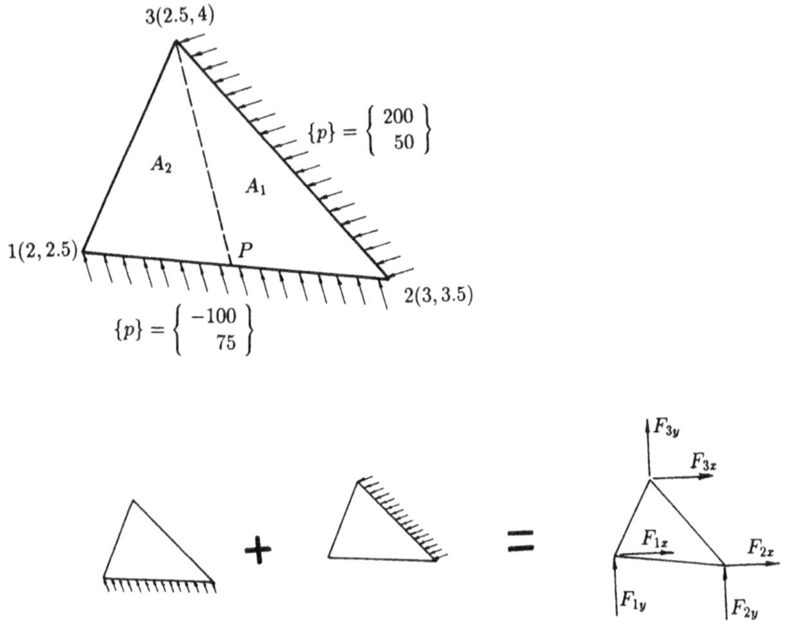


**FIGURE 8.9.** Element sides subjected to a general pressure load vector $\{p_x, p_y\}^T$.

find the equivalent element force vector (the vector of equivalent forces concentrated at the nodes).

*Solution* We observe that only surface forces are present at the boundary of the element.

$$\int [\mathbf{N}]^T \{\mathbf{f}_s\}\, dS = \int_{ij} (\cdots) + \int_{jk} (\cdots)$$

$$= \{\mathbf{f}\}_1^e + \{\mathbf{f}\}_2^e$$

$$\{\mathbf{f}\}_1^e = \int_{ij} \begin{bmatrix} N_1 & 0 \\ 0 & N_1 \\ N_2 & 0 \\ 0 & N_2 \\ N_3 & 0 \\ 0 & N_3 \end{bmatrix} \begin{Bmatrix} -100 \\ 75 \end{Bmatrix} dl$$

We need the values of $N_1$, $N_2$, and $N_3$ on the boundary $ij$. We note that on the curve $ij$, $y = (x + 0.5)$.

$$N_1 = \frac{A_1}{A} = (3.25 - 0.5x - 0.5y) = (3 - x)$$

$$N_2 = \frac{A_2}{A} = (-1.75 - 1.5x - 0.5y) = (x - 2)$$

$$N_3 = \frac{A_3}{A} = (-0.5 - x + y) = 0$$

Also,
$$dl = \sqrt{1 + y'^2}\, dx = \sqrt{2}\, dx$$

Note that the node $k$ has no influence on the force distribution on $i$ and $j$. In general the nodes on the curve and no others influence the force calculation.

$$\{\mathbf{f}\}_1^e = \int_2^3 \begin{bmatrix} (3-x) & 0 \\ 0 & (3-x) \\ (x-2) & 0 \\ 0 & (x-2) \\ 0 & 0 \\ 0 & 0 \end{bmatrix} \begin{Bmatrix} -100 \\ 75 \end{Bmatrix} \sqrt{2}\, dx$$

$$= \sqrt{2} \begin{Bmatrix} -50 \\ 37.5 \\ -50 \\ 37.5 \\ 0 \\ 0 \end{Bmatrix}$$

For the curve $jk$, we note $y = (6.5 - x)$. Also, we should expect $N_1 = 0$.

$$N_2 = \frac{A_2}{A} = (-1.75 - 1.5x - 0.5y) = (5 - x)$$

$$N_3 = \frac{A_3}{A} = (-0.5 - x + y) = (x - 4)$$

As before, $$dl = \sqrt{2}\,dx$$

$$\{\mathbf{f}\}_2^e = \int_2^3 \begin{bmatrix} 0 & 0 \\ 0 & 0 \\ (5-x) & 0 \\ 0 & (5-x) \\ (x-4) & 0 \\ 0 & (x-4) \end{bmatrix} \begin{Bmatrix} 200 \\ 50 \end{Bmatrix} \sqrt{2}\,dx$$

$$= \frac{1}{\sqrt{2}} \begin{Bmatrix} 0 \\ 0 \\ 200 \\ 50 \\ 200 \\ 50 \end{Bmatrix}$$

The element load vector is the sum of the two vectors.

## 8.8 ASSEMBLY OF THE GLOBAL STIFFNESS MATRIX

The various matrices calculated at the element level, namely the stiffness matrix $[K]^e$ and $\{f\}^e$ must all be assembled into one global stiffness matrix $[\mathbf{K}]$ and global force matrix $\{\mathbf{f}\}$ . The unknown global matrix of displacements $\{\mathbf{u}\}$ is related to the other two by the equation,

$$[\mathbf{K}]\{\mathbf{u}\} = \{\mathbf{f}\}$$

The process of assembly is based on the properties that

$$[\mathbf{K}] = \sum_e [K]^e \quad \text{and} \quad \{\mathbf{f}\} = \sum_e \{f\}^e.$$

These equations state that the stiffness of the entire assembly is the sum of the stiffness of the individual elements. The corresponding statement is true for the forces. This is true if the element stiffness are calculated in the global coordinate system. Likewise, the global force matrix is the sum of the element force matrices, if they are calculated in the global coordinate system. If the element stiffness matrices had been calculated in a local coordinate system, then they must be transformed back to the global coordinates, before the assembly stage. Figure 8.10 shows the scheme adopted for the assembly of the element matrices. Figure 8.10a shows a

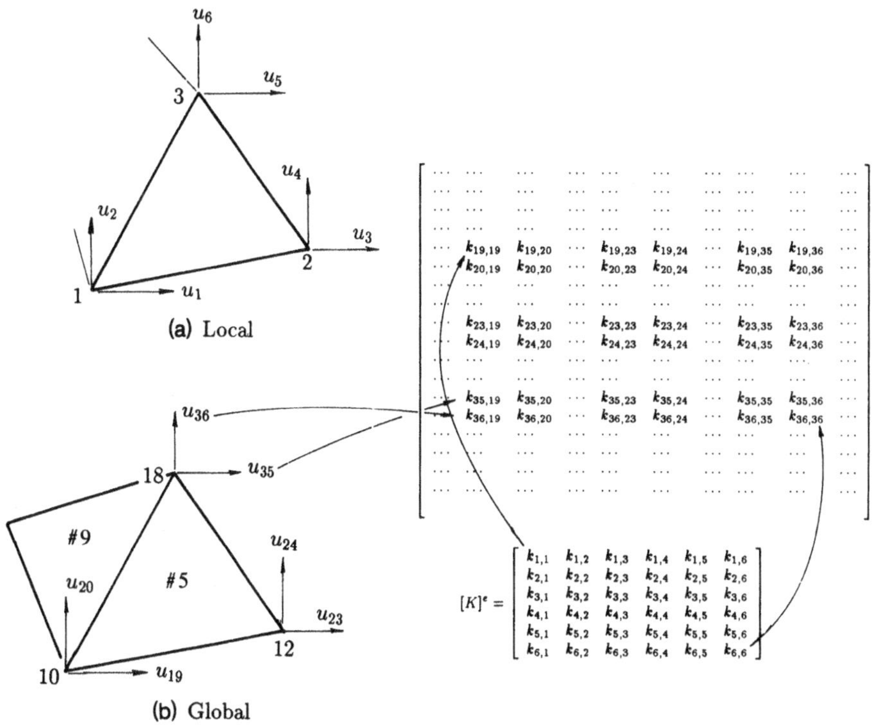


**FIGURE 8.10.** Assembly of global stiffness matrix. Note the correspondence between the global DOF number and the equation number in the stiffness matrix.

triangular element with internal numbering scheme 1, 2, and 3, and accordingly every entry in the element $[K]^e$ matrix is numbered using these internal identifications for nodes.

Let us suppose for illustration that the global numbering used for the nodes by the user happens to be 10, 12, and 18, as shown in Figure 8.10b. Because each node has two DOF, we count two per node. Hence we have the following.

Node 10 has the DOF numbers 19 and 20;
Node 12 has the DOF numbers 23 and 24;
Node 18 has the DOF numbers 35 and 36.

In accordance with the standard practice, we may refer to the global degrees of freedom as simply the equation numbers. This is appropriate as you recall that the DOF is simply the number of unknowns to be solved for.

Therefore, in the case shown in Figure 8.10b, the equation numbers must be 19, 20, 23, 24, 35, and 36. The internal numbering of the nodes for each element is just 1, 2, and 3. Thus, the internal equation numbers are 1, 2, 3, 4, 5, and 6. Since the global equation number is now available, the components of element stiffness are labelled with the corresponding global numbers. Thus, the correspondence between internal numbering and global numbering of the element stiffness are as listed below.

| Local subscript | | Is replaced by global subscript | | | |
|---|---|---|---|---|---|
| | | | $k_{1,1}$ | $\rightarrow$ | $k_{19,19}$ |
| | | | $k_{1,2}$ | $\rightarrow$ | $k_{19,20}$ |
| | | | $k_{1,3}$ | $\rightarrow$ | $k_{19,23}$ |
| 1 | $\rightarrow$ | 19 | $k_{1,4}$ | $\rightarrow$ | $k_{19,24}$ |
| 2 | $\rightarrow$ | 20 | $k_{1,5}$ | $\rightarrow$ | $k_{19,35}$ |
| 3 | $\rightarrow$ | 23 | $k_{1,6}$ | $\rightarrow$ | $k_{19,36}$ |
| 4 | $\rightarrow$ | 24 | $\cdots$ | $\cdots$ | $\cdots$ |
| 5 | $\rightarrow$ | 35 | $\vdots$ | | $\vdots$ |
| 6 | $\rightarrow$ | 36 | $k_{6,6}$ | $\rightarrow$ | $k_{36,36}$ |

Figure 8.10 also shows the position of the individual elements in the global stiffness matrix. The terms in the element stiffness matrix are simply posted to their corresponding positions as shown in this figure.

### 8.8.1 Assembly Procedure for Shared Nodes

Consider the element[6] number 5 adjacent to the element number 9 in Figure 8.10. Let us also assume that they share the edge 10 – 18. Element number 9 has its own $[K]^e$ matrix, of size $6 \times 6$. Of the 36 stiffness components, the ones that involve the shared edge 10 – 18 are $k_{19,19}$, $k_{19,20}$, $k_{19,35}$, $k_{19,36}$, $k_{36,19}$, $k_{36,20}$, $k_{36,35}$, and lastly $k_{36,36}$. Because, the subscripts in these are the DOF numbers of the nodes 10 and 18. Similar stiffness components are also present in the $[K]^e$ matrix of element number 5. However, their magnitudes are different, because of the different size of the triangle. However, in the global stiffness matrix, their positions are the same.

---

[6] Elements are identified by element numbers which are nothing but identification numbers. There is no significance or effect of these numbers on the FE calculations. However, they are useful as a data link when tracking the various element data in the assembly process and in the post-processing stage. If a frontal solver is used, then the element numbers are vital to the process of solution, although not for the solution itself.

Therefore, the procedure of assembly is simply to add the stiffness terms to those already present in [**K**] at the corresponding position.

To explain it further, let us return to Figure 8.10. Consider the term $k_{19,36}$ in the $[K]^e$ matrix of element number 5. Initially, in the global [**K**], there are no entries at the position (19, 36). Hence $k_{19,36}$ is simply posted to this position. Later on, while working with element number 9, we once again have a different $k_{19,36}$. We add this to whatever is already present in [**K**] at that location. If more elements share the same node, more terms are correspondingly added.

We can look at it in a different way. For a node, which is shared by many elements, there are as many entries in the global [**K**] matrix as the number of elements sharing that node. For example, if a node is shared by five elements, the position(s) in the global [**K**] corresponding to the node will be entered five times, once for each element. Each time an entry is made, it is added to the entry already present in the global matrix.

A similar exercise is performed also for the global force vector {**f**}. Usually, this vector is a single column of nodal forces. As in the case of the stiffness matrix, for each node the force vector is entered as many times as the number of elements sharing that node. However, there is a need to enter the force vector only for the elements which are subject to some form of loading.

When the exercise of assembling is done as described in the foregoing, we have the full global stiffness matrix, [**K**], and the global force matrix {**f**}.

## 8.9 THE NATURE OF THE GLOBAL STIFFNESS MATRIX—[K]

There are three important characteristics of the global stiffness matrix. They are,

1. it is banded,
2. it is symmetric, and
3. it is singular.

### 8.9.1 Cause of Bandedness

An example of a banded matrix is shown in Figure 8.11a and 8.11b. All the terms in such a matrix are crowded around the leading diagonal. One can define a semibandwidth for such matrices. In Figure 8.11a, the semi-bandwidth of the matrix is 3. All terms to the right of 3 terms from the diagonal

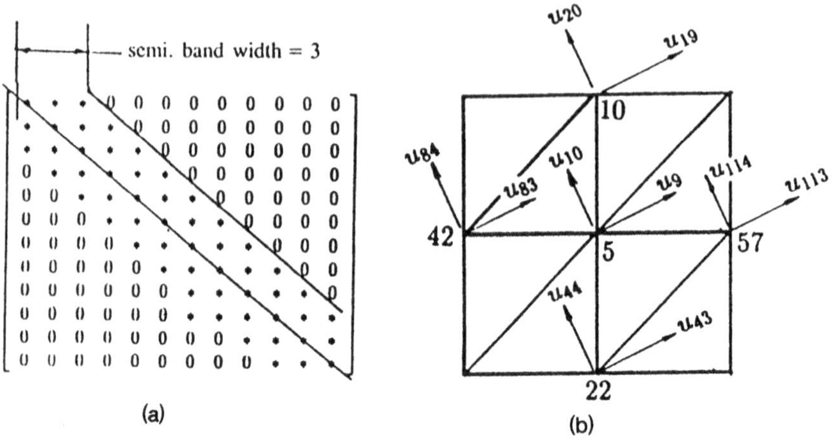


are zero. It does not mean that all terms inside the band are nonzero. There could be zeros inside the band, but all outside it are zero.

The bandedness is a consequence of the fact that a node is shared, after all by a finite number of elements. Consider the example in Figure 8.11b. Let us suppose that the nodes shown are a few that belong to a mesh containing 250 nodes. Assuming two DOFs per node, the global stiffness matrix must be of size 500 × 500. In the figure, the node 5 is seen to be connected with four other nodes, namely 10, 22, 42, and 57. For two DOFs per node the stiffness terms affecting node 5 are,

$$
\begin{array}{llll}
k_{9,9} & k_{9,44} & k_{10,9} & k_{10,44} \\
k_{9,10} & k_{9,83} & k_{10,10} & k_{10,83} \\
k_{9,19} & k_{9,84} & k_{10,19} & k_{10,84} \\
k_{9,20} & k_{9,113} & k_{10,20} & k_{10,113} \\
k_{9,43} & k_{9,114} & k_{10,43} & k_{10,114}.
\end{array}
$$

These terms are entered into the global stiffness matrix, and when all these entries are completed, the row numbers 9 and 10 are populated only up to the 114-th column, and not beyond. Consequently, all elements in [**K**] after that position are zero. This characteristic repeats itself for other nodes, too, of course, to various column positions. In the end, we have a banded matrix.

It is important to be aware of what contributes to the size of the bandwidth of the [**K**] matrix. We already have a clue in the above example. The rows 9 and 10 are populated to 114 positions, because node 5 is connected to node 57. In other words, of all the node numbers connected

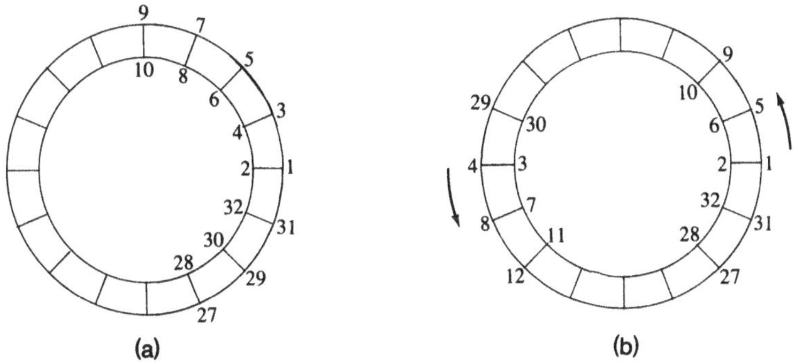


**FIGURE 8.12.** Different node numbering schemes lead to different bandwidths. The best one is that which minimizes bandwidth. In (a) the semibandwidth is 31 × Nodal DOF = 62. In (b), it is reduced to 27 × Nodal DOF = 54.

to node 5, the largest one is 57. Accordingly, the largest DOF number that goes with equations 9 and 10 is 114. Thus, row 9 is populated up to the 114th position. The semi-bandwidth in this example is (114 − 9 + 1) = 106.

> The bandwidth in the global stiffness matrix is directly dependent on the maximum node number difference in any element in the entire model.

$$\text{Semi-Bandwidth} = (\text{Max Node Number Difference} + 1) \times (\text{Degrees of Freedom per Node})$$

In Figures 8.12a and 8.12b, two different ways of node numbering are shown. The first one is produced by the user in the process of modeling. This has a semi-bandwidth of 62. The second one is produced by the computer using a special renumbering algorithm. In this case, the semi-bandwidth is 54, which is lower than that of the first case. Reducing the bandwidth is important for improving the computation time.

### 8.9.2 Cause of Symmetry of [K]

Symmetry implies that the [K] matrix is square, because symmetry exists only in square matrices. The term symmetry itself means

$$k_{i,j} = k_{j,i}.$$

By definition, the stiffness term $k_{i,j}$ is the force (or moment, or any kind of reaction) developed at $j$ due to a unit displacement applied at $i$.

Symmetry of the stiffness matrix implies that,

$$\left.\begin{array}{r}\text{Force developed at } j\\ \text{due to unit displacement at } i\end{array}\right\} = \left\{\begin{array}{l}\text{Force developed at } i \text{ due}\\ \text{to unit displacement at } j\end{array}\right.$$

Is this true? Yes it is, but only in certain cases. It is true in processes in which no energy is lost to the outside, such as in linear elasticity problems. For *conservative systems*, as they are called, this kind of reciprocity is guaranteed by the *Maxwell-Betti Theorem* in structural mechanics. For systems in which energy is dissipated, such as by friction or yielding, the stiffness matrix is not symmetric.

### 8.9.3 Singularity of [K]

By singularity we mean that the equation [**K**]{**u**} = {**f**} is not solvable as such. Because the matrix [**K**] has a zero determinant. The displacements referred to here denote the relative displacements between any two points in the body, which cause strains and stresses. It is well known that rigid body displacements do not cause strains. At this stage of the calculation, the support conditions of the structure have not yet been factored into the equation. Hence, a unique solution does not exist for displacements. Because in the absence of any kind of restraints, the solution for displacements is the sum of the strain-causing displacements *plus* an arbitrary rigid body motion. The arbitrary nature of the additive rigid body motion prevents a unique solution. This fact is represented by the singularity of the [**K**] matrix.

To avoid rigid body translation, at least one point on the body must be held fixed. In order to avoid rigid body rotation, at least three points, not all on a straight line, must be held fixed. However, in a real-life problem, usually many more supports are provided. Some of the supports may be rigid, some soft, and some others constrained to move on given surfaces.

The singularity of [**K**] actually begins at the element level. This can be seen from the element stiffness matrix in Example 8.4. Addition of the members in any column gives zeros, which means that the determinant of the element stiffness matrix is zero. All columns becoming zero denotes the complete absence of any kind of fixity of nodes. Assembly of such matrices should obviously result in a global matrix which has the same singularity.

(It may be mentioned in passing that a singularity may be caused inadvertently in the [**K**] matrix by errors in modeling. Common errors that can cause this trouble are (1) an inadequate number of restraints, and (2) dangling nodes, that is, nodes which had been defined in the model, but not connected to any elements at all. All software packages have provisions to find and eliminate such dangling nodes.)

## 8.10 DISPLACEMENT BOUNDARY CONDITIONS

The singularity of the global equation $[\mathbf{K}]\{\mathbf{u}\} = \{\mathbf{f}\}$ can be removed by applying proper displacement boundary conditions, also known as *kinematic boundary conditions*. Applying the displacement boundary conditions is physically equivalent to arresting the rigid body motions and hence the solution leads to the stress-causing displacements.

It should be noted that applied forces are also classified as boundary conditions in the mathematical sense. They are also known as *static boundary conditions*. From the FEA point of view, the static boundary conditions have already been accounted for in the calculation of element force vectors.

Let us suppose that at support point $S$, displacement is known, and the reaction $f_m$ is unknown. In general, it is not possible to know both of them a priori, except in statically determinate problems (in which case there is no need for FEM at all)! Figure 8.13a shows the assembled global equation, before application of boundary conditions. The $m$-th displace-

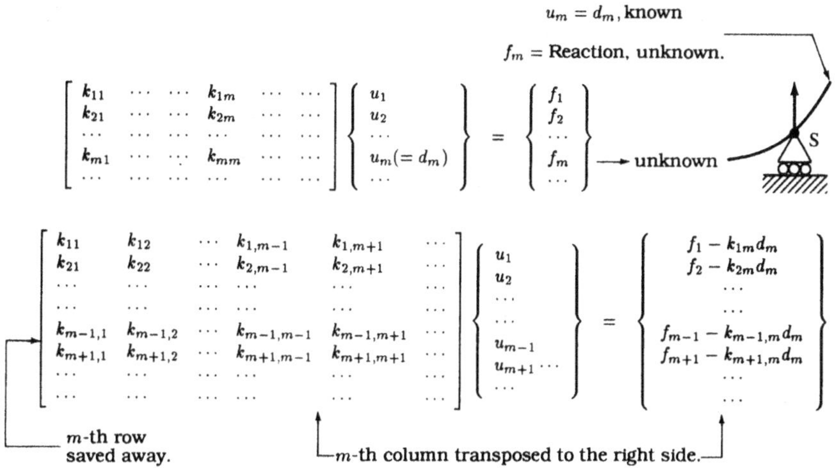


**FIGURE 8.13.** Matrix manipulations that are equivalent to the application of the displacement boundary conditions.

ment is known, so $u_m = d_m$. The corresponding reaction $f_m$ is not known. We may remove the $m$-th row and $m$-th column from the equation, after making the following modifications in the global equation.

- For any equation other than the $m$-th, we may write,

$$\sum_{j \neq m} k_{ij} d_j = f_i - k_{im} d_m \quad \text{no sum on } m. \tag{8.27}$$

  Equation 8.27 is simply equivalent to transposing the known terms from left side to the right side of all the equations. Thus, the resulting modified equation will be as shown in Figure 8.13b.
- Secondly, the $m$-th row, which is the equation containing the information about the reaction $f_m$, needs to be simply stored away in the memory or in a storage area. The coefficients needed are in the $m$-th equation, which simply reads

$$f_m = \sum_j k_{mj} u_j. \tag{8.28}$$

It may be noted that all the $k$'s in this equation are the stiffness components related to the $m$-th degree of freedom. The element matrices for those elements which share the node at the support $S$, contain the $k$'s. Also, since this equation needs all the $u$'s, the reactions can be calculated only after the displacements are solved.

## 8.11 SOLUTION OF THE UNKNOWN DISPLACEMENTS

### 8.11.1 The Gauss Elimination Method

The basic algorithm to solve for the unknown displacements is the Gauss elimination method. This method is discussed extensively in several texts on numerical methods for matrices [14]. There are two major steps in this algorithm. In the first step, called the *forward reduction*, the square matrix is reduced systematically to an *upper triangular form*. In this form the very last equation contains only one term, and hence it is the solution for the last degree of freedom. The second step begins here. By systematic substitution in the reverse direction—that is, starting from the last equation up—we can completely obtain the solution for the entire set of equations. It is called the *back substitution*.

An example of forward reduction and back substitution for a $6 \times 6$, nonsingular, symmetric matrix is shown in Example 8.6. Obviously, real life problems are not this small, and a more general procedure is needed to do the forward reduction, and back substitution.

**Example 8.6**

Solve the system of linear equations listed below, using the Gauss elimination method.

$$\begin{bmatrix} 1 & 1 & 1 & 1 & 1 & 1 \\ 2 & 3 & 2 & 1 & 3 & 1 \\ 2 & 3 & -2 & -2 & 3 & 1 \\ 1 & 2 & 1 & 1 & 2 & 2 \\ 2 & 3 & 2 & 0 & -3 & 0 \\ 2 & 3 & -2 & 1 & -3 & 2 \end{bmatrix} \begin{Bmatrix} x_1 \\ x_2 \\ x_3 \\ x_4 \\ x_5 \\ x_6 \end{Bmatrix} = \begin{Bmatrix} 1 \\ 1 \\ 0 \\ 1 \\ 2 \\ 5 \end{Bmatrix}$$

(Answer: $x_1 = \frac{1}{2}$; $x_2 = \frac{1}{3}$; $x_3 = -\frac{1}{2}$; $x_4 = 1$; $x_5 = -\frac{1}{3}$; $x_6 = 0$.)

*Solution*

***Forward Reduction:***
Let us first describe the forward reduction process. As mentioned above, this process converts the square matrix to the upper triangular form, that is, a matrix in which only the upper half is non-zero, and the lower half is *all* zeros. This is accomplished by linear combinations of rows. The following steps will clarify the technique.

*Step 1.* Perform the following operations.

$$(\text{Row } 2) - a_{21}\,(\text{Row } 1)$$

$$(\text{Row } 3) - a_{31}\,(\text{Row } 1)$$

$$(\text{Row } 4) - a_{41}\,(\text{Row } 1)$$

$$(\text{Row } 5) - a_{51}\,(\text{Row } 1)$$

$$(\text{Row } 6) - a_{61}\,(\text{Row } 1)$$

All of these operations can be briefly written as [(Row $i$) $- a_{i1}$(Row 1)], $i = 2, 3, 4, 5$, and 6. $a_{11}$ is called the pivot for this step.

$$\begin{bmatrix} 1 & 1 & 1 & 1 & 1 & 1 \\ 0 & 1 & 0 & -1 & 1 & -1 \\ 0 & 1 & -4 & -4 & 1 & -1 \\ 0 & 1 & 0 & 0 & 1 & 1 \\ 0 & 1 & 0 & -2 & -5 & -2 \\ 0 & 1 & -4 & -1 & -5 & 0 \end{bmatrix} \begin{Bmatrix} x_1 \\ x_2 \\ x_3 \\ x_4 \\ x_5 \\ x_6 \end{Bmatrix} = \begin{Bmatrix} 1 \\ -1 \\ -2 \\ 0 \\ 0 \\ 3 \end{Bmatrix}$$

*Step 2.* Perform the operations [(Row $i$) $- a_{i2}$(Row 2)], $i = 3, 4, 5$, and 6. For this step, the current $a_{22}$ is the pivot.

$$\begin{bmatrix} 1 & 1 & 1 & 1 & 1 & 1 \\ 0 & 1 & 0 & -1 & 1 & -1 \\ 0 & 0 & -4 & -3 & 0 & 0 \\ 0 & 0 & 0 & 1 & 0 & 2 \\ 0 & 0 & 0 & -1 & -6 & -1 \\ 0 & 0 & 4 & 0 & -6 & 1 \end{bmatrix} \begin{Bmatrix} x_1 \\ x_2 \\ x_3 \\ x_4 \\ x_5 \\ x_6 \end{Bmatrix} = \begin{Bmatrix} 1 \\ -1 \\ -1 \\ 1 \\ 1 \\ 4 \end{Bmatrix}$$

The diagonal element $a_{33} = -4 \neq 1$, and hence we divide the third row by $-4$, to obtain the following.

$$\begin{bmatrix} 1 & 1 & 1 & 1 & 1 & 1 \\ 0 & 1 & 0 & -1 & 1 & -1 \\ 0 & 0 & 1 & \frac{3}{4} & 0 & 0 \\ 0 & 0 & 0 & 1 & 0 & 2 \\ 0 & 0 & 0 & -1 & -6 & -1 \\ 0 & 0 & 4 & 0 & -6 & 1 \end{bmatrix} \begin{Bmatrix} x_1 \\ x_2 \\ x_3 \\ x_4 \\ x_5 \\ x_6 \end{Bmatrix} = \begin{Bmatrix} 1 \\ -1 \\ \frac{1}{4} \\ 1 \\ 1 \\ 4 \end{Bmatrix}$$

*Step 3.* Perform the operations [(Row $i$) $- a_{i3}$(Row 3)], $i = 4, 5$, and 6. The pivot is the current $a_{33}$.

$$\begin{bmatrix} 1 & 1 & 1 & 1 & 1 & 1 \\ 0 & 1 & 0 & -1 & 1 & -1 \\ 0 & 0 & 1 & \frac{3}{4} & 0 & 0 \\ 0 & 0 & 0 & 1 & 0 & 2 \\ 0 & 0 & 0 & -1 & -6 & -1 \\ 0 & 0 & 0 & 3 & -6 & 1 \end{bmatrix} \begin{Bmatrix} x_1 \\ x_2 \\ x_3 \\ x_4 \\ x_5 \\ x_6 \end{Bmatrix} = \begin{Bmatrix} 1 \\ -1 \\ \frac{1}{4} \\ 1 \\ 1 \\ 5 \end{Bmatrix}$$

*Step 4.* Perform the operations [(Row $i$) − $a_{i4}$ (Row 4)], $i$ = 5 and 6. The pivot is the current $a_{44}$.

$$\begin{bmatrix} 1 & 1 & 1 & 1 & 1 & 1 \\ 0 & 1 & 0 & -1 & 1 & -1 \\ 0 & 0 & 1 & \frac{3}{4} & 0 & 0 \\ 0 & 0 & 0 & 1 & 0 & 2 \\ 0 & 0 & 0 & 0 & -6 & 1 \\ 0 & 0 & 0 & 0 & -6 & -5 \end{bmatrix} \begin{Bmatrix} x_1 \\ x_2 \\ x_3 \\ x_4 \\ x_5 \\ x_6 \end{Bmatrix} = \begin{Bmatrix} 1 \\ -1 \\ \frac{1}{4} \\ 1 \\ 2 \\ 2 \end{Bmatrix}$$

As earlier, the diagonal element in the fifth row is rendered 1 by dividing by the current $a_{55}$.

$$\begin{bmatrix} 1 & 1 & 1 & 1 & 1 & 1 \\ 0 & 1 & 0 & -1 & 1 & -1 \\ 0 & 0 & 1 & \frac{3}{4} & 0 & 0 \\ 0 & 0 & 0 & 1 & 0 & 2 \\ 0 & 0 & 0 & 0 & 1 & \frac{1}{6} \\ 0 & 0 & 0 & 0 & -6 & -5 \end{bmatrix} \begin{Bmatrix} x_1 \\ x_2 \\ x_3 \\ x_4 \\ x_5 \\ x_6 \end{Bmatrix} = \begin{Bmatrix} 1 \\ -1 \\ \frac{1}{4} \\ 1 \\ -\frac{1}{3} \\ 2 \end{Bmatrix}$$

*Step 5.* Perform the operations [(Row $i$) − $a_{i5}$(Row $i$)], $i$ = 6, using the current $a_{55}$ for the pivot.

$$\begin{bmatrix} 1 & 1 & 1 & 1 & 1 & 1 \\ 0 & 1 & 0 & -1 & 1 & -1 \\ 0 & 0 & 1 & \frac{3}{4} & 0 & 0 \\ 0 & 0 & 0 & 1 & 0 & 3 \\ 0 & 0 & 0 & 0 & 1 & -\frac{1}{6} \\ 0 & 0 & 0 & 0 & 0 & -6 \end{bmatrix} \begin{Bmatrix} x_1 \\ x_2 \\ x_3 \\ x_4 \\ x_5 \\ x_6 \end{Bmatrix} = \begin{Bmatrix} 1 \\ -1 \\ \frac{1}{4} \\ 1 \\ -\frac{1}{3} \\ 0 \end{Bmatrix}$$

In the last step we have the upper triangular form. That is the non-zero elements are all above the leading diagonal, and the elements below the diagonal are *all* zero.

***Commentary***

*Certain generalizations are possible by observing the above steps. They are as follows.*

1. *For a matrix of size $n \times n$ the elimination steps have to be repeated $(n - 1)$ times.*
2. *At the end of each operation, for example the k-th, the column* $a_{kl} = 0$,

*for $l > k$. In other words, the diagonal element is non-zero, and all elements in that column, below the diagonal element become zero.*

3. *When performing the elimination with $a_{kk}$ as the pivot, the rows above the k-th are of no consequence.*
4. *When the diagonal element becomes zero, there is a stalemate. In a well-conditioned problem, this cannot happen. In FEM context, a zero diagonal element means that the problem has rigid body displacements and hence cannot be solved uniquely.*
5. *The generalization of the forward reduction process can be written in the following steps.*
   - *Perform [(Row $i$) − (Current $a_{ij}$) × (Row $j$)], $i = 1,\cdots,(n-1)$; and $j = (i+1),\cdots,n$. $a_{jj}$ is the pivot.*
   - *If after the above step, the diagonal element is not equal to unity, make it so by dividing it by the current $a_{ii}$.*
   - *If the diagonal element is zero, then terminate the procedure, since a unique solution does not exist.*

***Back Substitution***

The back substitution is an easy process. The last equation in the forward reduction is really a solution for the last variable (or, in the FEM language, it is a solution for the last DOF). Back substitution is simply solving the upper triangular form by starting with the last equation and proceeding upwards all the way. The following steps clarify the procedure.

$$-6x_6 = 0 \qquad x_6 = 0$$

$$x_5 - (1/6)x_6 = -\tfrac{1}{3} \qquad x_5 = -\tfrac{1}{3}$$

$$x_4 + 0.x_5 + 3x_6 = 1 \qquad x_4 = 1$$

$$x_3 + (3/4)x_4 + 0.x_5 + 0.x_6 = \tfrac{1}{4} \qquad x_3 = -\tfrac{1}{2}$$

$$x_2 + 0.x_3 - x_4 + x_5 - x_6 = -1 \qquad x_3 = \tfrac{1}{3}$$

$$x_1 + x_2 + x_3 + x_4 + x_5 + x_6 = 1 \qquad x_1 = \tfrac{1}{2}$$

The steps used in the above example are quite systematic and involve no heuristics. Hence the procedure can be easily programmed into the computer. The following rules may be seen to be the statements of the forward reduction and back substitution for any size of matrix.

### 8.11.2 Forward Reduction

The procedure used for forward reduction can be stated in formal algebraic terms for enabling programming. For well-conditioned problems (that is, ones with no surprises), the following cryptic statements summarize the forward reduction process. In the following, whenever we mention $a_{ij}$ or $F_i$, we mean their current value.

$$n = \text{Size of the matrix},$$

$$F_i = \text{Force vector}.$$

*Step 1.* If $a_{kk} \neq 1$, then make it so, by dividing the $k$-th row by $a_{kk}$.
*Step 2.* For all $i$ and $j$ from $(k+1)$ to $n$, replace

$$a_{ij} \text{ by } (a_{ij} - a_{ik}a_{kj}),$$

$$F_i \text{ by } (F_i - a_{ik}F_k).$$

### 8.11.3 Back Substitution

Back substitution, as mentioned earlier, is a sequential solution of the unknowns, starting from the last upwards to the first.

$$u_i = \text{Solution for the displacements}.$$

$$u_n = \frac{F_{nn}}{a_{nn}}$$

$$u_j = F_j - \sum_{l=j+1}^{n} a_{jl}u_l, \; j = (n-1) \text{ down to } 1.$$

The basic rules above are for a full square matrix. With modern computers and software, stiffness matrices of size 30000 × 30000 are quite common, and the number of calculations—or the so-called *operation count* —is prohibitively high. The operation count in the Gauss elimination scheme without any kind of modifications leads to prohibitively high demands on the memory, storage, and computing time. Evidently, therefore, we must adapt the Gauss elimination scheme to take advantage of the symmetric and banded nature of the [**K**] matrix.

In order to effect economy of space and time, the following artifices are used.

- Bandwidth minimization by renumbering nodes,
- Band matrix storage and band solver,
- Profile storage and profile solver,
- Use frontal solver, which is really an out-of-core solution. (The frontal solver is an alternative to the band solver and the profile solver.)

### 8.11.4 Bandwidth Minimization

It is easy to understand that higher the bandwidth, the higher the operation count in the solution. We have seen earlier that the bandwidth depends directly on the difference between the highest and lowest node numbers in any element over the entire model. But in building up a finite element model it is next to impossible for the user to number and renumber the nodes, to obtain the minimum bandwidth. Also, no general procedure exists to number the nodes to ensure minimum bandwidth. Minimization of bandwidth is a problem in the area of topology.

Most software packages let the user create a model and number the nodes in any arbitrary sequence. When the model is ready to run, the entire model is processed through a bandwidth minimization algorithm. This produces another set of internal node numbers and element connectivity, which is presumably very different from the one defined by the user. All calculations are performed for the internal node numbers, and converted back to the user's own numbering scheme. Thus the process of bandwidth minimization is transparent to the user. Two procedures that are popularly used by most software developers for automatic renumbering of nodes are the *Cuthill and McKee* (CM) algorithm and the *Gibbs, Poole, and Stockmeyer* (GPS) algorithm. The GPS algorithm is a more recent algorithm than the CM algorithm, and is employed in newer software.

### 8.11.5 Band Storage

It can be shown that when the Gauss elimination scheme is applied to a large banded matrix, the terms outside the band are inactive and are of no consequence in the overall calculation. Obviously therefore, it is an overhead to keep them in the memory. In the band storage, all the inactive zeros above the bandwidth are discarded, and are not stored at all.

**Table 8.2. Operation Counts for Gauss Elimination Scheme Performed on Regular and Banded Matrices.**

| | Un-modified Gauss Scheme | Band-Storage Algorithm |
|---|---|---|
| Number of elements in the global stiffness matrix | $(nd)^2$ | $(nd) \times B$ |
| Number of elements in the force matrix | $(nd)$ | $(nd)$ |
| Operation Count in the Gauss Scheme | $2(nd)^2 + \frac{1}{2}(nd)^3$ | $2(nd)B + \frac{1}{2}(nd)B^2$ |

Because of the symmetry, the portion of [**K**] that is below the leading diagonal can also be discarded. This is possible since all the elements below the leading diagonal have been stored *once* already. The resulting matrix is stored as a two-dimensional array.

In lieu of the elements discarded, a rule is to be set up to determine the correspondence between the positions of any element in the matrix in its original square format and those in its band matrix format. Once this correspondence is established, the Gauss elimination can be carried out on the band matrix. See Table 8.2 for the savings in the memory and operation counts by band storage. For a typical problem, the total DOF may be as high as 30,000 and the best estimate for the bandwidth (after optimizing it by renumbering algorithm) is about 750. That is, $(nd) = 30{,}000$ and $B = 750$. The savings in operation count is easy to see.

### 8.11.6 Profile Storage

Profile storage seeks to further rid the inactive zeros inside the band. In this method, we consider a column standing on the leading diagonal. The height of a column is given by the topmost nonzero entity in that column. The height of the column is referred to as the *profile* for that column. Note that the profile of each column may be different. In this respect profile storage is different from band storage, in which the bandwidth is just one number for the whole [**K**] matrix. Each column is stored to the extent of its profile. The order is from bottom of the profile to the top. The entire stiffness matrix is stored as a continuous one-dimensional array.

Again, as in band storage, a rule is set up to determine the one-to-one correspondence between the position of an element in the full square format, and in the profile storage format. Such a rule needs additional information about the profile size of each column. Therefore, in addition, another array is formed to keep account of the profile size of each column.

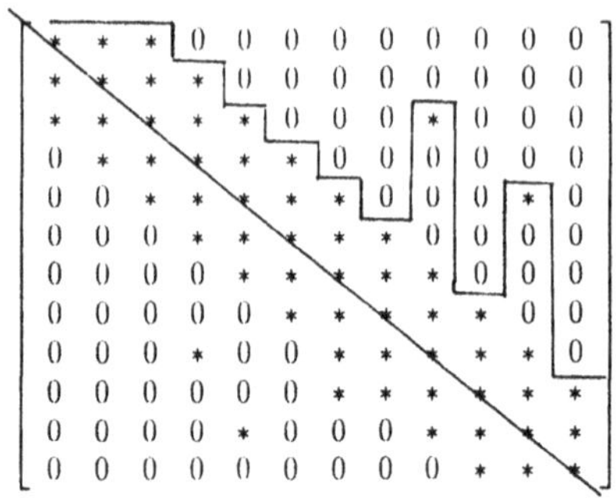

**FIGURE 8.14.** Profile or skyline storage procedure involves the stretching of the upper part of the stiffness matrix into a long one-dimensional array. The height of each column is stored as another array. The latter array helps in determining which entry in the stiffness matrix corresponds to which entry in the former array.

This scheme is shown in Figure 8.14. The Gauss elimination scheme is applied to the profile-stored stiffness matrix.

### 8.11.7 The Frontal Solver

The frontal solver, also called the *wave front solver,* is an ingenious method of solution. In contrast to the other two storage methods above, this method is not a concept of storage, but is a concept of *avoiding* storage. In fact, the global [K] matrix is *never* assembled in full. The basic idea underlying the frontal solution follows. See Figure 8.15.

Let us suppose that the $i$-th degree of freedom is being determined. It can be shown that in the Gauss elimination scheme, the only elements in the [K] matrix that affect the solution are those inside the shaded triangular area of the band. In the event that the solution of only the $i$-th DOF is sought, it is only necessary to have the elements inside the shaded area. Zienkiewicz further proposed that the global stiffness matrix need not be assembled in full! It is only necessary to assemble the elements in the shaded area in the core, as and when it is needed. Once the $i$-th DOF is solved in the memory—or in-core, so to say—the solution can move on the next DOF. Normally, a group of DOFs—as many as a few hundred in number—can be solved in-core at a time. When done, the next group is loaded into the memory, solved, stored away, and so on until the entire problem is completed. The appropriateness of the name wave-front for the method is now evident.

This method is extremely memory-efficient, and had been adopted by the early FEA software with advantage. The reader can visualize the

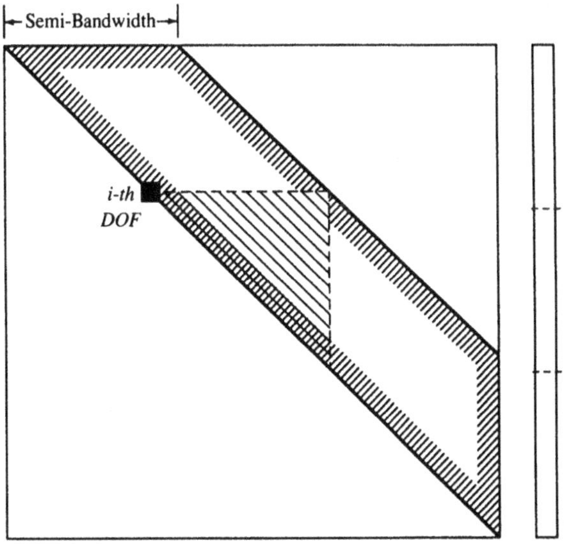


**FIGURE 8.15.** Wavefront algorithm depends on the fact that a DOF represented by the location at the black square, is affected by only those entries in the **[K]** matrix that are inside the shaded triangle.

enormous amount of bookkeeping involved in the procedure. The further variations of the frontal solver address the questions of how many DOFs can be solved in a computer of given memory size. These questions are relevant to solving large problems on relatively small computers.

However, in the late 1980s the price per bit of memory declined and size of the memory was no longer a factor in the choice of the solver algorithm. The profile solver, which is a fully in-core solver, came to be adopted in the more recent software. Its advantages are simplicity, and no bookkeeping requirements. Given enough memory in the computer, the difference between the profile solver and the frontal solver should disappear, since the profile solver is simply the frontal solver with just one wave containing the entire problem.

## 8.12 REACTIONS, STRAINS, AND STRESSES

We note that one talks about reactions at supports and restraints. Normally, the displacements at these points are known, but not the reactions. In Equation 8.28, we have a procedure for calculating the reactions. It must be recalled that the $k$s in this equation are stiffness components related to the $m$-th DOF. The reaction at a support is after all the force in the direction of the $m$-th DOF. It is easily seen that the reactions are simply calculated from Equation 8.28 by recalling the appropriate $k$s from the storage, and by using the appropriate $u$s from the solution database.

We have seen in Equation 8.21, that the strains can be obtained at any point inside an element using the **[B]** matrix for that element. If the solution is databased, then with the data on element connectivity, it is possible to obtain the nodal displacements relevant to each element. From this, by application of the **[B]** matrix at the appropriate point inside the element, we can calculate the strain at that point.

One step ahead, the stress at any point inside an element can be calculated straightaway from the strain by a simple multiplication by the **[D]** matrix for that element.

### 8.12.1 Uniqueness of the Computed Value of Strain and Stress

Because the calculation of strain needs the **[B]** matrix, which itself needs the interpolation functions, we observe that the value of strain at a point depends entirely on the functions $N_1, N_2, N_3, \cdots$. For a point inside an element, these functions are uniquely defined. However, at the nodes, this is not true, because nodes may be shared by more than one element. For a shared node, the interpolation functions depend on which element it is calculated from. So is the calculation of stresses. Thus, trying to calculate *nodal strains* and *nodal stresses* leads to non-uniqueness.

Thus, it is a common practice in all finite element software to calculate the strains and stresses at the element centroids, or at certain salient points inside the elements called the *Gauss points*. The resulting values are called the *element strains* and *element stresses*. It can be shown in the analysis of finite element methods, that the error in computation is minimum at the Gauss points.

For linear interpolation functions, displacements vary linearly inside an element. Consequently, strains are constant at all points inside the element. Extending the argument, we can say that quadratic interpolation functions lead to linearly varying strains and stresses, and so on.

### 8.12.2 How Then Are Unique Stress Fields Plotted on the Screen?

The non-uniqueness of the nodal stress is recognized in all software. The remedy for this is usually achieved in the following manner.

- Start with the element stresses, or Gauss point stresses as the correct ones.
- Do a spatial interpolation of the element or Gauss point stresses, and determine the stresses at any desired point—not just at the nodes.

It is readily noticed that there is no finite element related technique involved in this procedure, only a three-dimensional or two-dimensional interpolation.

## 8.13 POST-PROCESSING

Post-processing consists of all the steps necessary to create graphic outputs of a specific parameter, which is usually a stress component. All software vendors provide capabilities to view the stresses as a whole-field map. That is, the stresses are presented as a color map or a contour map superimposed on the problem geometry. Such a map is more meaningful to the engineer than is a tabular statement of stresses. This capability is supplemented by a variety of CAD-like features, which makes the post-processor module a vital adjunct to finite element analysis. A few of the features of a typical post-processor are listed below. Since these features provide the engineer with a convenient way of visualizing stress fields, software vendors constantly create more useful and attractive features.

*Subsets of Nodes and Elements:* It is possible to select a subset of nodes or elements for viewing. Such a selection can be made by specifying the range of node numbers, the range of element numbers, or the range of $x$, $y$, and $z$ coordinates. The coordinates themselves can be specified in a global or any local system.

*Zoom, Unzoom, View Angles:* Zooming in on a small area of interest is a common feature. Unzooming is also necessary. View angles can be chosen for seeing the details in the best way.

*Windowing:* The computer screen can be divided into several windows, and graphics can be plotted on them as though each one is completely independent of the others.

*Operations on the Subsets:* The selected subset of nodes and elements can be viewed in any arbitrary direction, and can be shaded. Also any specific output parameter like displacement stress or strain can be plotted on the subset. This may be considered as a way of sectioning the given geometry and looking at the stress on the section.

*Deformed Shapes:* The deformed shape of the object can be plotted either with or without the undeformed shape. Usually a scaling factor is automatically set up by the computer for visual clarity.

*Mode Shapes:* Mode shape is a term that refers to the displaced pattern while the part is in a state of vibration. Thus the mode shapes can be treated just as if they were the deformed shapes. Different modes of vibration correspond to different mode shapes and plotting them gives an insight into the vibration characteristics of the component.

*Animation:* The deformed shape is scaled down into a number of intermediate frames, say at 10%, 20%, 30% ··· of the full deflection. Assuming that there is enough memory available in the computer, these frames can be sequentially shown on the screen at reasonably high speed, so that it appears as an animated picture of the deformation. Its best use is intuition.

*Color-Fill and Color Contours:* A stress field can be mapped in terms of many contours, each contour representing a particular value. Alternatively, the region between two contours can be filled with a color. In the latter method, each color represents a *range* of stresses.

*Stresses on Deflected Shapes:* Stresses may by plotted on undeformed geometry or on the deformed geometry. In the latter case, two different kinds of information are seen on the same plot.

*Light Source Shading:* It is possible to assume a light source (or several light sources) is placed at different locations with respect to the object and to cast a particular color of light. The perception of depth and realism is greatly enhanced by this feature. In a number of cases errors in modeling can be detected by viewing a model with shading.

*Hidden Line Removal:* Lines hidden from view can be removed from plotting. This is a quick method of checking the model with its realistic appearance. However, this is not a final proof of the correctness of the model.

*View in a Local Coordinate System:* The model as well as stresses may be viewed in local coordinate systems. A pipe, for example, would look like a plain sheet of paper if viewed along the $\theta$-direction of a cylindrical coordinate system located at the center of the pipe. Stresses in this system automatically correspond to the hoop and axial components.

*Real-Time Rotation of Viewing Angle:* It is possible, in some powerful workstations and mainframe computers, to rotate an object on the screen much in the same way as if it were hand-held and viewed all around. Such real-time rotations are made possible by moving a pointing device (mouse).

*Path Operations:* Path operations is a generic term for taking a section of the model and plotting stress, strain, or displacement. The subset selection feature does not necessarily lead to a plane cross section. Path operations do. Hence some software packages provide this feature. A real-time interpolation is needed to calculate the quantities exactly on the line to obtain a plot of a prescribed output parameter (such as stress or strain). This is a very useful tool to plot the gradients near stress concentrations.

*Time-Based X-Y Plots:* When solving problems involving transients, one also needs a plot of the variation of an output parameter such as stress, acceleration, or displacement as a function of time. Since none of the above graphics is useful for this, many software packages provide a facility for a common $x$-$y$ plot of the parameter.

*Hard Copy Outputs:* Last but not the least is the hard copy. The finite element software has to provide the capability to use a printer or a plotter. Color plotters, ink-jet printers, color laser printers, and bubble jet printers are currently available, creating a problem of choice.

## 8.14 ISOPARAMETRIC ELEMENTS

### 8.14.1 Motivation

The discussions in the preceding sections were covered essentially for straight-edged triangular elements. However, when modeling curved boundaries in practical problems, especially with large sized elements, the straight edges of the elements do not model the curvatures well enough. Therefore, we need curved edges for the elements. To accomplish curved edges for elements we can have three nodes on each edge (not all on a straight line) and fit a parabola through them. The resulting geometry is shown in Figures 8.16a and 8.16b. However, recall that the several matrix calculations to be performed for such elements, involve integrals of the type

$$\int [\mathbf{B}]^T [\mathbf{D}][\mathbf{B}] t \, dA, \quad \text{and}$$

$$\int [\mathbf{N}]^T \{\mathbf{f}\} \, dl.$$

Obviously, from Figure 8.16 these integrations are impossible if performed in the Cartesian coordinate system.

To make such computations more formal and easier, the concept of *isoparametric elements* was introduced. This concept, advanced by Irons [8],

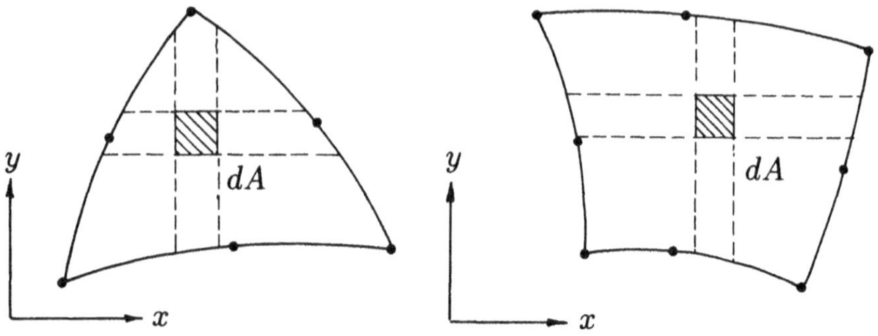


**FIGURE 8.16.** The curved edges of the elements make a general formula for integration of stiffness and force matrices impossible. One has to resort to numerical integration techniques.

is sufficiently general, and helps to create elements that are more complex in shape, in the degree of interpolation polynomials, and in the number of nodes per element. The steps involved in the formulation of an isoparametric element are as follows.

1. Define a unit-sized (triangular or square) element whose size usually extends over range of 0 to 1 or from $-1$ to 1, in the natural coordinate system.
2. It is possible to establish a one-to-one correspondence between points in the unit element and those in a real-life finite element. In the case of isoparametric elements, the displacement interpolation functions $N_i$ are also used to map the element geometry. In other words, the functions $N_i$ uniquely map the points in the unit element on to those of the real-life element. The formal definition of isoparametric elements results from this possibility. (See also Exercises 24 through 27.)

Elements which utilize the same set of functions $N_i$ for the interpolation of coordinates as well as displacements are called isoparametric elements.

3. Calculation procedures are written for the unit element, *using natural coordinates*. Because the real life element is defined in global Cartesian coordinates, the use of natural coordinates involves a conversion of scales at each step of the procedure. For example, the $[\mathbf{B}]$ and the $[K]^e$ matrices will have to be derived using additional matrices containing the coordinate conversion factors, as well as the area/volume magnification factors.
4. Step 3 above leads to integrals of the type $\int [\mathbf{B}]^T[\mathbf{D}][\mathbf{B}]|J|\,d\bar{V}$, for the

unit element (*and not for the real-life element*). Obviously, the **[B]** matrix is a variable, and cannot be derived as in Example 8.3. The **[D]** matrix, being a material property, does not change due to the coordinate transformation. Also, $d\bar{V}$ is the infinitesimal volume of the *unit element in natural coordinates*. For plane elements of uniform thickness, it is equal to $t\,d\bar{A}$, where $d\bar{A}$ is the area of the *unit element in natural coordinates*. $|J|$ is known as the *Jacobian Determinant*. It is the scale factor which converts the magnitude of $d\bar{A}$ (or $d\bar{V}$) to those of real life. Similar comments also apply for the calculation of force vectors.

5. The complexity of the element calculations has increased. However, they are modularized for a unit element, and hence can be computed for each element. For this reason, the isoparametric elements are sometimes also classified as *numerically integrated elements*.

By virtue of their generality, isoparametric elements can be applied easily to straight-edged elements.

### 8.14.2 The Isoparametric Linear Triangle

We may take one more look at the natural coordinates $L_1$, $L_2$, and $L_3$ of a point inside a triangular element. In Figure 8.17a, we see lines parallel to the sides of the triangle. Along each of the edges a coordinate is kept constant. When the line coincides with an edge, the value of the coordinate is zero, and as it moves towards a vertex, the value of the coordinate reaches 1. This way, we once again establish the uniqueness of the natural coordinates of any point inside the element.

Consider the equations of coordinate mapping as follows.

$$x = N_1 x_a + N_2 x_b + N_3 x_c \tag{8.29}$$

$$y = N_1 y_a + N_2 y_b + N_3 y_c \tag{8.30}$$

Evidently, these equations are identical in their form to the interpolation of displacements. For linear triangles, $N_i = L_i$. This maps the nodes 1, 2, and 3 in Figure 8.18a on to the nodes in Figure 8.18b. They also map edges on to edges and coordinate lines parallel to edges 12, 23, and 31 on to similar parallel lines. Only the size and orientation of the real-life element are different from those of the unit triangle. Because of the linear interpolation functions, straight lines in 123 map on to straight lines in *abc*. (For a different approach to proving this idea, see Exercise 8.24.)

Once the coordinate mapping is defined, the **[B]** matrix can be constructed. Recall that the **[B]** is a matrix of derivatives of $N_i$ with respect to

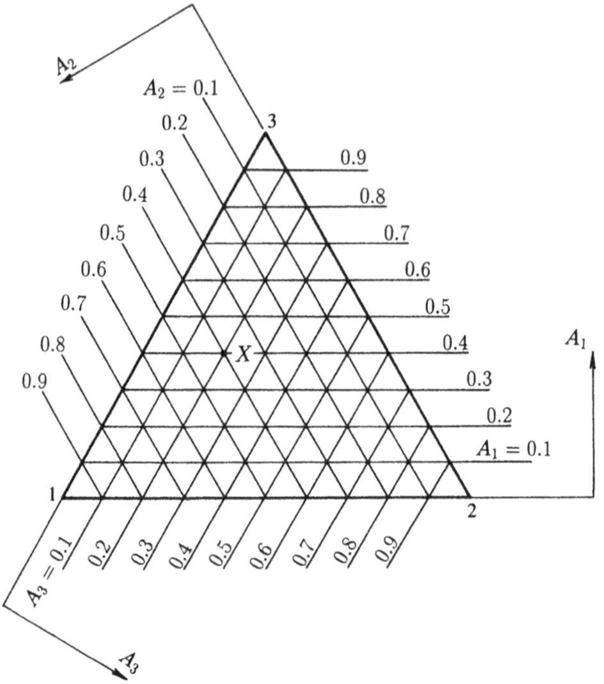


Natural Coordinates for the point $X$ : $(0.4, 0.4, 0.2)$

**FIGURE 8.17.** Illustration of natural coordinate lines.

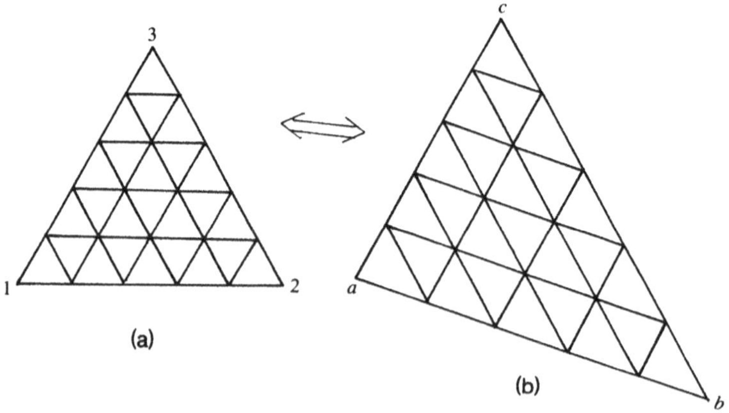


**FIGURE 8.18.** A unit-sized triangular element maps into a general straight-sided triangle.

coordinates. Since both coordinates and displacements are defined with respect to the unit triangle, the construction of the [**B**] matrix is somewhat complex. There is not much space in this book to discuss this in detail. We just call attention to the special nature of the calculation and refer the reader to one of the popular text books on finite elements, such as References 16, 15, or 17 for further details.

### 8.14.3 The Quadratic Triangle

The quadratic triangle has three corner nodes, and three midside nodes, and curved sides passing through all the three nodes on it, as shown in Figure 8.19b. It can be obtained by coordinate mapping of a unit-sized, straight-sided, six-noded triangular element, as in Figure 8.19a. In order to accomplish the mapping of straight lines onto parabolas, we need second degree interpolation functions, as below.

$$N_1 = L_1(2L_1 - 1) \quad N_2 = L_2(2L_2 - 1) \quad N_3 = L_3(2L_3 - 1) \tag{8.31}$$

$$N_4 = 4L_1L_2 \quad N_6 = 4L_2L_3 \quad N_6 = 4L_3L_1 \tag{8.32}$$

$$u = \sum_{i=1}^{6} N_i u_i; \quad v = \sum_{i=1}^{6} N_i v_i \tag{8.33}$$

$$x = \sum_{i=1}^{6} N_i x_i; \quad y = \sum_{i=1}^{6} N_i y_i \tag{8.34}$$

The functions $N_i$ above are chosen to satisfy certain basic requirements.

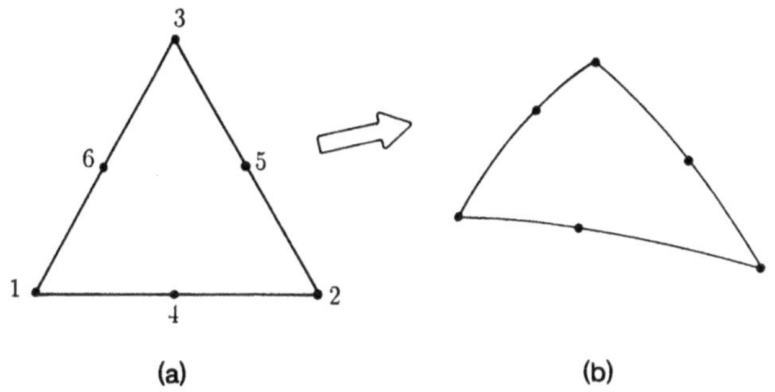


**FIGURE 8.19.** Quadratic triangle maps a unit-sized triangle onto one with curved sides. Mid-side nodes are used. Note also that the order of internal node numbering is important, since they are associated with a specific function $N_i$.

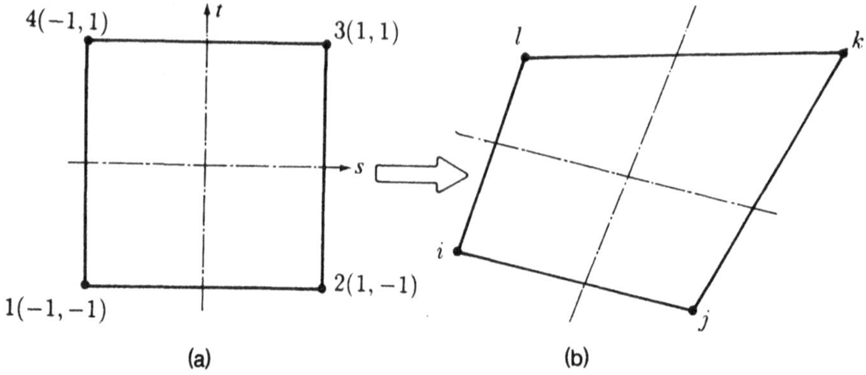


**FIGURE 8.20.** Linear isoparametric quadrilateral maps a unit square into a straight-sided general quadrilateral.

(See the commentary in Example 8.1.) The reader may verify that the equilateral triangle in Figure 8.19a maps into the real-life triangular element with curved sides as in Figure 8.19b.

### 8.14.4 The Linear Quadrilateral

The linear quadrilateral has four corner nodes and four straight edges, as shown in Figure 8.20. This can be formulated as a mapping of a four-noded square element with coordinates ranging from −1 to +1 units. The expressions for $N_i$ for this element are listed below. It can be verified that the unit square element in Figure 8.20a maps into the real life quadrilateral element shown in Figure 8.20b.

$$u = \sum_{i=1}^{6} N_i u_i; \quad v = \sum_{i=1}^{4} N_i v_i$$

$$x = \sum_{i=1}^{4} N_i x_i; \quad y = \sum_{i=1}^{4} N_i y_i$$

$$N_1 = \frac{1}{4}(1 - s)(1 - t); \quad N_2 = \frac{1}{4}(1 + s)(1 - t)$$

$$N_3 = \frac{1}{4}(1 + s)(1 + t); \quad N_4 = \frac{1}{4}(1 - s)(1 + t)$$

### 8.14.5 The Quadratic Quadrilateral

The quadratic quadrilateral in Figure 8.21a has four corner nodes, and four midside nodes. It may have curved boundaries. It is obtained as a

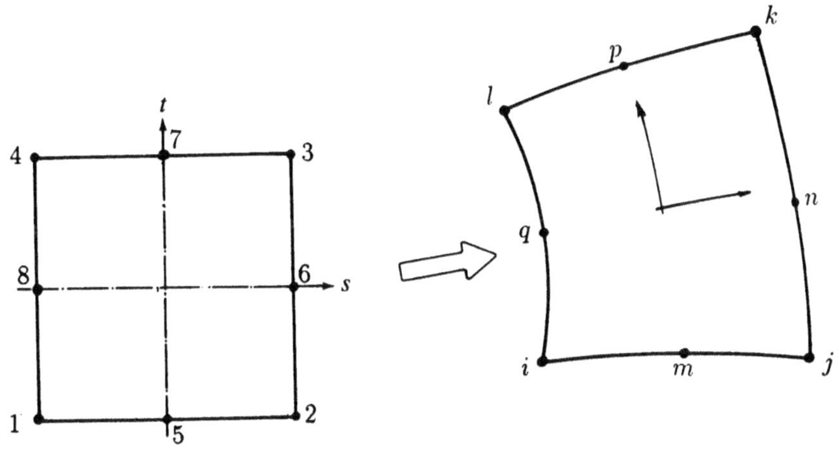


(a) Natural Coordinates

(b) Actual Spatial Coordinates.

**FIGURE 8.21.** Isoparametric quadratic quadrilateral maps a unit-sized square into a general, eight-noded, quadrilateral with curved sides.

mapping of an eight-noded square element, with natural coordinates ranging from $-1$ to $+1$, shown in Figure 8.21b. The unit element has midside nodes and straight boundaries. The interpolation functions $N_i$ are second-degree polynomials in the natural coordinates. Hence, they map straight sides of the unit element to parabolic sides of the real-life element. The equations relevant to this type of element are as follows.

$$u = \sum_{i=1}^{6} N_i u_i; \quad v = \sum_{i=1}^{8} N_i v_i$$

$$x = \sum_{i=1}^{8} N_i x_i; \quad y = \sum_{i=1}^{8} N_i y_i$$

$$N_1 = -\frac{1}{4}(1 - s)(1 - t)(1 + s + t);$$

$$N_2 = -\frac{1}{4}(1 + s)(1 - t)(1 - s + t)$$

$$N_3 = -\frac{1}{4}(1 + s)(1 + t)(1 - s - t);$$

$$N_4 = -\frac{1}{4}(1 - s)(1 + t)(1 + s - t)$$

$$N_5 = \frac{1}{2}(1 - s^2)(1 - t); \quad N_6 = \frac{1}{2}(1 + s)(1 - t^2)$$

$$N_6 = \frac{1}{2}(1 - s^2)(1 + t); \quad N_8 = \frac{1}{2}(1 - s)(1 - t^2)$$

The derivation of the [**B**] and $[K]$ matrices in all these cases involves some complexity, but in no case does it exceed the undergraduate level of engineering mathematics. The amount of calculation is high, but not the complexity. The interested reader is advised to refer to popular text books on finite elements cited earlier.

*Further, it must be noted that* [**B**] *never needs to be calculated as such at all. It enters the* $[K]^e$ *computation directly.*

## 8.15 THE GAUSS QUADRATURE

In the formulation of all the isoparametric elements, it is difficult to perform the integrations analytically. Hence, they are performed numerically. The natural coordinates for these elements vary from 0 to 1, or from $-1$ to 1. This enables us to utilize the Gauss numerical integration scheme, known in the literature as the *Gauss quadrature*.[7] According to this scheme, an integral in the range of $-1$ to 1 can be obtained by a weighted sum of the values' integrand at certain specific points, called the *Gauss Points*.

$$\int_{-1}^{1} f(x)\,dx = W_1 f(x_1) + W_2 f(x_2) + W_3 f(x_3) + \cdots + W_n f(x_n)$$

$$= \sum_{i=1}^{n} W_i f(x_i) \tag{8.35}$$

Note that the formula is approximate. The $W_i$ are called the *weight factors*, and $x_i$ are the Gauss points. The number of terms $n$ to which the summation is to be done is up to the user to decide. Each value of $n$ is associated with a set of $W_i$. Each value of $n$ leads to a particular accuracy of calculation. For quadratic functions, $n = 2$ gives the exact value. For

[7] The term quadrature of a curve refers to the finding an equivalent rectangle having the same area, and standing on the same base. Obviously, this is equivalent to finding the average ordinate, which in turn involves the computation of the integral.

**Table 8.3. Abscissae and Weight Coefficients of the Gaussian Quadrature Formula.**

$$\int_{-1}^{1} f(x)dx = \sum_{i=1}^{n} W_i f(x_i)$$

| $\pm x_i$ | $W$ |
|---|---|
| n = 1 | |
| 0 | 2.00000 00000 00000 |
| n = 2 | |
| 0.57735 02691 89626 | 1.00000 00000 00000 |
| n = 3 | |
| 0.77459 66692 41483 | 0.55555 55555 55556 |
| 0.00000 00000 00000 | 0.88888 88888 88889 |
| n = 4 | |
| 0.86113 63115 94053 | 0.34785 48451 37454 |
| 0.33998 10435 84856 | 0.65214 51548 62546 |

linear functions $n = 1$ gives the exact value. However, for functions like $f(x) = \sin x$, or $f(x) = \exp x$, more terms are needed. Most finite element formulations use linear and quadratic interpolation functions, and hence it is possible to use two terms in the summation and get accurate value.

Table 8.3 gives the weight factors to be used for Gauss quadrature as a function of the number of Gauss points used.

**Example 8.6**

Integrate $\int_{-1}^{1} x^2\, dx$ using the Gauss quadrature.

*Solution* We use a two-point scheme, and hence

$$\begin{aligned}\int_{-1}^{1} x^2\, dx &= W_1 f(x_1) + W_2 f(x_2)\\ &= 1 \times f(-0.5773503) + 1 \times f(0.5773503)\\ &= (-0.5773503)^2 + (0.5773503)^2\\ &= 0.33333 + 0.33333\\ &= 0.66667\end{aligned}$$

Analytical integration gives $x^3/3|_{-1}^{1} = 2/3 = 0.66667$. Hence, Gauss quadrature gave an exact answer.

**Example 8.7**

Calculate $\int_{-1}^{1} e^{-x}$ using Gauss quadrature.

*Solution* Using the two-point integration,

$$\int_{-1}^{1} e^{-x}\,dx = 1 \times f(-0.5773503) + 1 \times f(0.5773503)$$

$$= e^{-0.5773503} + e^{0.5773503}$$

$$= 2.34270$$

Using the three-point integration

$$\int_{-1}^{1} e^{-x}\,dx = 0.555556 \times f(-0.7745967) + 0.888889 \times f(0)$$

$$+ 0.555556 \times f(0.7745967)$$

$$= 0.555556 \times e^{-0.7745967} + 0.888889 \times e^{0}$$

$$+ 0.555556 \times e^{0.774596}$$

$$= (0.555556 \times 2.169717) + (0.888889)$$

$$+ (0.555556 \times 0.4608896)$$

$$= 2.9238889$$

Using analytical procedure,

$$\int_{-1}^{1} e^{-x}\,dx = -e^{-x}|_{-1}^{1}$$

$$= 3.08616$$

Evidently, the three-point integration gives better results in this case, with an error of only 5.25%.

### 8.15.1 Gauss Quadrature in Two and Three Dimensions

The Gauss quadrature can also be extended to two and three variables. that is two and three dimensions as well. In two dimensions, it has the form

$$\int_{-1}^{1}\int_{-1}^{1} f(x, y)\, dx\, dy = \sum_{i}\sum_{j} W_i W_j f(x_i, y_j)$$

$$\int_{-1}^{1}\int_{-1}^{1}\int_{-1}^{1} f(x, y, z)\, dx\, dy\, dz = \sum_{i}\sum_{j} W_i W_j W_k f(x_i, y_j, z_k)$$

**Example 8.8**

Evaluate the following integral over a square region, as shown in Figure 8.22.

$$I = \int_{-1}^{1}\int_{-1}^{1} x^2 y^3\, dx\, dy$$

*Solution* We may use the two-point integration first.

$$\begin{aligned} I &= \int_{-1}^{1}\int_{-1}^{1} x^2 y^3\, dx\, dy \\ &= W_1 W_1 f(x_1, y_1) + W_1 W_2 f(x_1, y_2) + W_2 W_1 f(x_2, y_1) + W_2 W_2 f(x_2, y_2) \\ &= f(-0.57735, -0.57735) + f(-0.57735, 0.55735) \\ &\quad + f(0.57735, -0.57735) + f(0.57735, 0.57735) \\ &= (-0.57735)^2 \times (-0.57735)^3 + (-0.57735)^2 \times (0.57735)^3 \\ &\quad + (0.57735)^2 \times (-0.57735)^3 + (0.57735)^2 \times (0.57735)^3 \\ &= (0.57735)^5 \times (-1 + 1 - 1 + 1) \\ &= 0 \end{aligned}$$

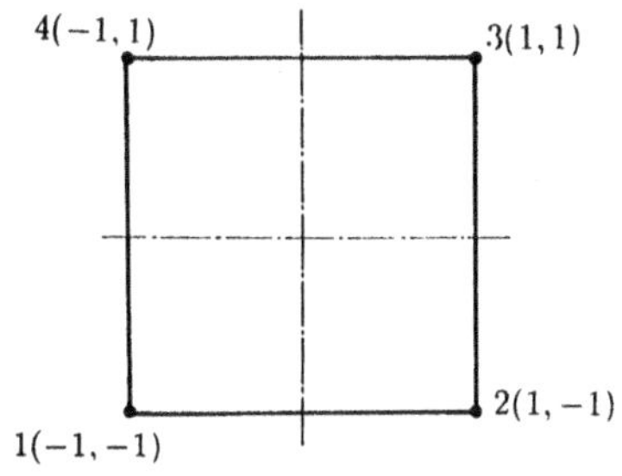


**FIGURE 8.22.** Integration over a square region.

By using the three-point integration,

$$I = \int_{-1}^{1}\int_{-1}^{1} x^2 y^3 \, dx\, dy$$

$$= W_1W_1f(x_1, y_1) + W_1W_2f(x_1, y_2) + W_1W_3f(x_1y_3)$$
$$+ W_2W_1f(x_2, y_1) + W_2W_2f(x_2, y_2) + W_2W_3f(x_2, y_3)$$
$$+ W_3W_1f(x_3, y_1) + W_3W_2f(x_3, y_2) + W_3W_3f(x_3, y_3)$$

Using the weight factors with $W_1 = W_3 = 0.555556$, and $W_2 = 0.888889$, and performing the summation, we obtain,

$$I = 0 \text{ exactly}$$

Doing the problem fully analytically, we get

$$I = \frac{x^3}{3}\bigg|_{-1}^{1} \cdots \frac{y^4}{4}\bigg|_{-1}^{1}$$
$$= 0$$

Once again, Gauss integration gives exact results in both cases.

## 8.16 CONCLUSION

This chapter describes the linear elastic behavior of a single continuous body. The chapter is elementary in nature, and is intended to give an insight into the real complexities of the finite element method. The real-life plastics problems involve nonlinear stress strain relations, buckling, impact resistance, minimum weight design, etc., none of which is exactly solvable with the amount of material covered. If exact relevance to plastics is sought, this chapter does not provide it. However, it alerts the engineer to the knowledge and care needed in using finite elements. Further it shows that aspects of economy such as the storage scheme and node renumbering are essential parts of FEM knowledge. The user of FEM software must become familiar with the several other points specific to the particular software package being used. Some of these special points are discussed in the next chapter.

**References**

1. Hrenikoff, A., "Solution to Problems in Elasticity by the Framework Method," *Journal of Applied Mechanics*, **8**:169–175 (1941).

2. Courant, R., "Variational Methods for the Solution of Problems of Equilibrium and Vibrations," *Bull. Amer. Math. Soc.* **49**:1–23 (1943).
3. Newmark, N. M., "Numerical Methods for Analysis of Bars, Plates and Elastic Bodies," in Grinter, L. E., (ed.), *Numerical Methods of Analysis in Engineering*, 1949, MacMillan, New York, NY.
4. Argyris, J. H., "Energy Theorems and Structural Analysis," *Aircraft Engineering*, **26**:(Oct-Nov 1954); **27**:(Feb-May 1955).
5. Turner, M. J., Clough, R. W., Martin, H. C., and Topp, L. J., "Stiffness and Deflection Analysis of Complex Structures," *J. of Aero. Sci.* **23**:(9)805–823, 854 (1956).
6. Clough, R. W., "The Finite Element Method in Plane Stress Analysis," *Proc. 2nd ASCE Conf. Electronic Computation*, Pittsburgh, PA. Sept 1960, pp 345–378.
7. Zienkiewicz, O. C. and Cheung, Y. K., "Finite Elements in the Solution of Field Problems," *The Engineer*, **220**:507–510 (1965).
8. Irons, B. M., "Engineering Application of Numerical Integration in Stiffness Method," *JAIAA*, **14**:2035–7, 1966.
9. Zienkiewicz, O. C., and Cheung, Y. K., *The Finite Element Method in Structural and Continuum Mechanics*, 1967, McGraw Hill, London.
10. Chu, S. L., "Analysis and Design Capabilities of STRUDL program," in Fenves, S. J., Perrone, N., Robinson, A. R., and Schnobrich, N. C., (editors), *Numerical and Computer Methods in Structural Mechanics*, 1973, Academic Press, New York, NY.
11. Noor, A., "Survey of Some Finite Element Software Systems," in Kardestuncer, H., and Norrie, D. H., (editors), 1987, *Finite Element Method Handbook*, McGraw Hill, New York, NY.
12. Tracey, D. M., "Finite Elements for Determination of Crack Tip Elastic Stress Intensity Factors," *Engineering Fracture Mechanics*, **3**:265–65, 1971.
13. Finlayson, B. A., *The Method of Weighted Residuals and Variational Principles*, 1972, Academic Press, New York, NY.
14. Golub, G. H. and Van Loan, C. F., *Matrix Computations*, 2nd Edn. 1989, John Hopkins University Press, Baltimore, MD.
15. Grandin, H., Jr., *Fundamentals of the Finite Element Method*, 1986, MacMillan Publishing Company, New York, NY.
16. Segerlind, L. J., *Applied Finite Element Analysis*, 2nd Edn. 1984, John Wiley & Sons, New York, NY.
17. Huebner, K. H., and Thornton, E. A., *The Finite Element Method for Engineers*, 2nd Edn. 1982, John Wiley & Sons, New York, NY.

## Exercises

**8.1.** Refer to any mechanical engineering design handbook for a helical spring design. State whether the following statements are true or false.

(a) The stiffness of the spring is proportional to the modulus of the material.

(b) The stiffness of the spring is reduced by reducing the number of turns.

(c) The stiffness of the spring is proportional to the mass of the spring.

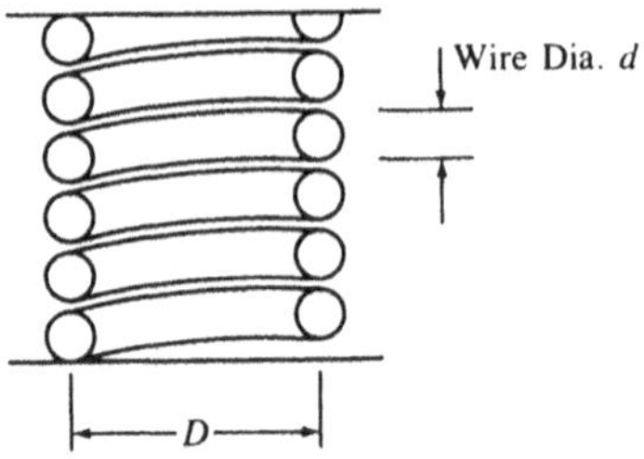


**EXERCISE 8.1.**

(d) If the coil diameter $D$ of the spring is reduced, so is the stiffness of the spring.

(e) If the wire diameter $d$ of the spring is decreased, so is the stiffness of the spring.

**8.2.** Two identical springs, each of stiffness $k$, work in parallel. That is, the loads get distributed, while they undergo the same compression. What is the equivalent stiffness of the combination?
Make an analogous statement for triangular finite elements.

**8.3.** Two identical springs, each of stiffness $k$, work in series. That is, the springs take the same load each, and the displacements get distributed. What is the equivalent stiffness of the combination?
Make an analogous statement for triangular finite elements.

**8.4.** True or False?

(a) The stiffness of a spring is the stress needed to produce a unit strain.

(b) The stiffness of a spring is the force needed to produce a unit elongation.

(c) The stiffness of a bar of unit length and unit area of cross section is numerically equal to the modulus of the material.

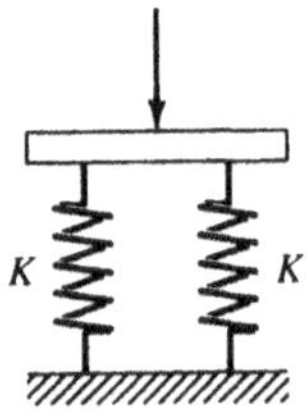


**EXERCISE 8.2.**

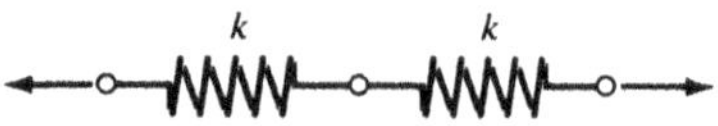


**EXERCISE 8.3.**

**8.5.** The stiffness of a bar of cross-sectional area $A$ is, which of the following? a. $AE$ b. $AE/L$ c. $L/AE$ d. $LE/A$ e. $LAE$ f. $LA/E$

**8.6.** A bar has stiffness $k$ in the axial direction. Use this information to write the stiffness matrix of a two noded bar element 12 shown in the figure. Each node has just one DOF, namely the axial displacement.

7. Two identical bar elements of stiffness $k$ each are connected in parallel to nodes 1 and 2. Write the assembled stiffness matrix.

**8.8.** Two identical bar elements 12 and 23 are connected in series at node 2, as shown in the figure. Write the assembled stiffness matrix.

**8.9.** Three bar elements 12, 23, and 34, all of the same material and area of cross section, are connected in series. Their lengths are in the ratio 2:2:3. The axial stiffness of bar 12 is equal to $k$. Write the stiffness matrix for the assembly.

**8.10.** Write the **[D]** matrix for plane stress conditions.

**8.11.** Write the **[D]** matrix for plane strain conditions.

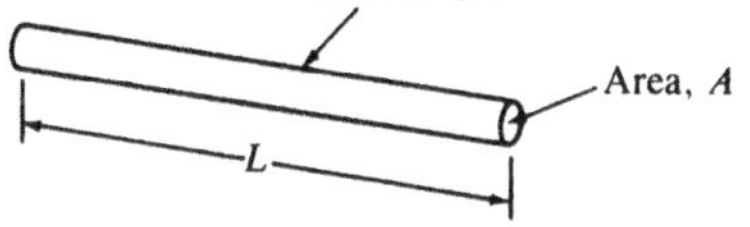


**EXERCISE 8.5.**

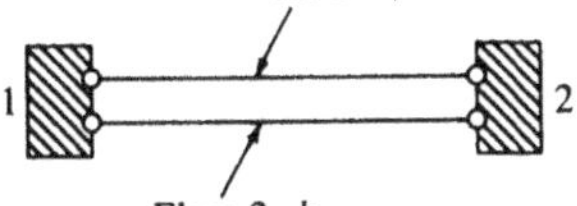


**EXERCISE 8.6.**

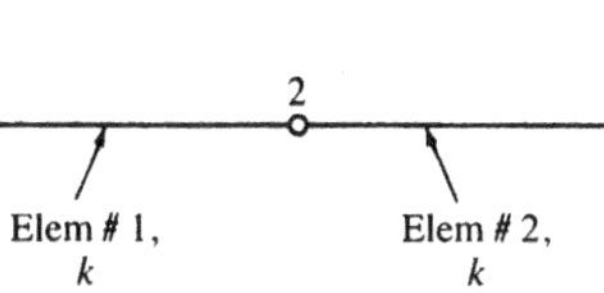


**EXERCISE 8.7.**




**EXERCISE 8.8.**

**8.12.** Write the [**D**] matrix for full three-dimensional elasticity.

**8.13.** Fact 1: The free end deflection of a cantilever beam is $PL^3/3EI$, where $P$ is the applied load.
Fact 2: A point in a beam deflects, as well as rotates.
Question: Is it correct to say that the stiffness of a beam element is $3EI/L^3$? Give reasons.

**8.14.** Two-dimensional beam elements are formulated directly from Euler's beam theory. It is seldom necessary to go through the interpolation functions and natural coordinates (although if you do, you are led to the same results). For the beam element shown, a rotation $\theta_1$ is applied at the end 1, while keeping all other displacements as zero. Find the forces (that is, forces as well as moments) developed at nodes 1 and 2.

**8.15.** For the beam element 12, a displacement $\delta_1$ is applied at node 1, while all the other displacements are kept zero. Find the forces developed at nodes 1 and 2.

**8.16.** Put together the results from Exercises 14 and 15 above, in the form of the element stiffness matrix $[K]^e$ for the beam element to fit into the following equation.

$$\begin{Bmatrix} F_1 \\ M_1 \\ F_2 \\ M_2 \end{Bmatrix} = [K]^e \begin{Bmatrix} \delta_1 \\ \theta_1 \\ \delta_2 \\ \theta_2 \end{Bmatrix}$$

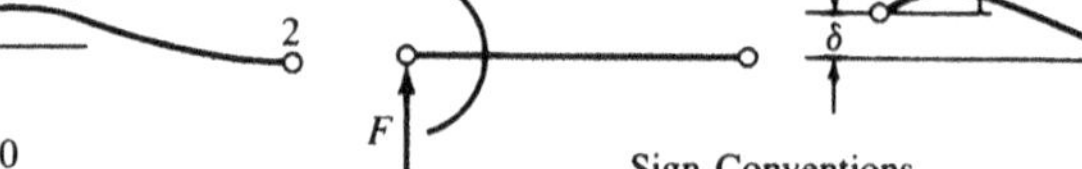


**EXERCISE 8.9.**

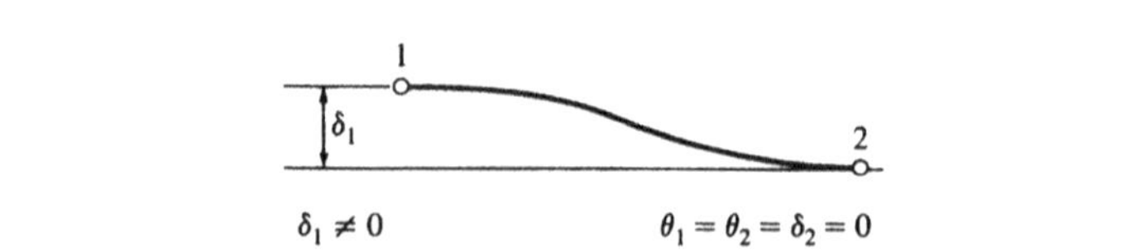


**EXERCISE 8.14.**




**EXERCISE 8.15.**

**8.17.** If you skipped Problem 13, go back and do it now.

**8.18.** The stiffness of a two-dimensional beam element is given by

$$[K]^e = \frac{EI}{L^3}\begin{bmatrix} 12 & -6L & -12 & -6L \\ -6L & 4L^2 & 6L & 2L^2 \\ -12 & 6L & 12 & 6L \\ -6L & 2L^2 & 6L & 4L^2 \end{bmatrix}$$

Use a one-element model, and the fixed-end boundary conditions at the node 1 to prove the well known results that the free end deflection $\delta$ and the rotation $\theta$ of an end-loaded cantilever beam are given by

$$\delta = \frac{PL^3}{3EI}$$

$$\theta = \frac{PL^2}{2EI}$$

**8.19.** What are the natural coordinates of the point $P$ in the diagram shown?

**8.20.** The interpolation functions satisfy a very important criterion, called the *inter-element compatibility*. By that it is meant that the displacements on an element boundary will be the same if calculated from inside another element sharing that boundary. This result is obtained by proving that the displacements of points along a boundary depends on the displacements of the nodes on that edge, and of no other edge. Prove it in the case of linear triangles, for which $N_i = L_i$.

**8.21.** Given the fact that the stiffness matrix of the triangular element 123 is given by

$$[K]^e = t(10^5)\begin{bmatrix} 25.70 & 5.36 & -23.80 & -0.41 & -1.92 & -4.95 \\ & 11.40 & 0.41 & -5.91 & 5.77 & -5.49 \\ & & 25.70 & -5.36 & -1.92 & 4.95 \\ & & & 11.40 & 5.77 & -5.49 \\ & & & & 3.85 & 0 \\ & & & & & 11.00 \end{bmatrix},$$

find the reaction at nodes 2 and 3 when node 1 is displaced in global

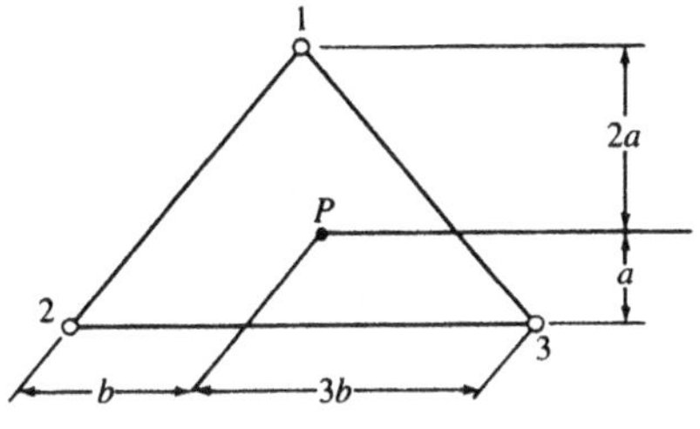


**EXERCISE 8.19.**

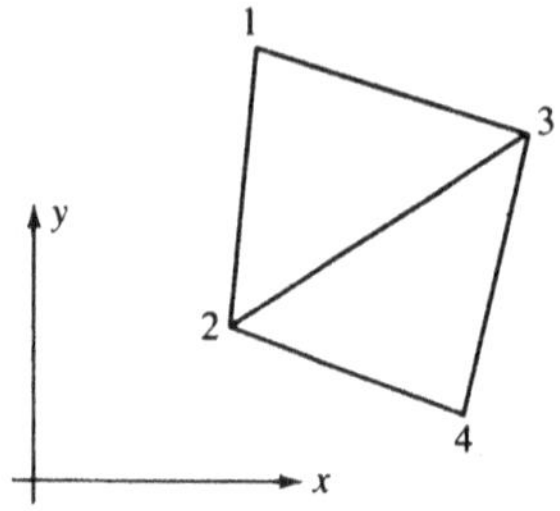


**EXERCISE 8.20.**

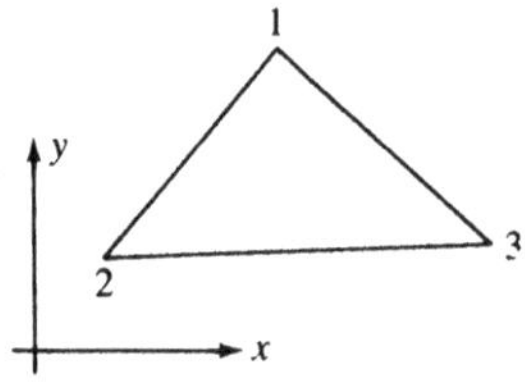


**EXERCISE 8.21.**

coordinates by (0.01**i** + 0.05**j**) units, while the other nodes are held fixed. Find also the displacement at the centroid. Take $t = 0.15''$.

**8.22.** For the previous problem, what are the forces at points 2 and 3, when the points 1, 2, and 3 are displaced by (0.05**i** + 0.07**j**), (0.04**i** + 0.05**j**), and (0.03**i** + 0.08**j**) units, respectively? Find also the displacement at the centroid.

**8.23.** For the same problem, assume that the triangle has linear distribution of displacements, and find the line(s) along which the $x$-displacement = 0.04 units.

**8.24.** Prove that in a plane triangular element, the point $P(x, y)$ inside the element, or on its boundary has coordinates given by

$$x = N_1 x_1 + N_2 x_2 + N_3 x_3$$

$$y = N_1 y_1 + N_2 y_2 + N_3 y_3.$$

(Hint: The coordinates of any point inside the triangular element are expressible as a linear combination of the nodal coordinates, such that

$$x = \lambda_1 x_1 + \lambda_2 x_2 + \lambda_3 x_3$$

$$y = \lambda_1 y_1 + \lambda_2 y_2 + \lambda_3 y_3$$

where, all $\lambda$'s are less than 1, and $\lambda_1 + \lambda_2 + \lambda_3 = 1$. Now we have three equations in the $\lambda$'s, and we can solve them preferably using the Cramer's rule. Observe that the expressions for $\lambda$'s are simply those of the $N$'s.)

**8.25.** Prove that the points $(x, y)$ given by

$$x = N_1 x_1 + N_2 x_2 + N_3 x_3$$

$$[K]^e = \begin{bmatrix} K_{11} & K_{12} & K_{13} & K_{14} & K_{15} & K_{16} \\ & K_{22} & K_{23} & K_{24} & K_{25} & K_{26} \\ & & K_{33} & K_{34} & K_{35} & K_{36} \\ & & & K_{44} & K_{45} & K_{46} \\ & & & & K_{55} & K_{56} \\ & & & & & K_{66} \end{bmatrix}$$

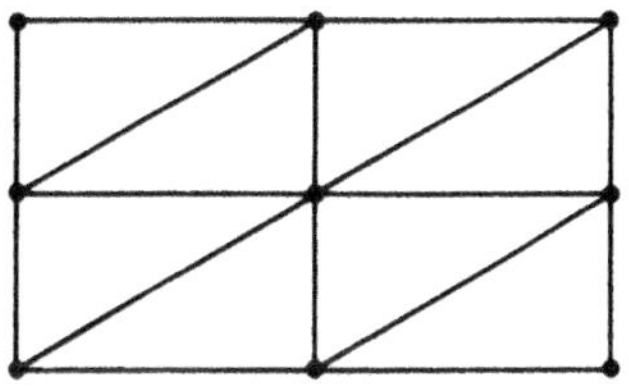

**EXERCISE 8.34.**

$$y = N_1 y_1 + N_2 y_2 + N_3 y_3$$

always lie inside the triangle. Given, $N_i = L_i$ and $(0 \leq L_i \leq 1)$.

**8.26.** For a quadratic triangle, verify that $\Sigma_{i=1}^{6} N_i = 1$. What is the physical meaning of the summation being equal to 1?

**8.27.** Verify that for a quadratic triangle, the lines parallel to the base map onto parabolas.

**8.28.** Verify that the interelement compatibility of displacements is satisfied by the quadratic triangles. (Refer to Problem 20.)

**8.29.** Verify the following for a linear quadrilateral.

(a) $\Sigma N_i = 1$.

(b) $N_i = 1$ at the $i$-th node and zero at the other nodes.

(c) Corner nodes map onto corner nodes, and straight edges map onto straight edges.

**8.30.** Verify that for a linear quadrilateral, the interpolation functions satisfy the "interelement compatibility" requirements.

**8.31.** Verify the following for a quadratic quadrilateral.

(a) $\Sigma N_i = 1$.

(b) $N_i = 1$ at the $i$-th node, and zero at all others.

(c) Corner nodes map onto corner nodes, and midside nodes map onto midside nodes.

(d) Straight edges map onto parabolic edges.

**8.32.** Verify that the quadratic interpolation functions for an eight-noded quadrilateral satisfy the interelement compatibility requirements.

**8.33.** When the elements are of identical shape and size, and have an identical $[K]^e$ matrix in the global coordinate system, then it is calculated just once and repeatedly used for the other elements. Use this idea to write the assembled global stiffness matrix for the four-element beam structure.

**8.34.** Given the $[K]^e$ matrix for the triangular element shown in the figure, assemble the global stiffness matrix of the eight-noded plate, modeled with triangular plane elements.

**8.35.** For Exercise 33, find the semibandwidth using the formula in Section 8.8. Compare it with the actual semibandwidth obtained.

**8.36.** Repeat Exercise 35 for the semibandwidth of the assembled stiffness matrix in Exercise 34.

# 9

# Guidelines for FE Analysis

## 9.1 CAPABILITIES OF A MODELING SOFTWARE

Finite element methods are applied to many different branches of engineering, including structural analysis, heat transfer, fluid flow, electromagnetics, mold analysis. Therefore the use of FEA needs to be guided by theoretical principles of the particular discipline, notwithstanding the computing power of the hardware. This chapter is a cursory review of the *use* of FEA in stress analysis.

The first step in any analysis is to model the problem. In this step, the geometry is created, meshed into finite elements, and the loads and other boundary conditions are applied. (There is, however, a step that precedes this, and that is nothing but the mental exercise of viewing the real-life problem as loads and displacements). Techniques of modeling a problem for finite element analysis have changed greatly in the late 1980s. The major factor in this change is the powerful modeling capabilities that go with the FEA software. In the absence of a modeling capability, the user will have to create the input file by building the structure, element by element. Evidently, this method is error-prone, time-consuming, and tedious. The modeling capability enables the user to think of the components in a problem in terms of its geometry rather than as finite elements. Consider, for example, the problem of finding the stress concentration around a circular hole in a square plate, Figure 9.1, subjected to a tensile load. In the absence of a modeler, one will have to build the model by defining nodes and elements, one at a time. On the contrary, modeling software lets the boundary be defined as circles and lines, and the region bounded by such curves is meshed into elements with a predefined size or

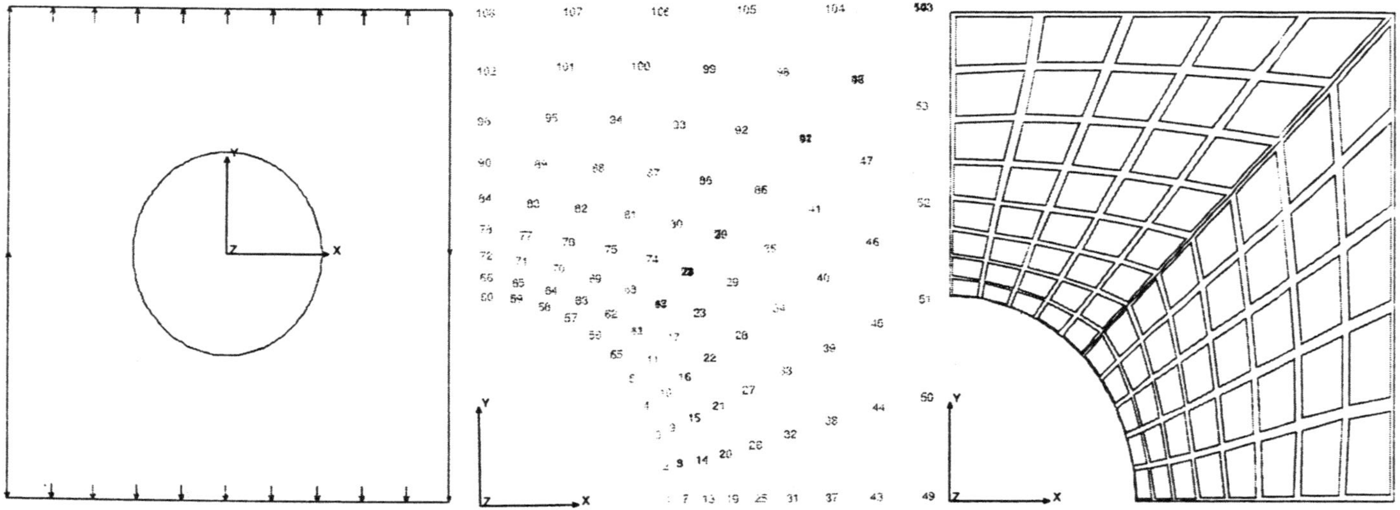

**FIGURE 9.1.** Before the modeler the modeling process involved explicit creation of each node and element. The modeler helps make this process fast and interesting.

population density distribution. Thus, there is no need to explicitly define the nodal coordinates and element connectivity. With higher complexities of the problem geometry, the benefits of a modeler become more evident. Once the geometry is defined, then the task of meshing it into finite elements is relatively an easier task. It is simply accomplished by identifying a zone and specifying the element density in some form. Also the shape of the element is to be defined, whether it is a one-dimensional element, a two-dimensional triangular element, or a quadrilateral element, etc.

Such modeling capabilities are characterized by vendors with such terms as geometric modeling, solid modeling, parametric modeling, etc. There are differences in each approach in terms of the algorithms and modeling power they lend to the user. We discuss below some of the common features that one needs in a modeling software. They are

- Creation of geometric details easily,
- Meshing a domain into elements, and
- Application of boundary conditions and submission of the problem for FE analysis.
- Accomplishment of the above steps preferably by interactive graphics.

***Modeling***

A modeling software enables the user to create the geometry in terms of the following hierarchy of *primitives*. (See Figure 9.2.)

1. Points or key points.
2. Lines or curves.
3. Surfaces.
4. Volumes.

None of the above is a finite element. Primitives can be assembled to make up an *entity*. For example, four key points in a plane can be used to define four lines, which, in turn can be used to define a rectangular surface. The surface is now an entity. In the limit, a primitive may also be an entity by itself.

A *key point* is a point in space—which may or may not belong to the final mesh—which is an important reference point in the construction of higher level entities. Examples of key points are the center of a circular arc, the location of a local coordinate system, the end points of a curve, or the intersection of two curves. In fact, any point that is used for building up the model can be termed a key point.

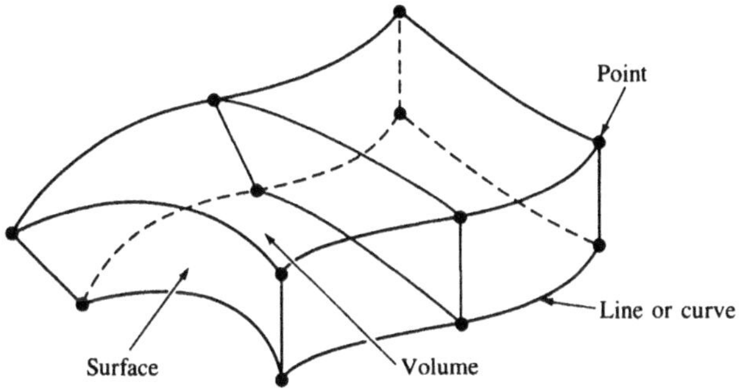


**FIGURE 9.2.** The basic primitives of modeling are unambiguous and conform to intuition.

Key points can be created by any of the following techniques:

- Mouse input—by grid-snapping, where possible.
- Typing in coordinates manually, where necessary.
- Reproduction of a predefined key point by one of the following methods.[1]
  - —FILLing between two predefined key points, with a given number of key points. The filling operation is coordinate system dependent. For example, take two points (10, 0) and (0, 10). Filling five nodes between them in the Cartesian coordinate system will give a straight line. The same operation performed in the polar coordinates will give an arc of a circle.
  - —MOVEing a key point, from one position to another position.
  - —RELOCATEing in a different coordinate system.
  - —COPYing a group of key points.
  - —REFLECTing or mirror-imaging a group of key points about a plane.
  - —GENERATEing a group of key points from another group by repeated but automatically incremented coordinates.
  - —SCALEing a group of key points, which is the same as COPYing but with a magnification factor applied to the pattern.

CURVEs are line segments created by two or more points. They may be

- straight lines,
- circles,
- parabolas, or other conic sections,
- Bezier curves and splines which are simply polynomials satisfying certain conditions,
- intersection of two surfaces.

[1] The terminology used below is technical. Similarity, if any, with the terminology used in any particular software is purely coincidental.

SURFACEs are two-dimensional entities, which are generally defined by

- defining three or four key points,
- defining three or four bounding plane curves,
- extruding a curve along a straight line,
- dragging a curve $S1$ along another curve $S2$,
- sweeping or rotating a curve about an axis.

See Figure 9.3 for a few examples.

VOLUMEs are three-dimensional chunks of space, which are generated by

- defining the bounding key points, usually eight of them,
- defining the bounding edges or curves, (assuming that the curves have well defined intersection points and that they do bound a volume),
- defining the bounding surfaces, (assuming that they properly bound a volume),

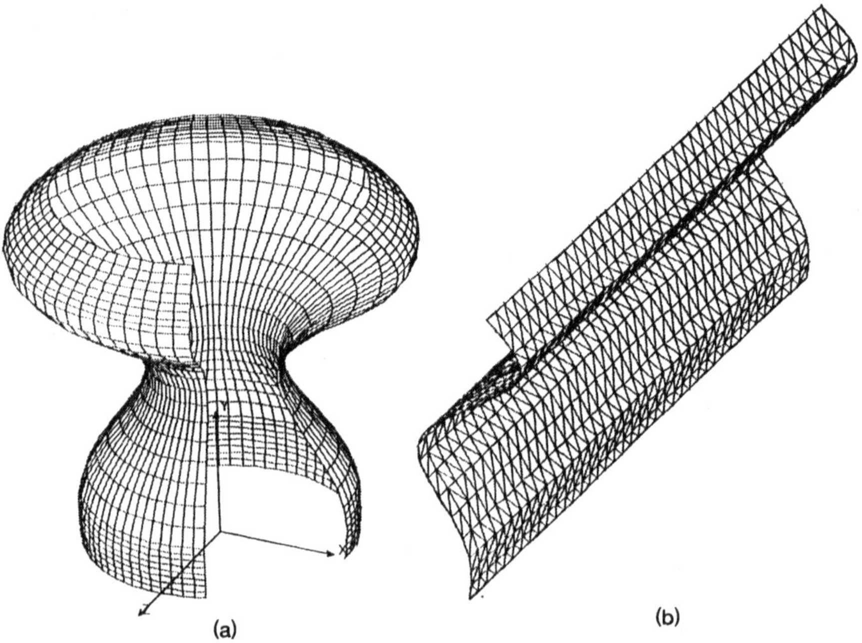


**FIGURE 9.3.** Some examples of creating a surface. Similar processes apply for volumes also.

- extruding a surface along a straight axis,
- dragging a surface along a curve,
- sweeping a surface about an axis.

All curves, surfaces and volumes can be created in any coordinate system, local or global. Every entity requires a minimum number of lower entities. For example, a straight line is a curve that will need two key points, and a circle or a parabola is a curve that requires three key points. Likewise, a volume needs at least four surfaces. Such requirements arise simply from the rules of geometry.

Any entity, be it line, surface, or volume, can be reproduced by any of the following operations, which are identical to those used for key points.

- Copy
- Move
- Relocate
- Rotate
- Reflect
- Scale (this operation shrinks or enlarges the size of a given entity by a specified factor, and places the new entity at a prescribed location and orientation)

All curves, surfaces and volumes can not only be built up, but also be broken down to more curves, surfaces and volumes. This feature may be helpful in selectively adjusting the element density at areas of interest.

Further, all of the above entities, can be subjected to *Boolean operations*. By Boolean operations, we mean "addition" of two objects, "intersection" of two objects or "negative addition" of two objects as explained below. In the examples to follow, common mathematical symbols are used. In an actual software package, these operations correspond usually to a sequence of commands, and not to just a single command.

Certain applications of finite element methods to biomechanics and plastics involve shapes which cannot be created using points, curves, surfaces and volumes. One can cite as examples the analysis of the human skull, optimal reshaping (or rehabilitation) of the spine, the analysis of an aircraft wing section, and the analysis of fancy plastic containers of cosmetics. The current practice is to input actual measurements at several locations, and to fit a curve or surface through such points. Complicated shapes like the skull are modeled using a technique called stereolithography. Actual cross sections are scanned by x-rays. These cross sections are digitized—the shapes become a part of the data base. Several cross sections are layered successively, and a Bezier surface is stretched over them in a manner that follows the contour exactly. This surface can be used for finite element meshing. Volume also can be generated in this manner. (Practicing plastics engineers will easily identify this technique being used in rapid prototyping.)

**Example 9.1 (Boolean Intersection):**

The curve of intersection of two surfaces $S1$ and $S2$ (when a well-defined one exists) can be obtained by using the operation,

$$S1 \cap S2$$

($S1 \cap S2$ denotes the set of points that belong to both $S1$ and $S2$). See Figure 9.4(a).

**Example 9.2 (Boolean Subtraction):**

A square plane figure with a hole can be obtained by supposing that it is obtained by "subtracting" a circular cut-out from the square. See Figure 9.4(b).

$$\text{Square plate} = S1$$

$$\text{A round solid bar} = S2$$

$$\text{Square plate with hole} = S1 \cap \widetilde{S2}$$

(The symbol $\widetilde{S2}$ denotes "what is not $S2$." Hence $S1 \cap \widetilde{S2}$ denotes a "subtraction" of $S2$ from $S1$).

**Example 9.3 (Boolean Addition):**

A large pipe with a small branch pipe (or nozzle) can be obtained by the following operations. See Figure 9.4(c).

$$\text{Large pipe with no holes} = S1$$

$$\text{Small pipe positioned properly} = S2$$

$$\text{The intersection curve} = (S1 \cap S2)$$

$$\text{Small pipe gets split into two surfaces} = S3, S4$$

$$\text{Large pipe gets split into two surfaces} = S5, S6$$

$$\text{Part of the small pipe to be retained} = S3$$

$$\text{Part of the large pipe to be retained} = S6$$

$$\text{Required model} = S3 \cup S6$$

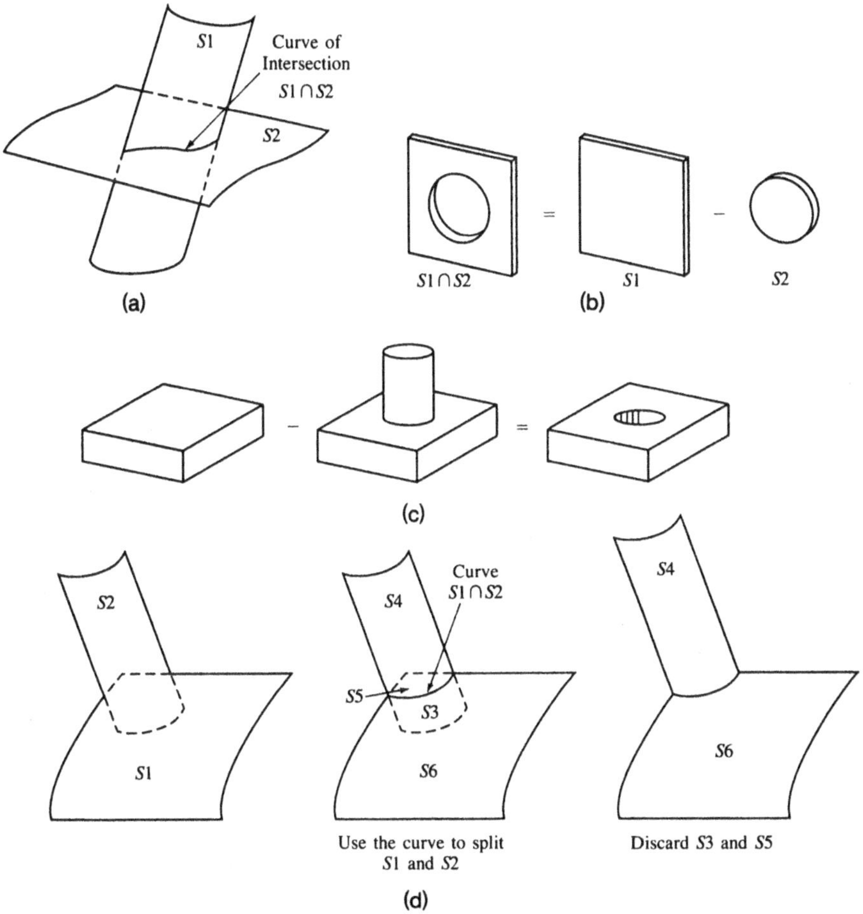


**FIGURE 9.4.** Examples of Boolean operations on geometric entities. These involve high volume computation and need strong database structure.

As before, the last step means the union, i.e., Boolean addition of $S3$ with $S6$.

Required in the process of solid modeling are several adjunct capabilities such as

- creation and display of grids, to assist in locating points,
- optional identification of key points, curves, surfaces, and volumes by labels,
- use of distinct colors for different entities and labels,

- creation and use of any number of local coordinate systems, of any type.
- blending of two surfaces with a fillet radius (actually, this is much more than an adjunct, and is also computation-intensive).
- a number of graphics-editing capabilities for points, curves, surfaces and volumes, which include but not limited to the following:
  —deleting,
  —re-creating,
  —viewing,
  —hard copy printing,
  —plotting,
  —windowing,
  —animation of displacements,
  —file-managing, such as saving work at intermediate stages,
  —restarting analysis from a previous work,
  —saving and redisplaying bit-mapped graphic images,
  —light source shading,
  —showing the start and end points of curves,
  —showing the directions of normals to the surfaces, by colors or arrows.
  —showing the directions of displacements and stresses by arrows if requested,
  —and a number of other useful features provided by the software developer.
- a capability to select or unselect a set of primitives or entities, any of its subsets, or its complements, and perform the above geometric or boolean operations. This capability needs a strong database structure for the software. Much of the speed and ease of modeling depends on this feature.

***Meshing:***
Once a curve or surface or volume is created in the modeler, the meshing is relatively easy. It is simply accomplished in three steps.

1. The first step is to identify the entity such as a curve (for meshing into one-dimensional elements), a surface (for meshing into two-dimensional elements) or a volume (for meshing into three-dimensional elements).
2. The second step is to specify the desired number of divisions, or equivalently, the number of nodes, along appropriate directions.
3. The third step is to specify the type of element desired, such as whether a triangle, quadrilateral, tetrahedron, or brick is desired.

*Adaptive mesh refinement* is a more recent feature of FEA software. In this process, the computer refines the mesh automatically depending on the results of an initial analysis. It does not involve user intervention, and several levels of refinement take place automatically until the error in the results is uniformly low over the whole model. In the absence of adaptive mesh refinement, the option available to the analyst is to zone the problem geometry into several entities, and adjust the mesh size inside such zones manually.

***Boundary Conditions:***
Boundary conditions are usually a set of displacement boundary conditions on a row of nodes, or a set of force/pressure boundary conditions along an edge or on a surface. The displacement boundary conditions may apply in a local coordinate system, which is typical of rollers and sliding supports, which may move along oblique planes or even curves. Likewise, the force/pressure boundary conditions also may have their local coordinate system.

Application of such boundary conditions is also a simple matter. One just needs to get into the proper coordinate system, and apply the pressures or loads on the appropriate key points, curves, or surfaces. That is, there is no need to work with the finite element mesh. What was applied to the solid model is automatically applied to the FE mesh.

## 9.2 DO'S AND DONT'S OF FEA

***General:***

- Avoid FEA if at all possible, if you know how to. FEA is not for the first-time analysis in the evolving design of a product. Several levels of the strength of materials approach are available for the engineer in sizing the parts. Where closed form solutions are available, they can help the engineer do a parametric study; FEA will not. Parametric studies give the real insight into the product design. Hence it is advisable that FEA be put off till the design is ready for a fine stress analysis.
- Organize and define your problem well. Get together the geometry, the materials, the material properties, geometric constants, the loads, the load cases, the constraints, and the unknowns for solution. Determine the units and dimensions at this stage.
- Review the geometry and plan an outline of how you are going to build the model, whether in the FEA pre-processor, or in a Modeler, or in a CAD. Plan also the stages in the input at which you will switch to a different material, different real constant, or different coordinate system. List all the local coordinate systems you wish to define and use. List the key points, surfaces, areas, and volumes to be generated and used.

Do this planning exercise whether you are going to use a CAD, a modeler, or a pre-processor.

- Remind yourself of the problem objective. The modeling depends on what it is. Figure 9.5 illustrates how a single product may be modeled in several ways, depending on the objective. Given below is a partial list of considerations.
  —Is the stress concentration factor to be found?
  —Are the stresses only to be found? If so, which stress? $x, y, z, 1, 2, 3$, Von Mises? In which coordinate system? Global? Local? Element coordinate system?
  —Is a through-thickness stress distribution needed?
  —Is the average stress to be found?

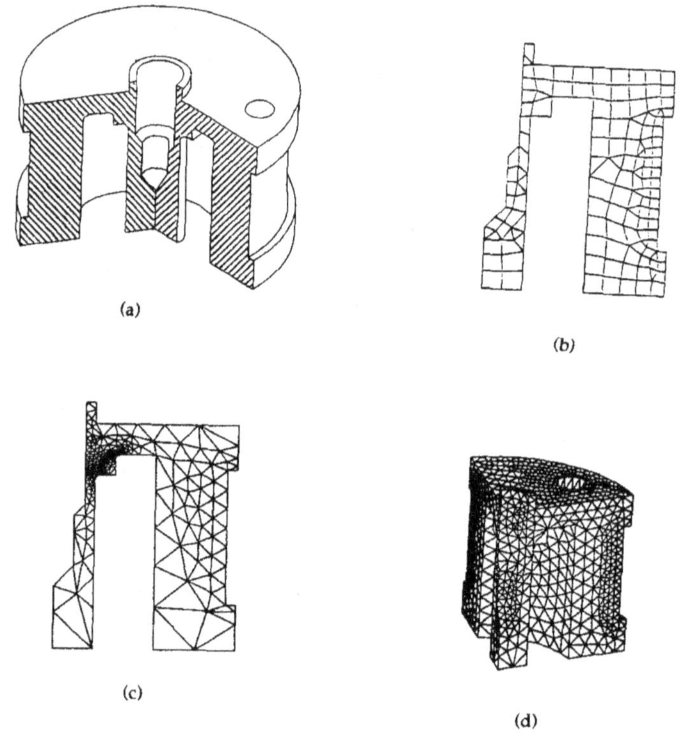


**FIGURE 9.5.** The spool in (a) may be modeled in several ways. All fillets may be ignored, as also the hole. In this case it becomes a simple axisymmetric model as in (b). Certain fillets may be included based on past experience with the product and the model would be like (c), only slightly more complicated than (b). Modeling in three dimensions as a quadrant may be justified if the hole is also to be included, in which case the model would be like (d). (*Courtesy*: *Swanson Analysis Systems*, *Houson*, *PA*.)

—Are strains needed too?
—Do you need the strain concentration factor too? (Think before you say yes. In a linear problem, this is trivially simple. In a problem with material nonlinearity, plasticity, or creep this can be quite complex. Some software packages simply cannot do this at all.)
—Do you want to find the long term effects using pseudo-elasticity?
—Is the final answer you seek, a second order effect, such as the in-plane stretching of a plate? Ordinary shell element formulations cannot handle this problem. However, there are special element formulations which have the capability to do this in a single pass. Consult your software manual.

- Visualize a problem in its smallest possible size. Can a subsystem be analyzed instead of a whole system? There is no particular merit in handling a problem in a large size. Cut down on its size commensurate with problem objectives.
- Ascertain the nature (primary or secondary) of the stress and its origin, so you will know whether the strains will grow or the stresses will relax.
- Determine the boundary conditions of the problem. It takes some practice to identify which ones are to be treated as applied displacements, and which ones as applied forces. In either case, the coordinate system in which the boundary conditions are to be applied must be identified well in advance. Be sure to determine how the forces are generated in your component: whether they are generated by some other part flexing and contacting the part being analyzed, etc. Be aware that contact forces existing between subassemblies are not easy to determine.
- Model an assembly of fastened components as individual components and replace the contacts as equivalent forces. Though some initial judgment is needed, this approach saves much woe. For fatigue analysis, however, this approach is not recommended. Because in a threaded assembly, all threads do not share the load equally. Special care is needed to calculate the real peak stress at the thread roots relevant for fatigue analysis.
- Consider the compatibility of your CAD and your FEA software. Passing data from one to the other is not always guaranteed for all kinds of problems. Furthermore, even after creating the geometry completely in a CAD software, you will always need to perform some "cleanup" tasks using the FEA software in order to complete the input for running.
- Remember that the results from two different FEA software packages need not, and will not, perfectly agree. (See Figure 9.6.) Differences may arise from the modeling, element types, element formulation, solution techniques, programming techniques, and the hardware accuracy con-

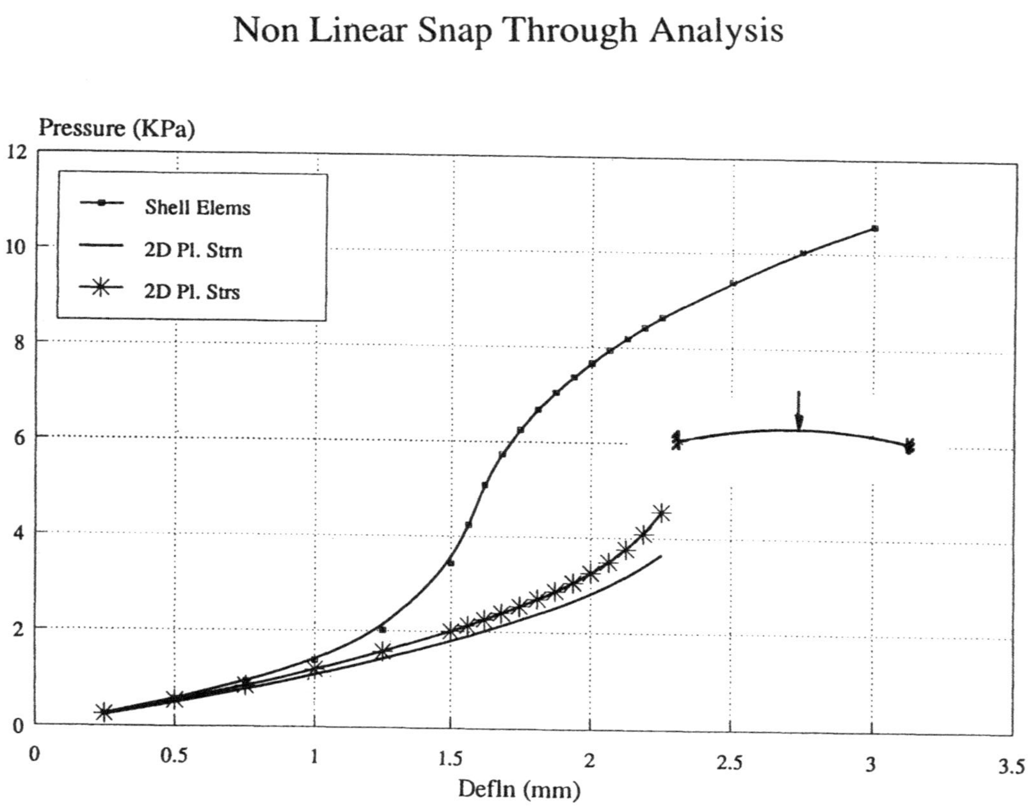


**FIGURE 9.6.** Three different modeling approaches used in a snap-through analysis predicted quite different behaviors, although the same software was used. Shell elements with and without in-plane stretching produce different results too.

straints. Reference 3 is a good source to learn about the different types of causes of errors.

- Remember that FEA solutions are just as good as the model. Reliable methodologies exist only for simple categories of problems, like the linear elastic problem. For thermoplastics the reliability suffers in other types of analyses, especially because of the void in the material data. There are no national standards for design techniques although there are standards stating performance requirements. Thus, the burden of reliable design rests heavily on the proper use of FEM and the interpretation of results.

### 9.2.1 Modeling and Post Processing

- Look into the suitability of an element type for your problem. A four-node quad element will do plane stress, plane strain, and axisymmetric problems, but not plate bending type problems involving out-of-plane moments. Plate or shell elements cannot adequately describe the strong through-thickness gradient of stresses (such as the thermal skin effects in the case of sudden cooling of a hot surface). Likewise, there are numerous situations with fiberglass designs, in which the composite shell elements offered by FEM software are not adequate. Think of how to stress analyze a repair patch of a composite pipe! The repair patch will have gradual variation in thickness, number of plies, and directions of fibers. This is quite difficult to model in many FEM software packages. (The so-called "composite wedge" element approximates to reality in stiffness, but does not approximate with respect to ply stresses, because wedge elements do not reflect the change in the number of plies).
- Study the general rules associated with the definition of your element types given in your software manual. In some old programs, quadrilateral elements may be numbered only in counterclockwise directions; and axisymmetric elements may be located only in the first and fourth quadrants, not in the second and third. In certain software, no nodes may be placed on the axis of symmetry of an axisymmetric problem. See Figure 9.7, in color insert, showing some spurious stresses that may be caused by these conditions.
- Determine if your problem involves orthotropic material properties, and if your element type has the capability for it.
- Ascertain the material property axes orientation with respect to the element coordinate system. It could complicate the problem input if the axes of orthotropy vary with location. In reinforced plastics, once again, there are many instances where continuous fibers are "blended" or

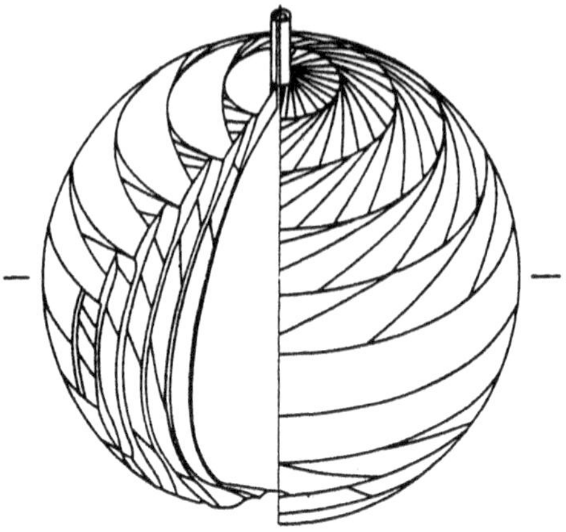

**FIGURE 9.8.** Filament wound vessels such as the above can present formidable problems of modeling. The material behavior is highly complex and may warrant the development of new element types.

"contoured" around openings. The stress concentration factors around these openings will involve elements with material axes changing in a continuous fashion! Figure 9.8 of a filament wound ball illustrates the point.

- Continuing the example considered above, of varying directions of orthotropy, it may a good idea to orient the mesh around a hole, roughly along the isostatics[2] and define material properties in the *element coordinate system*.
- Ascertain the up, down, top, and bottom sides of a plate (or shell) element. Note that the top or bottom of the element may not coincide with the real, physical, top or bottom of the problem. In fact, the modeling procedures may cause them to coincide at some places, and not to coincide at some others. Study the side effects of mirror-imaging, rotating or moving a whole group of elements to create new set of elements, which are the main source of this trouble.
- Continuing from the last point, it is advisable to study the powerful topological manipulation commands in greater detail by working out example problems. The clockwise and counterclockwise sense used in the element definition also defines its top or bottom. Would the powerful commands preserve the clockwise, counterclockwise sense of element connectivity? It is possible that the directions of load applications do not

[2]An isostatic is a line, such that the tangent at any point to the curve is along the principal stress direction.

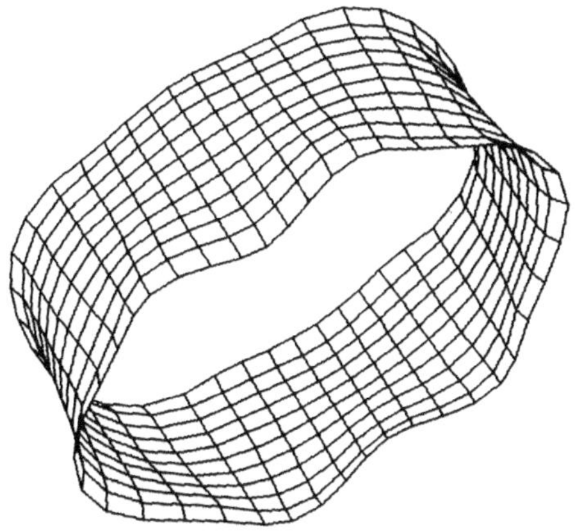

**FIGURE 9.9.** Unconvincing displacements result if the up and down sides of shell elements are taken care of. A quadrant of the cylinder was created at first, then the whole model was completed by two symmetric reflections.

correspond with the global directions you wanted. The finer software packages offer an option to reverse the clockwise-counterclockwise contortions. Use such options, if available, and as appropriate. Figure 9.9 shows unconvincing displacements in a pipe caused by pressure loads, which are supposedly internal.

- Determine the coordinate system in which the output will be written or viewed. Also determine the coordinate system for writing the output stresses, as well as for graphically displaying it. For every problem there is a certain coordinate system which is the natural one to work with, and in which intuition works best. For example, a circular plate clamped at its edges and uniformly loaded by pressure, will show symmetry, if at all, only in a polar coordinate system located at the center of the plate. See Figure 9.10 in color insert.
- List the nodes at which you need the reaction forces. In many software packages, the nodal reaction forces must be requested, and is not automatically printed. Also be sure why you want them and how you will use them. We may amplify this point. In a problem of shrink fit, distribution of nodal forces at the shrunk surface is a measure of the contact pressure distribution. In a problem of squeezing an O-ring, nodal reactions on the O-ring determine the total force needed to produce a prescribed squeeze.
- Do you have to use "master and slave" nodes? If yes, determine which one is the master node and which one is the slave node. Also a problem may involve coupling of nodes with each other—making them move together. Such needs have to be identified in advance.

- Obtain an estimate of the size of the problem, the storage space and computation time involved. Some software packages have an estimation routine for this purpose. The space used by FEA would be *greater during the solution process* than after it. The number of equations to be solved, and the bandwidth are both important factors, that determine the space and time requirements of a problem. Bandwidth is the stronger factor of the two. As a rule, triangular elements take up more space and time than quad elements; so also in three dimensions, tetrahedral elements take up considerably more space and time than do brick elements. See Reference 1.
- Activate the node renumbering routine, which is a capability almost all FEA software has. You are better off not depending on the default renumbering.
- Take care to know how the "parametric generation" of curves, surfaces, and volumes works in your FEA. Circles divided "equally" by you may not really turn out to be equal because of the special parametric algorithms. This is particularly important if you mix different techniques of node generation in the same problem and plan to merge them later on. Close examination of Figure 9.11 shows that synonymous-sounding commands can lead to different final models.
- Watch for the need to apply constraints such as roller supports, or points being allowed to slide on an inclined surface in a local coordinate system, etc.
- Consider whether you have more than one load case, and whether your FEA would let you list them as independent load cases. To qualify for

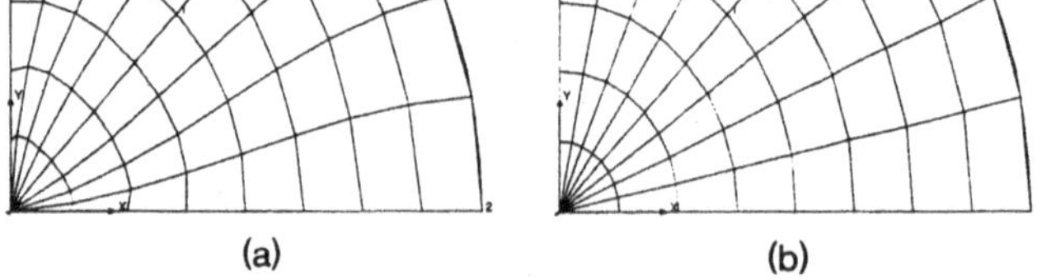


**FIGURE 9.11.** The exact nature of the mesh depends on the commands used. The different commands in a software have a shade of difference between them to avoid repetition. Note that lines are not straight in (a).

superimposition a *necessary but not sufficient* requirement is that support conditions must be identical in two different load cases. Note that a displacement boundary condition may not be treated as a load case in this sense. Also note that *any two load cases cannot always be superimposed.* Beam columns are a classic example of the case.

- Check carefully, visually *and otherwise,* the faces on which element pressures are applied. Did you get it where you wanted it? Take special care for three-dimensional models, since visualization is especially difficult in three dimensions.
- Did you know that in all FEA programs positive element pressure means compressive load?
- Remember to MERGE coincident nodes, if applicable.
- After construction of the model and before analysis, check the model carefully, specifically for the following points.
  —The geometry, from several view angles. Use dynamic/incremental rotation if available.
  —The element shapes. No element can be twisted excessively. See Figure 9.12.
  —The aspect ratio. Elements cannot be too long or too thin. The ratio of the longest distance between two nodes in an element to the shortest distance is called the aspect ratio. For a perfect square, it is $\sqrt{2}$, and for a cube it is $\sqrt{3}$. A ratio of 5 or less is generally accepted as a good number for most element types.

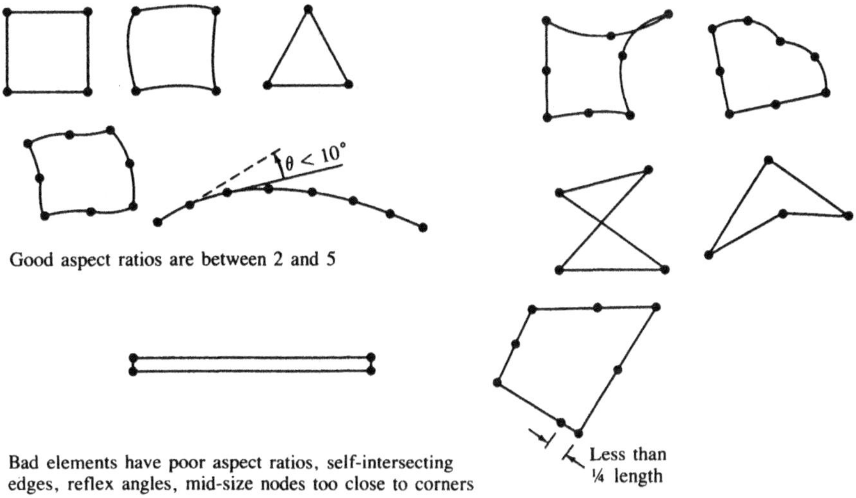


**FIGURE 9.12.** Only certain proportions of element geometry are allowable from the point of view of accuracy.

—The angles/solid angles at nodes. The angle at a node for a quad element is ideally 90°. However, it can be as low as 45° to 135°. Angles on either side of this range cause accuracy problems. Similarly, in a solid element, the angle between adjacent planes must be limited to the same range of values.
—Since four points need not lie on a plane, there is no guarantee that the four noded elements covering a curved surface, such as a sphere, will be in one plane. All software packages tolerate a moderate out-of-planeness of the element. Consult your software manual. In this connection, note that a triangular element will always be plane.
—Determine visually whether there are hollows, where an element is supposed to be. This can be found by using element coloring, element numbering, or shrinking the elements.
—Determine visually whether there is an overlap of two elements in the same space. This is more difficult to determine. There is no software that discovers such overlaps automatically, because such overlaps can be used as a legitimate trick in modeling for certain kinds of problems. Since none of the capabilities—element coloring, element numbering, element shrinking—can reveal this, you must wholly depend on the fuzzy overlap of element numbers on the screen at the same place, for all views. In any case, it will be discovered after the stress analysis, since all stress plots will show a "knot" in the stress field at that element.
—The applied forces, their locations, their directions; check their magnitudes in the text mode.
—The applied displacements, their locations, the coordinate system(s) used; check magnitudes in text mode.
—Check the top and bottom sides of plate and shell elements, if your software allows this option. Using color shading is a good idea for this.
—Determine whether there are unconnected and unrestrained nodes hanging loose in the air. (Finding them is an interesting problem in itself.) Delete them.
—Determine whether elements are connected to undefined nodes. Element plots can reveal this, since, in some software, such elements cannot be plotted at all, due to the nonexistence of the nodes. In some other software, it is possible that the undefined nodes are all placed by default at the global origin and elements connected to them are, in fact, plotted. However, such elements are easily visible to the eye, by their awkwardness.

- Consider the stress concentration in relation to whether it is "contained" or "extensive." Contained stress concentration is confined to small volumes and is surrounded by an elastic stress field. Contained stress concentration is benign from the point of view of cyclic application of loads, since it "shakes down" to elastic behavior. Living hinges are a good example of this point.

- Give special thought if you must do a shrink fit problem in FEA. One technique that works well is *not* to merge nodes at the interface, but to couple them along the normal, use fictitious thermal expansion coefficients for the outer part, and shrink it over the inner part by using a fictitious negative temperature increment. The fictitious values must be such that the product (reference length × expansion coefficient × temperature change) is equal to the interference. This technique is especially handy for those small FEA set-ups using only a linear analysis subset.
- Remember that text book formulas and analytical solutions have come a long way. It may not be wise to invest weeks of work to improve the accuracy by an amount which is less than the variance of the observed results.
- Limit the extent of writing the output to the areas of your interest, and not the entire problem.
- Plan the graphic outputs you want to view, save, and document/archive. Add clear text messages as necessary to the graphics for documentation.
- Review the output stress fields in depth. It takes some practice to interpret the whole field view of stresses. Consider stress concentrations with care. They may be called so only if the concentration decays over a short distance. If not, the component is likely to undergo gross yielding. Further, in such a situation, the possible conclusion is that the load is high enough to cause failure in one application. More views of the stress distributions may be needed to reach a conclusion.
- Consider secondary stress areas with care. Secondary stresses never occur in isolation, except by deliberate modeling. If you can isolate the primary and secondary components, then you know (roughly) which one relaxes and which one creeps.
- DON'T use a linear analysis if you have forces that depend on the deflected shape of the part (e.g., a fishing rod, or inflating a balloon).
- DON'T use FEA to answer the question: How strong is it? This question is ill-posed from the point of view of structural analysis, since it misses an important ingredient, namely the applied force. Rather, convert the question to: Is it safe under a load of ... lbs. acting at locations ...? This way, the problem is well-posed. One may argue—correctly, of course—that the strength of a structure is an inherent property of the configuration and material used, and that no more data should be required to answer the question. However, our point of view is from the technique to be used to answer the question.
- DON'T model several components all fastened together into one assem-

bly. Fastener loads are not easy to determine and control (e.g., an assembly of a container, a seal ring, and a snap-on cover).

- DON'T attempt to design a part to *fail* at a certain load. It is a high-technology problem. If you must, then the final proof is in a test, and not in analysis.
- DON'T go for fancy graphics such as
  —element coloring
  —hidden line removal
  —contouring of stresses
  —double light sources
  —shading
  —dynamic or incremental rotation
  all at one time. They take up time. Use them only for commercial presentation.
- DON'T try to avoid yield stress entirely. This is not a realistic objective. Consider the origin of the loads causing that stress and classify it as primary or secondary. Also consider whether the plastic zone is contained or extensive.
- Avoid large plasticity zones caused by primary stresses. This represents a combination of creep and plasticity. To handle such problems there are no reliable techniques for plastics. Plastics materials have not been systematically characterized for this combination of behaviors. Such behavior is extremely complex because of time dependence.

## 9.3 CURRENT DEVELOPMENTS

### 9.3.1 Singular Elements

Linear Elastic Fracture Mechanics (LEFM) involves crack tip stress fields which approach infinity as $1/\sqrt{r}$, where $r$ is the distance from the crack tip, leading to infinite stress. However, the pattern of approach to infinity is more important than is the stress magnitude (which is really finite in magnitude). The problem of determining the stress intensity factors ($K_I$, $K_{II}$, and $K_{III}$) while keeping the infinite stresses is solved by considering the displacements rather than stress. Theoretical treatments of singular elements have been available in the literature for over 15 years [4]. Commercial availability of the singular elements in personal computers started about five years ago. There are only a few software packages that offer the singular elements. The user must be aware that all materials do not behave the same way in the presence of cracks. The micromechanics of the fracture of plastics, composites, and metals are quite different from each other. Furthermore, the element can give only the stress intensity factor. It is not capable of carrying out a crack propagation study directly.

### 9.3.2 Coupled Problems

Coupled problems are those which involve entirely different engineering processes. Consider, for example, temperature field produced by work done to overcome friction. It is easy to say that work done against friction is transformed into heat. But it is very difficult to account for how much heat is produced, its distribution, how it flows in the body, and how much of it is dissipated. Such problems need a specially formulated FEM and cannot be handled by regular FEM packages. It is instructive to think that all problems in nature are really coupled problems.

### 9.3.3 Special Material Behaviors

Polymers and elastomers have prompted the formulation of special types of material behaviors. Viscoelasticity, hyperelasticity, rubber elasticity are a few types which are not generally available in FEM packages yet. Vendors of general purpose FEM packages have turned their attention to the needs of structural plastics engineering. The special material behavior types are becoming standard features in some of the packages. Hopefully, in a few years, say in the 1990s, most of the plastic behavior patterns can be handled in the FEM on personal computers.

Materials with different moduli in tension and compression need a special formulation too. Plastics and composites fall in this category. Such materials can be handled by the so-called Drucker-Prager constitutive relations. It is likely that these material types are available only in mainframe FEA software.

### 9.3.4 Moving Boundary Problems

Many ordinary phenomena in engineering are moving boundary problems. The melting of ice, the propagation of cracks, the combustion of solid fuels, the filling of a mold with a polymer melt are all examples of moving boundary problems. Since the application of an incremental load (thermal or mechanical) leads to a change in the shape of the problem domain itself, considerable computational effort is involved. The several checks involved are

- How is equilibrium satisfied in the new incremented geometry?
- How is the stress distributed for the load before increment and after?

- What will the new mesh look like for the possible shift in stress concentration areas?
- What happens to the elements that "melted" away?
- How are the new elements to be created in a mold to account for the additional sources of friction losses?

Moving boundary problems involve huge amounts of data movements and database operations. The reader with some experience of mold flow analysis programs can appreciate this point, especially when comparing running times of mold flow and stress analysis for the same model.

An objective of plastics design engineers is to quantify the molded-in stresses and to be able to superimpose them on the operating stresses. As yet, this objective is elusive. Because of the volume of computation in mold analysis, one has to compromise for low element population and the fineness of shrinkage stress concentration is lost right here. Further, the criteria to determine the element densities are different for mold flow and stress analysis.

### 9.3.5 *p*- and *h*-Versions of FEA-Adaptive Meshing

Of the many methods available to improve the accuracy of FE results, the *p*- and *h*-versions are currently very popular. The *p*-method redefines the order of the polynomial used in the element formulation. The *h*-method redefines the element size at the same order. Both may done concurrently and such an approach is called the *ph*-method. Both *p*- and *h*-methods are used to improve the user's initial model which is controlled by intuition rather than by an error estimate. In each of these methods, an *error norm* is calculated for each element. Those elements that contain errors higher than average are taken up for refinement by one of the two methods specified. Figure 9.13 (in color insert) shows a simple three-point bending specimen, with element mesh before and after *h*-refinement is shown.

The *p*- and *h*-refinements are optimal ways of accuracy improvement, since the element of judgment of stress gradient is eliminated and refinement is done only where needed, and to the degree needed. The process of redefining the mesh involves considerable computation. Existing load and displacement boundary conditions must be translated for the new mesh. Element load vectors are calculated again. The stiffness matrix is calculated again. The solution is performed all over again. Many of the recent software packages offer this feature as a standard.

**References**

1. Kelley, F. S., *Solid Modeling*, 1989, a manual published by Swanson Analysis Systems Inc., Houston, PA.

2. Meyer, C., (ed.) *Finite Element Idealization*, 1987, prepared by Task Committee on Finite Element Idealization of the Am. Society of Civil Engineers, Published by Am. Society of Civil Engineers, New York, NY.
3. Cook, R. D., Malkus, D. S., and Plesha, M. E., *Concepts and Applications of Finite Element Analysis*, 3rd Edn. 1989, John Wiley & Sons, New York, NY.
4. Tracey, D. M., "Finite Elements for Determination of Crack Tip Elastic Stress Intensity Factors," *Engineering Fracture Mechanics*, **3**, 255–65, 1971.

**Exercises**

**9.1.** The two element connections shown in the figure are not recommended. Explain why.

**9.2.** A square plate with a circular hole in the in-plane tension is modeled as a quadrant. The problem region can be divided into four-sided figures in two different ways as shown in the figure below. From the viewpoint of computational accuracy, which one would lead to a better mesh? Give reasons.

**9.3.** A lug taking up load from a snugly fitting pin was modeled as five regions as shown below. They were meshed, node-merged, and the problem was run with appropriate loads. The resulting displacement configuration is also shown below. Explain what could have gone wrong. What preventive measures would you suggest to avoid this problem?

**9.4.** The following is a node-numbering scheme for a circular ring. Assuming that the nodal degrees of freedom are two in-plane displacements, and one rotation, what is the maximum semibandwidth for the problem? Renumber the mesh to minimize the semibandwidth.

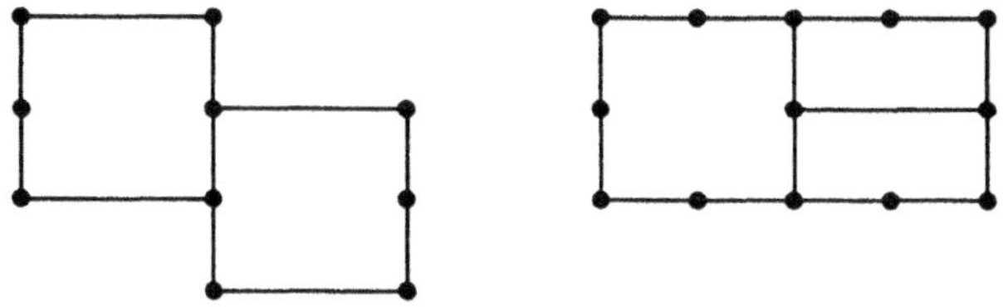

**EXERCISE 9.1.**

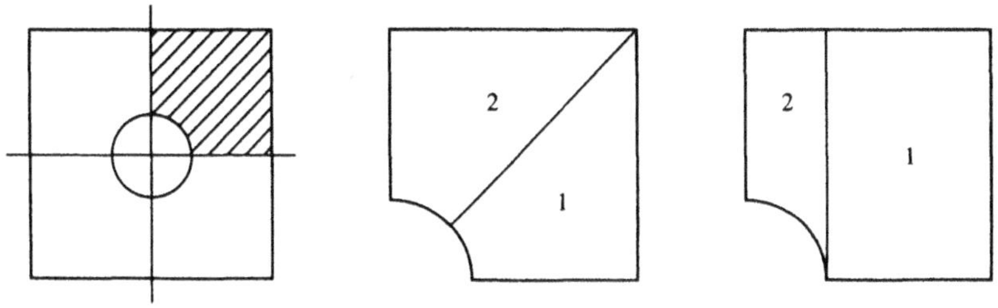


**EXERCISE 9.2.**

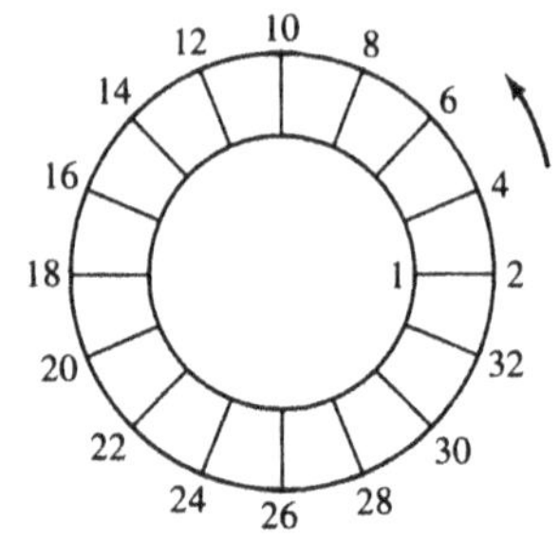


**EXERCISE 9.4.**

**9.5**. Reconstruct the steps involved in the modeling of the tire rim shown. Assume you have access to a high-ended solid modeling software package.

**9.6**. Outline the scheme of modeling the flange shown below. It is used on a straight pipeline, under pressure. No bending moments act on the pipe, and as such the flange sees only axisymmetric loads. All bolts may be assumed to be equally tight. The number of bolts is a multiple of 4. Solid elements are needed since the fillets are important for the model. If you are using symmetry boundary conditions, state them explicitly in terms of the specific degrees of freedom referred to a coordinate system located at the axis of the flange. Also explain a procedure that would use only a sector containing just one bolt hole in the model.

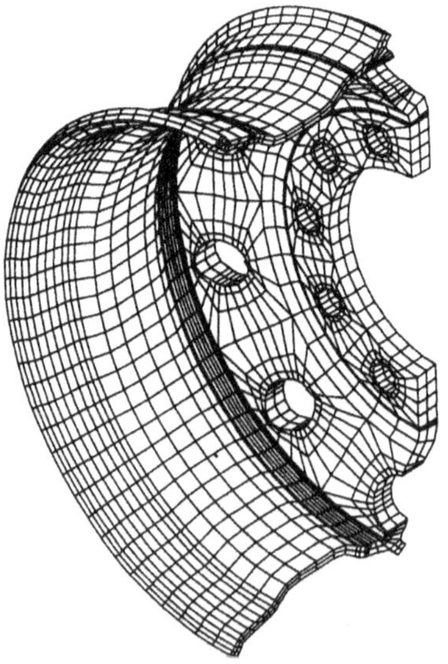

**EXERCISE 9.5.**

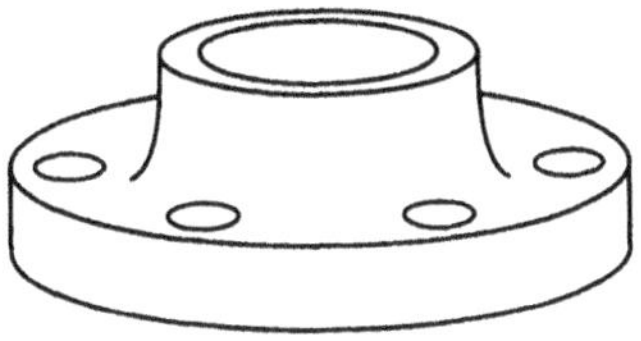

**EXERCISE 9.6.**

**9.7.** In a problem of meshing gears which are transmitting load, it is only necessary to analyze three teeth on each gear in mesh, since that is the maximum number of pairs of teeth that can possibly be in mesh. Assume that the geometry and horsepower transmitted are known. How would you model the contact forces? Remember that the contact forces at the three points are not equal for any position of the gear, and hence equal distribution of the forces would be incorrect.

**9.8.** A thick pipe with internal pressure is to be modeled using three-dimensional solid elements. The axial stress distribution in the pipe can be obtained in the following two ways.

- By holding fixed the node at the bottom of the pipe, and applying an element pressure at the top end of the pipe.

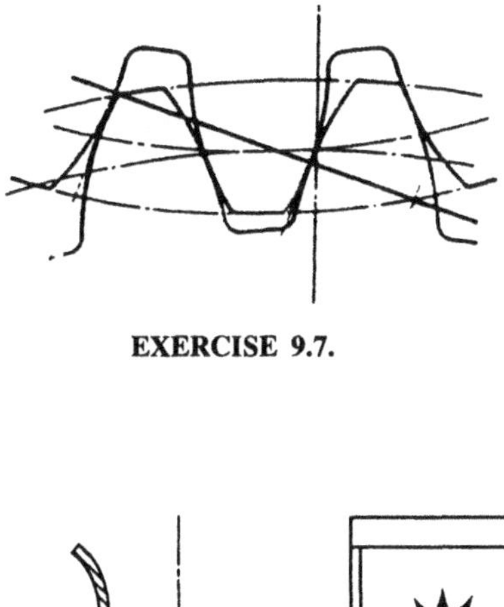

**EXERCISE 9.7.**

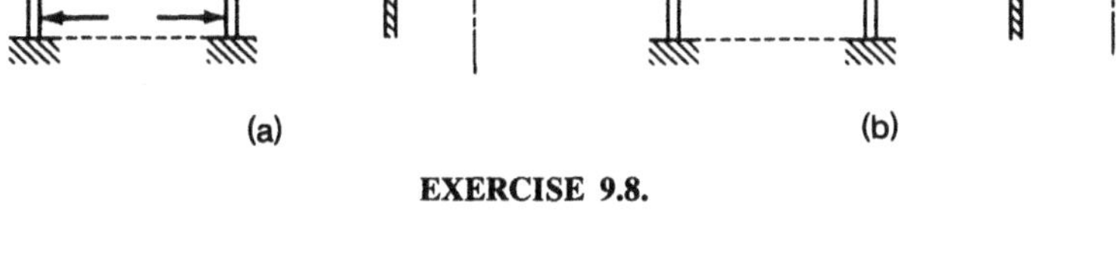


**EXERCISE 9.8.**

- By adding an extremely rigid disk-like closure at the end, holding the other end fixed, and applying the pressure all over the inside.

The first approach tends to show a flaring displacement at the top, whereas in the second approach the ends move axially. The latter case is closer to reality. Explain why a realistic looking load application in the first approach leads to an unrealistic displacement and an unrealistic closure would lead to a realistic displacement.

**9.9.** A tube of circular inner bore and a square outer profile is useful in certain industrial heat exchangers. They are normally welded with an adjacent tube at their "ears." In order to determine the effect of pressure, a single tube is taken and modeled as a quadrant. What boundary conditions are appropriate for the "ear" region?

**9.10.** The weight of a pipeline segment may be supported in a variety of ways, using clevis-hangers, rollers, spring-supported rollers, guided supports, rigid anchors, etc. Several ways are shown in the following diagrams. With respect to a coordinate system located at the center of the pipe at the support point, define the boundary conditions that each one of the supports represents.

**9.11.** In underground pipe applications, the cross section of the pipe surrounded by soil is taken as representative of the pipe. Hence a plain strain model is appropriate. In practice, the pipe is modeled as a three-dimensional beam element. For interelement compatibility, what type of solid element should be used for the soil? (Do not worry about the material model for the soil.)

**9.12.** When modeling a panel with ribs such as is shown below, two approaches are possible.

- Panel $S1$ is modeled using plates and the ribs $S2$ using beam elements with offset of neutral axis. This is by far the most popular technique used by engineers.
- Both panel $S1$ and ribs $S2$ are modeled using plate elements.

Show, that in the former procedure, interelement compatibility is not satisfied. Why then is it more popular?

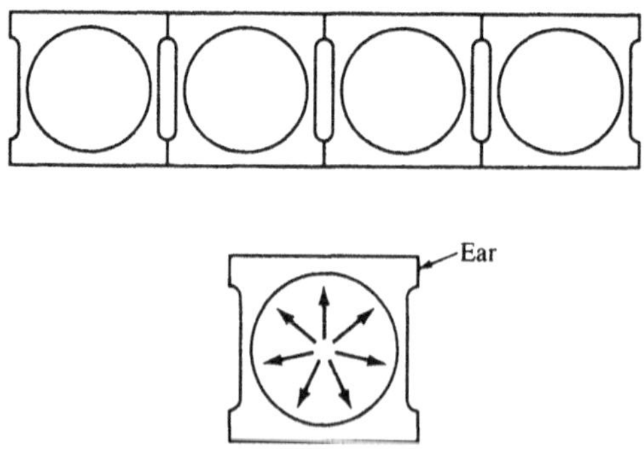


**EXERCISE 9.9.**

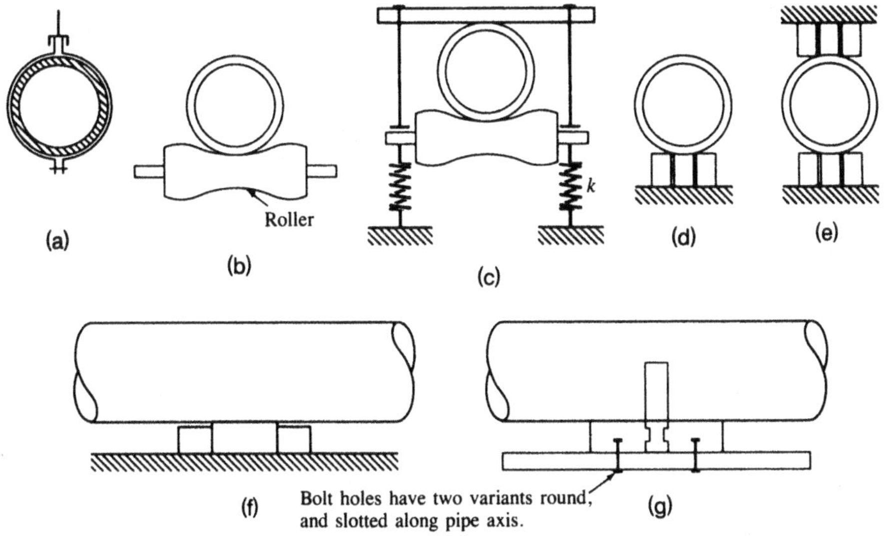


**EXERCISE 9.10.**

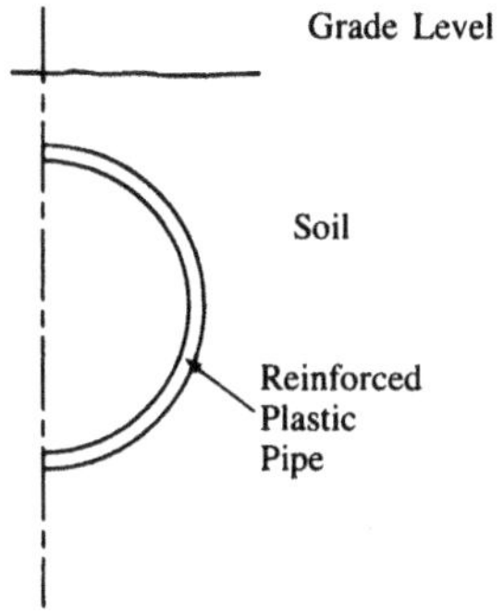


**EXERCISE 9.11.**

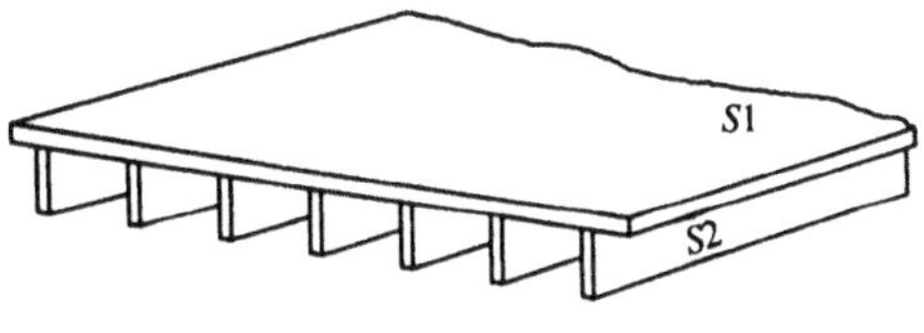


**EXERCISE 9.12.**

**9.13.** A polycarbonate tray with an array of stiffening ribs in a honeycomb-like structure is used as a paper tray in a business machine. How would you model the tray for a uniform load caused by a paper stack? Would you use a combination of plate-plus-offset beams or would you use plate elements throughout? Explain. (*No one answer is absolutely right. Just give your reasoning.*)

**9.14.** Imagine a very large beam structure, fabricated from thick metal plate 45 mm thick. It may be modeled as an assembly of thick plates as shown. Name at least one problem for which this model is (1) good, (2) not good.

**9.15.** The threads of a bolt exert loads on the threads of the mating nut in two different ways as shown. In the first figure, the load is simply an external force $F$. In the second figure, the load is applied by tightening the bolt against a spacer tube, thus forming a self-balancing force system. How can the two loading methods be modeled for finite element analysis?

**9.16.** Consider a garden hose connected to a water supply. When the faucet is opened the hose tends to whip about and stretch, thus indicating a longitudinal force in the pipe. Where does this come from? How would you calculate this force for inputting into a finite element analysis in order to simulate the axial stresses? Do you see an analog of this problem in the flow of melt in a mold?

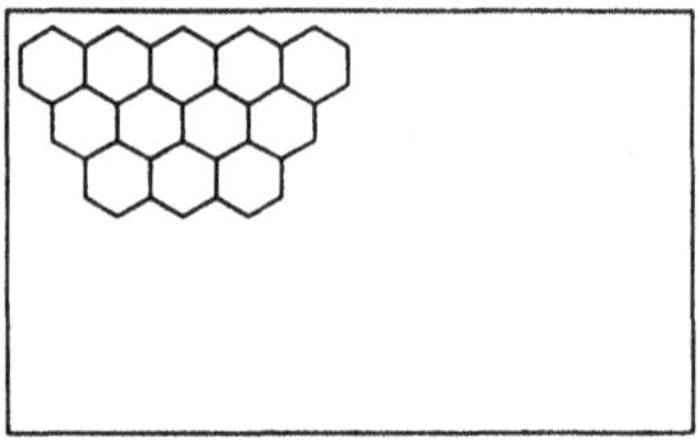

Reinforcement rib geometry at the bottom of tray.

**EXERCISE 9.13.**

**EXERCISE 9.14.**

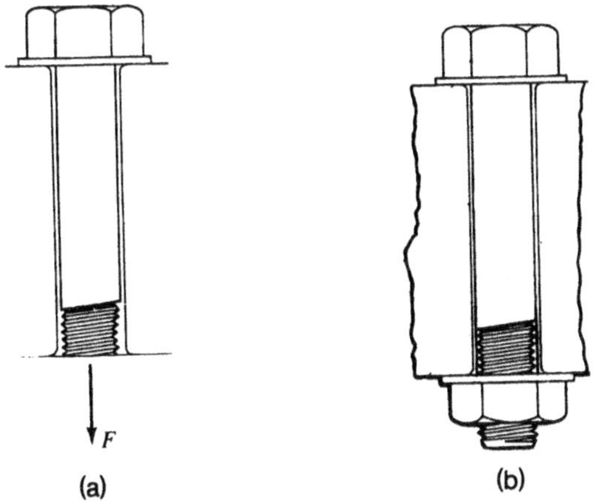


EXERCISE 9.15.

**9.17.** A rigid pipe bend is shown in the figure below. The pipe carries water flowing from an overhead tank out to the atmosphere. The pipe bend carries the force which arises from the momentum change of the water. This force is continuously distributed over the arc of the bend. Analyze the bend for the stresses due to momentum change. Assuming the flow quantity to be known, how would you model the force application?

**9.18.** A plate with uniform uniaxial applied stress was modeled using plane two-dimensional elements with equal nodal forces. According to elasticity and common sense the stress induced in the plate must be uniform and equal to the applied stress. The finite element output however does not agree with this. Explain what happened. The figure for this exercise appears in the color insert.

**9.19.** A number of equal sized rectangular blocks with a central hole are stacked side by side, and a tie rod is passed through the holes. The rod is tightened

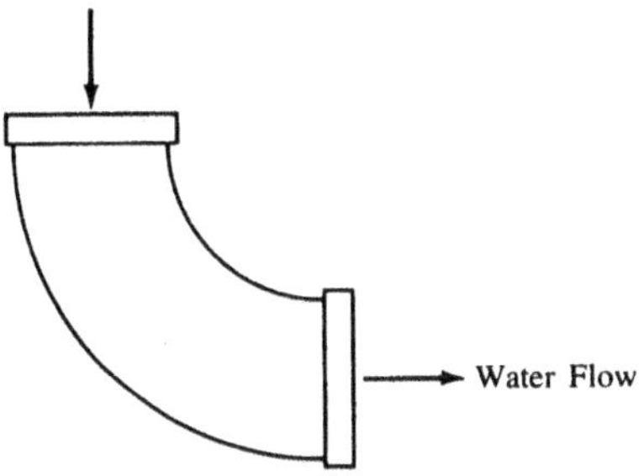


EXERCISE 9.17.

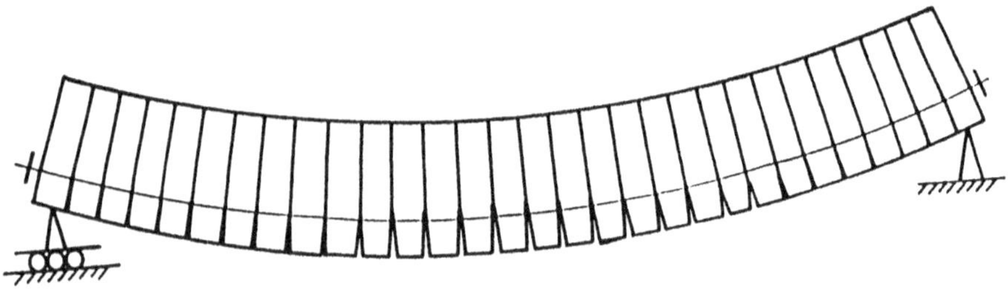

**EXERCISE 9.19.**

with nuts at the ends with a tension $T$. The assembly is supported at its ends and allowed to sag under gravity. It is common sense that as $T$ increases, the sag decreases, within certain limits of $T$. How can this behavior be coaxed in a finite element model?

**9.20.** In certain high technology applications, a very small pump is used. Its only function is to pump a certain fluid into a chamber during the first 10 minutes of operation. The life of the pump is just these 10 minutes. The impellers of the pump are made of plastic. You are investigating the use of an adhesive joint of the pump to the shaft, which rotates at about 30,000 rpm. Knowing nothing more about the product, what form of failure would you expect and where? What finite element model would achieve this end?

**9.21.** It was stated in Section 9.2.1 that the composite wedge element is unable to model the repair layups on composite pipes. Explain why? Find at least one more application where wedge elements would not do the job.

# Appendix 1

# Cartesian Tensor Analysis

## A1.1 ARE TENSORS FOR REAL?

Tensor analysis and calculus is a branch of mathematics that deals with coordinate transformation, and the appearance of the laws of physics in various coordinate systems. This appendix is a cursory review of the basics of tensor analysis. Additional reading is necessary for a fuller understanding. References listed at the end are only moderately complex since they discuss the application of tensors to elasticity theory.

Tensors are physically real quantities. They are a class of quantities which are more general than vectors. The fact that stress and strain are tensor quantities and not vectors gave rise to the name. The word *tensor* means *tension*.

Tensors use a notation that is intimidating to some. It is a system of subscript notation generalized to a great extent, so that a high degree of brevity is achieved. However, brevity is *not* its basic purpose. It is a way of transforming a set of quantities from one coordinate system to another. The expressibility of physical laws in tensor form is a guarantee of their coordinate independence and absoluteness. Fundamental laws of nature such as the laws of electromagnetism, conservation of energy, conservation of mass, etc. are expressible in tensor form. So is elasticity theory, which verifies its absoluteness. That is, elasticity theory is valid on earth and in space as well.

## A1.2 THE SUBSCRIPT NOTATION

Instead on naming the coordinate axes as $x$, $y$, and $z$, we just label them "first," "second," and "third." The coordinates are denoted by $x_1$, $x_2$, and

$x_3$, which may stand for Cartesian $x$, $y$, $z$ or cylindrical $r$, $\theta$, and $z$, or any curvilinear $u$, $v$, and $w$. In this appendix, we stick to the Cartesian coordinates of different orientations and positions. Thus, coordinates are denoted by

$$x_i, \qquad (i = 1, 2, \text{and } 3)$$

A *vector* $\boldsymbol{A}$ is only a first order tensor and needs just one subscript for its complete description. It has three components, $A_1$, $A_2$, and $A_3$. Also, the *unit vector* along the $i$-th axis is represented by $n_i$. Combining these two ideas, the vector itself can be represented by

$$\begin{aligned} \boldsymbol{A} &= A_1 n_1 + A_2 n_2 + A_3 n_3 \\ &= \sum_{i=1}^{3} A_i n_i \end{aligned}$$

The rules of addition and subtraction of vectors, namely adding or subtracting corresponding components, applies to tensors also, regardless of its order.

### A1.2.1 The Summation Convention

The *summation convention* adds to the brevity. Consider the dot product of two vectors $\boldsymbol{A} = A_1 n_1 + A_2 n_2 + A_3 n_3$ and $\boldsymbol{B} = B_1 n_1 + B_2 n_2 + B_3 n_3$. The dot product is given by

$$\boldsymbol{A} \cdot \boldsymbol{B} = \sum_{i=1}^{3} A_i B_i$$

We note that the subscript $i$ is repeated on the right side, denoting that like components are multiplied and added. Naturally, the repeated subscript is the one on which the summation is performed. The summation convention recognizes this and lets the summation sign be discarded.

When a subscript is repeated, it is understood that a summation is performed over that subscript. Thus,

$$A_i B_i = \sum_{i=1}^{3} A_i B_i = A_1 B_1 + A_2 B_2 + A_3 B_3$$

The repeated subscript, known as the *dummy subscript* disappears when the summation is performed. Any other symbol is just the same. Hence the name "dummy."

Matrices are two-dimensional arrays, and can be represented by two subscripts, the first one for the row number, and the second for the column. The matrix itself may represent a tensor or not. The rule of matrix multiplication is represented by

$$C_{ij} = \sum_k A_{ik} B_{kj}$$

Using the summation convention, this can be written as

$$C_{ij} = A_{ik} B_{kj}$$

The multiplication of a matrix by a column vector is represented as

$$C_i = A_{ij} B_j$$

### A1.2.2 The Kronecker Delta

An important tensor is the Kronecker Delta or the Unit Tensor. It is defined for any order greater than or equal to two. It is a tensor whose magnitude is equal to unity when two of its subscripts are the same. For the second order,

$$\begin{aligned} \delta_{ij} &= 1 \quad \text{if } i = j, \\ &= 0 \quad \text{otherwise.} \end{aligned} \tag{A1.1}$$

In the second (the lowest) order $\delta_{ij}$ is simply the unit matrix.

### A1.2.3 The Permutation Symbol

The permutation symbol of three subscripts is defined as follows.

$$e_{ijk} = \begin{cases} 1 & \text{if } ijk \text{ are even permutations of } 1, 2, 3 \\ -1 & \text{if } ijk \text{ are odd permutations of } 1, 2, 3 \\ 0 & \text{otherwise} \end{cases}$$

The permutation symbol has an alternating sign and is convenient in calculating the determinant of a matrix, or the curl of a vector. This symbol is used wherever there is rotation.

$$\mathrm{Det}[A] = \begin{cases} A_{11}A_{22}A_{33} + A_{12}A_{23}A_{31} + A_{13}A_{21}A_{32} \\ -A_{11}A_{23}A_{32} - A_{12}A_{21}A_{33} - A_{13}A_{22}A_{31} \end{cases}$$

The same equation can be expressed in subscript notation as,

$$\mathrm{Det}[A] = e_{ijk}A_{1i}A_{2j}A_{3k} = e_{ijk}e_{lmn}A_{il}A_{jm}A_{kn}$$

### A1.2.4 The Formal Definition of a Tensor

The important characteristic that identifies and defines a tensor is the rule it obeys under transformation of coordinates. Coordinates transform according to the rule

$$x'_i = L_{ij}x_j, \qquad \text{where } L_{ij} \text{ is the matrix of the direction cosines.}$$

The matrix $L_{ij}$ is central to the tensor definition.

If $A'_j$ are the components of a first order tensor in the $x'$ coordinates, and $A_i$ are its components in the $x$ coordinate system, then

$$A'_j = L_{ij}A_i \tag{A1.2}$$

This rule, identical to the vector transformation, does not tell much about tensors in particular. Consider the second and higher order transformation rules. For second order tensors (which stress and strain are),

$$A'_{ij} = L_{ik}L_{jl}A_{kl} \tag{A1.3}$$

$$B'_{ijk} = L_{iu}L_{jv}L_{kw}B_{uvw} \tag{A1.4}$$

$$T'_{ijkl} = L_{iu}L_{jv}L_{kw}L_{lx}T_{uvwx} \text{ and so on.} \tag{A1.5}$$

The second and higher order transformations really *define* the tensor. The rule for the second order tensor transformation is identical to what we learned as the Mohr's circle. Higher orders are generalizations.

### A1.2.5 The Stress Strain, and Elastic Moduli Tensors—The Quotient Law

From Chapter 1, we have a definition of stress as the density of force distribution $T_j$ on a small are $dA_i$. The force distribution vector $T_j$ is referred to sometimes as the *traction vector*. The definition of stress given in Chapter 1 in terms of traction vectors may be written in tensor notation as

$$\sigma_{ij} = \frac{\partial T_j}{\partial A_i}.$$

The fact that $\sigma_{ij} = \sigma_{ji}$ is a consequence of moment equilibrium of stress at a vanishingly small volume.

Strain as defined in Chapter 1 is not a tensor. However, the mathematical definition of strain is equivalent to halving the shear strains. The displacements $u$, $v$, and $w$ may be denoted as $u_i$, $(i = 1, 2, 3)$. In tensor notation, the definition of strain may be written in terms of displacements as

$$\epsilon_{ij} = \frac{1}{2}(u_{i,j} + u_{j,i}).$$

When $i = j$, the engineering and mathematical definitions of normal strains coincide, but not for shear strains. Also, the symmetry of strains is seen straightaway, since interchange of $i$ and $j$ leaves the definition unchanged.

Hooke's laws for isotropic materials can be expressed in tensor notation in terms of $G$ and $\lambda$ (usually called *Lame's constants*).

$$\sigma_{ij} = 2G\epsilon_{ij} + \lambda\epsilon_{kk}.$$

This is identical to the six Equations 1.23 and 1.24.

Stress and strains are tensors, because the rule of their transformation conforms to that of tensors. Initially, knowing nothing about the way they are related, we may assume the strain $\epsilon_{kl}$ and stress $\sigma_{ij}$ are linearly related. Hence, we may write the *generalized* Hooke's law in the form

$$\sigma_{ij} = E_{ijkl}\epsilon_{kl}$$

where the elastic moduli $E_{ijkl}$ is a fourth order tensor, which *contracts* to second order by multiplication and summation on the right side. There is no a priori reason to know that elastic moduli are tensors. The form of the equation relating them is the only clue. Guessing the higher order of the tensor from the form of the physical rules as above is called the *quotient rule*.

The four subscripts in three dimensions imply $3^4 = 81$ components of moduli. However, for three-dimensional orthotropic composites (which is the most general form for practical engineering purposes), the number of independent elastic constants reduces to only 9. As mentioned in Chapter 7, other physical principles such as symmetry, equivalence of left- and right-handed coordinate systems, etc. need to be exercised. Reference 2 is recommended in this context.

### A1.2.6 Differentiation and Equilibrium Equations

Partial differential coefficients with respect to a coordinate is represented by a comma, followed by the coordinate index. Thus

$$\frac{\partial \sigma_{ij}}{\partial x_k} = \sigma_{ij,k}$$

However, if $k = i$ or $j$, then a summation is indicated, in addition. Therefore,

$$\frac{\partial \sigma_{ij}}{\partial x_i} = \sigma_{ij,i} = \frac{\partial \sigma_{1j}}{\partial x_1} + \frac{\partial \sigma_{2j}}{\partial x_2} + \frac{\partial \sigma_{3j}}{\partial x_3}$$

Utilizing the notation, the equilibrium equations of Chapter 1 can be summarized by a single statement.

$$\sigma_{ij,j} + \rho g_i = 0, \qquad (i = 1,2,3)$$

Furthermore, the identities listed in Table A1.1 may be verified.

## A1.3 WHAT ARE CARTESIAN AND CURVILINEAR TENSORS?

Many text books in tensor analysis use the two terms "Cartesian" and "curvilinear" tensors. Many times, some knowledge of curvilinear coordinates and vector calculus in such coordinates is assumed.

What we discussed above are "Cartesian" tensors. The matrices $L_{ij}$ are the direction cosine matrices for transformation between Cartesian coordinates. If instead, we had a cylindrical or spherical coordinate system or any other system as our destination coordinate system, then the transformation matrix would change from point to point.

It is easy to expect that the local tangents to the *coordinate curves* will form the new coordinate axis. An equally valid system is the set of local *normals* to the curves. These two systems are not identical. Tensors in these coordinate systems will have to be distinguished by whether they are referred by the tangent coordinate system or the normal coordinate system. In the tensor analysis terminology, these are termed the *co-variant* (for the tangent coordinate system) and *contra-variant* (for the normal coordinate system) tensors. Very generally speaking, tensor analysis allows, *mixed* tensors also. Thus, a few components of a higher order tensor can be co-variant, and others contra-variant. The co-variant components are denoted as subscripts, and contra-variant components by superscripts.

As long as the transformation is done between Cartesian coordinate systems, there is no distinction needed. Co- and contra-variant tensors

**Table A1.1. Tensor Notation for Common Vector and Matrix Operations.**

| Quantities and Operations | | Tensor Notation |
|---|---|---|
| Scalar | : | $\phi$ |
| Vector | : | $A_i$ |
| Second Order Tensor | : | $A_{ij}$ |
| Matrix and Its Transpose | : | $A_{ij}$ and $A_{ji}$ |
| First Invariant of a Matrix | : | $A_{ii}$ |
| Determinant of a Matrix | : | $e_{ijk}e_{lmn}A_{il}A_{jm}A_{kn}$ |
| Dot Product of Vectors | : | $A_iB_i$ |
| Cross Product of Vectors | : | $e_{ijk}A_iB_j$ |
| Gradient of a Scalar | : | $\phi_{,i}$ |
| Divergence of a Vector | : | $A_{i,j}$ |
| Curl of a Vector | : | $e_{ijk}A_{k,j}$ |
| Gradient of a Vector | : | $A_{i,j}$ |
| Laplace Operator | : | $\phi_{,ii}$ |
| Biharmonic | : | $\phi_{,iijj}$ |
| Gauss Theorem | : | $\int_V A_{jklmn\cdots j} dV = \int_s A_{jklmn\cdots} n_i dS$ |
| General Gauss Theorem | : | $\int_V \nabla * \mathbf{A} dV = \int_s \mathbf{n} * \mathbf{A} dS$ where * denotes div, curl, or tensor product. |

represent one and the same set of components, and hence only subscripts are used throughout. Rigorous validation of the absoluteness of elasticity theory requires the use of curvilinear tensors.

**References**

1. Chou, P. C. and Pagano, N. J., *Elasticity, Tensor, Dyadic and Engineering Approaches*, 1967, D. Van Nostrand Company, Inc., Princeton, N.J.
2. Agarwal, B. D. and Broutman, L. J., *Analysis and Performance of Fiber Composites*, 2nd Edn. 1990, John Wiley & Sons, New York, NY.

# Appendix 2

# Methods in Beam Theory

Formulas for the deflection and stress in (statically determinate) beams are available as appendices and tables in several text books. See Reference 1 for the basic ones and Reference 2 for more complex formulas. The purpose of this appendix is to illustrate the use of the so-called *superposition method* for beams. The salient parameters needed in a beam problem are the bending moment ($M$), shear force ($V$), deflection ($\delta$), and slope of the deflected shape ($\theta$). All of these are related to distance $x$ along the axis of the beam by linear differential equations.

> The variables $M$, $V$, $\delta$, $\theta$ and consequently the bending stress $\sigma$ are all linearly additive. Thus, if two loads cause deflections $\delta_1$ and $\delta_2$, acting separately, then they cause a deflection ($\delta_1 + \delta_2$) acting concurrently. The only requirement is that the support conditions must remain same.

***Nomenclature***

| | |
|---|---|
| $I$: | Moment of area of cross section about its neutral axis |
| $z$: | Modulus of cross section $= I/y$ |
| $E$: | Young's modulus of the material |
| $M$: | Bending moment at a point |
| $R$: | Support reaction |
| $\sigma$: | Bending stress |
| $\delta, \theta$: | Deflection and slope of the beam |

**Example A2.1**

Find the deflection at the center of a simply supported beam loaded on only one half of its span.

**Table A2.1. Formulas for Bending of Beams.**

| Case | Loading | |
|---|---|---|
| 1 | 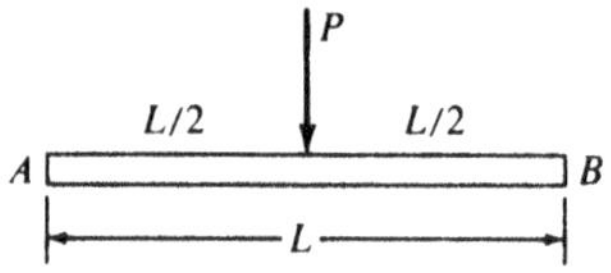 | $M_{max} = M \text{ at center} = \frac{PL}{2}$<br>$R_A = R_B = \frac{PL}{2}$<br>$\sigma_{max} = \frac{M_{max}}{z}$<br>$\delta_{max} - \delta \text{ at } A = \frac{PL^3}{48EI}$<br>$\theta_A = \theta_B = \frac{PL^2}{16EI}$ |
| 2 | 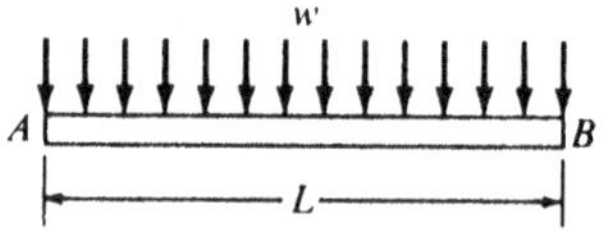 | $M_{max} = M \text{ at center} = \frac{wL^2}{8}$<br>$R_A = R_B = \frac{wL}{2}$<br>$\sigma_{max} = \frac{M_{max}}{z}$<br>$\delta_{max} = \delta \text{ at center} = \frac{5}{384}\frac{Wl^4}{EI}$<br>$\theta_A = \theta_B = \frac{wL^3}{24EI}$ |
| 3 | 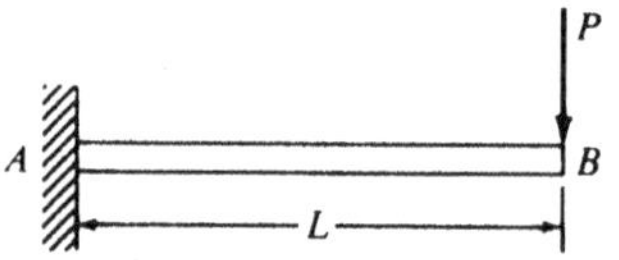 | $M_{max} = M_A = PL$<br>$R_A = P$<br>$\sigma_{max} = \frac{NM_{max}}{z}$<br>$\delta_{max} = \delta_B = \frac{PL^3}{3EI}$<br>$\theta_B = \frac{PL^2}{2EI}$ |
| 4 | 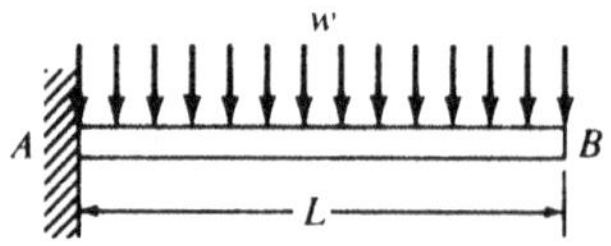 | $M_{max} = M_A = \frac{wL^2}{2}$<br>$R_A = wL$<br>$\sigma_{max} = \frac{M_{max}}{z}$<br>$\delta_{max} = \delta_B = \frac{wL^4}{8EI}$<br>$\theta_{max} = \theta_B = \frac{wL^3}{6EI}$ |

**Table A2.1.** (Continued)

| Case | Loading | |
|---|---|---|
| 5 | 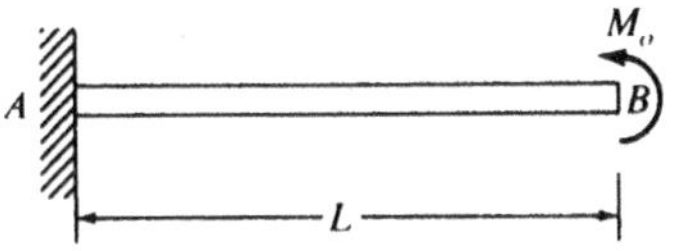 | $M$ is constant all over the length<br>$\sigma = \frac{M}{z}$<br>$\delta_{max} = \delta_B = \frac{M_o L^2}{2EI}$<br>$\theta_{max} = \theta_B = \frac{M_o L}{EI}$ |

*Solution* The diagrams show the logic involved.

$$\delta_c = \frac{1}{2}\,\frac{5}{384}\,\frac{wL^4}{EI} = \frac{5}{768}\,\frac{wL^4}{EI}$$

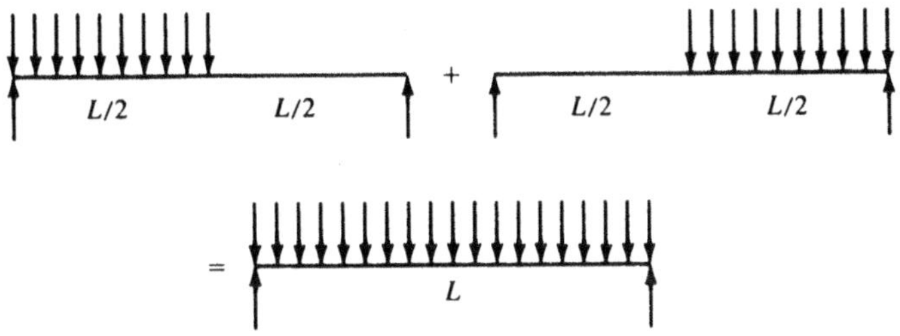


$$\sigma_c + \sigma_c = \frac{5}{384}\,\frac{wL^4}{EI}$$

**Example A2.2**

A support is added at the midspan of a uniformly loaded beam. Find the reaction at the middle support.

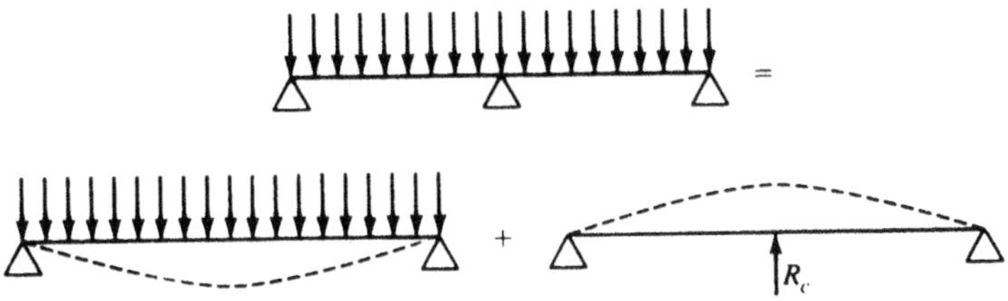


$$\delta_c = \delta_{c_1} + \delta_{c_2}$$

$$0 = \frac{5}{384}\,\frac{wL^4}{EI} + \left(\frac{-R_c L^3}{48\,EI}\right)$$

*Solution* Solving $R = \frac{5}{8}\,wL$. The support reactions at the ends are one half of the remaining load.

**Example A2.3**

Find the reaction of the prop placed at the end of a cantilevered beam with uniform load, $w$.

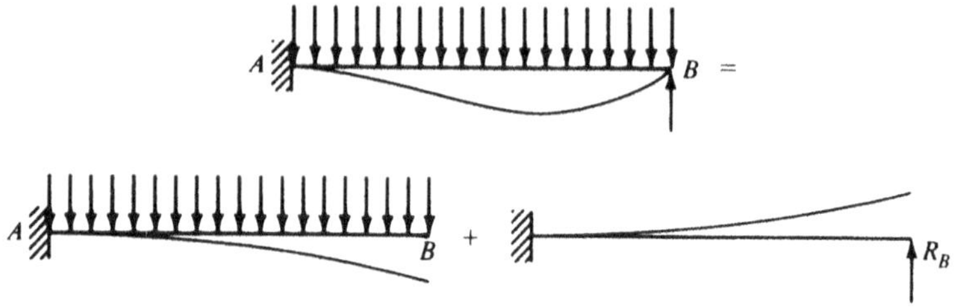


$$\delta_B = \delta_{B1} + \delta_{B2}$$

$$0 = \frac{wL^4}{EI} + \frac{-R_BL^3}{3EI}$$

*Solution* Solving, $R_B = \frac{3}{8} wL$.

**Example A2.4**

Find the end reactions and moments in a fully built-in beam, uniformly loaded.

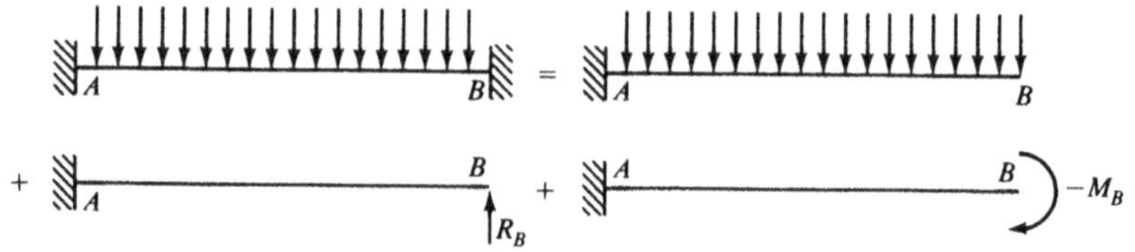


$$\delta_B = \frac{wL^4}{8EI} - \frac{R_BL^3}{3EI} - \frac{M_BL^2}{2EI} = 0$$

$$\theta_B = \frac{wL^3}{6EI} - \frac{R_BL^2}{2EI} - \frac{M_BL}{EI} = 0$$

*Solution* We treat this problem as a cantilever with additional reaction $R_B$ and a moment $M_B$ at the free end of a cantilever. By symmetry, we note that the reaction $R_B$ should be $(wL/2)$. Solving, $M_B = -wL^2/12$.

**Example A2.5**

Find the central deflection of a simply supported beam with uniform load, without reading it off the table.

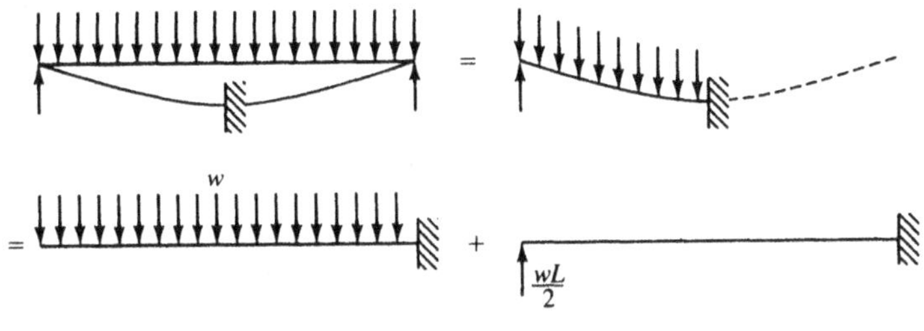


$$\delta = \frac{w(L/2)^4}{8EI} - \frac{(wL/2)\,L^3}{3EI}$$

*Solution* Based on the above idealization,

$$\delta = \frac{w}{8EI}\left(\frac{L}{2}\right)^4 - \frac{\frac{wL}{2}\left(\frac{L}{2}\right)^3}{3EI}$$

$$= \frac{wL^4}{EI}\left(\frac{1}{128} - \frac{1}{48}\right)$$

$$= -\frac{5}{384}\frac{wL^4}{EI}$$

With respect to the support point, the sign of the deflection gets reversed.

In all the above examples, the moments can be calculated by simple statics using the redundant forces (moments).

**References**

1. Gere, J. M. and Timoshenko, S. P., *Mechanics of Materials*, 2nd Edn., 1984, PWS Engineering, Boston, MA.
2. Young, W. C., *Roark's Formulas for Stress and Strain*, 6th Edn., 1989, McGraw Hill Book Company, New York, NY.

# Appendix 3

# Laplace Transforms

## A3.1 DEFINITION

The Laplace transform is a mathematical technique for transforming differential equations into algebraic equations. Usually, time dependent differential equations are transformed by this process into algebraic equations in a dummy variable $s$, whose physical meaning is undefined. As such it is a very useful tool for solving problems involving rate processes.

Let $f(t)$ be a given function for all $t \geq 0$. Its Laplace transform $\mathscr{L}[f(t)]$, denoted by $\bar{f}(s)$, is defined as,

$$\mathscr{L}[f(t)] = \bar{f}(s) = \int_0^{\infty} e^{-st} f(t)\, dt, \qquad \text{if } \bar{f}(s) \text{ exists.}$$

Conversely, if $\mathscr{L}[f(t)] = \bar{f}(s)$, then the inverse transform restores $f(t)$. It is written as $\mathscr{L}^{-1}[\bar{f}(s)] = f(t)$. The technique of inverse transformation of $\bar{f}(s)$ is simply to use a few properties of Laplace transforms, reduce it to well known standard forms, and then to look up a table of transforms. This procedure is the engineer's approach, however, and not the mathematician's.

## A3.2 TRANSFORMS OF ELEMENTARY FUNCTIONS

Taking transforms is simply an exercise in integral calculus. A few examples are listed below. Many others are given in Table A3.1 at the end of this appendix.

Transform of $f(t) = 1$.
Let $f(t) = 1$. $\mathscr{L}[1] = \int_0^\infty e^{-st} 1\, dt = 1/s$. Thus,

$$\mathscr{L}[1] = \frac{1}{s}$$

$$\mathscr{L}^{-1}\left[\frac{1}{s}\right] = 1.$$

Transform of $e^{at}$.

$$\mathscr{L}[f(t)] = \int_0^\infty e^{-st} e^{at}\, dt = \int_0^\infty e^{-(s-a)t}\, dt = \frac{1}{(a-s)}$$

Thus,

$$\mathscr{L}[e^{at}] = \frac{1}{(a-s)}$$

$$\mathscr{L}^{-1}\left[\frac{1}{(a-s)}\right] = e^{at}$$

Transform of $\cos \omega t$.

$$\mathscr{L}[\cos \omega t] = \int_0^\infty e^{-st} \cos \omega t\, dt = \frac{s}{s^2 + \omega^2}$$

Another way is to treat $\mathscr{L}[\cos \omega t]$ as the real part of the integral $\int_0^\infty e^{-st} e^{(iwt)}\, dt$.

$$\mathscr{L}[\cos \omega t] = \frac{s}{(s^2 + \omega^2)}$$

$$\mathscr{L}^{-1}\left[\frac{s}{(s^2 + \omega^2)}\right] = \cos \omega t$$

In the same manner it is possible to show that

$$\mathscr{L}[\sin \omega t] = \frac{\omega}{(s^2 + \omega^2)}$$

$$\mathscr{L}^{-1}\left[\frac{\omega}{(s^2 + \omega^2)}\right] = \sin \omega t$$

Transform of $t^a$, $(a \geq 0)$.

$$\mathscr{L}[t^a] = \int_0^\infty e^{-x}\left(\frac{x}{s}\right)^a \frac{dx}{s} \quad \text{by setting } st = x.$$

$$= \frac{1}{s^{a+1}} \int_0^\infty e^{-x} x^a \, dx = \frac{\Gamma(a+1)}{s^{a+1}}$$

where $\Gamma(\cdots)$ is Euler's gamma function, which is available in tabular form in several handbooks, such as Reference 2. In the context of Laplace transforms, its form, rather than its numerical value, is more helpful in obtaining the inverse transforms.

## A3.3 PROPERTIES OF LAPLACE TRANSFORMS

***Linear Additivity of Transforms***

The linear additivity of Laplace transforms is easy to verify.

$$\mathscr{L}[f(t) + g(t)] = \mathscr{L}[f(t)] + \mathscr{L}[g(t)]$$

For example,

$$\mathscr{L}[\cosh at] = \frac{1}{2}[\mathscr{L}[e^{at}] + \mathscr{L}[e^{-at}]]$$

$$= \frac{1}{2}\left[\frac{1}{s-a} + \frac{1}{s+a}\right]$$

$$= \frac{s}{s^2 - a^2}$$

***Laplace Transform of Derivatives of f(t)***

$$\mathscr{L}[f'(t)] = \int_0^\infty e^{-st} \frac{df}{dt} \, dt$$

$$= e^{-st} f(t)|_0^\infty + \int_0^\infty e^{-st} s.f.\, dt$$

$$= s\mathscr{L}[f] - f(0)$$

In general for the $n$-th derivative of $f(t)$, we can obtain the transform by repeated use of integration by parts.

$$\mathscr{L}[f^n(t)] = s^n\mathscr{L}[f] - s^{n-1}f(0) - s^{n-2}f'(0) - \cdots - f^{n-1}(0)$$

**Example**

Find $\mathscr{L}[t \cosh at]$.

*Solution* Normally, this is difficult to evaluate directly.

$$\begin{aligned}
f(t) &= t \cosh at; \quad f(0) = 0 \\
f'(t) &= at \sinh at + \cosh at; \quad f'(0) = 1 \\
f''(t) &= a^2 t \cosh at + 2a \sinh at \\
&= a^2 f(t) + 2a \sinh at \\
\mathscr{L}[f''(t)] &= a^2\mathscr{L}[f] + 2a\mathscr{L}[\sinh at] \\
&= s^2\mathscr{L}[f] - sf(0) - f'(0)
\end{aligned}$$

Transposing,

$$\begin{aligned}
\mathscr{L}[f] &= \left\{2a\left(\frac{a}{s^2 - a^2}\right) + 1\right\} \frac{1}{s^2 - a^2} \\
&= \frac{(s^2 + a^2)}{(s^2 - a^2)^2}
\end{aligned}$$

***Laplace Transform of Integral of f(t)***

Let $g'(t) = f(t)$, and $g(t) = \int_0^t f(\tau)\, d\tau$. Since

$$\mathscr{L}[f(t)] = \mathscr{L}[g'(t)] = s\mathscr{L}[g(t) - g(0)],$$

and since,

$$g(0) = \int_0^0 \cdots \equiv 0,$$

we obtain

$$\mathscr{L}[g] \equiv \mathscr{L}\left[\int_0^t f(\tau)\, d\tau\right] = \frac{1}{s}\mathscr{L}[f(t)]$$

**Example**

Find the inverse transform of $1/s(s^2 - 1)$.

*Solution*

$$\bar{g}(s) = \frac{1}{s}\bar{f}(s), \quad \text{where } \bar{f}(s) = (s^2 - 1)^{-1}.$$

$$g(t) = \int_0^t f(\tau)\,d\tau, \quad \text{where } f(t) = \mathscr{L}^{-1}\left[(s^2 - 1)^{-1}\right] = \sinh t$$

$$= \int_0^t \sinh \tau \,d\tau = \tau|_0^1$$

$$= (1 - \cosh t)$$

***The Shifting Property***

The "shifting" property of $\mathscr{L}$-Transform is a very important property, used freely in the examples of Chapter 5. It states that

$$\text{if } \mathscr{L}^{-1}[\bar{f}(s)] = f(t), \text{then, } \mathscr{L}^{-1}[\bar{f}(s - a)] = e^{at}f(t)$$

Since $\bar{f}(s) = \int_0^\infty e^{-st}f(t)\,dt$, it follows that

$$\bar{f}(s - a) = \int_0^\infty e^{-st}.[e^{at}f(t)]\,dt$$

Thus,

$$\mathscr{L}^{-1}[\bar{f}(s - a)] = e^{at}f(t)$$

**Example**

Find the inverse transform of $1/(s^2 + 2s + \omega^2)$.

*Solution*

$$\mathscr{L}^{-1}\left[\frac{1}{s^2 + 2s + \omega^2}\right] = \mathscr{L}^{-1}\left[\frac{1}{s^2 + 2s + 1 + \omega^2 - 1}\right]$$

$$= \mathscr{L}^{-1}\left[\frac{1}{(s + 1)^2 + \left(\sqrt{\omega^2 - 1}\right)^2}\right]$$

$$= \frac{1}{\sqrt{\omega^2 - 1}} e^{-t} \sin \sqrt{\omega^2 - 1}t$$

## A3.4 THE CONVOLUTION INTEGRAL

If $\bar{f}(s)$ and $\bar{g}(s)$ are the Laplace transforms of $f(t)$ and $g(t)$ respectively, then the product $\bar{h}(s) = \bar{f}(s)\bar{g}(s)$ is the transform of the *convolution* of $f(t)$ and $g(t)$, which is written as $f(t) * g(t)$, and defined as

$$f(t) * g(t) = \int_0^t f(\tau)g(t - \tau)\, d\tau$$

Its similarity to the BSP should suggest its utility to viscoelasticity. No proof is provided. See Reference 1 for a proof.

**Example**

Find the inverse transform of $1/(s - a)(s - b)$.

*Solution* Let $\bar{f}(s) = 1/(s - a)$ and $\bar{g}(s) = 1/(s - b)$. The answer is evidently the convolution of $f(t)$ and $g(t)$. Thus,

$$\mathscr{L}^{-1}\left[\frac{1}{(s - a)(s - b)}\right] = \int_0^t f(\tau)g(t - \tau)\, d\tau$$

$$= \int_0^t e^{-a\tau}e^{-b(t-\tau)}\, d\tau$$

$$= e^{-bt}\int_0^t e^{(b-a)\tau}\, d\tau$$

$$= \frac{e^{-bt} - e^{-at}}{(b - a)}$$

**References**

1. Kreyszig, E., *Advanced Engineering Mathematics*, 4th Edn., 1979, John Wiley & Sons, New York, NY.
2. Abramowitz, M., and Stegun, I., *Handbook of Mathematical Functions*, 1972, Dover Publications, New York, NY.

**Table A3.1. Laplace Transforms.**

| | $\overline{f}(s) = \mathcal{L}[f(t)]$ | $f(t) = \mathcal{L}^{-1}[\overline{f}(s)]$ |
|---|---|---|
| 1 | $1/s$ | $1$ |
| 2 | $1/s^n$($n$ is an integer) | $t^{n-1}/(n-1)!$ |
| 3 | $1/\sqrt{s}$ | $1/\sqrt{\pi t}$ |
| 4 | $1/s^n$($n$ positive, not an integer) | $t^{n-1}/\Gamma(n)$ |
| 5 | $\dfrac{1}{s-a}$ | $\epsilon^{at}$ |
| 6 | $\dfrac{1}{(s-a)^2}$ | $t\epsilon^{at}$ |
| 7 | $\dfrac{1}{(s-a)^n}$($n$ is an integer) | $\dfrac{1}{(n-1)!}t^{n-1}\epsilon^{at}$ |
| 8 | $\dfrac{1}{(s-a)^n}$($n$ positive, not an integer) | $\dfrac{1}{\Gamma(n)}t^{n-1}\epsilon^{at}$ |
| 9 | $\dfrac{1}{(s-a)(s-b)}, \quad (a \neq b)$ | $\dfrac{\epsilon^{at} - \epsilon^{bt}}{(a-b)}$ |
| 10 | $\dfrac{s}{(s-a)(s-b)}, \quad (a \neq b)$ | $\dfrac{a\epsilon^{at} - b\epsilon^{bt}}{(a-b)}$ |
| 11 | $\dfrac{1}{s^2+\omega^2}$ | $\dfrac{1}{\omega}\sin \omega t$ |
| 12 | $\dfrac{s}{s^2+\omega^2}$ | $\cos \omega t$ |

**Table A3.1. (Continued)**

| | $\bar{f}(s) = \mathcal{L}\|f(t)\|$ | $f(t) = \mathcal{L}^{-1}\|\bar{f}(s)\|$ |
|---|---|---|
| 13 | $\frac{1}{s^2 - a^2}$ | $\frac{1}{a}\sinh at$ |
| 14 | $\frac{s}{s^2 - a^2}$ | $\cosh at$ |
| 15 | $\frac{1}{s(s^2 + \omega^2)}$ | $\frac{1}{\omega^2}(1 - \cos \omega t)$ |
| 16 | $\frac{1}{s(s^2 + \omega^2)}$ | $\frac{1}{\omega^3}(\omega t - \sin \omega t)$ |
| 17 | $\frac{s}{(s^2 + a^2)(s^2 + b^2)}$ $(a^2 \neq b^2)$ | $\frac{1}{b^2 - a^2}(\cos at - \cos bt)$ |
| 18 | $\frac{s^2}{(s^2 + \omega^2)^2}$ | $\frac{1}{2\omega}(\sin \omega t + \omega t \cos \omega t)$ |
| 19 | $\frac{1}{s^4 + 4a^4}$ | $\frac{1}{4a^3}(\sin at \cosh at - \cos at \sinh at)$ |
| 20 | $\frac{s}{s^4 + 4a^4}$ | $\frac{1}{2a^2}\sin at \sinh at$ |
| 21 | $\frac{1}{s^4 - a^4}$ | $\frac{1}{2a^3}(\cosh at - \cos at)$ |
| 22 | $\sqrt{s-a} - \sqrt{s-b}$ | $\frac{1}{2\sqrt{\pi t^3}}(e^{bt} - e^{at})$ |

# Appendix 4

# Stress Intensity Factors for a Few Cases

In the following, solutions for stress intensity factors in mode $I$ fracture are presented for a few cases of loading. Use has been made of References 1, 2, 3, and 4. More complex cases are listed in References 5, 6, and 7. The cases listed here are broadly sufficient for a first estimate for several designs, as well as for the experimentalist. As stated in the main test, if a better approximation is needed, one may want to do a literature search for stress intensity factors.

Unless noted otherwise, plate thickness is taken to be unity. In all cases, check for yourself the dimensions of the expression for $K$ to determine how applied force/pressure is measured, per unit thickness or otherwise.

***CASE 1: Center Cracked Plate—Finite Width***

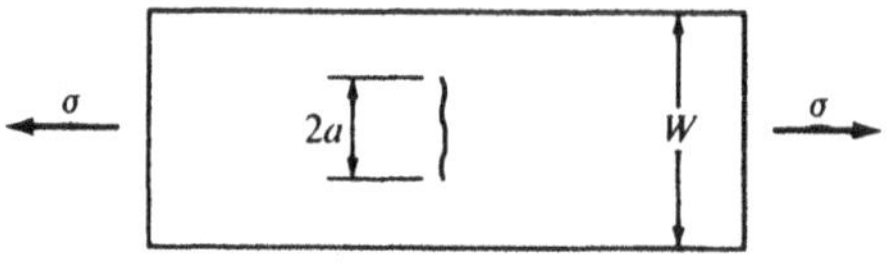


$$K = C\sigma\sqrt{\pi a}$$

where,
$$C = 1 + 0.256\left(\frac{a}{W}\right) - 1.152\left(\frac{a}{W}\right)^2 + 12.200\left(\frac{a}{W}\right)^3 \qquad \text{(A4.1)}$$

Or,
$$C = \sqrt{\sec\left(\frac{\pi a}{W}\right)} \qquad \text{(A4.2)}$$

Or, $$C = \frac{1}{\sqrt{1 - \left(\frac{2a}{W}\right)^2}} \tag{A4.3}$$

*CASE 2: Single Edge Notched Plate—Finite Width*

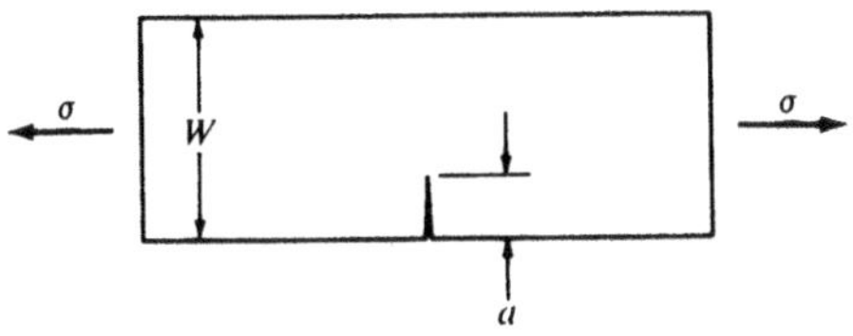


$$K = C\sigma\sqrt{\pi a}$$

where, $C = 1.12$ for small cracks. (A4.4)

Or, $$C = 1.12 - 0.231\left(\frac{a}{W}\right) + 10.55\left(\frac{a}{W}\right)^2 - 21.72\left(\frac{a}{W}\right)^3 + 30.95\left(\frac{a}{W}\right)^4. \tag{A4.5}$$

Equation A4.5 is valid for $0 \le (a/W) \le 0.6$.

*CASE 3: Double Edge Notched Plate—Finite Width*

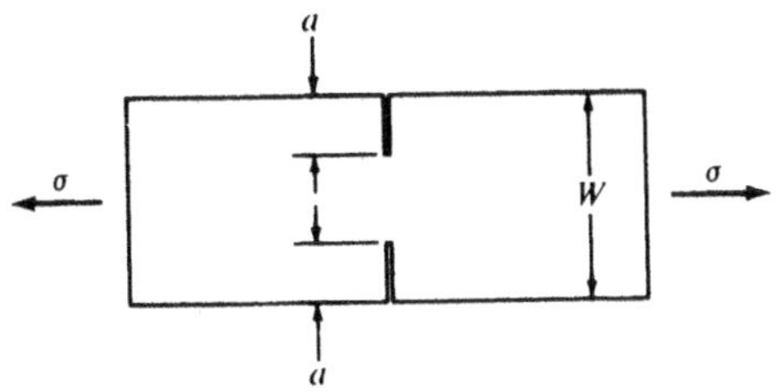


$$K = C\sigma\sqrt{\pi a}$$

where, $C = 1.12$ for small cracks. (A4.6)

Or,
$$C = \left(1 - \frac{a}{W}\right)^{-1/2}\left[1.122 - 0.561\left(\frac{a}{W}\right) - 0.205\left(\frac{a}{W}\right)^2 + 0.471\left(\frac{a}{W}\right)^3 - 0.190\left(\frac{a}{W}\right)^4\right] \qquad \text{(A4.7)}$$

***CASE 4: Crack Subjected to Pressure***

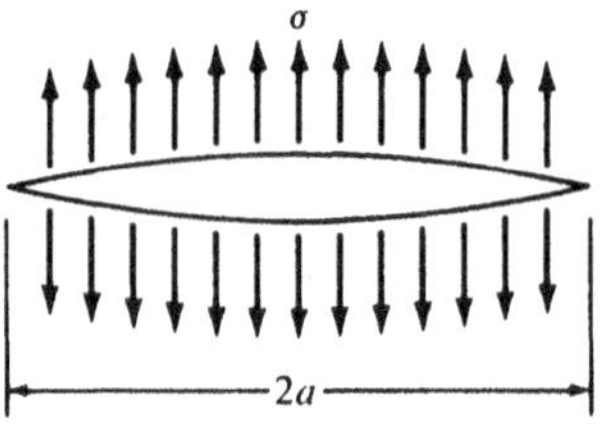


$$K = C\sigma\sqrt{\pi a}$$

$C$ can be taken to be the same as for a center cracked plate, as in Case 1 above.

***CASE 5: Crack Starting on Both Sides from a Small Hole***

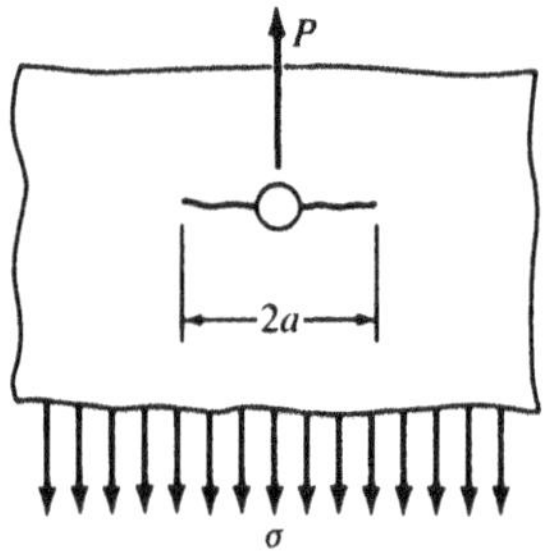


$$K = C\left(\frac{\sigma\sqrt{\pi a}}{2} + \frac{P}{2\sqrt{\pi a}}\right) \qquad \text{(A4.8)}$$

Note that $P$ is measured as force per unit thickness. $C$ is taken to be the same as for a center cracked plate, Case 1.

### *CASE 6: Crack Starting on One or Both Sides from a Large Hole*

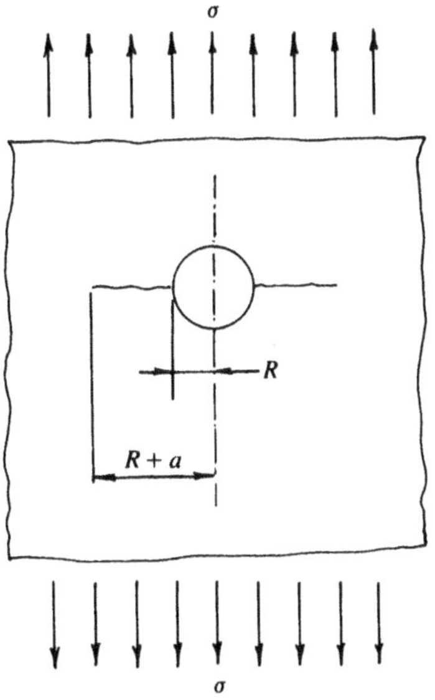


Case 6 (a)

$$K_1 = \sigma\sqrt{\pi a}\,F_1\left(\frac{a}{R+a}\right) \text{ for single crack.}$$

$$K = \sigma\sqrt{\pi a}\,F_2\left(\frac{a}{R+a}\right) \text{ for double crack.}$$

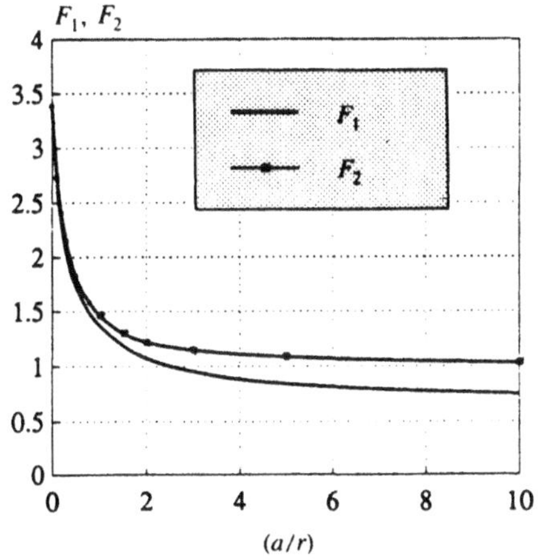


Case 6 (b)

***CASE 7: Notched Beam Loaded at the Center (Thickness = B)***

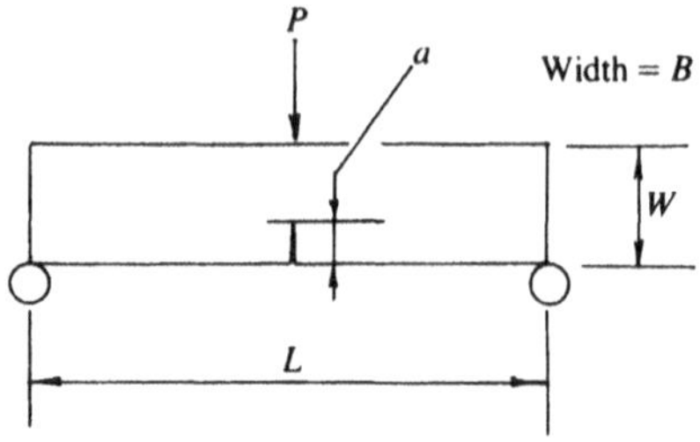


$$K = \frac{PL}{BW^{3/2}}\left[2.9\left(\frac{a}{W}\right)^{1/2} - 4.6\left(\frac{a}{W}\right)^{3/2} + 21.8\left(\frac{a}{W}\right)^{5/2} - 37.6\left(\frac{a}{W}\right)^{7/2} + 38.7\left(\frac{a}{W}\right)^{9/2}\right] \quad \text{(A4.9)}$$

***CASE 8: The Compact Tension Test Specimen (Thickness = B)***

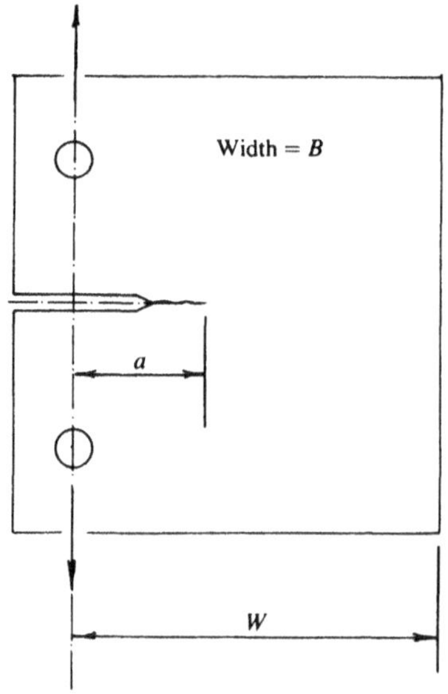


$$K = \frac{P}{B\sqrt{W}}\left[29.6\left(\frac{a}{W}\right)^{1/2} - 185.5\left(\frac{a}{W}\right)^{3/2} + 655.7\left(\frac{a}{W}\right)^{5/2} - 1017\left(\frac{a}{W}\right)^{7/2} + 63.9\left(\frac{a}{W}\right)^{9/2}\right] \quad \text{(A4.10)}$$

***CASE 9: Notched Beam with Bending Moments (Thickness = B)***

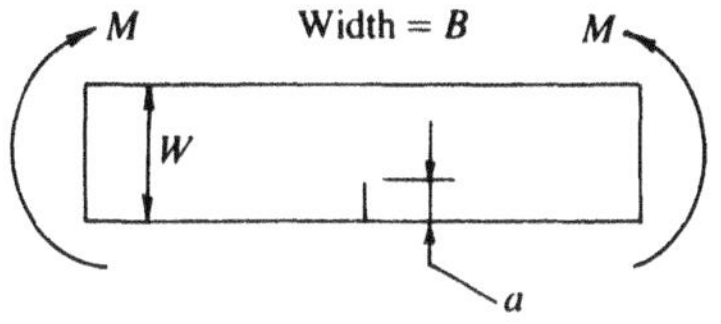


$$K = \frac{6M}{B(W-a)^{3/2}} \times g\left(\frac{a}{W}\right) \tag{A4.11}$$

The function $g(a/W)$ is given in the table below.

| $a/W$ | 0.05 | 0.1 | 0.2 | 0.3 | 0.4 | 0.5 | 0.6 (and higher) |
|---|---|---|---|---|---|---|---|
| $g(a/W)$ | 0.36 | 0.49 | 0.60 | 0.66 | 0.69 | 0.72 | 0.73 |

***CASE 10: Embedded Elliptic Crack in an Infinite Body—Uniform Tension Normal to Crack Plane***

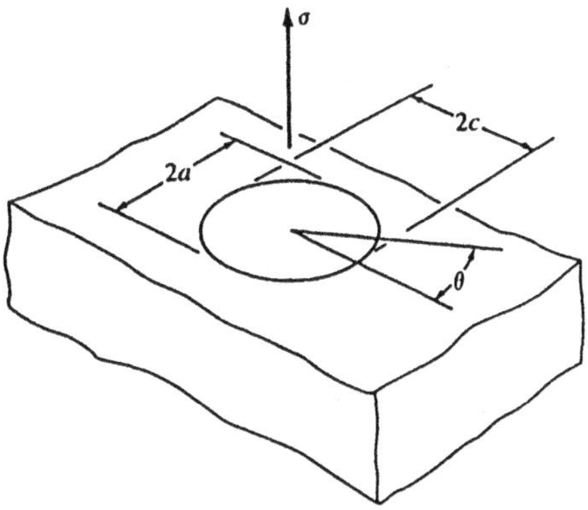


Note that the crack front is not a straight line, and hence the plane strain two-dimensional assumptions do not hold good. A full three-dimensional analysis is needed. Accordingly the $K$ varies with position on the crack front. The position of any point on the elliptic crack front is identified by the angle $\phi$ shown.

$$K = \frac{\sigma\sqrt{\pi a}}{\Phi_o}\left(\sin^2\phi + \frac{a^2}{c^2}\cos^2\phi\right)^{1/4} \tag{A4.12}$$

In the above, the parameter $\phi_o$ is the so-called elliptic integral defined by

$$\Phi_o = \int_0^{\pi/2} \left[ 1 - \left( \frac{c^2 - a^2}{c^2} \right) \sin^2 \theta \right]^{1/2} d\theta \qquad \text{(A4.13)}$$

This function is available in tabular form in several mathematical handbooks. *Handbook of Mathematical Functions* by Abramowitz and Stegun (Dover) is one such reference. More often its square $Q = \Phi_o^2$ is used in some equations in the form

$$K = \sigma \sqrt{\frac{\pi a}{Q}} \left( \sin^2 \phi + \frac{a^2}{c^2} \cos^2 \phi \right)^{1/4}$$

The function $Q$—called the Flaw Shape Parameter—is presented in Figure 6.22 for convenience.

***CASE 11: Elliptic Surface Crack (Thumb Nail Crack) on a Plate of Thickness t***

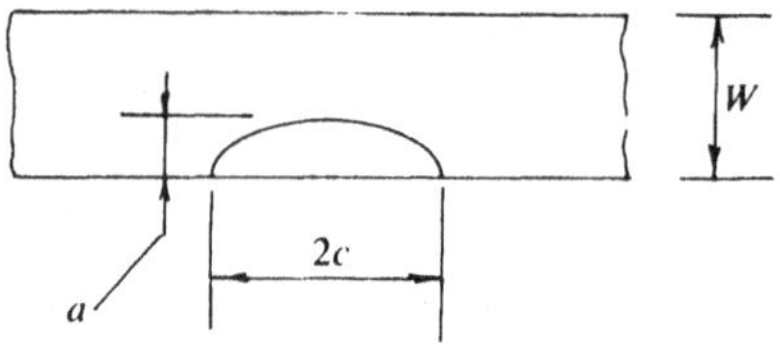


The maximum stress intensity $K$ for a thumb nail crack, occurring at $\phi = \pi/2$ is given by the following equation.

$$K = 1.12 M_K \sigma \sqrt{\frac{\pi a}{Q}} \qquad \text{(A4.14)}$$

where $Q$ is given in Figure 6.22 and $M_K$ is given by

$$M_K = 1.0 + 1.2 \left( \frac{a}{t} - 0.5 \right) \qquad \text{(A4.15)}$$

***CASE 12: Infinite Sequence of Evenly Spaced, Collinear Cracks***

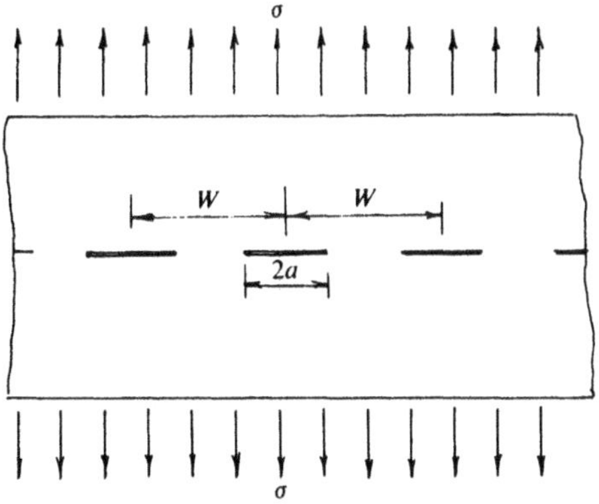


$$K = \sigma\sqrt{\pi a}\left(\frac{W}{\pi a}\tan\frac{\pi a}{W}\right)^{1/2} \tag{A4.16}$$

***CASE 13: Circumferentially Notched Round Bar***

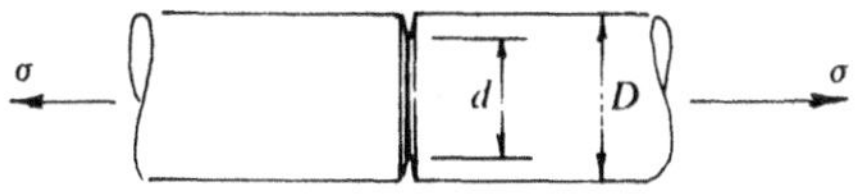


$$K = \frac{0.932}{\sqrt{\pi d^2}}P\sqrt{D} \tag{A4.17}$$

This equation is valid over the range $1.2 \le D/d \le 2.1$.

***CASE 14: Double Cantilever Beam (DCB)***

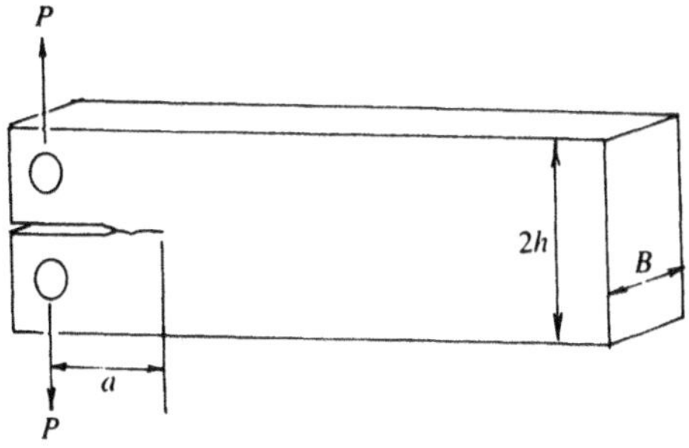


$$K = 2\sqrt{3}\,\frac{Pa}{Bh^{3/2}} \quad \text{(Plane stress)}, \tag{A4.18}$$

$$= \frac{2\sqrt{3}}{\sqrt{1-\nu^2}}\,\frac{Pa}{Bh^{3/2}} \quad \text{(Plane strain)}, \tag{A4.19}$$

### *CASE 15: Arc-Shaped Specimen*

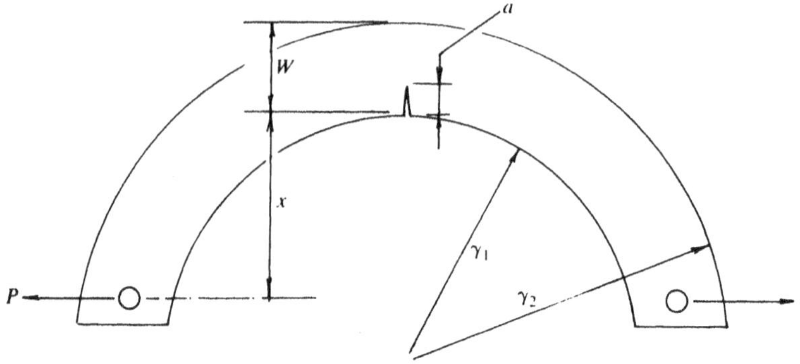


$$K = \frac{P}{B\sqrt{W}}\left[1 + 1.54\left(\frac{x}{W}\right) + 0.50\left(\frac{a}{W}\right)\right] \times\left[1 + 0.221\left\{1 - \left(\frac{a}{W}\right)^{1/2}\right\}\left(1 - \frac{r_1}{r_2}\right)\right] \times f\left(\frac{a}{W}\right) \tag{A4.20}$$

where,

$$f\left(\frac{a}{W}\right) = \left[18.23\left(\frac{a}{W}\right)^{1/2} - 106.2\left(\frac{a}{W}\right)^{3/2} + 389.7\left(\frac{a}{W}\right)^{5/2} - 582.0\left(\frac{a}{W}\right)^{7/2} + 369.1\left(\frac{a}{W}\right)^{9/2}\right] \tag{A4.21}$$

***CASE 16: Elliptic Crack at the Corner of a Pipe Branch***

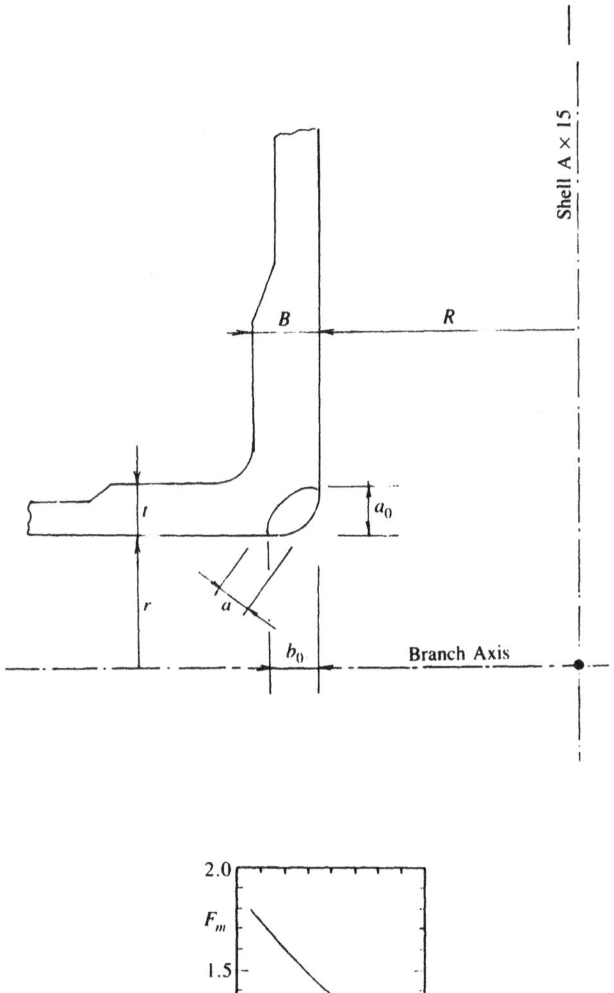

For pressure containers of cylindrical shape and with branch pipes, it is known that cracks in the longitudinal plane (that is, those subjected to the full hoop stress) are the critical ones. They are located in otherwise high stress areas too, namely the junction of the branch pipe and the vessel. The solution for $K$ for such cases is very important in pressurized container designs.

$$K = \sigma_h \sqrt{\pi a}\, F_m \left[1 + \left(\frac{rt}{BR}\right)^{1/2}\right] \tag{A4.22}$$

In the above equation $\sigma_h$ stands for the hoop stress in the vessel. The parameter $F_m$ is given in the adjoining figure.

**References**

1. Barsom, J. M. and Rolfe, S. T., *Fatigue and Fracture Control in Structures*, 2nd Edn. 1987, Prentice Hall Inc., Englewood Cliffs, NJ.
2. Broek, D., *Elementary Engineering Fracture Mechanics*, 3rd Edn., 1982, Martinus Nijhoff Publishers, Boston, MA.
3. Ewalds, H. L. and Wanhill, R. J. H., *Fracture Mechanics*, 1984, Co-published by Edward Arnold and Delftse Uitgevers Maatschappij, Delft, Netherlands.
4. Paris, P. C. and Sih, G. C., *ASTM STP 381*, American Society for Testing and Materials, 1964, Philadelphia, PA.
5. Sih, G. C., *Handbook of Stress-Intensity Factors—Stress Intensity Factor Solutions and Formulas for Reference*, 1973, Institute of Fracture and Solid Mechanics, Lehigh University, Bethlehem, PA.
6. Rooke, D. O. and Cartwright, D. J., *Compendium of Stress Intensity Factors*, 1974, Her Majesty's Stationery Office, London.
7. Tada, H., Paris, P. C., and Irwin, G. R., *The Stress Analysis of Cracks Handbook*, 1973, Del Research Corporation.

# Index*



*References to specific matrices appear after the letter z.

GPSR Compliance

*The European Union's (EU) General Product Safety Regulation (GPSR) is a set of rules that requires consumer products to be safe and our obligations to ensure this.*

*If you have any concerns about our products, you can contact us on ProductSafety@springernature.com*

In case Publisher is established outside the EU, the EU authorized representative is:

Springer Nature Customer Service Center GmbH
Europaplatz 3
69115 Heidelberg, Germany

**Batch number: 10761599**

Printed by Printforce, the Netherlands